Topics in
GEOGRAPHY

Kendall Hunt
publishing company

Cover image © Shutterstock, Inc.

www.kendallhunt.com
Send all inquiries to:
4050 Westmark Drive
Dubuque, IA 52004-1840

Copyright © 2022 by Kendall Hunt Publishing Company

ISBN 979-8-7657-7771-8

eBook ISBN 979-8-7657-7772-5

Published in the United States of America

Table of Contents

The Nature of Geography
By Tim Anderson

Most Americans relate the term *geography* to the rote memorization of trivial facts associated with various countries of the world: capitals, imports and exports, highest mountains, largest lakes, most populous cities, climates, and the like. This trivialized impression of the subject matter geography covers most likely stems from the fact that, until recently, most U.S. elementary schools and high schools did not teach geography as a distinctive subject.. Even today, it is often taught as a marginalized part of social science courses in elementary schools and civics or history courses in high schools. As a result, most Americans think of geography as a collection of trivial factoids rather than as a distinct academic subject such as history, economics, mathematics, or physics. But, within academia, geography is a thriving, diverse discipline with its own set of theories, research methods, terminology, and subject matter, and it has a unique way of looking at the world. Indeed, the Association of American Geographers, the largest professional society of geographers in the world, counts over 5,000 professors, students, and professionals as members.

So what do geographers do? What is their method of analysis? What questions do geographers ask about the world? These questions, as it turns out, are not easily answered. Indeed, any two geographers might answer them in very different ways, for geography is not unified by subject, but rather by method. This is yet another reason for the public's misconceptions about what geography is and, indeed, for misconceptions many academicians themselves have. Geography does not seem to fit into academe the same way that other subjects do because the discipline is not unified or defined by its subject matter. Rather, it is unified and defined by its method and approach, its mode of analysis. This mode of analysis (or approach or perspective) involves a **spatial perspective**. In this way, geography can be conceptualized as a way of thinking about the world, space, and places. This spatial (geographical) perspective involves the fundamental question about how both cultural and natural phenomena vary spatially (geographically) across the earth's surface. As such, it is possible to study the geography of almost any phenomenon that occurs on the earth's surface. Employing these ideas, we may define the academic discipline of geography as *the study of the spatial variation of phenomena across the earth's surface*. So, geographers do not study a certain thing, or subject. Instead, they study all sorts of things and subjects in a specific way—spatially.

The eighteeth-century German philosopher Immanuel Kant understood very well the unique position that geography held within academia. Kant argued that human beings make sense of and organize the world in one of three ways, and human knowledge, as well as academic departments in the modern university, is arranged or organized in the same way. First, Kant wrote, we make sense of the world *topically* by organizing knowledge according to specific subject matter: biology as the study of plant and animal life, geology as the study of the physical structure of the earth, sociology as the study of human society, and so on. Each of these fields is unified by its subject matter. Second, according to Kant, we make sense of the world *temporally* by organizing phenomena according to time, in periods and eras. This is the sole domain of the discipline of history. Finally, Kant argued that we make sense of the world *chorologically* (geographically) by organizing phenomena according to how they vary across space and from place to place. This is the sole domain of the discipline of geography. The disciplines of history and geography are similar in that they are both unified by a method rather than the study of a specific subject matter. The region (discussed below) in geography is analogous to the era or period in history— they are the main units of analysis in each of the respective disciplines.

Figure 1.1 represents a generalized model for understanding the nature of the academic discipline of geography. This figure illustrates the field's three main subfields: physical geography, human geography, and environmental geography. The description and analysis of patterns on the landscape unites each of these subfields. When most people think of the term landscape, they think of something depicted in a painting, or perhaps a garden. But, when geographers employ the term landscape, they are referring to the totality of our surroundings. In this sense, the **physical landscape** refers to the patterns created on the earth's surface by natural or physical processes. For example, tectonic forces create continents and mountain chains; Long-term climatic processes create varying vegetative realms; Wind and water shape and modify landforms. **Physical geography**, then, is the subfield of geography that is concerned with the description and analysis of the physical landscape and the processes that create and modify that landscape. The various subfields of physical geography, such as biogeography, climatology and geomorphology, are natural sciences, allied with such fields as botany, meteorology and geology, and, like these allied fields, research in physical geography is undertaken largely according to the scientific method.

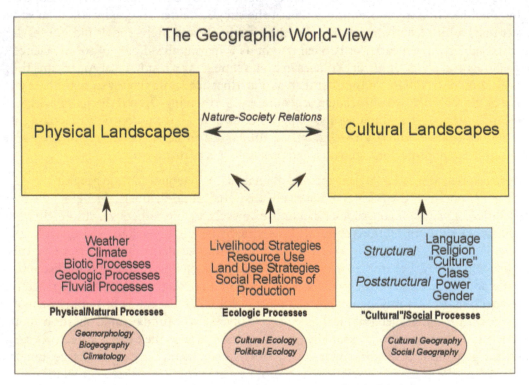

Source: Timothy G. Anderson

Figure 1.1   **The Geographic World-View**

Human geography is the subfield of geography that involves the description and analysis of cultural landscapes and the social and cultural processes that create and modify those landscapes. While physical geography concerns the natural forces that shape the earth's physical landscapes, human geography analyzes the social and cultural forces that create cultural landscapes. The **cultural landscape**, then, may be conceptualized as the human *imprint* on the physical landscape resulting from modification of the physical landscape by human social and cultural forces. Given the power and influence of human technology and institutions, many human geographers see human beings as the ultimate modifiers of the physical landscape. The various subfields of human geography, such as historical geography, economic geography, and political geography are social sciences, allied with such fields as history, economics, and political science. In addition to their own research methodologies, human geographers often employ research methods from these allied fields. Traditionally, human geography has pertained to how cultural processes such as religion, language, and worldview affect the cultural landscape. More recently, however, cultural geographers have begun to question traditional ideas about how cultural landscapes are created by focusing on so-called post-structural processes to reevaluate the nature of cultural landscapes. In such research, the cultural landscape is conceptualized as a stage upon which social struggles dealing with such concepts as race, class, power, and gender are played out. In this sense, the cultural landscape is the product of such struggles.

Although not all geographers recognize it as such, it can be argued that **environmental geography** is a third major subfield of geography. Environmental geography involves the interrelationships between humans and the natural environments in which they live, or nature-society relationships. Environmental geographers study the patterns created on the landscape by such interactions and the processes involved, such as social relations of production, cultural adaptations or maladaptations to particular environments, modifications to the environment wrought by human economies and technologies, and the use (or misuse) of natural resources.

Spatial Diffusion
By Dean Fairbanks

We can categorize most fields of knowledge as part of the humanities or as part of science (social and physical), two avenues of understanding the world that differ only in their medium. Those trained in the humanities view the world by expressing their deepest feelings and beliefs through the medium of canvas, clay, dance, theater, music, or language. Those trained in the sciences view the world by expressing their deepest feelings and beliefs by solving problems, exploring patterns and relationships, and testing ideas through objective observation (aka scientific method), and then building theories/models of how the world works.

The word **geography** literally means "description of the Earth." The goal of the geographical sciences is to explain how and why things, both physical and cultural, spatially and temporally differ over the surface of the Earth. Geography is the science that tries to describe, explain, experiment, understand, and predict the spatial distribution of Earth surface phenomena and the significance of those distributions to (1) an understanding of places and (2) an understanding of the relationships between society and natural phenomena. In short, geographers study "what is where, why, and so what." Physical geography differs from human or cultural geography only in the relative emphasis given to natural phenomena (features, processes, and materials of the atmosphere, hydrosphere, geosphere, and biosphere) compared with the phenomena of human society (features, processes, and materials created by humans— cultural elements). Geography, in general, is further distinguished from the other natural and social sciences by virtue of the attention geographers give to three major themes:

- Scale refers to the recognition that how we view and interpret phenomena depends on the spatial extent and the spatial resolution of the measurements we collect.
- **Integration** refers to the recognition that the landscape functions differently as a whole than one would predict by adding up the individual (parts) effects of climate, water, landforms, plants and animals, and human cultural elements.
- **Synthesis** refers to the attention geographers give to understanding the environmental opportunities (resources) and environmental constraints (hazards) upon human activities and upon human impacts on the environment.

The discipline of geography originated as a natural/physical science by Eratosthenes (2nd century B.C.) at the library of Alexandria. By the 18th century, scientific specialization emerged and physical geography (the study of the features and processes of the Earth surface) became a geographic sub-discipline. Physical geographers conducted "basic" research, which emphasized connective, integrative, and system perspectives that assembled Earth science knowledge from different disciplines. As geographers, they focused on emphasizing a variety of spatial and temporal scales as unique contributions to Earth science investigation.

While the contemporary geographical sciences include subfields that straddle the physical, social, and computer sciences, it is physical geography that specifically focuses on geography as an Earth system science. Its investigation is based on principles of the physical environment (natural laws), applies the scientific method of hypothesis generation and the development of theories, and emphasizes the importance of fieldwork, data collection, and quantitative analyses. Geospatial and temporal facts of the major Earth systems (atmosphere, hydrosphere, lithosphere, and biosphere) are sought and constantly revised, and then physical geographic theories are developed to interpret the facts.

Contemporary physical geographers view the Earth as a series of interconnected, interactive, and interlocking systems, which generate forces that produce the weather, climate, and landscapes encountered in our daily lives. Physical geography examines segments of geochemical, geophysical, and biological systems. An underlying focus is on the subsidiary systems in which energy and matter are transported, stored, and redeployed. What makes physical geography unique is the specific attention it imparts on facts, principles, and processes which form the foundations of living and non-living systems and that produce geospatial pattern—spatial arrangements, distributions, and organization—which then subsequently reinforces the processes. This allows for the special properties of scientifically measured spatial expressions of physical, chemical, and biological principles and functions to be interrogated via geospatial data (i.e., maps, global positioning systems, geographic information systems, satellite imagery), in the field physical measurement, laboratory analyses, and through computer modeling (**Figures 2.1**).

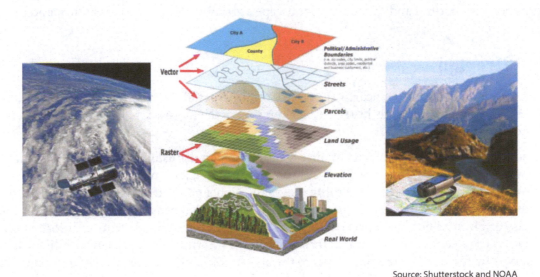

Source: Shutterstock and NOAA

Figure 2.1 Vector and raster data are two types of geospatial data that is used to build models in geographic information systems.

Physical geography covers topics as varied as the origin and development of living organisms and their evolution through time in response to the geospatial heterogeneity of Earth environments; and processes that transform and sculpt Earth materials through time to create landscapes. In order to approach these types of topics, physical geographers rely on physical laws. For example, physical geographers apply energy laws to explain the solar and terrestrial energy flows that set our atmosphere and hydrosphere in motion. The appropriate laws of motion then emerge to govern heat transfer across the Earth-atmosphere interface. The processes that shape the landscapes in the geosphere system—weathering, rivers, and desert formation—also rely on physical laws that apply across the Earth's surface.

Spatial Diffusion, Accessibility, and Connectivity

K.L. Chandler

The movement of people, things, and ideas can be geographically studied as mechanisms. Three mechanisms of change in both the natural and the cultural landscape are spatial diffusion, accessibility, and connectivity.

Spatial Diffusion

Every person, thing, or idea has a hearth, a point of origin. **Spatial diffusion** describes the scattering of a person, thing, or idea away from its hearth. It answers where the move started, what process facilitated the transfer, and where it stopped.

There are several different forms of spatial diffusion. **Expansion diffusion** is the spreading of an idea that increases in influence as its journey unfolds, either through contagious, hierarchical, or stimulus diffusion. **Contagious diffusion** occurs when something moves through a population from person to person, such as an infectious disease, viral video, or religious idea. The exposure happens randomly and rapidly, regardless of social class or physical borders. **Hierarchical diffusion** communicates concepts or information through a system, based on a command structure. Teachers present information to students; military leaders give orders to subordinates; employers assign tasks to the people who work for them. Fads and slang expressions are also adopted via hierarchical diffusion, introduced by influencers or particular groups before gaining traction with followers.

In **stimulus diffusion**, the idea spreads from its hearth to another location but morphs into something new along the way. Vodou, brought to the West Indies and the United States Gulf Coast by enslaved West African people, merged with the practices of Roman Catholicism, adding saints to religious rituals. McDonald's, founded in the United States, offers locally-inspired menu items and flavors, based on its locations in more than 100 countries around the world (**Figure 3.1**).

Relocation diffusion happens when people move from their hearth to a new place and bring their culture, traditions, or innovation with them. These practices may be shared with and embraced by the new community, as in the case of cuisine or an employer's implementing ideas from a new hire. In the realm of physical geography, air masses move across a landscape, generating storm activity as they travel through regions.

Perhaps, one of the most influential examples of spatial diffusion is colonialism, which occurs when a powerful, more developed country takes control of a particular country or region and controls its core functions and resources for the more developed country's benefit. For better or worse, colonialism became the predecessor of the contemporary idea of **globalization**, which has made the world more interconnected than ever

© OSORIOartist/Shutterstock.com

Figure 3.1   **Global contagion due to the Coronavirus is an example of contagious diffusion.**

before (especially through trade and cultural practices). It is incumbent upon citizens of the global community to interact with each other equitably and respectfully.

Accessibility and Connectivity

Accessibility indicates how easily something can move from one place to another and goes hand-in-hand with contagious diffusion. These concepts could pertain to forest fires, which spread quickly and without restraint through wooded areas until they reach an uncrossable zone, such as a major river, that renders the path inaccessible. Likewise, bodies of water have both granted and limited accessibility for human migration. For centuries, ocean voyages were an important means of travel and trade, but these ventures came at great economic cost and could be treacherous in unseasonable weather or dangerous storms.

Connectivity reveals the tangible and intangible ways things correlate with each other. Sometimes people, ideas, or objects that would naturally remain separate overlap, thanks to the connectivity of spatial diffusion. The Mississippi River begins in north-central Minnesota, making its way south through the center of the continental United States to the Mississippi Delta and the Gulf of Mexico. Particulates in the soil that have moved down the river in suspended stream load can be found at the Delta. Similarly, early people groups along the river's path found themselves in closer proximity to each other, as a result of the accessible water route. This nearness created opportunities for cultural diffusion and trade along the more than 2,000 miles between the headwaters of Minnesota's Lake Itasca to the port at New Orleans, a major power center for French colonialism and, later, the United States.

Geographic Regions
By Tim Anderson

Regions

The **region** is the primary unit of analysis for the geographer. A region is a bounded area of space within which there is homogeneity of a certain phenomenon or phenomena. Historically, the region is a central concept in geography that is employed to delimit and de ne spatial differentiation. Both physical and human geographers use the concept of region to describe and analyze places and subject matter. In this sense, a **culture region** is a place within which a certain culture is predominant. It should be noted, however, that ways in which regions are defied might differ from person to person; all types of regions are the product of human reasoning and as such are often cultural and socially constructed.

The geographers Robert Ostergren and Mathias Le Bossé identify three primary types of regions. **Instituted regions** are created by authorities within some kind of organization (such as governments and businesses) primarily for administrative purposes, such as planning, or for collecting data or revenue. The boundaries of such regions are usually clearly demarcated and agreed upon by almost everyone. In many cases, instituted regions consist of nested hierarchies of subregions. An example of this idea would be the nested hierarchy of federal, state, and local political boundaries in the United States and other federal states. **Denoted regions** are usually created by academics to reduce the complexity of the real world using systems of classification. Places in a denoted region are grouped together because they share one or more commonalities. Like all regions, they are the product of the person or persons who create them, and as such, their boundaries are often open to debate and critique.

Denoted regions can be subdivided into two different subtypes. **Formal denoted regions** are those that are rather easily identified using verifiable data. **Functional denoted regions** are areas in which the places in the region are all tied to a common central place by the movement of people, ideas, and things. A good example of a functional region is a trade area or the market area of a good or service. A **Perceptual region** is a region created informally, without official sanction by people within a community, or perhaps by those outside of a community. The boundaries of such regions are often imagined or internally perceived in the minds of the inhabitants based on real or perceived commonalities (such as language, religion, or ethnicity). By their very nature, perceptual regions are socially constructed, and therefore, subject to debate and continuous reinterpretation.

Maps, Projections, and Scale

K.L. Chandler

Maps and Projections

While the earth is not flat, most of the paper maps that depict its landforms are. These maps are called **projections**, and they demonstrate cartographers' (or map makers) attempts to transfer as much information as possible from a sphere to a flat surface. Three projections are frequently used: azimuthal, conic, and cylindrical. Each requires different mathematical formulas to plot gridlines that will correspond with landmasses, latitude, and longitude. Unlike a globe, which preserves Earth's *area, shape, direction, and distance*, a map projection can communicate only one or two features accurately, distorting the rest. For example, a projection that shows the correct land area tends to alter the land masses' shape and size. When choosing a projection to study, geographers need to understand cartographers' purpose when they designed various projections. Gerardus Mercator (1512-1594) created his map in 1569 as a navigational tool for sailors, who plotted their courses by hand. While its 90° angles at every latitude and longitude crossing did help seafarers plan how to get from one place to another, they also made land masses in the far north and south appear much larger. In contrast, an 1805 design by Karl Mollweide (1774-1825) offered a small-scale, equal-area map projection with meridian lines that would arc toward the poles, preserving the land masses' size but distorting shapes toward the edges of the map. (**Figure 5.1**) Which projection is correct? Both. Which project is incorrect? Both. Correctness depends on the purpose of what is being studied, thus it is critical to pick a projection based on its intention.

Map projections can mislead users with the type of information cartographers chose to include (or omit). Additionally, the way information is presented can show bias. Something as simple as the colors that shade certain areas can influence readers' impressions without their realizing it. Mercator's map, designed in the age of European colonialism, portrayed Europe to be much larger than it is in reality, using its vast land mass to exaggerate its importance and power on the world stage.

Types of Maps

Geographers commonly use maps that fall into the reference or thematic categories.

Reference maps can include physical, political, time zone, topographic, or road maps.

- Physical maps: landscape features, such as deserts, mountains, and rivers
- Political maps: formal boundaries of cities, countries, and continents

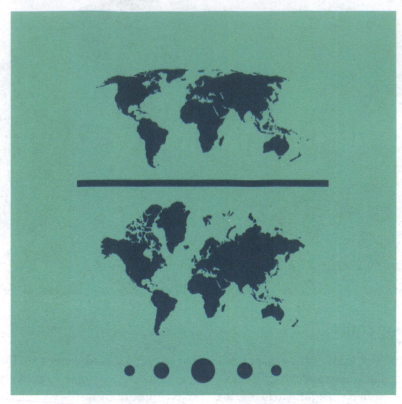

© Oleh Markov/Shutterstock.com

Figure 5.1 Mollenweide (top) and Mercator (bottom) projection maps.

- Time zone maps: time of day, as it relates to that of its adjacent zones
- Topographic maps: a detailed physical record of a land area, with features like mountains, valleys, and plains
- Road maps: primary and secondary roadways, including features of the physical and political landscape

Thematic maps can include climate, resource, socio-economic, or weather information.

- Climate maps: global climate zones or specific climate regions
- Resource maps: locations of natural resources
- Socio-economic maps: cost of living, income levels, literacy, and population density
- Weather information: barometric pressure, heat index, surface winds, or wind chill

In addition to map type and projection, a **scale** must be determined to indicate the distance on a map to the actual distance on the ground. A **large scale map**, such as a city map or a map of a theme park, shows a small area in great detail and includes features like streets, building footprints, and landmarks. A **small scale map**, such as a world map, shows a larger area in less detail. The measurement of the scale being used is then indicated on the map, often in a legend, as a representative ratio (e.g., 1 inch:100 miles), as a bar scale, or as a lexical scale (e.g., one inch will equal 100 miles).

The Scientific Method
By Terrence Bensel and Jon Turk

Why Study Environmental Science?

Almost every aspect of our daily lives is dependent upon and connected to the natural world around us. The food we eat, the air we breathe, and the water we drink all originate from the natural world. Perhaps less obvious, the products we use every day—such as fuel for our car, the clothes we wear, or our phones and electronic devices—all have their origins in the natural world. At the same time, our everyday actions and the use of these products, be it driving, eating, or throwing out the trash, all have an impact on the natural world and the environment on which we depend. The study of environmental science encompasses all of these relationships. At its most basic, **environmental science** is the study of how the natural world works, how we are affected by the natural world, and how we, in turn, impact the natural world around us.

Our fundamental dependence on the natural world makes the study of environmental science relevant to all of us. Environmental issues—including deforestation, ozone depletion, water pollution, and climate change—affect us all. These issues are also in the media now more than ever, and they are often at the center of heated political debates. Acquiring an understanding of the basic science behind these debates is thus an important part of becoming an educated citizen and forming your own opinion of the issues. Environmental science is also a highly **interdisciplinary** field of study, one that incorporates expertise, knowledge, and research methods from many different disciplines (including biology, chemistry, geology, economics, ethics, and history). For this reason environmental science can be connected to just about any major or field of study.

Note that there is a difference between environmental science and environmentalism. **Environmentalism** is a social and political movement committed to protecting the natural world. While many environmental scientists may also be environmentalists, scientists adopt a more objective approach to the issues they study. This approach is based in large part on the use of the **scientific method**, a scientific approach to research based on observation, data collection, hypothesis testing, and experimentation (see the upcoming feature for more). As a student, you are not required or expected to become an environmentalist, but as an educated citizen, you should learn to recognize the critical role played by the scientific method in forming our understanding of the environment and the environmental challenges we face. Such an understanding of the scientific method will help you develop critical thinking skills and enable you to weigh competing claims and arguments about environmental issues.

Consider This

How is environmentalism different from environmental science? Why might it be important to draw a distinction between these two concepts?

The Scientific Method

The scientific method is an approach whereby scientists observe, test, and draw conclusions about the world around us in a systematic manner rather than simply stating opinion. The scientific method consists of a series of five steps as illustrated in **Figure 6.1**. Scientists begin with simple observations of the world around us. They then form questions based on those observations. For example, an environmental scientist might observe the death and decline of numerous trees alongside a major highway and wonder what is causing this to happen. This leads to the third step, the formulation of a hypothesis or hypotheses that might explain the death of the trees. These hypotheses help scientists formulate predictions, specific statements that can be tested. In this case a scientist might predict that the death of the trees is caused by road salt running off the highway in the winter, or that an herbicide sprayed to control weeds on the side of the highway might also be impacting the trees.

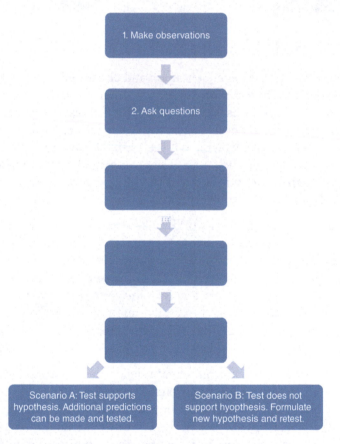

Figure 6.1   **The scientific method**

The scientific method is a five-step model used to observe, test, and draw conclusions scientifically.

All of these steps lead up to the final step of testing the predictions. In order to make a clear determination of what might be killing the trees, scientists devise experiments that attempt to hold conditions constant and then vary one variable at a time. In this case, scientists might identify four similar, small groves of trees that show no sign of stress or tree death. They might then expose one area to road salt applications, another to herbicide spraying, and a third to *both* road salt and herbicide, while the fourth area is left alone.

Note that regardless of the outcome of these experiments, scientists will typically still do two additional things. First, if the road salt and herbicide applications appeared to have some impact on the trees, they might refine their predictions to gain a better understanding of why this is happening. This might include adjusting the levels of road salt and herbicide application to see if they can better determine at what lev-els these become toxic. If the trees were not impacted by the road salt and herbicide applications, the scientists would be forced to revise their hypotheses or form new ones. Second, scientists typically seek to share their results with others, usually by presenting their research at scientific conferences and pub-lishing articles in professional journals. These presentations and papers are subject to analysis and scrutiny by other scientists, a process known as peer review. Scientists also have to explain the methods used in their research so that other scientists can run the same experiments, a process known as *replication*.

These two aspects of scientific research—peer review and replication—help to ensure the accuracy and legitimacy of the work that's done. Because the work of environmental scientists will often have an impact on policy debates, it's important to recognize just how the scientific method can shield scientists from claims of bias. Note that scientists don't really set out to "prove" anything; instead they observe, hypothesize, predict, test, and usually repeat. Demands by politicians for "scientific proof" before taking action to address an environmental issue are therefore problematic. Environmental policy should be informed by the best science available, and then combined with a consideration of other issues—such as ethical concerns, economic impacts, and consideration of the risks involved—to reach a decision.

Earth as a Grid
By Dean Fairbanks

An understanding of the Earth must start with an examination of its general shape. The Earth's shape is spherical, actually near spherical for as it rotates on its axis both gravity and centrifugal force combine to allow for a slight bulge at the equator.

Latitude and Longitude

The historical and modern confirmation of a near spherical planet we live on has allowed for the development of a reference system used for location awareness on the Earth's surface. This global address system is used and agreed upon by everyone as a means of identifying absolute location, and thereby allowing the accurate and near precise mapping of planetary phenomena.

Absolute location defines a point on the Earth's surface by specifying coordinates on a Cartesian mathematical grid (**Figure 7.1**).

The X and Y pair of a point define its location on the coordinate system. A coordinate system requires a point of origin (0,0), a set of prime axis reference lines, and a system of addressing any point within the system. These X, Y pairs represent longitude (X) and latitude (Y), our globally recognized address system for identifying the location of places and phenomena.

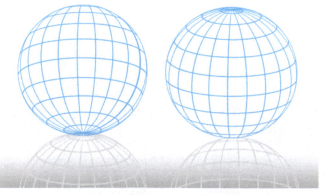

Image © Shutterstock, Inc.

Figure 7.1   A planar co-ordinate system using cartesian co-ordinates is the basis for atitude and longitude.

14

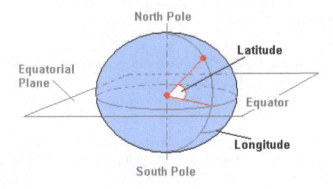

Figure 7.2   The equatorial plane divides the Earth's center.

- **Latitude** (north-south direction, or the Y-axis on a Cartesian grid) is an angular measure starting from the equator plane (0 degrees) of the Earth in relation to the center of the Earth sphere measured along lines that run east and west parallel to the equator called **parallels**. The image that this should conjure in your mind is that of slicing an onion, where the slices become smaller in diameter as one slices from the middle (equator) to the ends (poles). The north and south poles, respectively, represent 90 degrees and –90 degrees as measured from the center of our near spherical Earth (**Figure 7.2**).
- **Longitude** (east-west direction, or the X-axis on a Cartesian grid) is measured along lines that run north and south through the poles called **meridians**, and identifies the east to west location of a point on Earth along the equator by measuring the angular distance from the **Prime meridian** (0 degrees). The image that this should conjure is that of slicing wedges of an orange, that is, slicing down from pole to pole. The prime meridian (0 longitude) is internationally recognized as running through Greenwich, England, with the opposite side of the circle representing the International Date Line (180 longitude).

Latitude and Longitude

To understand a world geographic "addresses" we can use what was stated earlier about latitude and longitude to see that the Earth, being spheroid, is divided into 360°. Therefore, using the analogy of a clock: 1° = 60' (minutes) and 1' = 60" (seconds). Let's look at a pair of coordinates:

Table 7.1	
Degrees (°), Minutes (')	**Decimal number**
Traditional Method	**Computer Age Method**
37° 45′ N, 122° 26′ W	37.75, -122.433
34° 03′ N, 118° 14′ W	34.05, -118.233

The first "address" above is for the city of San Francisco, California, and it states that San Francisco is 37° (degrees) and 45' (minutes) north of the equator (0° line), and 122° (degrees) and 26' (minutes) west of the Prime meridian (0° longitude line). The other address coordinate is for Los Angeles, California, and it shows that this location is a little closer to the equator at 34° and 03' north of the equator, and a little closer to the Prime meridian at 118° and 14' west. In contrast to other global addresses, if you were given a location for somewhere in Chile the latitude would have a negative in front of the number or an "S" for south of the equator after the complete latitude depiction of degrees, minutes, and sometimes seconds. Likewise, if you were given a location for somewhere in Iraq, while it would be north of the equator, its longitudinal position would be east of the Prime meridian and therefore be a positive number or have an "E" after the complete longitude depiction.

Knowing a few basic rules about working with latitude and longitude values can allow you to calculate the distance from one location to another. Looking at the longitude values again for San Francisco and Los Angeles we can calculate how far apart they are in a west-east direction. To do this, we subtract 122° 26' from 118° 14' and get 4° 12'.

- **Rule 1:** Since 1° = 85 km (53 miles); we can therefore calculate the distance by multiplying and thus get 357 km (223 miles). Remember, however, that while you should get 340 km (212 miles) for the 4°, there are 12' to work with. Therefore, since 1° = 60', then 12'/60' = 0.2; or in other words 12' is 20% of the full 60' and you should therefore multiply 0.2 with 85 km (53 miles) to derive the fractional distance (17 km or 11 miles) and then add it to the result from the whole degrees multiplication (340 + 17 = 357 km).
- **Rule 2:** This rule involves carrying over values, recalling that 1° = 60'. If the values had been slightly different, such as 122° 14' from 118° 26', then you would not bring over 19 to subtract in the minutes column but rather 60', thereby reducing 122° to 121° and adding 60' to 14' to get 74', which then is subtracted from 26'. The answer is now 3° 48'.

The above example was shown to you in the traditional manner of degrees, minutes, and sometimes we have seconds as the coordinate pair system. But we live in the computer age so we commonly convert the traditional system to a decimal degree number, which is notably simpler to work with. The conversion between the two forms is:

$$\textbf{Decimal = Degrees.(Minutes/60 + Seconds/3600)}$$

Looking at the example table again you can see that the decimal number is much easier to work with as the minutes and seconds are converted, summed, and then added as the decimal to the degrees. In addition, there is no need for a letter denoting compass direction from the equator or Prime meridian, instead just as in the Cartesian coordinate system the X and Y coordinates are denoted as positive or negative depending on their relation to the origin (0,0).

How is any of this applied in everyday life? Following are some examples. We can use geographic coordinates to search for places in Google Earth. U.S. Federal Law (the Patriot Act) requires your cell phone to have a global positioning system (GPS) device to locate your position in case you dial a 911 call. Next time you travel on an airplane look in at the cockpit (without looking suspicious). There you will see a large digital display of the aircraft's current latitude–longitude location with a corresponding latitude–longitude location for its final destination or for the landmarks it passes over along its flight path. The aircraft's computers are calculating the air miles using latitude–longitude as you fly to your holiday destination.

Great Circles and Small Circles

A discussion of the Earth's near-spherical shape and the devised coordinate system for determining absolute location allows for the discussion of great circles and small circles. If we consider a sphere and a plane, there are two ways that the plane may intersect the sphere:

- Either the plane does not pass through the center of the sphere: in this case the circle of intersection is called a small circle. In relation to latitude, every parallel is a small circle except the equator.
- Or the plane does pass through the center of the sphere: in this case the circle of intersection is as large as possible (its radius is the same as the radius of the sphere) and is called a **great circle**. In relation to latitude, the equator is the only parallel that is a great circle, while an infinite number of longitude meridians are great circles.

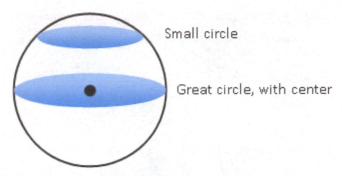

Small circle

Great circle, with center

Figure 7.3   Only a great circle will pass through the center of the sphere.

On a sphere, the shortest distance between two points is measured along the surface arc route of a great circle. This would be the shortest flying distance an aircraft would fly from say Chicago, Illinois (41° 53' N, 87° 37' W) to Rome, Italy (41° 53' N, 12° 30' E). Note that even though both these locations are on the same latitude, flying along that parallel is not the shortest distance (which represents a small circle. Only the equator as a parallel is a great circle). Thus, the shortest distance between these two locations is to fly a great circle route (see image).

It's Time for Time–Time Zones

We use temporal concepts all the time, yet seldom think about the origin of these terms. Time, as a measurement based on Earth motions, may be considered on an annual (yearly) or diurnal (daily) basis. The time of year is determined by the Earth's position in its orbit as it **revolves** around the sun. The time of day is determined as the Earth **rotates** (spins on its axis). Here's an easy way to remember the difference: One day is one rotation, while one year is one revolution.

Your position on Earth relative to the position of the sun in the sky determines the local time of day where you are. The longitudinal meridian on the Earth that is turned directly toward the sun is at the noon position. The meridian directly on the opposite side of the Earth is at the midnight position, while all other meridians fall somewhere between midnight and noon. All times of the day exist simultaneously, somewhere on Earth, and your time simply depends upon where you are at the moment. As you rotate with the Earth, you will experience all 24 hours of the day including each minute and each second of the day. Time begins at Greenwich, UK on the prime meridian, also known as **Greenwich mean time** (GMT) or **universal coordinated time** (UCT).

There is a relationship between our global grids (outlined earlier) and temporal constructs by remembering that there are 360 angular degrees in a circle, 60 minutes in a degree, and 60 seconds in a minute. As there are 360° in a complete rotation, which takes approximately 24 hours, the Earth rotates by (360° / 24 hours) = 15° every hour. Because there are 60 minutes of time in one hour, and the Earth rotates 15° per hour, we can determine that it takes the Earth (60 minutes / 15°) = 4 minutes to rotate by 1°.

Standard time zones have been established almost worldwide to avoid the confusion inherent with local solar time. Solar time differs by 4 minutes for every one degree of longitude. With standard time, the same time is assumed to exist throughout a zone that spans 15° of longitude. When crossing from one standard time zone into another, time changes by one hour. For example, in the adjacent zone to the east, the standard time is one hour later, while in the adjacent zone to the west it is one hour earlier. The zones are centered on controlling meridians which are 15° multiples: 0°, 15°, 30°, 45°, 60°, 75°, 90°, 105°, 120°, 135°, 150°, 165°, and 180°. Every point in a time zone assumes the same time as its controlling meridian. However, the zone boundaries are drawn in irregular shapes for convenience and/or political considerations.

Image © Shutterstock, Inc.

Figure 7.4   Standard time zones span 15° of longitude.

The **International Date Line** (IDL) falls in the middle of the Pacific Ocean where its inconvenience is minimized. So, you can understand why Greenwich, UK wasn't such a bad idea for a prime meridian origin. When one crosses the IDL the day and date must be changed, regardless of the hour of the day. This is necessary to keep the time observed on the other side of the line. Here's how the IDL works: When crossing from the west side of the IDL going eastward (toward the United States) 24 hours is repeated so that you gain a day: that is, Friday becomes Thursday. When crossing westward (toward Japan) 24 hours is skipped; you lose a day and Monday becomes Tuesday. The same hour exists on either side of the Date Line, but on different days and dates. Standard time zones are handled as before. Note that the −12 zone and the +12 zone each span only 7.5° and together make the 15°.

Modern Geographic Technology

K.L. Chandler

Modern Geographic Technology

Modern geographic study employs more than just maps that were drawn by cartographers hundreds of years ago. Computers and technology, satellites, and imagery offer a more detailed, accurate picture of the world than has ever been available before. This improved technology has even corrected previously inaccurate data: Mt. Everest, once estimated to reach 29,002' above sea level, is now recognized to stand 29,032'.

Global Positioning System

Developed by the United States military as a strategic system in 1978, the **global positioning system (GPS)** is a constellation of satellites that orbit above Earth and send radio wave signals to receivers around the globe. Ground stations monitor the satellites, tracking them, uploading updated navigational data, and adjusting their clocks as needed (**Figure 8.1**).

© nmedia/Shutterstock.com

Figure 8.1   Planet Earth with satellite, rendered in detailed view, as seen from space. Space debris can endanger the equipment.

GPS combines three capabilities: *positioning, navigation, and timing*. Positioning identifies an accurate location and orientation. Navigation establishes a path from a current to a desired position (relative or absolute) and corrects for course, orientation, and speed. Timing synchronizes complex systems to achieve precise time. Modern navigation systems combine these factors with map data and other information, such as weather and other criteria, to help users develop a detailed understanding of an area.

The United States is committed to maintaining the availability of at least twenty-four operational GPS satellites 95% of the time. In January 2021, twenty-seven operational satellites were in the GPS constellation, not including decommissioned, on-orbit spares. Calculating a precise location requires four GPS satellites' working together: three to determine a position on Earth and one to adjust for error in the receiver's clock. The technology can be found in everything from smartwatches to bulldozers and shipping containers.

Remote Sensing

Remote sensing is the science of acquiring information about an object, land area, phenomenon, or process without making physical contact with it. It began in the 1840s when balloonists took photographs of the ground with the newly invented camera. Aerial photography became a valuable reconnaissance tool in wartime, eventually maturing into today's satellite and drone imagery.

Landsat, a joint program between NASA and the United States Geological Survey (USGS), provides the longest continuous space-based record of Earth's land in existence. It uses lower resolution than other imagery but still shows changes in Earth's landscape and allows scientists to track human-scale processes, like urban growth.

Weather satellites monitor Earth's weather and climate. Polar-orbiting satellites travel north-south orbits, passing over the poles to observe the same spot on Earth twice each day – once in daylight and once at night – to provide imagery and atmosphere data over the entire planet. Geostationary satellites orbit above the Equator, spinning at the same rate as Earth, and focus constantly on the same area, enabling them to photograph Earth at the same location every thirty minutes.

Infrared (IR) satellite imagery is like a temperature map that detects the radiation emitted by Earth's surface, atmosphere, and clouds. The satellite detects heat energy in the infrared spectrum. It also shows objects like clouds, water, or land surfaces, based on the object's temperature.

Geographic Information Systems

A **geographic information system (GIS)** is a computer system that gathers, manages, and analyzes data about a unique location. Two different types of data: *vector and raster*, are collected, **geocoded** (assigned a latitude and longitude), and mapped. Vector data is a type of spatial data with geographic boundaries that represents individual real-world features on the landscape: trees, rivers, houses, and roads. Raster data is a type of spatial data with geographic boundaries, organized in computerized rows and columns that displays information that is continuous across an area, such as a panoramic aerial photograph of the grasslands. Computer modeling programs layer individual data files, discovering variable, workable solutions to geographic problems, such as where to open the next coffee shop or how to determine the length and breadth of a volcano's debris field (**Figure 8.2**).

Figure 8.2   CALIFORNIA, USA – example of vector data used in GIS.

Living with Earth's Natural Hazards

By Ingrid Ukstins and David Best

Topic 9

Does it seem like Earth is a dangerous place to live? A hazard by definition is "a possible source of danger." Among some of Earth's natural hazards are hurricanes, tornadoes, earthquakes, tsunami, floods, mudslides, wildfires and volcanic eruptions (**Figure 9.1a–c**) When these possible events do occur and impact people, they become disasters, such as the 2005 Hurricane Katrina, the 2011 tsunami in Japan and the 2017 tornadoes in east Texas. These all resulted in widespread destruction and loss of life (**Figure 9.2a–b**).

© AMF Photography/Shutterstock.com

Figure 9.1a   Severe flooding occurred in and around Houston, Texas, as the result of Hurricane Harvey in late August 2017.

These scenes of unprecedented devastation have been etched in our minds and have made us increasingly aware of our human vulnerability to the awesome power of nature. However, these events are the result of natural processes involving physical, chemical and biological mechanisms and forces. When the processes occur, they become hazards for people and create disasters.

Source: NASA.

Figure 9.1b　Hurricane Chris off the east coast of the United States in July 2018.

Worldwide we are facing **natural disasters** that have an increasingly large impact on people. Data from the United Nations Office for the Coordination of Humanitarian Affairs outlined the global effects of natural disasters for 2016. In that year the number of deaths caused by the documented 342 disasters was 8,733, making it the second lowest in the past decade. The number of people affected by these natural disasters was 564 million and the economic damage was estimated to be $154 billion (USD).

natural disaster: The loss of life, injuries, or property damage as a result of a natural event or process, usually within a more local geographic area.

Hydrological disasters, mainly floods, and meteorological disasters (cyclonic storms) accounted for 80 percent of the events, while climatological (drought) and geophysical (earthquakes) comprised the remaining 20 percent.

United States Geological Survey.

Figure 9.1c　Kilauea resumed erupting on the island of Hawaii in May 2018, destroying more than 600 homes.

FEMA, photo by Andrea Booher.

Figure 9.2a Destroyed neighborhoods along the Gulf coast in Biloxi and Gulfport, Mississippi, after Hurricane Katrina struck the region in late August 2005.

National Weather Service.

Figure 9.2b Tornado damage was severe in east Texas during an outbreak on April 29, 2017.

Since 2006, China, the United States, India, Indonesia, and the Philippines are the top five countries most frequently affected by natural disasters. Coincidently, the first four of these are also the four most populated countries on Earth. In 2016 the greatest number of deaths was due to floods (4,731) while storms and earthquakes ranked second and third, killing 1,797 and 1,315 people, respectively. Such large numbers may appear abstract and difficult to conceptualize, but they are a harsh reality for many families who have lost loved ones, had their homes destroyed, or have watched their investments and economic future be destroyed by a natural disaster.

But what can we do? Natural hazards are natural phenomena, but their transition into disasters or catastrophes is often a result of the organization, distribution, and behavior of our society. Natural disasters know no political boundaries. Smoke from forest fires can make air unbreathable in neighboring nations, ash plumes from volcanoes can disrupt air traffic across the globe and emerging diseases and global warming impact the entire planet. Whether an extreme hazard event becomes a natural disaster depends on our ability to predict, prepare, and mitigate. Most decision makers and policymakers agree that the key to reducing the vulnerability and risk of human populations to natural hazards is the integration of disaster preparedness, **mitigation**, and prevention measures into future policy development. Yet funding patterns, an indicator of real priorities, show that disaster relief—not reduction or prevention—tops the list of disaster-management funding, which leads to continued losses and suffering after we recover and a disaster strikes again.

> **mitigation:** The act of making less severe or intense; measures taken to reduce adverse impacts on humans or the environment.

Fortunately, public awareness of the losses in human lives and property from natural disasters is starting to grow, mostly through media images of devastation. Once we understand and accept the need for a change in our life styles, such as recycling or seat-belt use, our willingness to make the change increases. But we still have a long way to go if we are to focus on hazard preparedness and prevention. Better information and public education about the natural environment and Earth processes are essential to reducing losses. Therefore, the focus of this book is to help develop an understanding of Earth's natural processes through the study of relationships between people and their environment. The information on natural hazards presented in this book is of practical value to people in making choices of where to live and understanding what hazards might be present so they may make the most informed decisions.

From Natural Hazards, to Disasters, to Catastrophes: Interactions with Natural Processes

In describing our interaction with natural processes we use the terms natural hazard, disaster, and catastrophe. A **natural hazard** is a potential event that could have a negative impact on people and their property resulting from natural processes in the Earth's environment. Some of these events include the possibility of occurrence of earthquakes, hurricanes, tsunami, floods, volcanic eruptions, droughts, landslides, and coastal erosion. These events are the result of natural Earth processes that have been operating over the lifetime of our planet; therefore they will continue to happen, regardless of whether or not humans are exposed to them. Only when these processes threaten humans and their property do they become hazards, and if they occur they can become a disaster. A natural disaster is the loss of life, injuries, or property damage as a result of a natural event or process. A natural disaster could be as small as an individual's house damaged in a wildfire with losses handled through a private insurance carrier or a local community devastated by flooding that could require assistance from government agencies. A **natural catastrophe** is considered a massive disaster often affecting a larger region and requiring significant amounts of time and money for recovery. A catastrophe can be so large that it devastates a metropolis such as the impact of Hurricane Katrina that requires large amounts of disaster relief from outside the community, which can stifle the economy of an entire country.

> **natural hazard:** An event or phenomenon that could have a negative impact on people and their property resulting from natural processes in the Earth's environment.
>
> **natural catastrophe:** A massive natural disaster often affecting a large region and requiring significant amounts of time and money for recovery.

Types and Characteristics of Natural Hazards

Natural hazards, and the disasters or catastrophes that result from them, come in a variety of forms. Natural hazards of terrestrial origin can be divided into three different categories: geologic, atmospheric (or hydro-meteorological), and environmental (or biological) (Table 9.1). These originate from the flow of energy and matter contained on or in our planet. Examples include geologic events such as earthquakes and volcanic eruptions, atmospheric events like hurricanes and tornadoes, and environmental events like wildfires and diseases. Natural hazards of extraterrestrial origin include the possibility of asteroid or comet impacts and solar-related events such as geomagnetic storms.

Table 9.1 Categories of Natural Hazards and Examples

TERRESTRIAL HAZARDS

Geologic	Atmospheric	Environmental
• Volcanic eruptions	• Global warming	• Wildfires
• Earthquakes	• Tropical cyclones	• Biological diseases
• Landslides	• Storms	• Insect infestations
• Land subsidence	• Tornadoes	• Environmental pollution
• Tsunamis	• Droughts	
• Coastal erosion	• Lightning	
• Floods	• Blizzards	

EXTRATERRESTRIAL HAZARDS

- Impacts from asteroids and comets
- Solar flares
- Geomagnetic storms

Hazards can also be categorized as **rapid-onset hazards** which expend their energy very quickly, such as volcanic eruptions, earthquakes, floods, landslides, thunderstorms, and lightning. These hazards can develop with little warning and strike rapidly. These are the hazards we hear about the most because of their violent effects and real-time media coverage. In contrast, **slow-onset hazards**, such as drought, insect infestations, disease epidemics, and climate change take years to develop and are usually neglected by the media as not newsworthy until long after the effects have started.

rapid-onset hazard: A hazard that develops with little warning and strikes rapidly. They expend their energy very quickly, such as volcanic eruptions, earthquakes, floods, landslides, thunderstorms, and lightning.

slow-onset hazard: A hazard that takes years to develop such as drought, insect infestations, disease epidemics, and global warming and climate change.

Other types of hazards include man-made *technological hazards* that originate in accidental or intentional human activity, such as oil and chemical spills, building fires, plane crashes, and acts of terrorism. *Anthropogenic hazards* are also caused by humans, but they affect our environment and ecosystem, which eventually affects us in the long run. These include things such as pollution and deforestation, that lead to global warming, ozone layer destruction, increased magnitude of storms and landslides, and other effects. We will explore some of the anthropogenic hazards that are linked to natural hazards and disasters.

Often, several hazards and processes can be linked to a single event and occur at the same time or be triggered by the main event. These multiple events must be taken into account in preparing in advance for a hazard. For example, along with strong winds, hurricanes are associated with intense precipitation that can cause flooding, erosion along the coast, and landslides on inland hill slopes. Volcanic eruptions can cause lahars (mudflows) and flooding, and if the volcano erupts or collapses in the ocean, it can cause a tsunami.

Natural hazards also frequently occur in a series, meaning they follow an initial onset. Each subsequent hazard acts upon the effects of the preceding event. For instance, a hurricane can down enormous stands of timber, an event that could result in an disastrous outbreak of detrimental insect populations and lead to increased wildfire risk in future years from the volume of dead wood. Likewise, during long periods of drought, wildfires that burn standing or downed timbers can denude hillsides of vegetation and thereby trigger soil erosion, flooding, and landslides.

Natural Hazards in the United States: What's at Stake?

The extraordinary natural, climatic, and geographic diversity of the United States exposes people to a wide range of natural hazards throughout the nation. Natural hazards such as earthquakes, volcanic eruptions, wildfires, storms, and drought strike nearly every part of the United States, exacting an unacceptable toll on life, property, natural resources, and economic well-being (**Figure 9.3**). Each year, natural hazards cause hundreds to thousands of deaths and cost billions of dollars in disaster aid, disruption of commerce, and destruction of homes and critical infrastructure. The cost of major disaster response and recovery continues to rise, and property damage from natural hazards events doubles or triples each decade. The following discussion is a short summary of some common natural hazards in the United States intended to help illustrate the need for understanding these events and reducing their occurrence and effects.

Severe Weather

Severe weather events such as tornadoes, hurricanes, storms, and heat waves strike many areas of the United States each year (**Figure 9.4**). Due to changes in population demographics and the establishment of more complex weather-sensitive infrastructures, the United States is more vulnerable to severe weather events today than in the past. In 2017 the estimated cost of disasters in the United States was $300 billion.

Over the past 30 years, coastal population growth has quadrupled, along with accompanying property and infrastructure development, to the extent that more than 75 million people now reside along hurricane-prone coastlines. At least 1,836 people lost their lives in Hurricane Katrina (**Figure 9.5**) and in the subsequent floods, making it the deadliest hurricane since the 1928 Okeechobee hurricane. The storm was also responsible for an estimated $81.2 billion (2005 U.S. dollars) in damage, making it the costliest natural catastophe in our history.

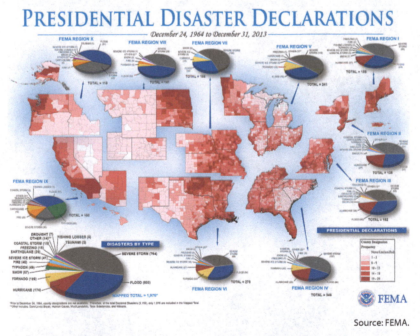

Source: FEMA.

Figure 9.3   Presidential Disaster Declarations in the United States by county from 1964 to 2013 reflect the regional geographic distribution and human impacts of earthquakes, floods, hurricanes, tornadoes, severe storms, and wildfires.

© Minerva Studio/Shutterstock.com.

Figure 9.4 The tornado that hit Moore, Oklahoma, in 2013 was 27 km long, 2 km wide and had winds over 320 km per hour. It was classified as an EF5—of about 1000 tornadoes to hit the US each year, only one on average reaches EF5 status.

Source: NASA.

Figure 9.5 Hurricane Katrina near peak strength on August 28, 2005. This developed into the most catastrophic storm to hit the United States in more than 80 years.

Tornadoes are more common in the United States than anywhere else in the world, with an average of 1,000 reported annually nationwide, resulting in an average of 80 deaths and over 1,500 injuries every year. The tornadoes that struck the Oklahoma City area on May 3, 1999, were some of the most devastating in history, destroying over 2,500 structures and causing over $1 billion in damage (**Figure 9.6**).

FEMA, photo by Andrea Booher.

Figure 9.6   Several supercell thunderstorms produced more than 70 tornadoes that struck central Oklahoma, southern Kansas, and northern Texas on May 3, 1999. Forty-six people were killed and more than 8,000 homes and businesses were damaged or destroyed, at a cost of more than $1.2 billion.

Rapid-onset hazards such as hurricanes and tornadoes strike with deadly force, but slow-onset hazards can also be deadly. Such is the case with heat waves. In July 1995, a heat wave in Chicago killed 739 people. From June to mid-August 2003 a severe heat wave struck continental Europe, killing more than 30,000 people.

Wildfires

Globally, wildfires, also known as wildland fires, engulfing millions of acres of land and encroaching on residential neighborhoods, have become all too familiar scenes (**Figure 9.7**). Since 2000 wildfires have burned

FEMA, photo by Mark Wolfe.

Figure 9.7   Lake City, Florida, May 15, 2007. The Florida Bugaboo Fire raged out of control in some locations. The U.S. Department of Homeland Security's Federal Emergency Management Agency (FEMA) authorized five Fire Management Assistance Grants between March 27 and May 10, 2007, to help Florida fight fires in 16 counties.

an average of 6.5 million acres in the United States annually, and in recent years the nation has spent over $2 billion annually for fire suppression. This does not include costs to communities and individuals in terms of structural losses or economic disruptions. The extreme fire seasons from 2000 to 2007 saw the largest areas burned by wildfires in the United States since the 1960s. The 2003 California fires burned over 743,000 acres (3007 km²), destroyed 3,300 homes, and killed 26 people. The October 2007 California wildfires were a series of wildfires in which at least 1,500 homes were destroyed and over 500,000 acres (2,000 km²) of land burned from Santa Barbara County to the U.S.–Mexico border (**Figure 9.8**). Two days into the fires, approximately 500,000 people from at least 346,000 homes were under mandatory orders to evacuate, the largest evacuation in the region's history. Nine people died as a direct result of the fire; 85 others were injured, including at least 61 firefighters.

NASA http://rapidfire.sci.gsfc.nasa.gov/subsets/?AERONET_La_Jolla/2007297/AERONET_La_Jolla.2007297.aqua.250m.jpg

Figure 9.8   NASA satellite photo from October 24, 2007, showing the active fire zones and smoke plumes in southern California from Santa Barbara County to the U.S.–Mexico border.

Wildfires are not restricted to the drier climates of the western United States. The 2007 Bugaboo Scrub Fire, a wildfire in the southeastern corner of the country, raged from April to June, ultimately became the largest fire in the history of both Georgia and Florida. A thick, sultry smoke from the fires blanketed the city of Jacksonville, Florida and the entire area of northeast Florida, and southeast Georgia for many days, reducing visibility and causing many health concerns.

Drought

Drought is a complex, slow-onset, nonstructural-impact natural hazard that affects more people in the United States than any other hazard, with annual losses estimated at $6 to $8 billion. Compared to all natural hazards, droughts are considered the leading cause of economic losses. The 1987–1989 drought cost an estimated $39 billion nationwide and was at one time the greatest single-year natural disaster in U.S. history (Hurricane Katrina topped that amount in a single event). In May 2014, 40 percent of the continental United States was in drought, the highest percentage recorded. Growing population, a shift in population to drier regions of the country, urbanization, and changes in land and water use have increased the magnitude and complexity of drought hazards.

Earthquakes

Each year the United States experiences thousands of earthquakes. An average of seven earthquakes with a magnitude of 6.0 or greater (enough to cause major damage) occur each year. About 75 million people in 39 states face significant risk from earthquakes, and earthquakes remain one of the nation's most significant natural hazard threats. The last major earthquake to strike a large urban area was the Northridge, California, earthquake that occurred on January 17, 1994 (**Figure 9.9**). Sixty-one people died as a result of the earthquake, and over 7,000 were injured. In addition, the earthquake caused an estimated $20 billion in damage and $49 billion in economic loss, making it one of the costliest natural disasters in U.S. history.

FEMA News Photo.

Figure 9.9 The January 17, 1994, Northridge, California, earthquake. Numerous highways were damaged and approximately 114,000 residential and commercial structures were damaged and 72 deaths were attributed to the earthquake. Damage costs were estimated at $25 billion.

Volcanoes

The United States is among the most volcanically active counties in the world with about 170 active or dormant volcanoes (**Figure 9.10**). During the past century, volcanoes erupted in Washington, Oregon, California, Alaska, and Hawaii, devastating thousands of square kilometers and causing substantial economic and societal disruption and loss of life. Since 1980, 45 eruptions and 15 cases of notable volcanic unrest have occurred at 33 volcanoes, producing lava flows, debris avalanches, and explosive blasts that have invaded communities, swept people to their deaths, choked major rivers, destroyed bridges, and devastated huge tracts of timber forests. Volcanic ash plumes ejected into the atmosphere can be a costly and serious danger to aircraft. A Boeing 747 sustained $80 million in damages when it encountered ash from Mount Redoubt in Alaska during a 1989 eruption (**Figure 9.11**).

Floods

Floods are the most frequent natural disaster in the United States and can be caused by hurricanes, weather systems, and snowmelt (**Figure 9.12**). Failure of levees and dams and inadequate drainage in urban areas can also result in flooding. Nearly 75 percent of federal disaster declarations are related to flooding, which, on average, kills about 140 people each year and causes $6 billion in property damage. In 1993, flooding in the Mississippi Basin resulted in an estimated $12 to $16 billion in damages across nine states in the Midwest. An increase in population and development in river floodplains and an increase in heavy rain events over the past 50 years have gradually increased the economic losses from floods.

Landslides

Landslide hazards occur and cause damage in all 50 states (**Figure 9.13**). Severe storms, earthquakes, volcanic activity, coastal wave attacks, and wildfires can cause widespread slope instability. Each year landslides in the United

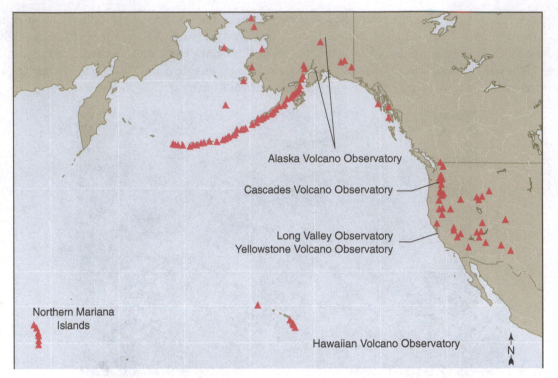

© Kendall Hunt Publishing Company.

Figure 9.10 The U.S. Geological Survey (USGS) is responsible for monitoring the Nation's 170 active volcanoes (red triangles) for signs of unrest and for issuing timely warnings of hazardous activity to government officials and the public. This responsibility is carried out by scientists at the five volcano observatories operated by the USGS Volcano Hazards Program and also by state and university cooperators.

Alaska Volcano Observatory/United States Geological Survey, Photo by Cyrus Read.

Figure 9.11 Redoubt Volcano has been active since a major eruption in 1989.

FEMA, photo by Andrea Booher.

Figure 9.12 Volunteers filled more than 300,000 sandbags in one day in order to build levees along the Red River in Fargo, North Dakota, as the river crested at 41 feet in late March 2009.

FEMA, photo by Adam DuBrowa.

Figure 9.13 Roads in Cougar, Washington, were destroyed in February 2009 by landslides and mudslides produced by a series of several winter storms that came off the Pacific Ocean.

States cause $1 to 2 billion in damage and more than 25 fatalities. The May 1980 eruption of Mount St. Helens caused the largest landslide in history—large enough to fill 250 million dump trucks. Human activities and population expansion, however, are major factors in increased landslide damage and costs. Wildfires also contribute to increased landslide and mudslide activity as they reduced the vegetation that holds the surface in place.

Disease Epidemics

While disease outbreaks lack the rapid-onset aspect of other disaster discussed above, they potentially present an even greater threat to the United States population. The West Nile virus epidemic of 2002 demonstrates how outbreaks can become a public health emergency. In 2002, a total of 4,156 cases were reported, including 284 fatalities, not to mention the deaths of hundreds of thousands of birds and mammals. The cost of West Nile-related health issues was estimated at $200 million and exposed vulnerabilities in the U.S. health care system. It also served as both a warning and a wake-up call for the nation.

Human Population Growth: Moving Toward Disaster

Complicating all natural hazards issues is the rapid growth of our world population (**Figure 9.14**), along with the goal of having a better standard of living which is heavily tied to natural resource use. Remember that a natural hazard only becomes a disaster when it affects people where they live or work. The increase we see compared to past natural disaster occurrences is due mostly to expansion of our population into hazardous areas which is occurring at a fast rate. The biggest population increases have been in cities and their suburbs (**Figure 9.15**). In the United States between 1950 and 2018, the urban population grew from 64 percent of the total population to 81 percent. The worldwide urban rate is only 54%, making the United States a very urbanized country. Densely populated cities can be easy targets for natural catastrophes as witnessed in New Orleans in 2005. Mexico City, with over 20 million people, is a megacity in a region of earthquake and volcanic hazards. The southern and western regions of the United States (areas prone to drought, wildfires, hurricanes, earthquakes, and mudslides) are expected to grow by 32 and 51 percent, respectively, by the year 2050.

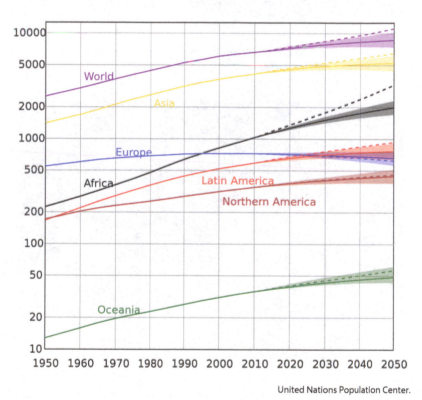

United Nations Population Center.

Figure 9.14 Population growth curve showing different parts of the world. North America and Europe maintain a fairly constant growth while Asia and Africa have much higher growth rates.

The increase in our world population by over 5 billion people during the past 200 years is without precedent in human history (**Figure 9.16**). The world population is estimated to have been only about 5 million people 10,000 years ago, but the nearly flat population curve began to rise about 8,000 years ago when agriculture and domestication of animals began to replace a hunter-gatherer culture. For thousands of years people lived in sparsely settled rural areas and growth rates were still very low, so population increased slowly but steadily to about 400 million by the Middle Ages (1100 to 1500). A sharp drop in population occurred when the Black Death struck Europe in the mid-1300s, but by 1700 the world population rebounded to about 650 million. Since that time, **exponential growth** in world population reached the world's first billion by the early 1800s, 2 billion by 1930, 3 billion by 1960, and 6 billion by 2000 (**Figure 9.17**). It is estimated that world population will reach 9.8 billion by 2050.

Figure 9.15 Cities throughout the world have grown rapidly in the past fifty years, using many more resources than in the past.

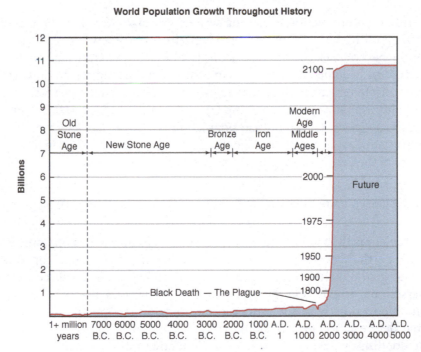

Source: Population Reference Bureau; and United Nations, *World Population Projections to 2100* (1998). © 2006 Population Reference Bureau.

Figure 9.16 Human population growth curve. Source: As shown and http://www.prb.org/Publications/GraphicsBank/ PopulationTrends.aspx.

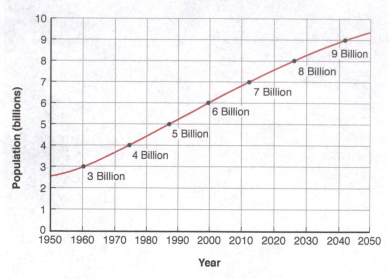

World Population: 1950–2050

Source: U.S. Census Bureau, International Data Base, August 2006 version.

Figure 9.17 **Latest projections of world population from the U.S. Census Bureau showing that world population increased from 3 billion in 1959 to 6 billion by 1999, a doubling that occurred over 40 years. The Census Bureau's latest projections imply that population growth will continue into the twenty-first century, although more slowly. The world population is projected to grow from 6 billion in 1999 to 9 billion by 2042, an increase of 50 percent in 43 years.**

exponential growth: Growth in which some quantity, such as population size, increases by a constant percentage of the whole during each year or other time period; when the increase in quantity over time is plotted, this type of growth yields a curve shaped like the letter J.

The rapid increase in human population is causing serious shortages of resources, including oil, food, and water, and is heavily degrading the environment. Some resources are renewable, such as water, whereas many others such as fuels (oil, natural gas, coal) and minerals (ores for copper, iron, etc.) are not. From population and natural resource data it becomes clear that it is impossible in the long run to support exponential population growth with our finite resources. Scientists are worried that it will be impossible to supply resources and a high-quality environment for the billions of people who will be added to the planet in the twenty-first century. Some scientists suggest that the present population is already above our planet's **carrying capacity**, which is the maximum number of people the Earth can hold without causing environmental degradation that reduces the ability of the planet to support the population.

carrying capacity: The maximum population size that can be regularly sustained by an environment.

Compounding the problem is the fact that technologically advanced societies consume greater quantities of resources that are only partially deemed essential, with the remaining consumed for convenience or luxury. Therefore, to reach our goal of achieving a higher standard of living, our resource consumption increases substantially per person, not to mention the vast quantities of wastes generated. Human links to environmental degradation, due to our thirst for using more resources, can greatly increase the risks of natural hazards. For example, deforestation for wood production and land development (also a resource) can lead to more landslides and flooding. Therefore, population growth and resource consumption leads to more loss of life and property from individual hazard events, as well as more hazards or higher-magnitude hazards due to resource development.

One of the obvious ways to reduce natural disasters and catastrophes would be population control, for which there is no easy answer. Left unabated, some scientists predict that population growth will take care of itself through wars (mostly over resources such as land, fuel, minerals, water, etc.) and other catastrophes such as famine, disease, and ecosystem collapses. Others believe we will find better ways to control population through increased education, improved regulation of resources and space, and being environmentally friendly.

Impacts from Natural Hazards: Human Fatalities, Economic Losses, and Environmental Damage

Human Fatalities

Increasing world populations in urban areas and coastal regions has resulted in more people living in hazardous areas. As a result, more than 255 million people globally were affected by natural disasters each year between 1994 and 2003. During the same period, these disasters claimed an average of 58,000 lives annually, with a range of 10,000 to 123,000.

The five deadliest disasters since the year 2000 are listed in Table 9.2 and the 10 deadliest historical disasters are listed in Table 9.3. Because many of these events occur in areas that have inadequate communication and reporting capabilities, the fatality numbers are often not precise and can vary. Notice that as deadly as the 2004 Asian disaster was, it was not the deadliest in human history. Also notice that the greatest disasters occurred where human population density is high in a belt of poorer countries running from China and Bangladesh through India, northwestward into Iran and Turkey. Population growth, urbanization, and the inability of poor populations to escape from the vicious cycle of poverty make it more likely that there will be a continued, and increased, number of people who are vulnerable to natural hazards.

Economic Losses

The deaths and injuries caused by natural disasters are what grab our immediate attention, but there are economic losses as well that are increasing at a staggering rate (**Figure 9.18**). Since 1980 there have been hundreds of natural disasters in the world resulting in over $500 billion dollars in damage and thousands of deaths. The economic cost associated with natural disasters has increased 14-fold since the 1950s.

The list of most expensive events is dominated by hurricane storm events along coastal regions (Table 9.4). Tables 9.2 and 9.4 show that there is a disparity between the number of deaths in poorer countries and economic loss among the wealthy countries. The poorer, underdeveloped countries suffer increasing numbers of deaths, whereas developed countries suffer greater economic losses. In both the developed and undeveloped countries, population growth once again has led to the increases, with more people living in more dangerous places. However, in developed countries the number of deaths has not increased due to better prediction, forecasting, and warning systems, as well as safer buildings. Therefore, disasters cause the loss of property and lives wherever they occur, but more highly developed countries usually lose more property while poorer countries always lose more lives. Scientific predictions and evidence indicate that global climate change will

Table 9.2 **Five Deadliest Natural Disasters, 2000 to 2017**				
Date	**Event**	**Region**	**Overall Losses**	**Fatalities**
December 26, 2004	Earthquake, tsunami	South Asia		230,000
January 12, 2010	Earthquake	Haiti	$14 billion	160,000
May 2, 2008	Cyclone	Myramar	$4 billion	138,000
May 12, 2008	Earthquake	Sichuan, China	$29 billion	88,000
October 8, 2005	Earthquake	Kashmir, Pakistan	$5 billion	>87,000

Table 9.3 Ten Deadliest Natural Disasters

Rank	Event	Location	Date	Death Toll (Estimate)
1	1931 Yellow River flood	Yellow River, China	Summer 1931	850,000–4,000,000
2	1887 Yellow River flood	Yellow River, China	September–October 1887	900,000–2,000,000
3	1970 Bhola cyclone	Ganges Delta, East Pakistan	November 13,1970	500,000–1,000,000
4	1938 Huang He flood	China	1938	500,000–900,000
5	Shaanxi earthquake	Shaanxi Prov-ince, China	January 23,1556	830,000
6	1839 India cyclone	Coringa, India	November 25,1839	3,000,001
7	1642 Kaifeng flood	Kaifeng, Henan Province, China	1642	300,000
8	2004 Indian Ocean earthquake/tsunami	Indian Ocean	December 26,2004	225,000–275,000
9	Tangshan earthquake	Tangshan, Chi-na	July 28,1976	242,000*
10	1138 Aleppo earth-quake	Syria	1138	230,000

* Official government figure. Estimated death toll as high as 655,000.

increase the number of extreme events, creating more frequent and intensified natural hazards such as floods, storms, hurricanes, and droughts.

Environmental Damage

Natural hazards not only affect humans, but also affect and shape the environment and biodiversity. In August 1992 Hurricane Andrew had a strong economic effect not only on the city of Homestead, Florida, but also on the ecosystem of the nearby Everglades and other coastal waters. Despite the destruction in Homestead, timber losses during the winter storms of 1993–1994 cost insurers more than the total payout from Hurricane Andrew. The drought in 1988 in the Midwest and the 1993 floods both had impacts on riverine ecosystems as well as downstream coastal ecologies of the Gulf of Mexico, by changing water salinity and the addition of toxins and other pollutants washed out upstream by the floods. The floods also accelerated the spread of invasive species, like the Zebra mussels, in the Mississippi watershed. These few examples show that strategies for protecting renewable resources and biodiversity must be taken into account when looking at the effects of natural hazards.

Human Vulnerability and Natural Hazards

Each year natural hazards are responsible for causing significant loss of life and staggering amounts of property damage, and they continue to adversely affect millions of people every day. Statistics published by the International Strategy for Disaster Reduction, a subsidiary of the United Nations, show that there is an

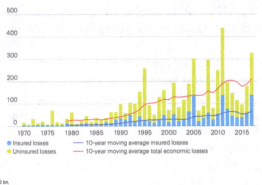

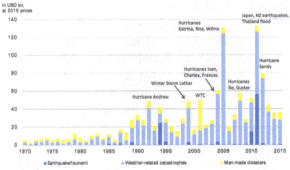

Source: Swiss Re Institute.

Figure 9.18 **Global overall (economic) and insured losses.**

increasing trend of natural disasters, especially over the last several decades. However, scientists don't see Earth as becoming more violent; instead, because of population and land management trends in our society, humans are increasingly becoming more *vulnerable* to the natural hazards, and more people are being impacted by these events when they do occur. A summary of some reasons for the increase in disaster and catastrophe losses includes:

Table 9.4 **Natural Disasters 1980-2014, 10 Costliest Natural Disasters Ordered by Insured Losses**

Date	Loss Event	Region	Overall Losses	Fatalities
2011	Tohoku earthquake	Japan	$235 billion	16,000
1995	Kobe earthquake	Japan	$100 billion	6,434
2005	Hurricane Katrina	USA	$108 billion	1833
1988	North American drought	USA	$78 billion	>10,000
1980	Heat wave and drought	USA	$55 billion	>1,700
2012	Superstorm Sandy	USA	$50 billion	286
1994	Northridge earthquake	USA	$42 billion	57
2004	Chuestsu earthquake	Japan	$34 billion	68
2010	Maule earthquake	Chile	$31 billion	525
2008	Sichuan earthquake	China	$29 billion	87,419 dead or missing

- Global population growth. In 1960, for example, there were 3 billion people living on Earth, and by 2000 there were 6 billion—a doubling in only 40 years. In 2011 the Earth's population reached 7 billion people.
- Increased settlement and industrialization of hazardous lands subject to floods, landslides, hurricanes, earthquakes, wildfires, volcanoes, and other hazards.
- Concentration of population and values in conurbations (when towns expand sufficiently that their urban areas join up with each other) and the emergence of numerous megacities—even in hazardous regions (e.g., metropolitan Tokyo, with more than 34 million inhabitants).
- The increased demand for natural resources and rising standard of living in nearly all countries of the world produces growing accumulations of wealth and the need for increased insurance. Losses thus escalate in the event of a disaster or catastrophe.
- The vulnerability of modern societies and technologies in the form of structural engineering, lifelines (water, sewer, electrical, etc.), transportation services, and other networks that are both fragile and costly to repair when damaged by natural hazards. Such effects are mostly due to lack of scientific understanding or education and lack of awareness of the hazards.
- Global changes in environmental conditions, climate change, water scarcity, and loss of biodiversity, mostly due to human activities and interactions with natural hazards.
- Globalization of the world economy now makes us all vulnerable to disasters wherever they occur, for example, by shutting off supplies of goods.

Therefore, natural hazard events are only classified as disasters or catastrophes when people or their properties are adversely affected. An earthquake in the Gobi Desert or gales in the Antarctic are not natural disasters if they do not have any impact on human such as or property. Conversely, a natural event that is considered to be a normal scale event in some places, such as the natural flooding of a stream, can quickly become a disaster in a region that is densely populated and poorly prepared.

Studying Natural Hazards: Fundamental Principles for Understanding Natural Processes

People are usually surprised to learn that every day the Earth experiences earthquakes, volcanic eruptions, landslides, floods, fires, meteor impacts, extinctions, and storms. Natural disasters are often perceived as being "acts of God," with little causal relationship to human activities. However, a growing body of evidence points to the effects of human behavior on the global natural environment and on the possibility that certain types of natural disasters, such as floods, may be increasing as a direct consequence of human activity. Natural hazards, which the geologic record shows have been continuously shaping our planet for hundreds of millions of years (and operating for billions of years), are not problems to be solved but are a critical part of how the Earth functions. Ecosystems and individual species have evolved to coexist with these hazards. Natural hazards become disasters only when natural forces collide with people, structures, or other property. Understanding how the Earth works and our relationship with it is the first step in reducing the impacts of natural hazards. The following concepts serve as a basic framework for our study.

Before a Natural Hazard Strikes

When disasters strike, we find that areas that had properly prepared for the hazard suffered fewer losses and recovered more quickly than those that did not have preparedness programs in place. Being prepared requires seven steps: (1) hazard process research and development, (2) hazard identification, (3) risk assessment, (4) risk communication, (5) mitigation; (6) prediction, and (7) preparedness, described as follows:

1. **Hazard process research and development.** The science activities dedicated to improving understanding of the underlying processes and dynamics of each type of hazard. This includes fundamental

and applied research on geologic, meteorological, epidemiological, and fire hazards; development and application of remote-sensing technologies, software models, infrastructure models, and organizational and social behavior models; emergency medical techniques; and many other science disciplines applicable to all facets of disasters and disaster management.

Knowledge of how hazards operate is the fundamental first step in understanding how to reduce impacts to humans and the environment. Basic research leads to fundamental breakthroughs in understanding natural processes, such as the theory of plate tectonics that greatly improved our knowledge of volcanic and earthquake hazards. Long-term scientific studies of coupled ocean–atmosphere oscillations enabled NOAA in 1997 to predict a powerful El Niño during 1997–1998 that produced large rainfall-induced landslide events in California. NOAA and USGS worked with the California Office of Emergency Services to prepare landslide hazard maps showing former landslide deposits and debris-flow source areas along with dangerous rainfall thresholds for the San Francisco Bay area. This prediction allowed individuals and emergency responders to prepare for the extreme hazard event. No loss of life occurred from landslides in this area, as compared to 25 deaths in comparable storms from an El Niño event in 1982.

2. **Hazard identification.** Determining which hazards threaten a given area. This includes understanding an area's history of hazard events and the range of severity of those events. The continuous study of the nation's active faults, seismic risks, and volcanoes is included in this category, as are efforts to understand the dynamics of hurricanes, tornadoes, floods, droughts, and other extreme weather events.

   Since natural disasters are repetitive events caused by natural processes, we can identify potential hazards and assess their probability of occurring in the area. Some events may not occur more than once in a person's lifetime or even for several generations. Also, many events such as volcanic eruptions—with hundreds to thousands of years elapsing between eruptions—are not on the same timeline as humans (average life expectancy of 70 years), thus we often do not see the danger. Therefore we become complacent by not knowing there is a hazard until disaster strikes. It is important, therefore, to take our working knowledge of natural hazards (from hazard process research) and identify their potential occurrences. Also known as *hazard assessment*, hazard identification consists of determining *where* and *when* hazardous events occurred in the past and how they affected the area.

   Hazards vary widely in their power, and the extent to which hazards impact an area is partly a function of their **magnitude**, the energy released, and the interval between occurrences, or frequency. All other factors being equal, larger-magnitude events occur less frequently but cause more damage than those of a smaller magnitude. The **recurrence interval** is how frequently, on average, a hazard event of a certain magnitude occurs. This information is important and useful to planners and public officials responsible for making decisions in the event of a possible disaster.

> **magnitude:** The size or scale of an event, such as an earthquake.
>
> **recurrence interval:** (1) A statistical expression of the average time between events equaling or exceeding a given magnitude. (2) The average time interval, usually in years, between the occurrence of an event of a given magnitude or larger.

Lessons and Challenges

Today we hear more about natural hazards and have seen more vividly the negative effects these hazards have on people, property, and the environment. We cannot totally avoid natural hazards, but how we perceive and react to them determines our destiny. We can act to minimize and reduce their impacts. After all, natural hazards do not become disasters or catastrophes unless the people and communities they touch are not prepared to deal with them. The previously described strategy of reducing the impacts from inevitable natural hazards can be strengthened as we learn more about natural processes and work together to mitigate their hazards. We still have a long way to go if we are to focus on preparedness and prevention rather than the usual bandage solutions.

Our general approach to disaster management has remained reactive, focusing on relief, followed by recovery and reconstruction. Prevention planning or community preparedness has been poorly funded and has not been a major policy priority. Relief remains media-friendly, action oriented, easy to quantify (e.g. tons of food distributed, number of family shelters shipped), and readily accountable to donors as concrete actions in response to a disaster. With the increase in magnitude of disaster impacts, mostly in poorer developing countries, concern is rising over inadequate preparedness and prevention. Natural disasters create serious setbacks to the economic development of countries. This has been proven time and time again, particularly in the last twenty years with the devastation caused by Hurricane Mitch in Central America; the Yangtze River floods in China; earthquakes in Turkey, Iran, and Indonesia; the Japanese earthquake and tsunami; and Hurricanes Katrina and Sandy in the United States. All of these events diverted possible economic development funds toward reconstruction.

One of the biggest challenges is to change our culture from one of continually responding to and recovering from a natural disaster to one of learning to live with natural hazards through mitigation efforts, which are more cost efficient. The USGS and the World Bank calculated the worldwide economic losses from natural disasters in the 1990s could have been reduced by $280 billion if $40 billion had been invested in disaster preparedness, mitigation, and prevention strategies. Unfortunately, many policies concerning hazards are driven by politics and special interests.

Policymakers must understand that mitigating hazards is prudent both politically and economically. Developers, companies, and local governments often allow or encourage people to move into known hazardous areas. Developers, real estate companies, and some corporations are reluctant to admit to the presence of hazards for fear of reducing the value of land and scaring off potential clients. So the adage "buyer beware" applies here. Many local governments consider any news about hazards to be bad for growth and economy and resist restrictive zoning for fear of losing their tax base. Even developers and private individuals view restrictive zoning as infringement on their property rights and doing as they please with their property. But having good land-use practices can lead to an improved quality of life, new recreation opportunities using open-space lands, and a stronger tax base in the long run.

Instead of mitigating hazards, natural disasters continue to kill and inflict human suffering as well as destroy property, economic productivity, and natural resources. Financial assets are diverted from much needed investments in our future—such as research, education, and the reduction of crime, disease, and poverty. Inaction today regarding natural hazards compromises our safety, economic growth, and environmental quality for future generations. Natural disasters greatly jeopardize our goal of sustainable development, which meets the needs of the present without compromising the ability of future generations to meet their needs. Sustainable development is usually understood to require (1) economic growth, (2) protection of the environment, and (3) sustainable use of ecological systems. However, a fourth criterion of equal importance is that sustainable development must be resilient to the natural variability of the Earth and the solar system. By variability, we mean natural hazards events such as earthquakes, tsunamis, hurricanes, and so on that brutally interrupt our societies and ecosystems. These events have not merely punctuated Earth's history as much as they have defined it, and so understanding them is the first step in our success in coping with them.

The Four Spheres
By Ingrid Ukstins and David Best

Topic 10

We live on a planet that is complex and dynamic, and has been in a continuous state of change since its origin about 4.6 billion years ago. The present day landscapes and features we observe and live with are still changing as the result of interactions among Earth's many internal and external components. In studying the Earth, it is clear that the planet can be viewed as a system of interacting parts or spheres (or layers). The atmosphere, hydrosphere and cryosphere, biosphere, and lithosphere, and all of their components, can be defined separately, but each is continuously interacting as a whole called the **Earth system** (**Figure 10.1**). Geologists attempt to understand the past, present, and future behavior of the whole Earth system by taking a planetary approach to the study of global change.

Source: NOAA.

Figure 10.1 **View of the Great Lakes region of the United States, where the spheres on Earth interact constantly.**

43

The Earth's surface is a major boundary in the Earth system where the four spheres intersect, where energy is transferred, and therefore, where we find volcanoes, landslides, earthquakes, severe weather, and long-term climate changes that impact us. The dynamic Earth system is powered by energy from two sources, one external and the other internal. Energy from the Sun drives the external processes that occur in the atmosphere, hydrosphere, and at Earth's surface of the lithosphere. Processes such as weather, climate, ocean circulation, and erosion of the land are driven by the Sun's energy. Earth's interior portion of the lithosphere is the second source of energy. Heat remaining from when the planet formed along with heat produced continuously from decay of radioactive elements, powers the internal processes to produce volcanoes, earthquakes, and mountains. This constant motion of the Earth's spheres will continue as long as the energy sources remain.

Earth's Spheres

Atmosphere

Earth is surrounded by a thin envelope of gases called the **atmosphere**. The gases consist of a mixture of primarily nitrogen (78 percent), oxygen (21 percent) and minor amounts of other gases, such as argon (less than 1 percent), carbon dioxide (0.035 percent) and water vapor. These gases are held near Earth by gravity but thin rapidly with altitude. Almost 99 percent of the atmosphere is concentrated within 30 kilometers of Earth's surface and forms an insignificant fraction of the planet's mass (less than 0.01 percent). Despite its modest dimensions, the atmosphere plays an integral role in planet dynamics. It supports life for both animals that need oxygen and plants that need carbon dioxide. It supports life indirectly by controlling climate, acting as a filter, and being a blanket that helps retain heat at night and shield us from the Sun's intense heat and dangerous ultraviolet radiation. The atmosphere is also responsible for transporting heat and water vapor place to place on Earth, with solar heat being the driving force of atmospheric circulation. Energy is continuously being exchanged between the atmosphere and the surface. This interchange influences the different effects of weather and climate. The atmosphere is divided into four layers (**Figure 10.2**) in which different factors control the temperature.

Layers of the Atmosphere

The **troposphere** is the first layer above the surface and contains half of the Earth's atmosphere and about 75 percent of the total mass. This is where all plants and animals live and breathe. Virtually all of the water vapor and clouds exist in this layer and almost all weather occurs here. The air is very well mixed and the temperature decreases with altitude. Air in the troposphere is heated primarily from the ground up because the surface of the Earth absorbs energy and heats up faster than the air does. The heat is spread through the troposphere because the air is slightly unstable. At the tropopause the steady decline in temperature with altitude ceases abruptly and the cold air of the upper troposphere is too dense to rise above this point. As a result, little mixing occurs between the troposphere and the stratosphere above. The tropopause and the troposphere form the lower atmosphere.

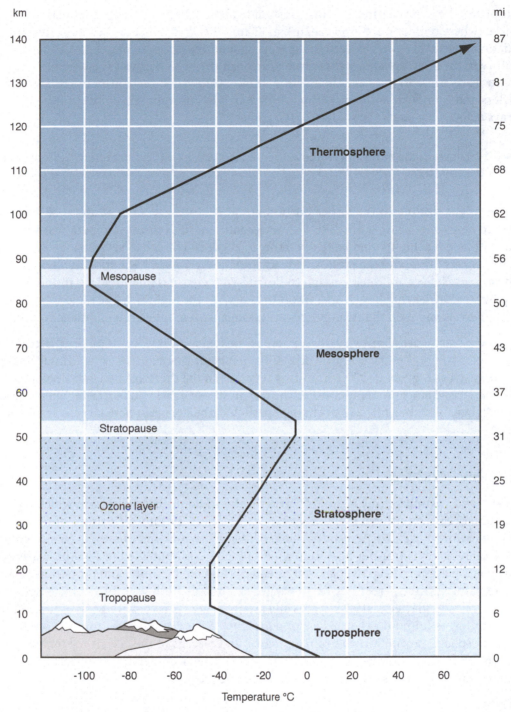

Figure 10.2 Earth's atmosphere consists of four layers. The troposphere has the greatest effect on activity at the surface. The solid line shows the average temperature; note the variation of temperature with increased altitude. The ozone layer shields the surface from incoming ultraviolet radiation.

Source: NOAA.

troposphere: The lowest part of the atmosphere that is in contact with the surface of the Earth. It ranges in altitude above the surface up to 10 or 12 kilometers.

The next layer is the **stratosphere** where many jet aircraft fly because it is very stable. It extends to about 55 km above the Earth and this layer plus the troposphere make up 99 percent of the total mass of the atmosphere.

In the stratosphere, the temperature remains constant to about 20 km and then it increases with altitude to the stratopause (where temperature levels again) at 50 km. The stratosphere is heated primarily from above by solar radiation instead of from below as in the troposphere. The presence of ozone (O_3) causes the increasing temperature in the stratosphere by absorbing ultraviolet rays which causes the temperature in the upper stratosphere to reach those found near the Earth's surface. Ozone is more concentrated around an altitude of 25 kilometers and is important in protecting life on Earth by absorbing much of the harmful high-energy ultraviolet rays before they reach the surface of the planet.

stratosphere: The level of the atmosphere above the troposphere, extending to about 50–55 kilometers above the Earth's surface.

Ozone concentrations decline in the upper portion of the stratosphere, and, at about 55 km above Earth, temperature decreases once again with altitude. This second zone of declining temperature is the mesosphere where the atmosphere reaches its coldest temperature of about 100°C due to very little radiation absorption. The mesosphere starts just above the stratosphere and extends to 85 km high until reaching the mesopause. The gases in the mesosphere are thick enough to slow down meteorites hurtling into the atmosphere, where they burn up by friction, leaving fiery trails, "shooting stars," in the night sky. The regions of the stratosphere and the mesosphere, along with the stratopause and mesopause, are called the middle atmosphere.

The thermosphere starts just above the mesosphere and extends to 600 km high. The thermosphere contains auroras and is where the space shuttle orbits. The temperature increases with altitude due to the Sun's energy and can reach as high as 1,727°C. The air is very thin in the thermosphere but small changes in solar energy can cause large changes in temperature. The thermosphere also includes the region of the atmosphere called the ionosphere, which is filled with charged particles. The thermosphere layer is known as the upper atmosphere.

Hydrosphere and Cryosphere

The **hydrosphere** and **cryosphere** include all of Earth's water in liquid and solid form which continually circulate among the oceans, continents, and atmosphere. The most prominent feature of the hydrosphere is the global ocean, which covers 71 percent of the Earth's surface and contains about 97 percent of its water (**Figure 10.3**). Ocean currents transfer heat from areas of high heat input like the equator to areas of low heat input like the poles, helping to keep the planet at an equilibrium temperature, and the oceans also alter global climate. The remaining three percent of Earth's water is found as freshwater in the hydrosphere and cryosphere and is in ice caps, glaciers, lakes, streams and groundwater. This freshwater is responsible for creating many varied landforms found on the surface of our planet. Water evaporates from the oceans and moves through the atmosphere, precipitating as rain and snow, and returns to the oceans in streams, groundwater, and glaciers. This movement of water over Earth's surface weathers, erodes and transports rock material and deposits it as sediment thus constantly modifying the landscape.

hydrosphere: The part of the Earth composed of water including clouds, oceans, seas, lakes, rivers, underground water supplies, and atmospheric water vapor.

cryosphere: The frozen water part of the Earth system, including ice caps and glaciers.

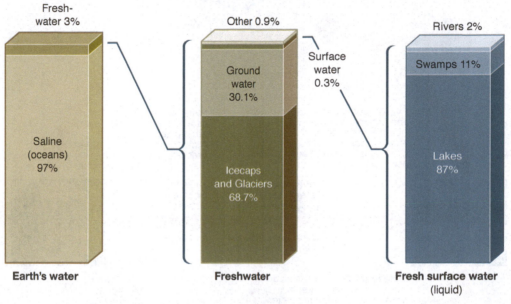

Distribution of Earth's Water

Fresh-water 3%

Saline (oceans) 97%

Earth's water

Other 0.9%

Ground water 30.1%

Icecaps and Glaciers 68.7%

Freshwater

Surface water 0.3%

Rivers 2%

Swamps 11%

Lakes 87%

Fresh surface water (liquid)

Figure 10.3 Most of Earth's water is found in the oceans. About two-thirds of the relatively small amount of freshwater is confined to cold environments.

Biosphere

The **biosphere** represents all life on Earth that exists in a thin layer in the uppermost lithosphere, the hydrosphere, and the lower atmosphere (**Figure 10.4**). Plants and animals depend on the physical environment for the basics of life and are affected by Earth's environment. Organisms breathe air, need water, and live in a relatively narrow temperature range. Land organisms depend on soil, which is part of the lithosphere. Ultimately, the biosphere strongly influences the other spheres. Essentially, the present atmosphere has been produced by chemical activity of the biosphere, mainly through photosynthesis by plants. Organisms control some of the composition of the oceans; for example, many organisms extract calcium carbonate from seawater for their bones and shells, and when they die, this material settles to the seafloor and accumulates as sediment that lithifies to limestone. Also, the biosphere forms Earth's natural resources of coal, oil, and natural gas. Therefore, many of the rocks in the Earth's crust can have their origins traced back to some form of biological activity.

biosphere: The living and dead organisms found near the Earth's surface in parts of the lithosphere, atmosphere, and hydrosphere. The part of the global carbon cycle that includes living organisms and biogenic organic matter.

Atmosphere

Atmospheric gases and
precipitation contribute to
weathering of rocks.

Evaporation, condensation,
and precipitation transfer
water between atmosphere
and hydrosphere, influencing
weather and climate and
distribution of water.

Plant, animal, and human
activity affect composition of
atmospheric gases.
Atmospheric temperature and
precipitation help to determine
distribution of Earth's biota.

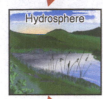

Hydrosphere

Plants absorb and transpire water.
Water is used by people for domestic,
agricultural, and industrial uses.

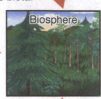

Biosphere

Water helps determine abundance,
diversity, and distribution of organisms

Plate movement affects size,
shape, and distribution of
ocean basins. Running water
and glaciers erode rock and
sculpt landscapes.

Organisms break down rock
into soil. People alter the
landscape. Plate movement
affects evolution and
distribution of Earth's biota.

Heat reflected from land surface affects
temperature of atmosphere. Distribution
of mountains affects weather patterns.

Lithosphere
(plates)

Convection cells within the mantle
contribute to movement of plates
(lithosphere) and recycling of
lithospheric material.

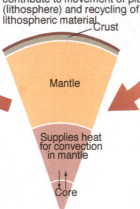

Crust

Mantle

Supplies heat
for convection
in mantle

Core

Figure 10.4 Simplified diagram of Earth's subsystems (atmosphere, hydrosphere, biosphere, and lithosphere) and their interactions showing examples of how material and energy is cycled throughout the Earth system.

Lithosphere

Beneath our feet are the rocks and soil that make up the solid part of our planet known as the **lithosphere**, or solid Earth. The solid materials of Earth are separated into layers based on chemical composition and physical properties. The compositional layers consist of the hot, central core, surrounded by a large mantle that comprises the majority of Earth's volume, and a thin, cool **crust** at the surface. The layers separated by different physical properties include the **inner core**, **outer core**, **asthenosphere**, and **lithosphere** (**Figure 10.5**). The characteristics of these layers are controlled by an increase in density, temperature, and pressure with depth.

lithosphere: The soils, sediments, and rock layers of the Earth including the crust, both continental and beneath the ocean floors.

crust: The rocky, relatively low density, outermost layer of the Earth.

inner core: The solid central part of Earth's core.

outer core: The liquid outer layer of the core that lies directly beneath the mantle.

asthenosphere: The uppermost layer of the mantle, located below the lithosphere. This zone of soft (plastic), easily deformed rock exists at depths of 100 kilometers to as deep as 700 kilometers.

lithosphere: The outer layer of solid rock that includes the crust and uppermost mantle. This layer, up to 100 kilometers (60 miles) thick, forms the Earth's tectonic plates. Tectonic plates float above the more dense, flowing layer of mantle called the asthenosphere.

Layers of the Lithosphere

The crust is a compositionally defined layer of the Earth and is the outermost and thinnest layer of the planet. It can be divided into two types based on the composition. The continental crust is thick (20–70 km) and consists of a wide variety of rock types but has an overall composition similar to granite, with an average density of 2.7 g/cm³. Continental rocks are dominated by the elements silicon, oxygen, potassium, aluminum and iron. The **oceanic crust** is thin (5–10 km) and consists of dark, igneous basalt that contains more iron and magnesium than granite, thus making oceanic crust denser (3.0 g/cm³) than continental crust. Both continental and oceanic crusts are cooler relative to the layers below, and thus consist of hard, strong rock that is rigid and behaves as a brittle material. The chemical makeup of the crust overall is dominated by eight main elements (Table 10.1).

oceanic crust: That part of the Earth's crust of the lithosphere underlying the ocean basins. It is composed of basalt and has a thickness of about 5 km.

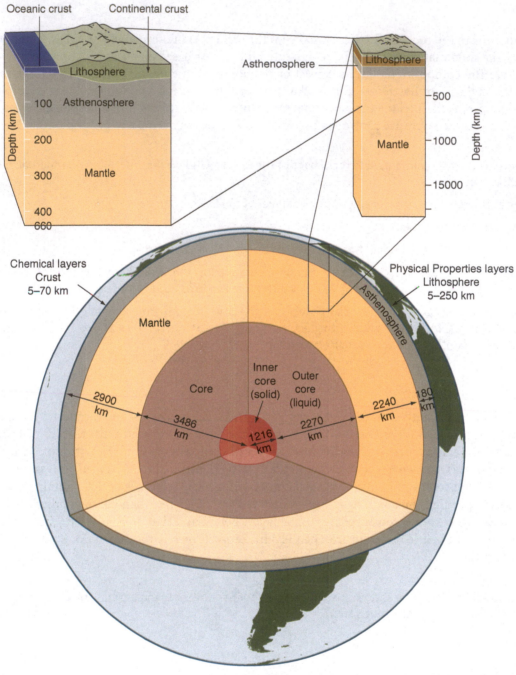

Figure 10.5 Internal structure of the lithosphere. Notice that the lithosphere includes both the crust and the uppermost layer of mantle that overlies the asthenosphere.

The compositional layer of the **mantle** is the middle layer between the crust and core and occupies about 83 percent of Earth's volume. It is denser than the crust with a density ranging from about 3.3 g/cm³ in its upper part to about 5.7 g/cm³ near the contact with the outer core. The composition is mostly of peridotite which is a dark, dense igneous rock containing **silicate** minerals (compounds of silicon and oxygen atoms) with abundant iron and magnesium like olivine and pyroxene. Although the chemical composition is uniform throughout the mantle, the temperature and pressure increase with depth. Temperatures near the top of the mantle approach 1000°C, increasing to about 3,300°C near the mantle-core boundary. At these temperatures the mantle should be molten rock. However, the high pressure from the thickness of overlying rocks prevents the mantle rocks from melting. This is because rocks will expand as much as 10 percent when they melt and the high pressures make it difficult for rock to expand and therefore inhibit melting. If the combination of temperature and pressure effects is close to the rock's melting point, the rock remains solid but loses its strength which makes it weak and plastic. If the temperature rises, or the pressure decreases, the rock will begin to melt and form magma (molten rock). These temperature and pressure changes with depth cause the strength of the rocks to vary with depth and create layering within the mantle, as well as in the core below.

mantle: The layer of the Earth below the crust and above the core. The uppermost part of the mantle is rigid and, along with the crust, forms the 'plates' of plate tectonics. The mantle is made up of dense iron and magnesium rich (ultramafic) rock such as peridotite.

silicate: Refers to the chemical unit silicon tetrahedron, SiO_4, the fundamental building block of silicate minerals. Silicate minerals represent about one third of all minerals and hence make up most rocks we see at the Earth's surface.

Table 10.1 Chemical Composition of Earth's Crust

Element	Percent by Weight	Percent by Volume
Oxygen (O)	46.6	93.8
Silicon (Si)	27.7	0.9
Aluminum (Al)	8.1	0.5
Iron (Fe)	5	0.4
Calcium (Ca)	3.6	1
Sodium (Na)	2.8	1.3
Potassium (K)	2.6	1.8
Magnesium (Mg)	2.1	0.3
All other elements	1.5	0.3
TOTAL	100	100

From *Physical Geology* by Dallmeyer. Copyright © 2000. Reprinted by permission of Kendall Hunt Publishing Co.

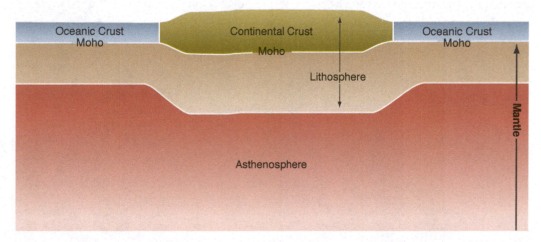

Figure 10.6 Structure of the outer Earth based on mechanical strength. The outer portion of Earth's layers includes oceanic and continental crustal material, along with a rigid lithosphere and a hotter, plastic asthenosphere.

The uppermost mantle below the crust is relatively cool and its pressure is low, which are similar physical conditions to the crust and produce hard, strong rocks as well. The rigid physical properties of both the crust and upper mantle cause the two to combine and form an important outer layer, based on the physical properties of the rocks, called the lithosphere (rock sphere). The lithosphere, consisting of the crust (either continental or oceanic) and a thin layer of the uppermost upper mantle, averages 100 km thick but is as much as 200 km thick in some continental regions (**Figure 10.6**).

The asthenosphere (weak sphere) is the zone in the upper part of the mantle, about 250 km thick, that is more ductile or plastic than the rest of the mantle. The temperatures and pressures in this zone cause the rocks to lose much of their strength and become soft and plastic, like warm putty, and they behave like a ductile layer underlying the brittle lithosphere. Partial melting within the asthenosphere generates magma (molten rock) which can rise to the surface to form volcanic eruptions. This is because the magma is molten and less dense than the mantle rock from which it was derived.

Below the lithosphere and asthenosphere is the mantle which is a solid and forms most of the volume of the Earth's interior. Because the high pressure at these depths offsets the effect of high temperature, the rocks are stronger and more rigid than the overlying asthenosphere. Despite the stronger nature of the mantle, it never gets as strong as the lithosphere and remains capable of flowing slowly over geologic time.

The **core** is the innermost region of Earth and has an average calculated density of 11 g/cm³ and comprises about 16 percent of the planet's total volume. The composition is considered to be predominantly metallic iron and some nickel. While the core is compositionally uniform and is a single layer of the Earth based on composition, it is subdivided into two zones based on very different physical properties: the inner core is solid and the outer core is a molten liquid. Near the center, the core's temperature is estimated to be nearly 7000°C, hotter than the Sun's surface, and pressures are 3.5 million times that of Earth's atmosphere at sea level. This extreme pressure is responsible for keeping the inner core solid despite being hotter than the outer core.

core: The innermost layer of the Earth, made up of mostly of iron and nickel. The core is divided into a liquid outer core and a solid inner core. The core is the densest of the Earth's layers.

The combination of the rigid lithosphere lying over the non-rigid asthenosphere has extremely important geologic implications for natural hazards. The lithosphere is broken into numerous individual slabs called **plates** that slide over the asthenosphere. This motion is the reason that plate tectonics exists on Earth. Movements of the plates create earthquakes, volcanism, and mountain uplifts.

plate: Slab of rigid lithosphere (crust and uppermost mantle) that moves over the asthenosphere.

Topic 11

Earth's Seasonability
By Dean Fairbanks

Earth depends on energy from the sun. The amount of solar energy we receive, however, varies with latitude, distance from the sun, and time of year. The length of time between sunrise and sunset (**day length**) changes throughout the year, and it's different depending on the latitude. Finally, Earth's seasonality is linked to all the above, but primarily to the tilt of the Earth's polar axis in relation to the plane of the ecliptic (sun's equatorial plane). Thus, some latitudes experience extreme differences between summer and winter, while at other latitudes, there's almost no difference between the seasons at all.

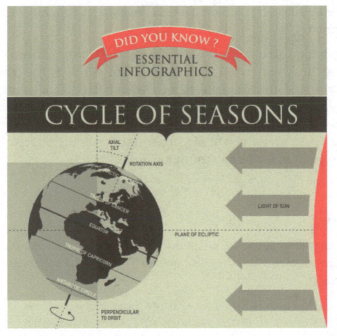

Image © Shutterstock, Inc.

Figure 11.1 The amount of solar energy received on Earth varies on latitude, season and distance between the Earth and the Sun.

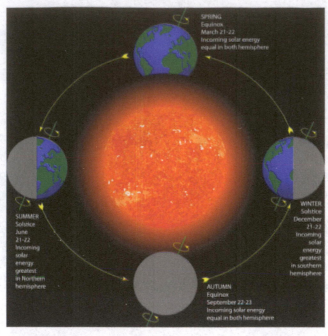

Image © Shutterstock, Inc.

Figure 11.2 The Earth's revolution around the Sun creates the seasons.

The Earth has a slightly lopsided orbit around the sun. The Earth is closest to the sun (about 147.3 million km away) on approximately January 3 during **Perihelion** (Southern hemisphere summer) and farthest from the sun (about 152.1 million km away) on approximately July 4, during **Aphelion** (Northern hemisphere summer). As a result of the orbital geometry, the southern hemisphere summer is slightly more intense than the northern hemisphere summer.

When discussing the distribution of solar radiation, we define it by the term **insolation**. Incoming solar radiation is that energy that reaches a horizontal plane at the Earth's surface. The tropics, between 23.5° N and 23.5° S, receive more concentrated insolation due to the Earth's curvature or its **sphericity**. In fact the tropics receive 2.5 times more energy than the poles annually. The energy arrives parallel at the Earth from the sun, causing a direct interaction at the tropics (90 degree angle), but the sun's rays becoming increasingly oblique in their angle of incidence with increasing latitude. Thus, at the poles the surface area receiving insolation is larger and more diffuse, in contrast to the concentration at the tropics.

Earth's Seasonality

Seasonality has two meanings: The seasonal variations in the sun's position above the horizon and seasonal variations in day length. The following are the components to describing sun's geometry and movement from a point on the Earth's surface:

- Sun's position above the horizon is called its **altitude** measured in degrees.
- When the sun is directly overhead (90 above horizon) is termed its **zenith**.
- The location on the Earth's surface where the zenith occurs and insolation is maximum is termed the **subsolar point**.
- The sun's **declination** is the latitude where zenith occurs. This latitude annually migrates between 23.5° N and 23.5° S for a full 47° range of declination. This region is defined as the tropics.

For an observer at any latitude the sun's altitude at solar noon changes throughout the year. In the northern hemisphere the lowest angle occurs on December 21 and its highest angle on June 21. This is the exact opposite for the southern hemisphere.

Figure 11.3 Even early human cultures understood seasonality. Stonehenge in the United Kingdom is a 5000 year old testament to our co-evolvement with the Earth's seasonal rhythms.

Seasonality, or the march of the seasons, occurs because the Earth is **tilted 23.5°** on its axis. This tilt, which is constant, combines with the Earth's **revolution** around the sun to derive winter and summer seasons. As the Earth revolves around the sun, it does so on the same plane, the plane of the **ecliptic**. The axis maintains alignment during the orbit around the sun—the same alignment relative to the plane of ecliptic is termed **axial parallelism**. Synthesizing the tilt, revolution, axial parallelism, and sphericity gives rise to the northern hemisphere receiving the most extra sunlight on June 21 (summer **solstice**) since the north pole is tilted toward the sun; and the southern hemisphere receives the least sunlight on December 21 (winter solstice) with the north pole being tilted away from the sun. Opposite situations apply for the southern hemisphere. On the **equinoxes**, the Earth is neither tilted toward or away from the sun thus there is equal night and days on March 21 and September 21. No matter what the Earth's position might be during its revolution, it is always 50% in day and 50% in night as it rotates on its axis every 24 hours. The circle of illumination, or **terminator**, marks the separation. On the equinoxes the circle of illumination passes directly from pole to pole. In the northern hemisphere summer, the circle of illumination passes through the 66.5° N parallel (arctic circle) to the 66.5° S parallel (Antarctic circle) which causes the north Polar Regions to experience 24-hr day light. In summary, the seasons are linked to **day length, which is a function of the latitude**. These seasonal differences are caused by the Earth's tilt (major contributor), revolution, rotation, axial parallelism, and sphericity.

Seasons, Weather, and Climate

K.L. Chandler

The **atmosphere** is the layer of gases that surround the planetary body of Earth, held in place by gravity. A geographer's focus usually remains within Earth's closed system; however, a discussion of Earth's atmosphere naturally involves its sun, the source of all energy on Earth. The solar energy (from the sun) heats Earth's surface, oceans, and atmosphere.

Earth's position in its daily rotation and yearly revolution determines the duration and amount of sunlight that reaches any particular place on Earth. The sun's direct rays vary daily, due to Earth's revolution, but fall only within the area that surrounds the Equator, between the Tropic of Cancer and Tropic of Capricorn, also called the **Tropics**. The areas north of the Tropic of Cancer and south of the Tropic of Capricorn receive only indirect rays from the sun.

The Four Seasons

Earth's 23.5° tilt and position along its yearly revolution around the sun governs how many hours of daylight different points on Earth receive throughout the year. It also drives the seasons, which are marked by two equinoxes and two solstices. An **equinox** occurs when the sun is at its zenith directly over the Equator. This position causes a nearly equal amount of daylight and darkness at all latitudes, although the amounts of solar radiation emitted are not even. In the Northern Hemisphere, the vernal equinox in March is the beginning of spring, and September's autumnal equinox starts the fall season. The Southern Hemisphere's equinoxes are opposite with the autumnal equinox in March, and the vernal equinox in September.

Solstices also happen twice a year, when the sun's direct rays are at their highest (summer) or lowest (winter) point over the Tropics. Ancient astronomers described them as the days that the sun gave the impression of standing still. A **solstice** is the point of Earth's orbit when the north or south hemisphere sees the most hours of daylight, while the other hemisphere sees the least. The hemispheres' respective polar regions experience either complete light or a day of total darkness. Those days are referred to as the summer solstice – occurring in June for the Northern Hemisphere and December for the Southern Hemisphere – and the winter solstice, occurring in December for the Northern Hemisphere and June for the Southern Hemisphere (**Figure 12.1**).

© Honza Krej/Shutterstock.com

Figure 12.1 Locations in the mid-latitudes experience all four seasons, while areas in the Tropics and Poles are usually described as having only two seasons. The Tropics have a wet season and a dry season, while the Poles have winter and summer.

Weather and Climate

Weather is the condition of the atmosphere at a given time and place in relation to temperature, wind, moisture, and amount of solar radiation. Weather is localized and specific – it refers to what is happening right here, right now. Meteorology studies the atmosphere and how it affects the weather in a specific area on a day-to-day basis. People have monitored daily and seasonal weather changes since the earliest civilizations, noting patterns and speculating about what might happen next. Technological advances over the centuries have increased the precision of information gathering; today, satellites and computers provide modern weather models with such precise information that a three-day temperature forecast now is nearly accurate as a one-day forecast was thirty years ago.

Many people use the terms *weather* and *climate* interchangeably, but they represent markedly different occurrences – if weather is described as an event, then climate is a trend. **Climate** is the average weather in a given area over many years. Some scientists use 30 years as a minimum threshold to establish climate, but 100 years is a more favorable time span to truly understand what can be predicted for an area. Put simply, climate is what should be happening right now.

Earth Processes, Cycles and Heat Transfer

By Ingrid Ukstins and David Best

As a system, remember that Earth is an assemblage or combination of interacting components that form a complex whole. The system is driven by the flow of matter and energy through the components in order to try to reach equilibrium. Each of Earth's major spheres, the atmosphere, hydrosphere, biosphere, and geosphere, are usually described and studied as separate entities, each containing numerous interacting smaller systems; however, each sphere and their components interact with each other continuously, exchanging both matter and energy.

Several energy and material (matter) cycles are fundamental to our understanding and study of natural processes and hazards connected to the dynamic movements of matter and energy. These include the **rock cycle**, the **hydrologic cycle**, and the **biogeochemical cycle**. Keep in mind that during the course of these cycles, matter is always conserved, that is, it is neither created nor destroyed, but continuously changes form. The cyclical exchange of matter and energy occurs between storage reservoirs within the spheres and on a wide range of time scales. This cyclic movement of matter and energy plays a large role in producing natural disasters when we find ourselves in the path of their motions.

rock cycle: The sequence of events in which rocks are formed, destroyed, and reformed by geological processes. Provides a way of viewing the interrelationship of internal and external processes and how the three rock groups relate to each other.

hydrologic cycle: The cyclic transfer of water in the hydrosphere by water movement from the oceans to the atmosphere and to the Earth and return to the atmosphere through various stages or processes such as precipitation, interception, runoff, infiltration, percolation, storage, evaporation, and transportation.

biogeochemical cycle: Natural processes that recycle nutrients in various chemical forms from the environment, to organisms, and then back to the environment. Examples are the carbon, oxygen, nitrogen, phosphorus, and hydrologic cycles.

Materials

The materials that make up the different Earth spheres, and what energy performs work on, is known as matter, which is defined as anything that has mass and occupies space. Therefore, the atmosphere, water, plants, animals, minerals, and rocks are composed of matter made up of atoms of different elements. Matter exists in three states: solids, liquids, and gases. The cycling of this matter is important in Earth's processes.

Energy

Natural processes responsible for natural disasters (such as erosion, atmospheric circulation, or plate tectonics) occur on or within the Earth and require energy for their operation of performing work on matter. Energy is defined as "the ability to do work," which can be constructive or destructive. Work itself can mean moving something, lifting something, warming something, or lighting something. Energy exists in two states: stored and waiting to do work (**potential energy**), or actively moving (**kinetic energy**). Potential energy is so named because this stored energy or work has the *potential* to change the state of other objects when released. The kinetic energy of an object is the energy it possesses because of its motion and is an expression of the fact that a moving object can do work on anything it hits and thus it quantifies the amount of work the object could do as a result of its motion. There are many different forms of energy that can be transformed from one to the other:

- **Heat energy** is exhibited by moving atoms and the more heat energy an object has, the higher its temperature, which is a measure of its kinetic energy. When gasoline is burned in a car's engine, the heat energy is converted to kinetic energy that moves the car.
- **Radiant energy** is energy carried in the form of electromagnetic waves such as visible light, ultraviolet rays, and radio waves. Most of the Sun's energy reaches Earth in this form (especially light) and is converted to heat energy.
- **Nuclear energy** is energy released from an atom's nucleus. Energy is released by several processes including radioactive decay, where an unstable radioactive nucleus decays spontaneously into a new nucleus, and nuclear fusion where multiple nuclei join together to form a new element nucleus and heat-producing nuclear reactions where two nuclei merge to produce two different nuclei.
- **Elastic energy** is potential energy formed through deformation (stretching, bending, or compressing) of an elastic material like a rubber band. When energy is released, it is converted to kinetic energy and heat (by friction). Earthquakes exhibit a release of energy stored in elastically deformed rocks that eventually break.
- **Electrical energy** is produced by moving electrons through material and is often converted into heat energy. Lighting storms are good examples.
- **Chemical energy** is produced by breaking or forming chemical bonds and this type of energy is usually converted to heat.
- **Gravitational energy** is produced when an object falls from higher to lower elevations. As the object falls, the energy can be converted to kinetic energy or heat.

potential energy: The energy available in a substance because of position (e.g., water held behind a dam) or chemical composition (hydrocarbons). This form of energy can be converted to other, more useful forms (for example, hydroelectric energy from falling water).

kinetic energy: The energy inherent in a substance because of its motion, expressed as a function of its velocity and mass, or $MV^2/2$.

Earth's major cycles and processes are powered by energy from two sources. The Sun drives external processes that occur in the atmosphere, hydrosphere, cryosphere, biosphere and surface of the lithosphere. Climate, weather, ocean circulation, and erosional processes are the result of energy from the Sun being transformed into heat energy which powers them. The second source of energy is derived from the Earth's interior. Heat left over from the Earth's formation along with heat produced from the constant decay of radioactive elements powers the internal processes driving plate tectonics and forming volcanoes, earthquakes, and mountain ranges.

Energy from the Sun

The Sun is the most prominent feature in our solar system and contains approximately 98 percent of the total solar system mass (**Figure 13.1a–b**). The Sun is composed of hydrogen (about 74 percent of its mass, or 92 percent of its volume), helium (about 25 percent of mass, 7 percent of volume), and trace quantities of other elements. The Sun's outer layer is called the photosphere and has a temperature of 6,000°C and a mottled appearance due to the turbulent eruptions of energy at the surface. Solar energy is created deep within the core of the Sun, where the temperature (15,000,000°C) and pressure (340 billion times Earth's air pressure at sea level) are so intense that nuclear reactions take place. Here, nuclear fusion produces energy by fusing hydrogen atoms together to form atoms of helium. During nuclear fusion, some of the mass from the hydrogen atoms is expelled as energy and carried to the surface of the Sun by convection (see following discussion), where it is released as radiant energy. Energy generated in the Sun's core takes about a million years to reach its surface, but every second, 700 million tons of hydrogen are converted into helium and 5 million tons of pure energy are released. The Sun has been producing energy since the solar system formed 4.6 billion years ago, and it is estimated that solar energy production will continue at this level for another 5 billion years.

International Astronomical Union.

Figure 13.1a Earth is the third planet from the Sun. When the size of Earth is compared to that of the Sun, more than one million Earths could fit inside the Sun and 109 Earths could be placed across its equator.

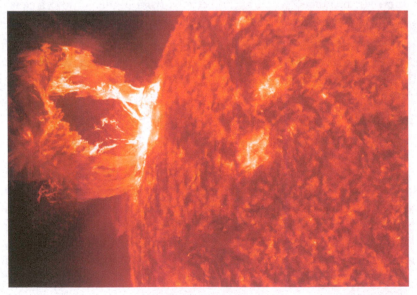

Source: NASA.

Figure 13.1b Solar flares are sudden bursts of energy that occur at the Sun's surface. Major flares can interfere with communications on Earth.

The radiant energy from the Sun travels in many different wavelengths (**Figure 13.2**) through space, with about 43 percent being visible light, 49 percent near-infrared, and 7 percent ultraviolet (UV) solar radiation received on Earth. Some of the energy is immediately reflected back into space (about 40 percent) by the atmosphere, clouds and the Earth's surface (mostly the oceans). Some of the energy is converted to heat and is absorbed by the atmosphere, hydrosphere, and lithosphere. Heat drives the hydrologic cycle, causing evaporation of the oceans and circulation of the atmosphere, and, with gravity, rain to fall on the land and run downhill. Thus energy from the Sun is responsible for such natural disasters as severe weather and floods.

In the biosphere, some energy is absorbed and stored by plants during photosynthesis and used by other organisms, or is stored in fossil fuels such as coal and petroleum. This stored energy in the biosphere is released through burning to form heat energy, which also releases the stored carbon that produces carbon dioxide—a greenhouse gas that is partially responsible for global warming issues.

Energy from Earth's Interior

Energy in the form of heat is generated within the Earth's interior and contributes only about 0.013 percent of the total energy reaching the Earth's surface (the other 99.987 percent is from the Sun). However, this smaller amount of energy that flows to the Earth's surface is responsible for driving plate tectonics, which creates mountains and cause earthquakes, and for melting of rocks to create magmas that result in volcanism. Heat energy is produced from three energy sources within the Earth's interior:

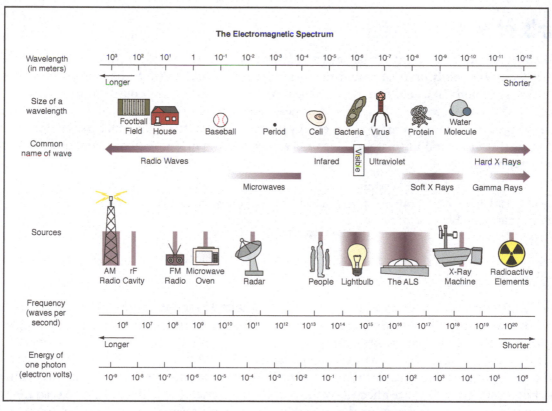

Source: NASA.

Figure 13.2   The electromagnetic spectrum has an extremely wide range of wavelengths. Visible light is only a small portion of the spectrum.

Radioactive decay continuously produces heat energy from radioactive isotopes such as ^{235}U (Uranium), ^{232}U, ^{232}Th (Thorium), and ^{40}K (potassium). Radioactive isotopes have unstable nuclei that break down (decay) to a more stable isotope and expel subatomic particles such as protons, neutrons, and electrons from the radioactive parent atom that interact with surrounding matter (in this case rocks).

> **radioactive decay:** Natural spontaneous decay of the nucleus of an atom where alpha or beta and/or gamma rays are released at a fixed rate.

Gravitational energy and **impact energy** produced a tremendous amount of heat during the initial formation of the Earth. Gravitational energy developed as the Earth grew in size by accretion in the original solar nebula. The deepening burial of material caused an increasing gravitational pull that continued to compact the interior and convert gravitational energy to heat. At the same time in Earth's early history, tremendous numbers of space objects (asteroids, planetessimals, and comets) were hitting the Earth. The impacts of these objects converted their energy of motion (kinetic energy) to heat. The large quantity of heat produced by the combination of these two sources did not easily escape to the surface due to the poor conductivity (see following heat transfer discussion) of rock material. The sum of heat energy generated from radioactive decay, impacts, and gravity is slowly rising to the surface of our planet through plate tectonics, and along its journey it is causing the continents to drift and ocean basins to form.

> **gravitational energy:** The force of attraction between objects due to their mass and is produced when an object falls from higher to lower elevations.
>
> **impact energy:** Cosmic impacts with a larger body convert their energy of motion (kinetic energy) to heat.

Methods of Heat Transfer

Any body, whether a solid, liquid, or gas, has a temperature associated with it. This thermal energy can be transferred to another body or its surroundings by the process of **heat transfer**. The movement occurs by heat moving from a hot body to a cold one through the processes of conduction, convection, or radiation, or any combination of these (**Figure 13.3**). Heat transfer never stops; it can only be slowed down. Earth's internal and external processes are powered by heat energy moving from hotter areas (such as the Earth's interior) to cooler ones (the crust and ultimately into space) and so it is important to understand how heat can move through materials.

> **heat transfer:** Heat moving from a hot body to a cold one through processes of conduction, convection, or radiation, or any combination of these

Conduction

Conduction transfers heat directly from atom to atom. It works best in solids where the atoms are closely packed together, but also transfers heat from solids to liquids or gas. Heat energy causes atoms to vibrate against each other, and these vibrations pass from high-temperature areas (rapid vibrations) to low-temperature areas (slower vibrations), causing them to heat up and vibrate faster. Heat from the Earth's interior moves through the solid crust by this mode of heat transfer, but very slowly. Each substance has a different heat conductivity based on how easily the atoms are affected by the heat energy. Metals are highly conductive, so they transfer heat energy by heating up and cooling down rapidly which is why we use metal pots for cooking—heat is transferred quickly to the contents inside. Rocks are much less conductive and so heat up and cool down more slowly, which helps heat build up inside the Earth.

> **conduction:** Heat transfer directly from atom to atom in solids.

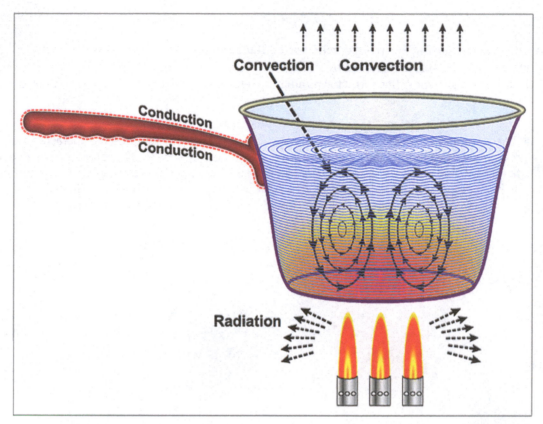

© Fouad A. Saad/Shutterstock.com

Figure 13.3 Convection, conduction, and radiation often occur together. These are three ways in which heat is transferred through objects and space.

Convection

Convection transfers heat by being carried along by circulation by the material that was heated. Since heat energy moves with the material, the materials that are able to move are mostly liquids and gases. We are familiar with convection of air in our houses when we heat them in the winter by a heat source (whether by a wood stove, radiator, or heater). The warmed, less-dense air rises to the ceiling as a moving current, which then cools to a denser air mass that returns to the floor to be heated again. A complete circulation of hot air rising and cold air sinking is called a **convection cell**. Convection is an important method of moving energy through Earth materials. Inside the Earth, the mantle heats up by this method as well as through the flow of magma to the surface. On the surface, heat is transferred in the atmosphere and oceans by this mode that influence climate and weather.

convection: (1) (Physics) Heat transfer in a gas or liquid by the circulation of currents from one region to another; also fluid motion caused by an external force such as gravity. (2) (Meteorology) The phenomenon occurring where large masses of warm air, heated by contact with a warm land surface and usually containing appreciable amounts of moisture, rise upward from the surface of the Earth.

convection cell: Within the geosphere it is the movement of the asthenosphere where heated material from close to the Earth's core becomes less dense and rises toward the solid lithosphere. At the lithosphere-asthenosphere boundary heated asthenosphere material begins to move horizontally until it cools and eventually sinks down lower into the mantle, where it is heated and rises up again, repeating the cycle.

Radiation

Radiation transfers heat using electromagnetic waves. Radiant electromagnetic energy from the Sun (produced by nuclear reactions) is transferred by this method. On Earth radiative heat transfer is responsible for warming the oceans and atmosphere, and for reradiating heat back into space to keep our planet at a constant average temperature.

radiation: Energy emitted in the form of electromagnetic waves. Radiation has differing characteristics depending upon the wavelength. Because the radiation from the Sun is relatively energetic, it has a short wavelength (ultra-violet, visible, and near infrared) while energy radiated from the Earth's surface and the atmosphere has a longer wavelength (e.g., infrared radiation) because the Earth is cooler than the Sun.

Structure of the Atmosphere

By Dean Fairbanks

The air we breathe, we take for granted. The atmosphere is crucial to life. It filters the energized shortwave radiation of gamma and x-rays in the upper levels on the edge of the virtual vacuum of space, down to the lower stratospheres role of filtering a large proportion of ultraviolet (UV) rays. In the lowest layer, the troposphere, visible light is transmitted. In the troposphere resides the majority composition of elemental gases and most weather processes that interact with biotic life and landscape surface processes. Yet, the atmosphere is incredibly thin—imagine a standard class globe with one coat of varnish for a scaled comparison.

Composition, Energy Interactions and Heat

A profile of the atmosphere can best use composition, temperature changes, and function as examination criteria. In order to cover these criteria atmospheric pressure must be explained. Atmospheric pressure is driven by gravity. Air density (number of molecules/cm3) decreases with height in the atmosphere because earth's gravity pulls the molecules to the surface. Thus, the atmosphere at higher altitudes, for instance above 10,000 ft. (3038 m), is less dense than at sea level, and this can have real effects on sports games like baseball. Power hitters can hit a baseball farther at Coors Field in Denver, Colorado (mile-high stadium) than at Fenway Park in Boston, Massachusetts because the ball is rubbing against fewer air molecules and thus experiencing less frictional drag. The air density in Denver is 15% less than at Boston.

Air pressure is the force exerted on you and the surface by the weight of air molecules. A **barometer** is used to measure air pressure at any given moment for a particular place. Using mercury in a glass column closed on one end and placed in a mercury-filled reservoir, the pressure of the atmosphere is measured in millibars (mb) or in inches of mercury (29.92 inches of mercury is equal to 1013 mb experienced at sea level). The weight of air literally pushes mercury up the column. Atmospheric pressure **decreases exponentially** with altitude, from 1013.25 mb at sea level to almost 0 mb at 50 km. Because we live at the surface we experience air pressure all around us as a force of 1 kg per square centimeter (14.7 pounds per square inch). At the elevation of Mt. Everest (29,028 ft./8848 m) the air pressure is 300 mb, while at the typical flying height of commercial jets (34,000 ft./10,363 m) the air pressure is only 200 mb. At these elevations mountain climbers require oxygen tanks and aircrafts are pressurized. What would the mercury column's height be on top of Mt. Everest? Say the elevation is ~8 km, so 75% of air is below you, so 0.25*29.92 inches is ~7.5 inches or 300 mb.

The atmosphere technically extends to 32,000 km from the surface if we include the magnetosphere (Exosphere), however the principal atmosphere stops at 480km. The structure of the atmosphere is analogous to the layers of an onion, with four layers with distinct properties and functions. We can characterize vertical profiles according to three criteria: compositions, functions, and relationships to temperature (**Figure 14.1**). The top of the principle atmosphere is at 480 km called the thermopause in descending order are the following layers:

- **Thermosphere**—beginning at 80 to 480 km altitude is the virtual vacuum of space. This is the region of the auroras.
- **Mesosphere**—between 50 and 80 km altitudes is the last region of any appreciable gas density. At this layer the highest clouds can form—**noctilucent clouds** represent crystals of ice formed on smoke particles from burnt up meteors.
- **Stratosphere**—between 18 and 50 km altitudes is a region that lacks dust and water vapor, but still contains the vestige of a stable thin gas density of ~10% of the total gas molecules present.
- **Troposphere**—this is the region of the atmosphere that contains 90% of the gas molecules present in the atmosphere. Most weather processes take place here, and thus this is the layer that is most associated with physical and biotic surface interactions.

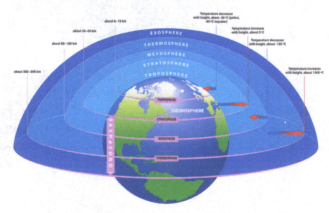

Figure 14.1   **The layers of the atmosphere.**

In terms of gas and solid compositions, the lower atmosphere (below 80 km) is very well mixed, with the exception of O_3 (ozone), H_2O, and CO_2 (carbon dioxide). The air in the troposphere is basically a mixture of nitrogen and oxygen with some trace gases. This mixture makes for an average density of air of 1.3 km/m3. The atmosphere is not static over time, but instead is constantly evolving based on its interactions with the physical and biotic surface. The present air composition represents the net sum of many processes, both biological and chemical, operating over millions of years. The evolution of biological life is an important factor; if earth were lifeless, its atmospheric composition would be very different. For example, there would be very little oxygen. Its current composition is composed of 78% nitrogen (N_2), 20.94% oxygen (O_2), 0.93% argon (Ar), 0.040% CO_2, with the remainder being trace gases. Most of the atmospheric nitrogen was and still is produced by volcanoes; in fact the tremendous amount of volcanic activity in the first 4 billion years set its dominance in the atmosphere. The earth's atmosphere in the distant past did not contain any oxygen, 2.5 billion years ago. Given the rise of photosynthesis in plants, the greening of the planet led to the uptake of CO_2 which was then broken down and combined with water to create simple sugars for plants to grow on and a release of O_2 as a waste product. Plant life is the primary producers of oxygen through the uniqueness of the photosynthetic gas exchange process. Finally there is argon, an inert (not reactive) noble gas that is not involved in biological processes but is instead associated with the planet's oceans. Oceans cover 72% of the planet's surface, where the salty seawater and the earth's crust that they cover contain potassium-40, which through its radioactive decay creates argon-40 which escapes to the atmosphere.

The relationship of temperature with atmospheric height is controlled by pressure (air density) and air composition, and therefore is not straightforward. Keep in mind that 99% of the atmosphere is contained in the troposphere and stratosphere. To understand the contradictory nature of temperature we will focus on these two principle layers. Between these two layers is a transition zone called the tropopause, as the layers are not strictly distinct in their adjacency to each other. The tropopause, while on average is at 18 km altitude, actually varies in elevation with latitude. The tropical tropopause is twice as high (26 km) as the polar tropopause (10 km) because of vigorous surface heating at the low latitudes. The relationship is such that the tropopause height is proportional to the mean tropospheric temperature. The relationship of temperature with altitude is opposite between the two layers:

- In the troposphere, the air temperature decreases with height until the tropopause where it stops and remains constant.
- In the stratosphere the air temperature increases with height until it decreases again in the mesosphere.

Why does air temperature decrease with height in the troposphere? Warm surface heated air rises because pressure decreases with height and this rising air expands and does work (pushing on the air around it) by releasing heat and thus cooling itself. The lapse rate is the rate of temperature decrease with altitude. The **environmental lapse rate (ELR)** on average is −6.5°C per 1000 m. As for the air temperature switching the relationship to an increasing one, this is based on the composition of the air in the stratosphere—ozone. Ozone in the stratosphere absorbs UV radiation from the sun thus increasing its temperature and re-radiates the energy at longer thermal wavelengths (**Figure 14.2**).

FORMATION OF OZONE

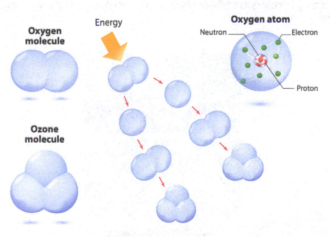

Image © Shutterstock, Inc.

Figure 14.2 Seasonal variations in surface air temperature (Courtesy: Department of Geography, University of Oregon, supported by the US National Science Foundation, 2000).

The atmosphere also has a filtering effect. The thermosphere and the stratosphere have a functional role in blocking certain forms of radiation. However, the atmosphere as a whole is transparent to visible and infrared radiation. The thermosphere in its functional role is called the ionosphere as particles become charged as they block gamma and x-rays from reaching the surface. The stratosphere in its functional role is called the ozonosphere as this is where ozone is concentrated (90% of atmospheres ozone). Ozone is important as it filters out UV radiation (**Figure 14.3**). This role became prominent when in the late 1970s it was discovered that the ozone layer was being destroyed over the poles and most notably over Antarctica. The ozone hole was created by human-manufactured halocarbon refrigerants, solvents, propellants, and foam-blowing agents (CFCs, HCFCs, freons, and halons) (**Figure 14.4**). When CFCs reach the stratosphere they are exposed to UV rays and break down into free chlorine, which breaks down ozone. As this destruction of zone is faster than its natural regeneration, the amount of ozone is decreasing. The Montreal Protocol (1987) required countries to cut-off CFCs by 1999. Many developing countries were wary of signing, with China and India only responding to cuts in 2012. Since the protocols policy intervention and the development of alternatives to CFCs, the ozone hole is repairing itself slowly through the natural process of ozone regeneration. Finally, another function of the atmosphere was curiously discovered in the 1950s called Schumann resonance. Schumann resonances are the name given to the resonant frequency of the earth's atmosphere, between the surface and the densest part of the ionosphere. Schumann predicted that the earth's atmosphere would resonate certain electromagnetic frequencies, a frequency of 7.83 Hz which is a wavelength of 38,000 km. This is about the circumference of the earth, which is why its atmospheric cavity resonates at that frequency. There is preliminary research to show that the Schumann resonance has a relationship to human psychobiology, which might make sense since we evolved with it (See documentary video: **Beings of Resonance** for more details).

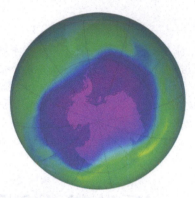

Figure 14.3   Satellite instruments monitor the ozone layer, and we use their data to create the images that depict the amount of ozone. The blue and purple colors are where there is the least ozone, and the greens, yellows, and reds are where there is more ozone. (**NASA 2006**)

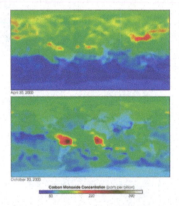

Figure 14.4   The false colors in these images represent levels of carbon monoxide in the lower atmosphere, ranging from about 390 parts per billion (dark brown pixels), to 220 parts per billion (red pixels), to 50 parts per billion (blue pixels). (**NASA**)

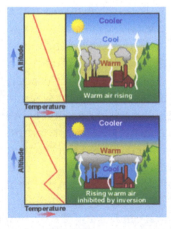

Figure 14.5   Air pollution in a temperature inversion. (**US EPA**)

While the focus has been on the role of gas elements in the atmosphere, we cannot understate the role of air pollution (**Figure 14.5**). First, at the surface the atmosphere mixes rapidly from east to west with full mixing of pollution plumes circling the earth within 10 days. The majority of these pollution plumes occur in the mid-latitudes of the northern hemisphere associated with industrial processes, while in the southern hemisphere they are most associated with slash and burn agriculture. The bottom line is that within a 10-day span we are polluting each other, whereas mixing between north and south is much slower (~1 year).

In the troposphere were 95% of the air pollution occurs much of this is from combustion particles (natural and anthropogenic), bacteria, dust, sea salt, algae, pollen, and so on. These represent **aerosols**, tiny solid or liquid particles suspended in the atmosphere. They affect visibility, the amount of incoming solar radiation, air chemistry, and cloud formation. There are several natural factors that affect air pollution levels: local and global winds; topography where surrounding mountain ranges can trap pollutants in a valley; and temperature inversions where the normal vertical mixing and dilution of polluted air is inhibited. Normally air at the surface is warmed and rises, as this happens the ELR states that the air temperature decreases (6.5°C/1000 m). An atmospheric layer can form where the temperature decrease with height is much less than normal and even reverses to warming again, thus trapping colder (less buoyant) air below warmer (more buoyant air). This air layer effectively caps the rising air and its pollutants. These trapped aerosols with the addition photochemical smog, caused by nitrogen oxides and volatile organic compounds emitted by vehicles and industry chemically reacting with sunlight lead to the characteristic yellow-brown coloration made famous in Los Angeles, Mexico City, Beijing, as well as all modern cities (**Figure 14.6**).

mage © Shutterstock, Inc.

Figure 14.6 Air polluted skyline of Mexico City at dawn. Temperature inversion and the fact that the city is built in a very large volcanic crater leads to these air pollution problems.

Weather and Climate
By Ingrid Ukstins and David Best

Severe Weather is a Worldwide Killer

In early March of 2019, severe thunderstorms ripped through Alabama, Florida, Georgia and South Carolina, generating 34 confirmed tornadoes including one that cut a swath through Lee County, eastern Alabama, killing at least 23 people, including 7 from one family. Many more still remain missing or unaccounted for after the storm, and authorities are using drones with heat-seeking cameras to help locate survivors who are trapped by debris. This was an EF-4 tornado that had wind speeds of 275 km/hour (170 mph), was more than 1.5 km wide (1 mile), and had a track on the ground of at least 39 km (24 miles) long. America is one of the most severe weather-prone countries on Earth, and being prepared for our increasing vulnerability to extreme weather is critical for health and safety. No community is storm-proof, but advanced planning, education and awareness can help save lives.

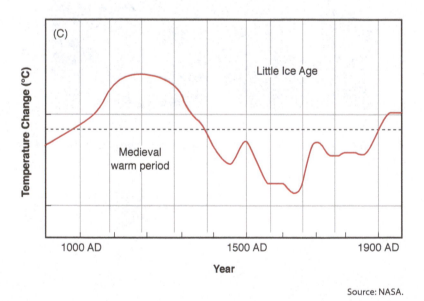

Source: NASA.

Figure 15.1a Changes in global temperatures between 1400 and 1800 produced the Little Ice Age, a time when temperatures were lower and glaciers advance. The dashed line shows the baseline average temperature.

Earth's weather and climate are strongly influenced by the oceans, which cover 71 percent of Earth's surface and contain about 97 percent of Earth's water. The oceans along with the atmosphere serve as key reservoirs for much of the water in the hydrosphere. The atmosphere gets the majority of its heat and moisture from the oceans, which control weather patterns and climate. The Sun provides the energy that heats the ocean and drives the hydrosphere.

One purpose the oceans serve is to regulate temperature in the troposphere—the lower part of the atmosphere that contains most of our weather—clouds, rain and snow. The atmosphere in turn provides energy through winds to create waves and to help move currents below the surface. Weather that is created by the interaction of the oceans and the atmosphere can become a problem, both on land and at sea. Severe weather, such as hurricanes, tornadoes, severe thunderstorms, and winter blizzards, generates violent winds and heavy precipitation that can produce loss of property and lives.

© Everett-Art/Shutterstock.com.

Figure 15.1b   The painting Ice Skating in a Village, by Hendrick Avercamp, shows Dutch villagers slipping and sliding on ice on a Dutch canal. Cold conditions during the Little Ice Age may have been due to a period of low solar activity.

Weather and Climate

People often confuse the terms *weather* and *climate*. **Weather** is the short-term condition of the atmosphere and how it behaves, mainly with respect to its impact on life and human activities. Often measured over periods of hours, days, or even weeks, these conditions are usually localized or confined to a small region of the country. There are a lot of components to weather. It includes sunshine, rain, cloud cover, winds, hail, snow, sleet, freezing rain, flooding, blizzards, ice storms, thunderstorms, steady rains from a cold front or warm front, excessive heat, heat waves and more, according to the National Weather Service. Temperature, amount of precipitation, wind speeds, and humidity are associated with the weather. The ever-changing, complex nature of the atmosphere makes it difficult to forecast weather conditions very far into the future.

weather: The composite condition of the near earth atmosphere, which includes temperature, barometric pressure, wind, humidity, clouds, and precipitation. Weather variations over a long period create the Climate.

Climate describes the atmospheric conditions over long periods of time, ranging from a few years to thousands or even millions of years. It is the average of weather over both time and space. These persistent conditions can have a major effect on a region, a continent, or even the entire globe. In the geologic past there are numerous examples of long-lasting ice ages or warming trends that affected large areas of the Earth's surface. In relatively recent history, areas of Europe experienced a Little Ice Age, which lasted from about 1400 to 1800 (**Figure 15.1a**). During this period, glaciers increased in size, and temperatures dropped, which affected farming and crop production (**Figure 15.1b**).

climate: The long-term average weather, usually taken over a period of years or decades, for a particular region and time period.

Weather and climate are intertwined with the atmosphere and the hydrosphere in complex ways. Varying amounts of water in the oceans change the climate and thus affect weather. Changes in solar radiation patterns over long periods of time can increase or decrease global temperatures, which can cause changes in the amount of ice held in polar regions. Changes in the concentration of different atmospheric gases such as carbon dioxide can also alter global temperatures.

Climate Factors and Climate Zones
By K.L. Chandler

Climate Factors

Climate is the average weather in a given area over many years. The minimum threshold to establish climate is 30 years, although some scientists prefer to use a criterion of 100 years of temperature, precipitation, and various weather phenomena to generate more precise predictions. Five factors can impact the climate of a given area:

1. Latitude and the sun
2. Proximity to water
3. Altitude and Elevation
4. Landforms
5. Prevailing winds and ocean currents

Latitude and the Sun

Latitude has the greatest effect on a region's climate. The amount of solar radiation (sunlight) an area receives varies according to its latitude, the angle of inclination from the sun, and the number of daylight hours an area receives.

The Tropics receive direct solar radiation for an average of twelve hours a day, resulting in warmer surface temperatures near the Equator. On the December solstice, the Tropic of Capricorn receives thirteen hours and thirty-five minutes of daylight but only ten hours and forty-one minutes on the June solstice. The inverse is true for the Tropic of Cancer.

The Arctic and Antarctic Circles, in the polar regions, receive the least amount of solar radiation in their respective winters because they experience darkness for thirty days or longer, depending on the location's vicinity to the pole. The amount of solar radiation increases dramatically in each hemisphere's summer months, with some areas experiencing continuous sun for thirty days or longer, depending on the vicinity to the pole. Although the polar regions receive more solar radiation in their summer months than the Tropics do, the indirect angle of inclination does not transmit the same warmth felt in the Tropics (**Figure 16.1**).

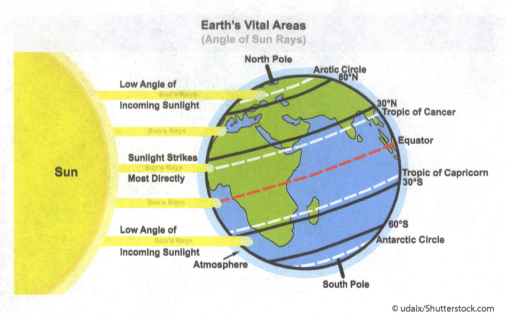

Earth's Vital Areas
(Angle of Sun Rays)

© udaix/Shutterstock.com

Figure 16.1 Direct solar radiation is experienced only between the Tropics of Cancer and the Tropic of Capricorn. North and south of the tropics all sunlight is indirect.

Proximity to Water

Large bodies of water absorb a significant amount of the sun's radiation, essentially functioning as a heat-retaining solar panel. This water heats and cools more slowly than the landmass it surrounds, causing a moderating effect on seasonal temperatures. Coastal regions, also called **maritime locations**, enjoy a more temperate climate, due to the water's moderating influence, warming the air in winter and cooling it during summer months. **Continental locations**, found in inland regions, do not experience water's insulating effect on climate, so temperature fluctuates more broadly across these large landmasses, producing hotter summers and more frigid winters.

Altitude and Elevation

As **altitude** (the vertical position of an object above sea level or land) increases, atmospheric pressure and air temperature drop. High-elevation locations have colder air, which holds less water vapor, than the warmer air, which holds more water vapor, found in low-elevation locations. Rising air has less pressure on it, which allows it to expand. The expansion process of ascending air causes a decrease in temperature, which is known as **adiabatic cooling**. Not to be confused with altitude, **elevation** is the vertical distance between the local surface of Earth and global sea level. The **lapse rate** is the rate at which air temperature falls as altitude increases in the troposphere. For every 1,000-foot increase in elevation, the temperature drops 3.6°F on average.

Landforms

Variations in **topography** (landforms that comprise an area's physical surface features), such as mountains, affect climate by causing shifts in air movement and evaporation rates, initiating changes in a region's temperature, relative humidity, and precipitation. When a moving air mass encounters a topographic barrier, it is forced to ascend and pass over it. This process, called **orographic lift**, initiates change in the air mass: the temperature drops, inversely increasing the relative humidity, and, if conditions are right, precipitation may occur on the **windward side** of the mountain (the side that faces the wind), often supporting an environment of lush, green vegetation (see **Figure 16.2**). Orographic lift occurs on the windward side of the mountain. In mid-latitude regions, the windward side is the western side of the topographic barrier, where the prevailing winds blow from the west.

As the now-drier air mass moves to the **leeward side** of the mountain (the side protected from the wind), it descends and becomes warmer. As a result, the leeward side's weather is arid (dry), sometimes desert-like. The same mountain that caused precipitation to fall on one side now blocks it from the other, creating a **rain shadow**, an area with relatively little precipitation, due to the effect of a topographic barrier. Tall mountain ranges that run parallel to coastlines, like the Cascades in North America or the Andes in South America, form a physical barrier that separates different biomes.

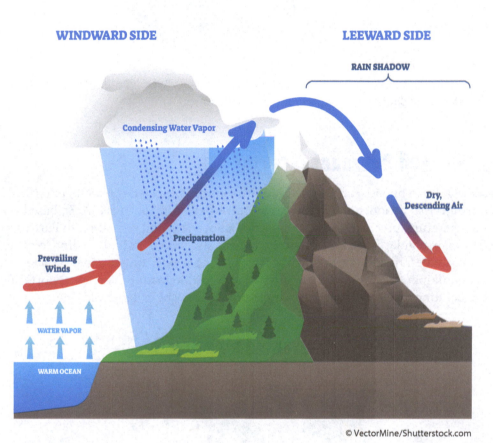

© VectorMine/Shutterstock.com

Figure 16.2   An image of orographic lift, also known as orographic effect.

Prevailing winds and ocean currents

The **prevailing winds** (winds that blow from a specific place or during a certain season) move masses of air around Earth in an attempt to balance atmospheric temperatures. These temperatures will always vary, due to Earth's rotation, the angle of inclination from the sun, and the amount of solar radiation to which it is exposed. *Trade winds, prevailing westerlies, and polar easterlies* are the main three prevailing wind belts that create drag on the ocean and act as a major influence, creating a consistent ocean current pattern. **Trade winds** blow from the east in the north and south latitudes near the Equator. **Prevailing westerlies** blow from the west in the middle latitudes, 30-60°N/S. **Polar easterlies** blow from the east near the North and South Poles. Warm ocean currents push toward the poles as cold ocean currents push toward the Tropics in a perpetual warming and cooling process. In addition to continental shape, tides, ocean basins, and mountains affect ocean currents' movement in both local and regional spaces and surface and deep water sources (**Figure 16.3**).

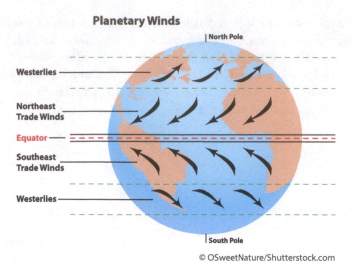

Planetary Winds

North Pole

Westerlies

Northeast
Trade Winds

Equator

Southeast
Trade Winds

Westerlies

South Pole

© OSweetNature/Shutterstock.com

Figure 16.3 Planetary wind directions and atmospheric circulation map.

Climate Zones and Subzones

In 1884, Russian-German botanist and climatologist Wladimir Köppen (1846-1940) published the first iteration of his climate classification system, which recognized five climate zones (A-E), based on the relationship between temperature, precipitation, elevation, bodies of water, and vegetation in Earth's various biomes (**Figure 16.4**). His map, regularly updated by modern scientists, uses colors and shading to differentiate major zones and their subzones. In addition to the five climate zones, an H climate zone was later used to identify a highland or mountainous region over 5,000' above sea level. Mountainous areas support different zones of temperatures depending on elevation, thus multiple climates may be experienced.

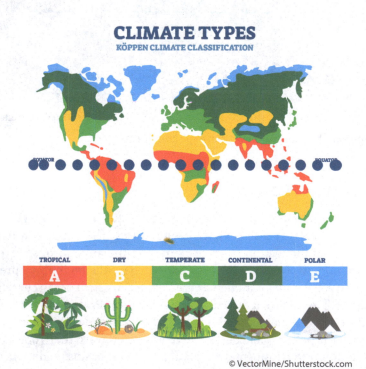

© VectorMine/Shutterstock.com

Figure 16.4 Five climate zones were first recognized based on precipitation, temperature, elevation, vegetation, and bodies of water.

Table 16.1 Köppen Climate Classification System

A 0°-25°	type of climate	precipitation	season
	tropical	more than 59" per year	dry season wet season
	climate subzones		
	• Af, Am: humid tropical	• As, Aw: seasonally humid tropical	

B 20°-35°	type of climate	precipitation	season
	dry	less than 10" per year	cold season hot season
	climate subzones		
	• BSh, BWh: arid hot	• BSk, BWk: arid cold	

C 30°-50°	type of climate	precipitation	season
	temperate	28-156" per year	four seasons
	climate subzones		
	• Csa, Csb, Csc: mediterranean • Cfb, Cwb: marine west coast	• Cfa, Cwa: humid subtropical • Cfc, Cwc: subpolar oceanic	

D 35°-70°	type of climate	precipitation	season
	continental	8-60" per year	four seasons
	climate subzones		
	• Dfa, Dsa, Dwa, Dfb, Dsb, Dwb: humid continental	• Dfc, Dsc, Dwc, Dfd, Dsd, Dwd: subarctic	
	The D climate zone does not appear in the Southern Hemisphere.		

E 70°-90°	type of climate	precipitation	season
	polar	less than 10" per year	two seasons
	climate subzones		
	• ET: tundra	• EF: ice cap	

H	type of climate	precipitation	temperature
	highlands	varies by region	varies by altitude

Solar Radiation and the Atmosphere
By Ingrid Ukstins and David Best

The Atmosphere

Earth is the only planet in the solar system with an atmosphere that can sustain life and it is composed of many gases. In the lower atmosphere, the troposphere extends up to about 7 to 10 km above Earth's surface at the poles and 17 to 18 km high above the equator, and consists primarily of nitrogen (78%) and oxygen (21%). Argon makes up slightly less than one percent of the atmosphere. Other gases present in very small amounts include carbon dioxide, methane, and nitrous oxides. Depending on atmospheric conditions, water vapor can make up as much as four percent of the atmosphere when the air is saturated. Water content varies with air temperature.

We measure the amount of water present in air in terms of the saturation level. **Relative humidity** (measured as a percent) indicates the amount of saturation present as measured against total saturation (100%) at a given temperature. Total saturation results in condensation of moisture as dew and precipitation (either rain or snow). A relative humidity of 50 percent tells us that the air is holding one half of the amount it would hold if totally saturated at a given temperature.

> **relative humidity:** The percentage of moisture present in the air as measured against the amount it can hold at a given temperature and pressure to be saturated.

A cross section of the atmosphere shows that it is layered (**Figure 17.1a**). Heat in the lowest portions of the troposphere helps drive the weather that occurs here. Turbulence is generated as wind blows over the Earth's surface, and by warm air rising from the land as it is heated by the Sun. This turbulence redistributes heat and moisture. Air becomes less dense with increasing altitude, as there are fewer gas molecules present. There is also less gravitational attraction to hold the molecules close to the surface.

As altitude increases, there is less air pressure pushing down on the surface. The decrease in temperature with height is a result of the decreasing pressure. For example, the air outside an airplane flying at 10 km is about 242°C. In the troposphere, the layer nearest the surface, air temperatures decrease steadily by about 6.5°C per kilometer to about 280°C at an altitude of 18 to 20 kilometers. The tropopause is the boundary separating the troposphere and the overlying stratosphere, in which the ozone layer is found. This layer, consisting of ozone (O_3), is the primary line of defense for harmful ultraviolet (UV) radiation that is produced by the sun. The increase in temperature with height in the stratosphere occurs because of absorption of ultraviolet (UV) radiation from the sun by the ozone. Above this layer is the stratopause, which lies below the mesosphere,

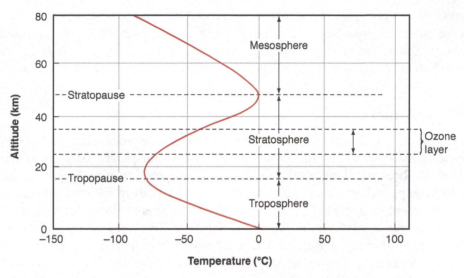

Source: NASA.

Figure 17.1a A simplified cross section of Earth's atmosphere up to 60 km. Human activity and weather are confined to the troposphere. The solid, curved line shows the change in temperature as altitude changes.

ISS022E062672

Source: NASA.

Figure 17.1b This photo of the space shuttle Endeavour orbiting Earth shows the orange troposphere, whitish stratosphere and blue mesosphere layers.

the layer that extends to an altitude of about 50 to 55 kilometers. Here the temperature again decreases with height, reaching a minimum of about –90°C. Although other layers of the atmosphere lie above the mesosphere, they do not have any significant effect on the climate and surface conditions. These upper layers do play a role in protecting Earth from extraterrestrial objects.

Solar Radiation and the Atmosphere

Electromagnetic radiation from the sun has a wide range of wavelengths. Part of this spectrum of energy includes visible light. However, a much wider range of energy bombards Earth, including many wavelengths that are harmful. Certain gases in the atmosphere interact with selected wavelengths in different ways. As mentioned earlier, ozone shields the lower atmosphere and Earth's surface from 97 to 99 percent of the

Sun's UV radiation, which damages the genetic material of DNA and is related to some types of skin cancer. Destruction of the ozone layer has occurred by the release of man-made organic compounds, including chlorofluorocarbons (CFCs), into the atmosphere. Ozone reacts readily with CFCs and is destroyed in the chemical reaction. Thus UV radiation can then penetrate the atmosphere and reach the surface.

Effects of Volcanic Activity

The eruption of volcanoes often involves expulsion of gases and other pyroclastic material (**Figure 17.2**). Sulfur gases are a component in explosive volcanic eruptions and can be transported to the upper atmosphere where they are able to reflect short wavelength solar radiation, which produces a cooling effect. The combination of sulfur gases and the pyroclastic dust can produce a significant drop in temperature that is equivalent to a nuclear winter. Recent examples of volcanic eruptions that altered the atmosphere and the amount of incoming radiation include the 1982 eruption of El Chichón in Chiapas, Mexico, and the 1991 eruption of Mount Pinatubo in the Philippine Islands. More than 20 million tons of sulfur dioxide were thrown into the atmosphere, causing global temperatures to drop 0.5°C for a two-year period following the Pinatubo event. In 1912 the eruption of Novarupta Volcano on the Katmai Peninsula of Alaska produced more than twice the pryoclastics and gases of Mount Pinatubo. When measured in the context of geologic time, major eruptions similar to these can occur rather frequently. One major eruption every 100 or 200 years adds great volumes of gases to the atmosphere, and water to the hydrosphere.

United States Geological Survey.

Figure 17.2   **Gases and other pyroclastic materials produced by volcanoes contribute significantly to affecting the amount of sunlight that can reach the surface.**

In the geologic past, several episodes of massive flood basalt eruptions sent countless millions of tons of gases into the atmosphere. The flood basalts of Siberia that formed about 250 million years ago, the Deccan Traps in India that erupted 66 million years ago and the Colombia River Basalts which erupted from 17 to 14 million years ago in the Pacific Northwest of the continental United States all had a disastrous effect on living organisms. The continual eruption of basalts associated with seafloor spreading adds gases directly into sea water, thereby changing the chemistry of the largest body of water on Earth.

Greenhouse Gases

Once solar radiation penetrates the atmosphere and strikes the surface, some of the energy is absorbed by the land and water, while some of it is reflected back into space. Gases and dust particles also reflect radiation back into the upper atmosphere, as well as clouds and ice.

When this energy is reflected, there is a change in the wavelengths of the radiation. Longer wavelength infrared waves are absorbed by moisture and gases in the middle and upper troposphere (**Figure 17.3**). This absorption of heat by these gases causes the temperature of the atmosphere to rise. Radiation is also re-reflected back toward the surface. This trapped energy causes a global warming effect, similar to that that takes place in a greenhouse, and results in a land and ocean surface temperature that is on average 14 degrees C warmer than it would be without this process. It is important to note that this greenhouse effect is a natural process and helps contribute to the habitability of Earth; it is the increase in **greenhouse gases** since the Industrial Revolution that has caused a change in the composition of the atmosphere and increased the magnitude of the greenhouse effect.

> **greenhouse gases:** Atmospheric gases, primarily carbon dioxide, methane, and nitrous oxide restricting some heat-energy from escaping directly back into space.

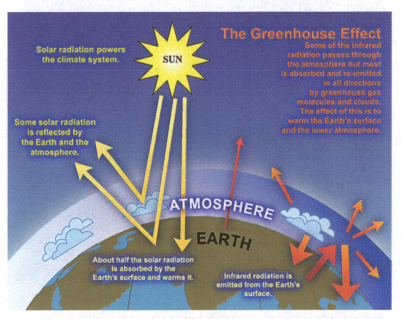

Source: IPCC—Intergovernmental Panel on Climate Change.

Figure 17.3   **Energy from solar radiation is divided between reflected and reradiated energy.**

Several gases contribute to this **greenhouse effect**. The primary gases are carbon dioxide, water vapor, methane, and ozone. Because so much water is present on Earth's surface, it is easily heated, and rises to become a significant contributor to the warming process. However, carbon dioxide, a gas that is generated by the oxidation (burning) of fossil fuels that contain carbon, is also a key greenhouse gas. Over the past 200 years, there has been a 40 percent increase in the presence of carbon dioxide in the atmosphere. This has contributed to the warming of the atmosphere.

> **greenhouse effect:** The heating that occurs when gases such as carbon dioxide trap heat escaping from the Earth and radiate it back to the surface.

Methane (CH_4) is the simplest of the hydrocarbons. It is a byproduct of the decay of organic material, and it also forms during the digestive process of organisms. Estimates show that as much a 20 percent of all methane is produced by cattle. Methane is also released during the processing and transportation of petroleum products.

Reducing the Presence of Carbon

The role that carbon dioxide plays in enhancing the greenhouse effect, and to air pollution, is evident to many scientists. If the amount of carbon can be reduced, the end result would be a better environment. The federal government, in conjunction with many private industries, is working on the process of **carbon sequestration**, which involves removing carbon from the environment or reducing or eliminating its presence. The increased use of hybrid vehicles decreases the demand for petroleum fuels and reduces the amount of carbon dioxide put into the atmosphere. The rapidly increasing use of wind turbine technology to generate electricity reduces the need for coal-fired power plants (**Figure 17.4**) and reduces the production of carbon dioxide.

> **carbon sequestration:** The storage or removal of carbon from the environment or the reducing or elimination of its presence.

Source: David M. Best.

Figure 17.4   This wind turbine farm in southern California is part of a larger plan nationwide to increase the production of electricity from clean, existing resources.

Atmospheric Energy Interactions

By Dean Fairbanks

Atmospheric Energy Balance

The wavelengths of radiation the surface receives from the sun span roughly 0.3 to 3 mm (visible light to far infrared), while the wavelengths emitted from earth span roughly 5 to 50 mm. The overall energy balance model of radiation can be described as follows:

Table 18.1

Input	Output
Solar radiation to earth	Earth's infrared emission to space
Concentrated in shorter wavelengths: UV, visible, and shortwave infrared	Longer wavelengths: thermal infrared (heat energy)
0.3 to 3 mm	5 to 50 mm

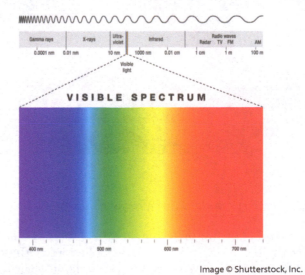

Image © Shutterstock, Inc.

Figure 18.1   The light that is visible to humans on the electromagnetic spectrum, called the visible spectrum, is quite narrow ranging from only 380 to 750 nanometers.

This figure compares the emission from the earth with the amount of solar energy that reaches the top of the earth's atmosphere (**Figure 18.1**). Notice that now, the emission from the earth is larger than the energy received from the sun, for wavelengths longer than about 5 microns. The exact cross-over point depends on the earth's emission temperature. The average surface temperature is a bit warmer than the emission temperature shown in this graph, so the cross-over point for surface emission occurs at a shorter wavelength. This slightly fuzzy cross-over point is used to define two principal wavelengths: shortwave (SW) and longwave (LW).

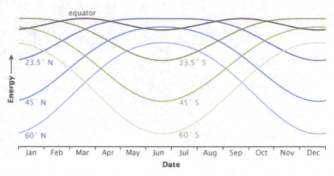

NASA illustration by Robert Simmon.

Figure 18.2   **Difference in peak energy at various latitudes.**

The **peak energy** received at different latitudes changes throughout the year. **Figure 18.2** shows how the solar energy received at local noon each day of the year changes with latitude. At the equator (gray line), the peak energy changes very little throughout the year. At high northern (blue lines) and southern (green) latitudes, the seasonal change is extreme.

The **total energy** received each day at the top of the atmosphere depends on latitude. In **Figure 18.3** the highest daily amounts of incoming energy (pale pink) occur at high latitudes in summer, when days are long, rather than at the equator. In winter, some polar latitudes receive no light at all (black). The southern hemisphere receives more energy during December (southern summer) than the northern hemisphere does in June (northern summer) because earth's orbit is not a perfect circle and earth is slightly closer to the sun during that part of its orbit. Total energy received ranges from 0 (during polar winter) to about 50 (during polar summer) megajoules per square meter per day (NASA).

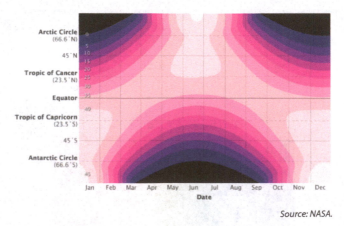

Source: NASA.

Figure 18.3   **Latitudinal location determines the total energy received each day.**

The term "balance" refers to the *net* of two opposing fluxes—think of radiation from the sun to the earth and radiation leaving the earth. The earth emits longwave radiation continuously (24 hours/day, 365 days/year), whereas the surface receives shortwave primarily visible light when rotated to face the sun. As was explained in Section 1, the daylength changes seasonally and latitudinally, but at any time only 50% of the earth's surface is in sunlight. The atmosphere is mostly transparent to shortwave visible radiation, but the surface receives less than the amount first entering the top of the atmosphere at the thermopause 480 km altitude. The loss can be attributed to the solar radiation being modified as it passes through the atmosphere and interacts with gas molecules, aerosols, and water vapor (cloud droplets). The following solar energy interactions/modifications and pathways present the exchange of energy between the sun, atmosphere, and earth:

- **Transmission**—passage of electromagnetic radiation EMR through a medium (air or water). The thicker the medium, the lesser the transmission. The atmospheric transmission changes based on the wavelength.
- **Scattering**—the interaction of atmospheric gas molecules and aerosols with radiation as it passes through the atmosphere. This interaction does not change the wavelength of radiation. The direction of scattering is random, and some radiation is redirected back to space. Scattering **diffuses** the light, eliminating shadows— think of a partly cloudy day, as opposed to **direct** radiation (have shadows). There are two types of scattering, depending on the size of the scattering particles:
  - **Rayleigh**—wavelength > particle size. Rayleigh scattering gives the atmosphere its blue color. Blue light is the shortest wavelength scattering off individual gas atoms and molecules. The type of scattering also leads to red sunsets because the low sun angle causes the visible light to come through a thicker atmosphere and thus the scattering is stretched to the oranges and reds.

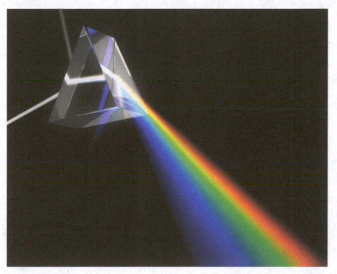

Image © Shutterstock, Inc.

Figure 18.4 Prism refracting white light.

  - **Mie**—wavelength < particle size. This type of scattering gives the color white to clouds since the H2O droplets are large enough to scatter the entire visible spectrum. From color theory, red, green, and blue mixed give white (**Figure 18.4**).
- **Refraction**—represents the change in speed and direction of solar radiation as it passes from one medium to another (e.g., from the empty vacuum of space to our atmosphere, or from air to water). The incoming angle of the incident ray is bent to a new angle in the new medium, becoming a refracted ray.
- **Reflection**—when radiation is bounced back to space at the same angle it arrived at. The incident ray and reflected ray have the same incident and reflected angle. Think of a mirror or chrome on car. The reflective ability of a surface/material refers to its **albedo**. Albedo is determined by the color of the material where darker colors are less reflective then lighter ones. The angle of the incoming radiation to a surface also makes a difference where lower angles produce higher albedos. The texture of the surface also plays a role, where smoother surfaces have higher albedo. In the atmosphere clouds (white) account for roughly two-thirds of planetary albedo, forcing shortwave radiation to be reflected back to space.

January 2013

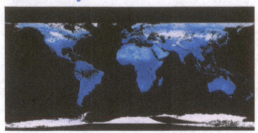

July 2013

Source: NASA.

Figures 18.5–18.6 **Maps showing albedo effect.**

These maps show albedo on a scale from 0 (no incoming sunlight being reflected) to 0.9 (nearly all incoming light being reflected). Darker blue colors indicate that the surface is not reflecting much light, while paler blues indicate higher proportions of incoming light are being reflected. Black areas indicate "no data," either over ocean or because persistent cloudiness prevented enough views of the surface. Our planet's brightest surfaces (highest albedos) are ice caps, glaciers, and snow-covered ground. Deserts also have high albedos. Forests have low albedos, especially boreal forests during summer months (**Figures 18.5–18.6**).

- **Absorption**—assimilation and conversion of radiation by molecules into a different form. *This interaction does change the wavelength of the radiation*—shortwave visible re-radiated as longwave thermal. This absorption and re-radiation is exactly what ozone does in the stratosphere. Atmospheric dust (aerosols) and water droplets also participate in absorption.

The exchange of energy between earth and atmosphere can be expressed as longwave re-radiation, heat transfers. Shortwave energy is converted to longwave energy (thermal heat), which can be absorbed by the atmosphere; some is re-radiated back—water vapor acts as an insulator.

In synthesis only about half of the radiation at the top of the atmosphere is absorbed at the earth's surface. Globally, taken as a whole, the earth is in radiative equilibrium where the surface is net positive and the atmosphere is net negative. The two balance one another out. Looking at the energy budget by latitude illustrates the second law of thermodynamics. Surplus incoming heat energy gain at the equatorial and tropical regions is transported poleward where polar energy is in deficit. This energy gradient is dissipated by meridional transport that drives the general circulation of the atmosphere and oceans. If the second law of thermodynamics did not work and thus no energy transfers to the poles, they would be a lot colder (greater than 25°C cooler than average) and the equator would be much warmer (greater than 14°C warmer than average).

Net shortwave radiation = shortwave down - shortwave up. Net longwave radiation = longwave down - longwave up. **Net radiation** = net shortwave

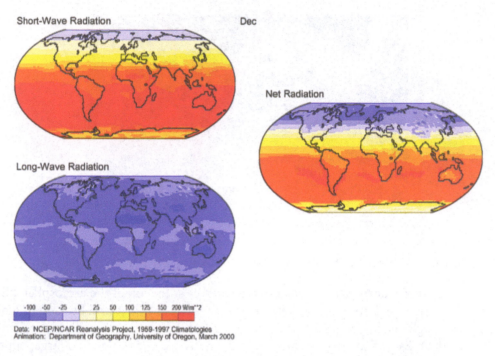

Source: Department of Geography, University of Oregon, supported by the US National Science Foundation, 2000

Figure 18.7   **Net radiation is the balance between short-wave and long-wave radiation.**

Greenhouse Effect

Ultimately, what has been described is the role the atmosphere plays in its composition and function as a **greenhouse** for the planet. The atmosphere absorbs heat energy and delays transfer of that heat from earth into space, acting as an insulator. The major components causing this delay of heat and re-radiation back to the surface are water vapor (60%), carbon dioxide (25%), ozone (8%), and other trace gases such as nitrous oxide, methane, and CFCs make up the remainder of the total. What needs to be stressed here is the tremendous role of clouds—water vapor insulating the planet delaying the effects of the second law of thermodynamics, but also the powerful effect of CO2, when as was pointed out earlier that it only makes up 0.040% of the atmosphere's total composition. CO2 is a very powerful greenhouse gas indeed. Clouds are also powerful but play two different roles depending on their altitude. When clouds are high in the atmosphere, such as cirrus clouds (thin-wispy and cold), they act like clear air transmitting most shortwave insolation to the surface, but they absorb and re-radiate longwave radiation both out to space and back to the earth's surface. Their cold nature, however, causes less energy radiated to outer space than would be without the cloud, so they have a large cloud greenhouse forcing—enhancing atmospheric greenhouse warming. In contrast, low clouds (think typical rain clouds) reflect most shortwave insolation to space (net albedo forcing), and what shortwave does get absorbed is equally emitted to space leading to atmospheric cooling (**Figure 18.7**).

To emphasize the power of the **greenhouse effect**, imagine the earth's average temperature without the greenhouse effect. The average black body radiative equilibrium temperature without an atmosphere would be −17°C; however, the enhanced greenhouse effect gives earth an actual average temperature of 15°C. The difference is the greenhouse effect.

The atmosphere radiates the equivalent of 59% of incoming sunlight back to space as thermal infrared energy, or heat. Where does the atmosphere get its energy? The atmosphere directly absorbs about 23% of incoming sunlight, and the remaining energy is transferred from the earth's surface by evaporation (25%), convection (5%), and thermal infrared radiation (a net of 5–6%). The remaining thermal infrared energy from the surface (12%) passes through the atmosphere and escapes to space.

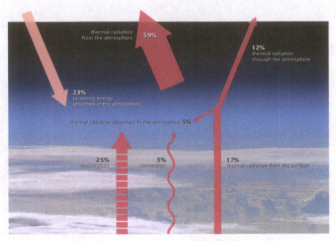

NASA illustration by Robert Simmon.

Figure 18.8 Large amounts of heat is radiated back to space.

On average, 1370 watts per square meter of solar energy arrives at the top of the atmosphere. Earth returns an equal amount of energy back to space by reflecting some incoming light and by radiating heat (thermal infrared energy). Most solar energy is absorbed at the surface, while most heat is radiated back to space by the atmosphere (**Figure 18.8**). Earth's average surface temperature is maintained by two large, opposing energy fluxes between the atmosphere and the ground (right)—the greenhouse effect.

Earth Surface Energy Balance

The majority of Section 1 so far has looked at global scale processes to balance the earth's energy budget: insolation entering the earth's atmosphere, interactions in the atmosphere and only a general discussion of surface processes. At each level, the amount of incoming and outgoing energy, or net flux, must be equal. Since roughly 29% of incoming shortwave insolation is reflected back to space that leaves 71% to be absorbed by the atmosphere (23%) and the land/oceans (48%). That 48% that the ocean and land surfaces absorb must be gotten rid of for the energy budget at the earth's surface to balance. The processes operating at the surface specifically refer to latent heat, emission of thermal infrared radiation as sensible heat, and more specific considerations of conduction and convection.

Image © Shutterstock, Inc.

Image © 2014 Shutterstock, Inc.

Figures 18.9–18.10 Controls on latent heat and sensible heat: Which surface will partition much more energy into latent heat than sensible heat (and thus feel cooler)?

The exchange of energy between earth and atmosphere can be expressed as longwave re-radiation, thermal infrared heat transfers, and latent heat transfers. Shortwave energy is converted to longwave energy (thermal heat), which can be absorbed by the atmosphere; some is re-radiated back —water vapor acts as an insulator. Conduction and convection are the main heat transfer processes:

- **Conduction**—is transfer of energy (heat) through molecule-to- molecule collisions. This is what happens when you grasp a handle to a pan heated on the stove and it is hot to the touch. On the earth's surface heat passes from the ground to the air because it is cooler; air, however, is a poor conductor.
- **Convection**—is transfer of energy (heat) through vertical mixing of molecules (gas or liquid). The surface heating of air causes it to expand, and lower density leads to air rising.

Then there is the latent heat of water, pertaining to the heat involved in water phase changes. In atmospheric **latent heat** transfers, energy is used to convert water into water vapor. The heat is retained as latent heat. When vapor turns back into liquid water, heat is released in the atmosphere. This is different from **sensible heat**—the heat you measure with a thermometer, what you can sense (i.e., when you burn your hand on the stove).

Table 18.2

← Heat Energy Released←				
Solid Water	**←Freezing**	**Liquid Water**	**←Condensation**	**Water Vapor**
	Melting→		**Evaporation→**	
→Heat Energy Absorbed→				

The following set of animated images provides a monthly look at global sensible heat flux (direct heating, a function of surface and air temperature), latent heat flux (energy that is stored in water vapor as it evaporates, a function of surface wetness and relative humidity), net radiation flux, and the overall change in surface heat storage. The heat storage = net radiation - latent heat flux – sensible heat flux. It is best to focus on what is happening in the oceans and on the land surfaces, especially those over large deserts, the tropics, and large boreal forest lands. Then focus on the question posed above with regards the two landscapes and how heat is partitioned.

Positive values for sensible and latent heat flux represent energy moving toward the atmosphere, negative values represent energy moving away from the atmosphere. Positive values for change in heat storage represent energy moving out of storage, negative values represent energy moving into storage.

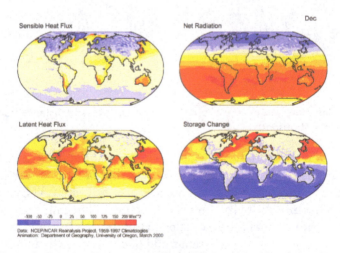

Figure 18.11   Heat storage = net radiation – latent heat flux – sensible heat flux.

Atmospheric Movement
By Ingrid Ukstins and David Best

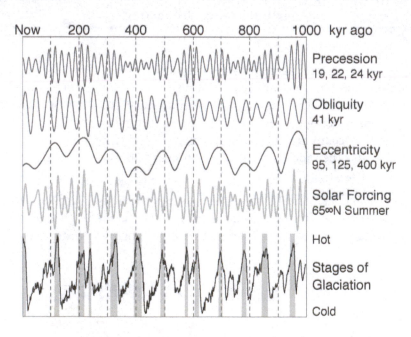

Source: NOAA National Centers for Environmental Information.

Figure 19.1 Milankovitch cycles show how changes in the Earth's orbit, its tilt and distance from the Sun affect the climate on the scale of thousands to tens of thousands of years.

As the Earth rotates on its axis, there are times when the angle of tilt of the axis ranges from 22.1 degrees to 24.5 degrees. Currently Earth is tilted 23.5° from the vertical. The period of change is roughly 41,000 years. The planet's tilt reached a maximum in 8,700 BCE, we are currently about halfway between the maximum and minimum, and will reach a minimum in 11,800 CE. Increased tilt means that each hemisphere's summer will get more incoming solar radiation as the pole is tipped towards the sun, and less solar radiation in winter as the pole is tipped away from the sun. Our current trend of decreasing tilt and the changes in solar energy reaching the poles promotes warmer winters and cooler summers. As Earth rotates, it also tends to wobble on its axis, in the same way a spinning top begins to wobble as its rotation rate decreases. This wobble, also termed **axial precession**, has a period of about 26,000 years. It is caused by the tidal forces from the Sun and the Moon on the solid Earth. The net effect of all these astronomic factors occurring together is that their period, together with their maximum influences, corresponds to times when active glaciations happened, due to major changes in the amount of solar radiation striking Earth over time (**Figure 19.1**).

> **axial precession:** The wobble that occurs when a spinning object slows down.

The Coriolis Effect

Earth rotates on its axis roughly once every 24 hours, or one day. The Earth spins on its axis from west to east, this rotation produces an effect on the movement of fluids on or near the surface perpendicular to that, in a north-south direction. The effect is termed the **Coriolis Effect**, after an early nineteenth-century French mathematician who proposed its existence. The effect is that any moving object in the northern hemisphere moves to the right (clockwise) and an object in the southern hemisphere moves to the left (counterclockwise) (**Figure 19.2**). There is actually no physical force involved, because it is just the ground moving a different speed than an object in the air. For the planet's atmosphere, it means that the winds appear to be deflected counter-clockwise in the Northern Hemisphere and clockwise in the Southern Hemisphere. The effect is zero at the equator. The magnitude of the deflection increases toward the poles.

> **Coriolis Effect:** An imaginary force that appears to be exerted on an object moving within a rotating system. The apparent force is simply the acceleration of the object caused by the rotation. Along the equator, there will be no such rotation.

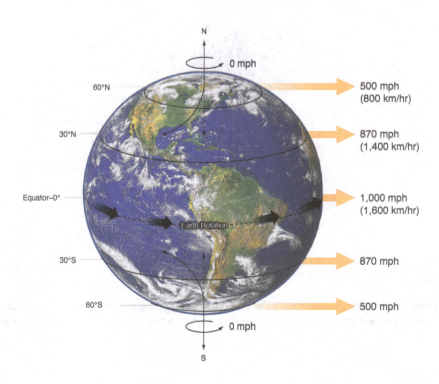

© Sandy Stifler/Shutterstock.com. Adapted by © Kendall Hunt Publishing Company.

Figure 19.2 **The highest velocities of Earth's rotation are at the equator; the slowest are at the poles.**

Atmospheric Circulation

The atmosphere is a fluid and thus moves readily. Several factors produce movement in the troposphere. Differential heating of the surface of the Earth and the overlying air generate warm and cool **air masses** that then move across the surface. The upward motion of warm air into the upper troposphere is caused by convection. The equatorial regions receive the greatest amount of solar radiation (**Figure 19.3**). Rising, warm air moves toward the poles in the upper portion of the troposphere. This air contains large amounts of water vapor derived from the evaporation of ocean waters near the equator. The area near the equator, where warm air is rising, is one of low pressure because of the suction effect of the rising air.

> **air mass:** A large body of air of considerable depth which are approximately homogeneous horizontally. At the same level, the air has nearly uniform physical properties, especially temperature and moisture.

Colder temperatures in the upper troposphere cause condensation of this water vapor, which returns to Earth as precipitation. The air in the upper troposphere is now cold and depleted of moisture, so it begins to sink in a region near 30° to 35° north and south of the equator. This region is called the subtropical high-pressure zone. High pressure exists in those areas where the cooler, dry air is descending back to the surface. As the air descends, it is compressed and begins to heat up, so the surface air is warm and dry. These conditions form many of the mid-latitude deserts on Earth, such as the deserts of North America, the Middle East, and Saudi Arabia in the northern hemisphere, and in Australia in the southern hemisphere. These descending winds are also deflected by the Coriolis Effect and produce the prevailing westerlies.

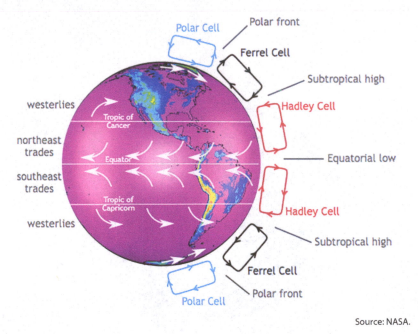

Source: NASA.

Figure 19.3 Large amounts of solar radiation near the equator cause heating of the surface and atmosphere. Rising air then moves poleward where it descends in the subtropical regions near 30 degrees north or south of the equator.

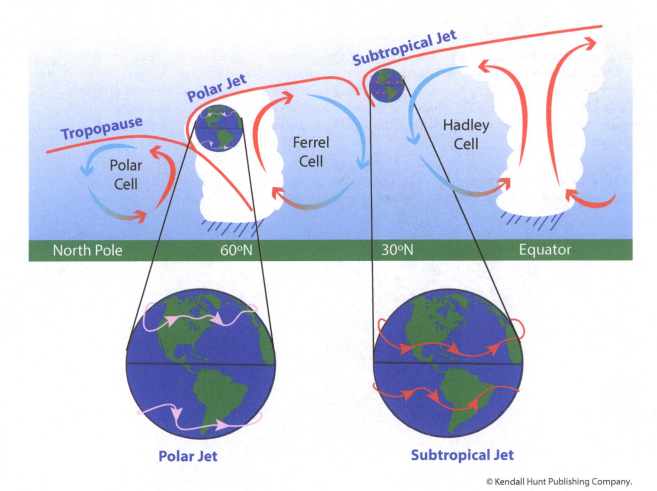

© Kendall Hunt Publishing Company.

Figure 19.4   Jet stream diagram.

Air Masses

Large-scale movement of the atmosphere often involves expansive air masses. North America is affected by these masses, as they often move down from regions around the North Pole or off the Pacific Ocean. Polar air masses are typically cold and dry, while those coming off the Pacific and Gulf of Mexico are warm and moist. General movement is from west and northwest to east and southeast. These directions are driven by Coriolis forces and the prevailing west-to-east **jet streams** that traverse the country (**Figure 19.4**). Occasionally these two different air masses will collide; the result is often very unstable weather conditions.

jet stream: A high-speed, meandering wind current, generally moving from a westerly direction at speeds often exceeding 400 kilometers (250 miles) per hour at altitudes of 15 to 25 kilometers (10 to 15 miles).

Low-Pressure Conditions

In an area where less dense air rises due to heating, the upward force generates an area of low pressure (similar to a vacuum cleaner). This lower pressure causes the air to move from areas of high pressure into the lower pressure area. In the northern hemisphere this rising air rotates in a counterclockwise manner (**Figure 19.5**). Because of the Coriolis Effect, low pressure rotates in a clockwise manner in the southern hemisphere. Low pressure systems are termed **cyclones**. This rotational system is very evident when we examine hurricanes and other cyclonic storms.

cyclone: An area of low atmospheric pressure having is counterclockwise circulation in the northern hemisphere and clockwise motion in the southern hemisphere.

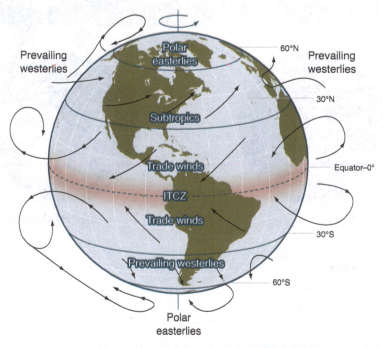

© Kendall Hunt Publishing Company.

Figure 19.5 Circulation of Earth's atmosphere consists of a series of belts of air that produce the trade winds, westerlies, and polar easterlies.

High-Pressure Conditions

Whenever air has been cooled, it becomes denser and sinks, thus producing a higher pressure region on the surface. The winds generated by the descending air mass move outward from the center in a spiral fashion. These areas of high pressure rotate in a clockwise fashion in the northern hemisphere (counterclockwise in the southern hemisphere) and are termed **anticyclones** (**Figure 19.6**).

anticyclone: An area of high atmospheric pressure having clockwise circulation in the northern hemisphere and counterclockwise motion in the southern hemisphere.

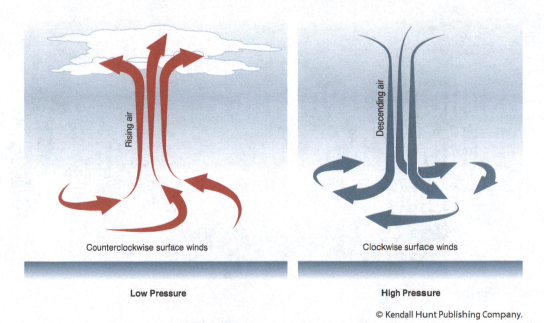

Counterclockwise surface winds

Clockwise surface winds

Low Pressure

High Pressure

© Kendall Hunt Publishing Company.

Figure 19.6 Rising air produces a low pressure condition on the surface in the northern hemisphere, producing a counterclockwise rotation; high pressure produced by colder, descending air generates clockwise winds at the surface.

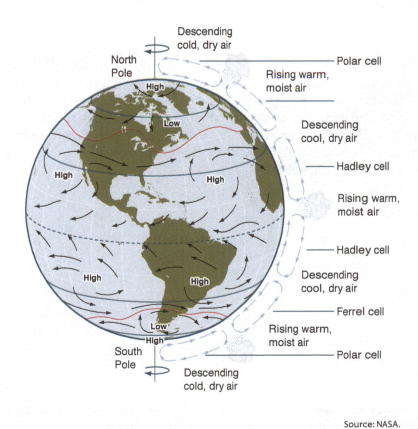

Source: NASA.

Figure 19.7 Hadley cells move warm, moist air from the equator to about 30° north and south latitudes, where it descends and produces high pressure. Ferrel cells lie between Hadley cells and Polar cells and move warm air to higher latitudes and shift cold air toward the subtropics.

Atmospheric Dynamics: Atmospheric Motion

By Dean Fairbanks

Atmospheric Motion

Atmospheric motion is controlled by the combination of the following forces:

- **Pressure-gradient force**
- **Coriolis force**
- **The geostrophic wind**
- **The gradient wind**
- **Friction**

The air above an object exerts a force per unit area upon that object and we call this pressure. The variations in pressure lead to the development of winds, and thus the development of weather. We measure pressure using a barometer, which is why atmospheric pressure is referred to as barometric pressure. Also, reviewing from earlier discussion, pressure is measured in millibars (mb), the unit of pressure found on weather maps (**Figure 20.1**). The average pressure at sea level is 1013.25 mb. Pressure decreases with increasing altitude since the number of molecules above a surface is less with height. On a map a line connecting points of equal pressure is called an **isobar.**

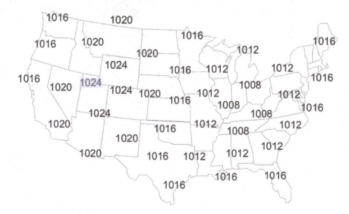

Figure 20.1 This map shows the sea-level pressures for various locations over the contiguous United States. The values are in whole millibars. (**NOAA**)

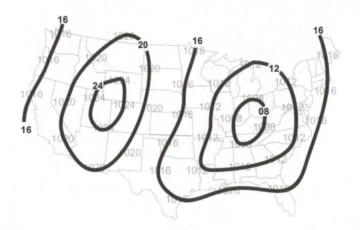

Figure 20.2 Isobars are usually drawn for every 4 mb, using 1000 mb as the starting point. Traditionally, only the last two digits are used for labels. For example, the label on the 1024 mb isobar would be 24. A 1008 mb isobar would be labeled 08. **(NOAA)**

A constant pressure surface is a surface in the atmosphere where the pressure is equal everywhere along the surface. For example, a 100 mb surface, everywhere along that surface would measure 100 mb, and since pressure decreases with height, the 100 mb surface is higher in the atmosphere than the 500 mb surface. Thus, when identifying a specific pressure surface, we are defining a location (height) in the atmosphere that refers to that specific pressure. For example, 1000 mb = ~100 m, 500 mb = ~5,000 m, and 100 mb = ~16,000 m. The height of a pressure surface varies with temperature. For example, if the temperature is the same at two different locations (columns of air) then the height of say the 700 mb surface would be the same at both locations. However, if one of the columns of air cooled and the other warmed then the height of the 700 mb surface would change in each column. Colder air contracts lowering the 700 mb height and warming air expands, raising the height of the 700 mb surface. Thus, areas of colder temperature will have pressure surfaces with a lower height than if the same pressure surface was located in warmer air. The fundamentals of pressure, isobars for mapping, and the concept of the pressure surface allow for the explanation of high- and low-pressure centers:

- **High pressure**—where pressure is measured to be the highest relative to its surroundings. A high-pressure region represents the center of an anticyclone and is indicated on weather maps by a blue "**H**." In the northern hemisphere at the surface, winds tend to flow clockwise and outward away from a low-pressure region, while in the southern hemisphere, winds flow counterclockwise around a high-pressure region. In a high-pressure region, the air is sinking thus compressing and therefore suppressing the development of clouds. A region of high pressure is a region with clear skies.
- **Low pressure**—where pressure is measured to be the lowest relative to its surroundings. A low-pressure region represents the center of a cyclone and is indicated on weather maps by a red "**L**." In the northern hemisphere at the surface, winds tend to flow counterclockwise and inward toward a low-pressure region, while in the southern hemisphere, winds flow clockwise around a low-pressure region. In a low-pressure region, the air is rising thus expanding and therefore favoring the development of clouds. A region of low pressure is a region with cloudy skies.

Pressure maps are created hourly from weather stations and are used to locate areas of high and low pressure. A map of isobars is also useful for identifying pressure gradients, which are identified by lose or tight packing of the isobars. Stronger winds are associated with stronger pressure gradients (**Figure 20.2**).

Isobars can be used to identify "Highs" and "Lows." The pressure in a high region is greater than the surrounding air and flows clockwise in the Northern Hemisphere. The pressure in a low region is lower than the surrounding air and flows counter-clockwise in the Northern Hemisphere (NOAA) (**Figure 20.3**).

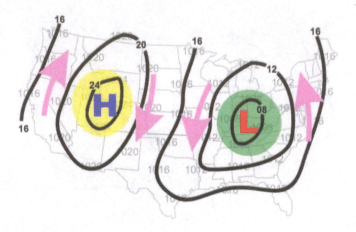

Figure 20.3 Isobars also identify high and low pressure.

The change in pressure measured across a given distance is called a pressure gradient. Wind is the horizontal flow of air in response to differences in air pressure, these pressure differences are due to uneven solar heating at the surface. The relationship between the **pressure-gradient force** (PGF) and wind is positive linear. For example, a 4× increase in the PGF corresponds to a 4× increase in wind speed. Pressure-gradient wind is the movement of air that occurs along pressure gradients from high to low pressure (**Figure 20.4**).

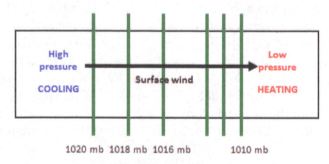

Figure 20.4 The PGF is directed from high to low pressure. The tighter the isobars of pressure the faster the wind speed.

Once air is set in motion by the PGF it undergoes an apparent deflection from its path. The **Coriolis force** (CF) is the deflection of winds due to the earth's rotation. Deflection is to the right in the northern hemisphere and to the left in the southern hemisphere. The amount of deflection air makes is related to the speed the air is moving and the latitude. Faster winds are deflected more than slowly blowing winds. The rotation of the earth is fastest at the equator and zero at the poles. The deflection of winds increases leaving the equator to the poles. The CF is zero right at the equator. The following movie illustrates the process.

In the mid-latitudes, the PGF and the CF are directly balanced. This leads to air moving not from high to low pressure but between the two, parallel to the isobars as a **geostrophic wind**. These winds tend to be in the upper troposphere around the 500 mb level (**Figure 20.5**).

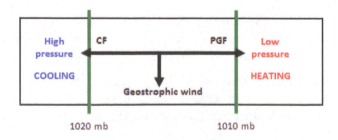

Figure 20.5 Air movement between the isobars is geostrophic wind.

The geostrophic wind is found in the upper troposphere. It is caused by a balance between the PGF and the CF that makes the wind flow between the isobars.

The truth of the matter is that geostrophic winds are rare since they exist only in areas with no friction and where isobars are straight. In reality, isobars are always curved and are rarely evenly spaced. This changes the winds from geostrophic to what is really a **gradient wind**. These winds blow parallel to the isobars but are no longer balanced by the PGF and the CF. Where isobars are curved, centrifugal and centripetal forces act upon the wind to maintain a flow parallel to the isobars. This curved path is called the gradient wind. What this does to the air movements in the upper troposphere is develop a twist to the air movements allowing for high and low-pressure areas to develop a rotary motion as the wind flows parallel to the isobars. In the northern hemisphere, high-pressure areas rotate clockwise (anticyclone) and low-pressure areas rotate counterclockwise (cyclone). They are opposite rotations in the southern hemisphere due to the left hand deflection by the CF (**Figure 20.6**).

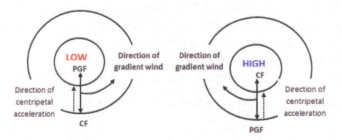

Figure 20.6   Where isobars are curved, centripetal forces act upon the wind to maintain a flow parallel to the isobars. The curved path is called a gradient wind, and is illustrated for the northern hemisphere.

Finally there is the effect of **friction** on air movements. Frictional drag from the earth's surface modifies the balance between PGF and the CF. Friction decreases wind speed but also changes wind direction. The difference in terrain conditions directly affects how much friction is exerted. For example, hills and forests force the wind to slow down and/or change direction much more, opposed to blowing across a flat expanse of terrain. Friction causes the wind to cross isobars at an angle. This decreases with height, which means wind changes direction with height. The height at which winds are no longer affected by surface friction is 1–2 km altitude, which is where the winds are nearly geostrophic or gradient. Since friction adds a countering force to the CF and cross isobars at an angle, the winds tend to spiral out of high-pressure areas and spiral into low-pressure areas (**Figuer 20.7**).

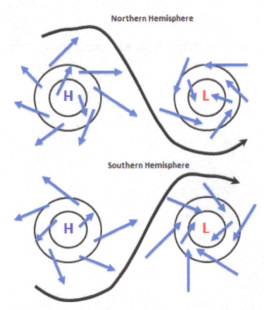

Figure 20.7   Wind flows from high-pressure area anticyclones into low-pressure area cyclones. The friction force causes winds to move across isobars at angle.

Global and Local Winds

Winds, by convention, are designated based on the direction they come from, not the direction they are going. This is true for the labeling of global winds, but not so for local winds as we shall see. At the global scale there are three major winds, one at the equator, one at the mid-latitudes (30–60°), and the other at the polar regions (60°):

- **Eastern trade winds**—at the equator air is heated and rises, leaving low-pressure areas behind. The air moves north and south of the equator to about the 30° parallel where the air begins to cool and sink, thus compressing and becoming high-pressure areas. The majority of this cooling air moves back to the equator, and the remainder flows to the poles. The Coriolis effect on both sides of the equator curves (converges) the air coming back to the equator to the west. These winds are warm and blow nearly continuously from the east to the west. These are the same winds that early European explorers used to sail across the Atlantic Ocean to the new world. Thus the "trade" winds.
- **Prevailing westerlies**—The winds that move toward the poles appear to curve to the east. The westerlies are responsible for much of the weather movements across the United States and Canada.
- **Polar easterlies**—As the atmosphere over the poles cools, this air sinks and spreads over the surface flowing away from the poles. The CF turns the air to the west.

All of these global winds are a direct demonstration of the CF (**Figure 20.8**).

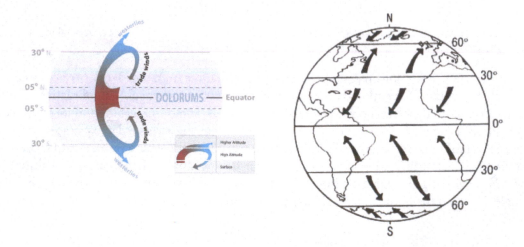

Figure 20.8   Atmospheric circulation and the Coriolis effect create global wind patterns including the eastern trade winds and the prevailing westerlies. (**NOAA**)

Atmospheric Dynamics: Global and Local Winds

By Dean Fairbanks

Global and Local Winds

Local winds are what we are most used to in our everyday lives on the surface. These local winds are a result of uneven surface heating. Two of the most common types of local wind pairs include land and sea breezes, and mountain and valley winds.

- **Land and sea breezes**—are most often experienced on warm sunny days during spring and summer when the temperature on the land is normally higher than the temperature of the water. Remember that the specific heat of water is much higher than for land. Land is able to heat up rapidly while large bodies of water like a large lake or ocean are able to absorb more insolation than land without warming. The long wave thermal heat radiated back into the atmosphere via convection from the land creates a low-pressure condition of rising warm air that creates an elevation rise in the pressure surfaces in the upper atmosphere. The cooled air descends on to the cool sea causing a high-pressure area. High pressure flows to low pressure, thus during the morning and daytime a sea breeze called an **onshore wind** occurs because land temperatures rise more rapidly during the day. In the late afternoon and nighttime, a land breeze condition

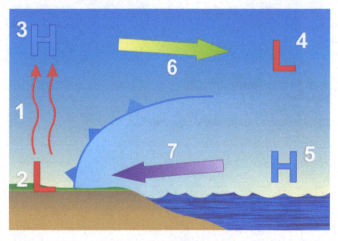

Figure 21.1   The warmed air, with its decreased density, begins to rise (1). The rising air creates a weak low-pressure area due to a decrease in air mass at the surface (2). Typically, from 1,000 to 1,500 m above this low pressure, as the air cools, it begins to collect resulting in an increase in pressure, creating a "high" (3). These differences in pressures over land, both at the surface and aloft are greater than the differences in pressures over water at the same elevations (4 and 5). Therefore, as the atmosphere seeks to reestablish the equal pressure both onshore and offshore, two high-pressure to low-pressure airflows develop; the offshore flow aloft (6) and surface onshore, called the sea breeze (7). (**NOAA**)

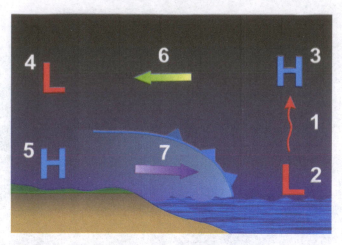

Figure 21.2   At night, the land temperature falls to below that of the ocean and becomes less dense. Therefore it begins to rise (1). The rising air creates a low-pressure area due to a decrease in air mass at the surface (2). As the air cools, it begins to collect resulting in an increase in pressure, creating a "high" (3). These differences in pressures over the water, both at the surface and aloft are greater than the differences in pressures over land at the same elevations over land (4 and 5). Therefore, as the atmosphere seeks to reestablish the equal pressure both onshore and offshore, two high-pressure to low-pressure airflows develop; the onshore flow aloft (6) and surface offshore, called the land breeze (7). (NOAA)

generates an **offshore wind** because the waterbody cools less rapidly than the land. Thus the lower temperature land develops a high pressure and the higher temperature water bodies create a low pressure reversing the wind flow. This is why surfers like to go surfing in the morning and early daytime since the onshore breeze is helping develop the waves. Surfing is much poorer in the late afternoon with the offshore winds now blowing against the waves and ruining their riding structure (**Figures 21.1–21.2**).

- **Mountain and valley winds**—incoming radiation heats the valley bottoms and sides during the day causing warm air to rise up mountain slopes at the same time that colder denser air settles down on the warm valley air also forcing the warm air to move up the mountain slopes. This is a valley wind (also known as an **anabatic wind**) because it flows up the mountain slope out of the valley. At night, the process reverses as cold, denser air at higher elevations drains into the valleys. This is a mountain wind (also known as a **katabatic wind**) (**Figure 21.3**).

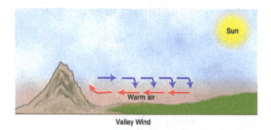 

Figure 21.3   Valley wind condition versus a mountain wind condition. (NASA)

Global Circulation Models

At the simplest the low latitudes are warmer than the higher latitudes. This energy deficit for which the second law of thermodynamics is most apparent should result in a large convection cell under the following path (first figure on right). Warm air rises over the equator from strong insolation heating. Air then moves polewards to sink, this is then drawn back to the low pressure. This would be the case if the earth did not rotate, was not tilted, and did not have a mixture of water and landmasses. Because the earth is a rotating, tilted sphere with large water bodies and land masses, the global circulations are much more complicated (**Figure 21.4**).

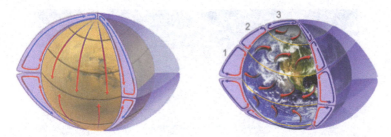

Figure 21.4 Earth did not rotate, had no tilt, or had no water then excess heat from the tropics would move to the poles and back again. **(NOAA)**

There are actually three circulations or cell circulations between the equator and the poles because the earth's rotation leads to zonal (latitudinal) flow: the Hadley cell, the Ferrel cell, and the polar cell. It all starts at the equator where a low-pressure belt is created called the intertropical convergence zone (ITCZ). The ITCZ is where air flow converges to create the easterly trade winds due to the Coriolis effect. Combined with strong convectional uplift from a large amount of insolation this creates a convergence/convection zone or the equatorial trough (**Figure 21.5**).

1. **Hadley cell**—Equatorial low pressure (ITCZ) and heating of air rises, with poleward movement in the upper atmosphere. The poleward-flowing air is deflected to the right in the northern hemisphere to become southwesterlies. The air descends at the 30°N/S latitude as a high-pressure cell. Note that high pressure is associated with dry/hot weather and at the 30° latitude resides the majority of the world's deserts. The pressure gradient flows the air back to the equator. This forms a convection cell that dominates tropical and subtropical latitudes.

2. **Ferrel cell**—A mid-latitude atmospheric circulation cell covering the westerlies (30–60° N/S). In this cell the air flows poleward and eastward near the surface and equator-ward and westward at higher altitudes. At the 50–60° N/S latitude a low-pressure cell is created.

3. **Polar cell**—Air rises and travels toward the poles to which it descends forming a polar high-pressure cell (60–90° N/S). Surface winds in the polar cell are easterly (polar easterlies).

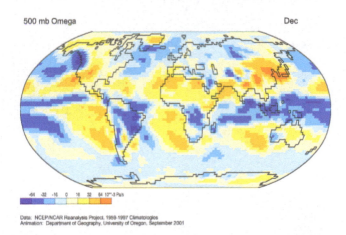

Figure 21.5 The 500 mb vertical velocity indicates areas of large-scale rising (blue) and sinking (orange) motion of air. (Source: Department of Geography, University of Oregon, supported by the US National Science Foundation, 2000)

Newer global circulation models have changed the relative importance of the three convection cells in each hemisphere with the influence of the jet streams and Rossby waves:

- **Jet streams**—There are four principal jet streams on the earth (two for each hemisphere). A polar jet (30–50° N/S) and the subtropical jet (20–30° N/S). They are strong and regular winds which blow in the upper atmosphere about 5–7 miles above the surface. They are generally 1–3 miles thick an flow 100–300 km/h. In the northern hemisphere the polar jet flows eastwards and the subtropical jet flows westwards. The polar jet stream influences storm tracks over much of the United States and Europe in the winter (**Figure 21.6**).

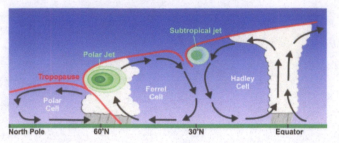

Figure 21.6 Cross section of convection cells identifying jet streams.

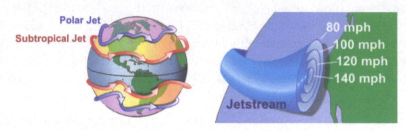

Figure 21.7 The meandering nature of the jet stream.

- **Rossby waves**—These are meandering rivers of air formed by the westerly winds. There are three to six waves in each hemisphere formed by major relief barriers (Rocky mountains, Himalayas), thermal differences, and uneven land-sea interface (**Figure 21.7**).

Topic 22

Air Pollution
By Ingrid Ukstins and David Best

Air Pollution

Clean air is necessary for life to survive and thrive on the planet. We have seen that the atmosphere near the surface consists of 78 percent nitrogen, 21 percent oxygen, and about 1 percent of many lesser gases. Oxygen is the component critical for animal life. However, when the air (and atmosphere) is contaminated with pollutants such as nitrous and sulfurous oxides and dangerous hydrocarbon volatiles, the air is rendered harmful (**Figure 22.1**). Pollution can come from numerous sources and can include particulate matter such as dust and ash from volcanoes and wildfires. Anthropogenic sources include coal-fired power plants, factories, vehicles that operate on fossil fuels, and even dry-cleaning establishments.

A secondary effect of air pollutants is the formation of haze that can reduce visibility around large cities and in wilderness areas and in national parks. In northern Arizona a large coal-fired generating station is located only 25 km (15 miles) from the northeastern boundary of Grand Canyon National Park. Until several years ago, a similar type of station located to the west of the park contributed significantly to lower visibility in the regions of national parks in northern Arizona and southern Utah. The Mohave Generating Station in Laughlin, Nevada, was shut down in December 2005 rather than have its operators face the expense of installing scrubbers and other pollution control equipment to reduce its emissions.

The Clean Air Act, originally passed in 1963, has been amended several times to insure healthy air in the United States. The EPA is entrusted to insure that clean air standards are met by monitoring the amounts of air pollutants across the country.

Figure 22.1   Pollutants from industrial smokestacks enter the atmosphere without any reduction in the particulates or gases generated through the manufacturing process.

Topic
23

Ocean Circulation
By Dean Fairbanks

Large-Scale Ocean Circulation

The large-scale global winds drag via friction along the surface of oceanic waters at the equator and the mid-latitudes. We know from the discussion earlier that the winds flow east to west at the equator and west to east at the mid- latitudes. Thus surface waters are flowing in these similar directions at these latitudinal zones. Combine these flows with continental landmass configuration creating large ocean areas (Pacific, Atlantic, and Indian) then the process of large-scale ocean circulation called **gyres** are created. In the northern hemisphere gyres flow clockwise and in the southern hemisphere they flow counterclockwise.

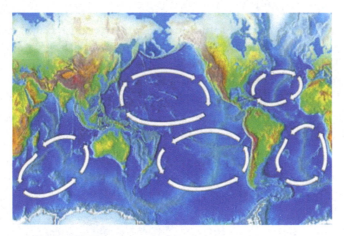

Figure 23.1   The five major ocean gyres. **(NOAA)**

Specifically speaking the surface circulation includes five major subtropical gyres, a descriptive term applied to current flows that generally form a continuous or semi-continuous loop; two subpolar gyres; the Antarctic circumpolar current; and the Arctic Ocean current system. Each major gyre is further subdivided into equatorial currents and boundary currents (run along the continents). The five major gyres includes: North Pacific gyre, South Pacific gyre, North Atlantic gyre, South Atlantic gyre, and the South Indian gyre (**Figure 23.1**).

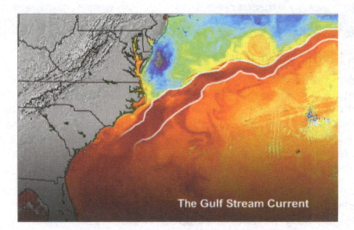

The Gulf Stream Current

Figure 23.2   A very strong western boundary current is the Gulf Stream. **(NOAA)**

There is a common rule to understanding the influence of gyres on the continental landmasses their waters influence, think of latent heat from earlier discussions. This rule holds for both hemispheres.

- **Left sides** of continents = warm water currents
- **Right sides** of continents = cold water currents

If you are interested in spending your summer vacation swimming in a warm ocean, then Florida up to North Carolina is your best destination, not Washington to California where a wetsuit will be required.

The North Atlantic gyre to which the Gulf Stream current is linked drives one of the largest oceanic circulation systems on the planet (**Figure 23.2**). A circulation system that acts like an oceanic conveyor belt starting with the Gulf Stream and covering all the world's oceans as a deep ocean current. Once the Gulf Stream has delivered its warmth via latent heat exchange heating the air temperatures off of northern Europe the water is ready to sink. The water is now is colder after releasing lots of heat and saltier because of evaporation, and therefore it sinks. The Gulf Stream system and the sinking of the water in the North Atlantic are so strong that it drives a deep water current called **thermohaline circulation**. This global circulation system takes 1000 years to complete one cycle (**Figure 23.3**).

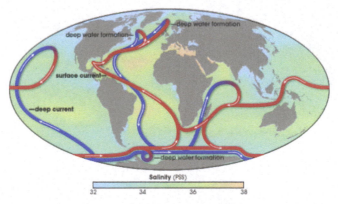

Figure 23.3   The ocean thermohaline circulation system is a slow, three-dimensional pattern of flow involving the surface and deep oceans around the world. **(NASA)**

NORTH SEA BREEZES—A MIGHTY "POWERFUL" WIND

Windmills have been around at least since the 9th century (in Iran) and prominently part of the European landscape since the 12th century. While it is typical to think of the Dutch windmill as the classic European windmill we really have to look at the last 120 years to the invention of the wind turbine to turn the rotational energy derived from the wind against the sails to electricity. It's the Scottish that created the first wind turbine but at the same time in the late 1800s, it's the Danish that perfected their use. The wind power map shown for Europe should tell you why. The North Sea region of Europe has a lot of wind power potential. That onshore electricity generation potential is nearly **52 Terawatts**! This would potentially cover an area of 4,895,560 km2.

The continuous development of onshore wind farms is an important feature of the European transition towards an energy system powered by distributed renewables and low-carbon resources. There is enough wind power to produce the equivalent of 1 Megawatt per 16 European citizens – a supply that would be sufficient to cover the global all-sector energy demand from now through to 2050.

Source: Enevoldsen, P., Permien, F-H., Bakhtaoui, I., von Krauland, A-K., Jacobson, M., Xydis, G., Sovacool, B.K., Valentine, S.V., Luecht, D., and G. Oxley (2019). How much wind power potential does Europe have? Examining European wind power potential with an enhanced socio-technical atlas. Energy Policy, 132: 1092-1100.

Role of Land Masses on Global Circulation

By Ingrid Ukstins and David Best

The mean height of land above sea level is 840 meters, and ranges from –418 meters at the Dead Sea to 8,848 meters at the top of Mount Everest. An idealized circulation pattern for Earth does not take into account land masses. On several continents there are many mountain ranges that disrupt the flow of air in the atmosphere. The long, relatively linear stretches of mountains such the Rocky Mountains of North America and the Andes Mountains of South American stretch for thousands of miles in a north-south direction. These impedances cause the flow of air to be altered, and thus change the weather and climate associated with the theoretical flow patterns. The Himalaya Mountains of Asia are the highest on Earth, reaching 8,850 m above sea level. Air that strikes these peaks is driven higher into the upper troposphere where the moisture is concentrated and returned to Earth as snow. These features also disrupt the normal flow of air around the globe. **Figure 24.1** shows global atmospheric patterns as seen in the clouds. The line of clouds along the equator in the tropical eastern Pacific and Atlantic Oceans is typical of that area where the air is heated, rises, and condenses to form clouds.

Source: NASA.

Figure 24.1 This MODIS image—moderate resolution imaging spectroradiometer—are instruments on NASA's Terra and Aqua satellites that acquire new images of the Earth';s entire surface every 1 to 2 days.

Role of Water in Global Climate

Water can exist in one of three states (or phases)—as a liquid, as a solid (ice), or as a gas (water vapor). When water changes from one state to another, heat is either absorbed or released, as seen in **Figure 24.2a and 24.2b**. The amount of heat, measured in calories, needed to change 1 gram of water from one phase to another can be measured. For water to be transformed from a liquid to a gas, the liquid must absorb 600 calories to evaporate. This is termed the **latent heat of vaporization**. When liquid water freezes, heat is released (80 cal per gram). This is the **latent heat of fusion**. In a case where the liquid phase is bypassed (the transformation of ice directly into water vapor), 680 calories are absorbed. In this instance, the process is termed **sublimation**. The **latent heat of condensation** (600 cal per gram) is associated with the change from water vapor to a liquid. This is a primary source of energy in cyclonic storms, as countless billions of grams of water vapor at high altitudes condense to form liquid water (as rain). This heat adds to unstable atmospheric conditions and also helps fuel the storm.

latent heat of vaporization: Heat stored in water vapor as as it changes states from a liquid to a vapor.

latent heat of fusion: Heat released when water freezes to form ice.

sublimation: The process that changes a solid into a gas, bypassing the liquid phase.

latent heat of condensation: Heat released when water vapor absorbs heat to be transformed to water.

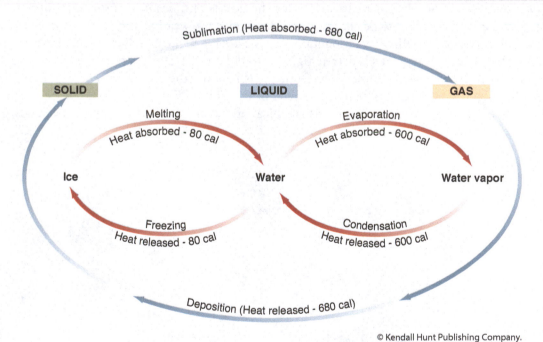

© Kendall Hunt Publishing Company.

Figure 24.2a   Heat is absorbed as water changes from a solid to a liquid and then to a gas. Heat is given off when it changes from a gas to a liquid and then a solid. The number of calories shown are those needed to change the state for 1 gram of water.

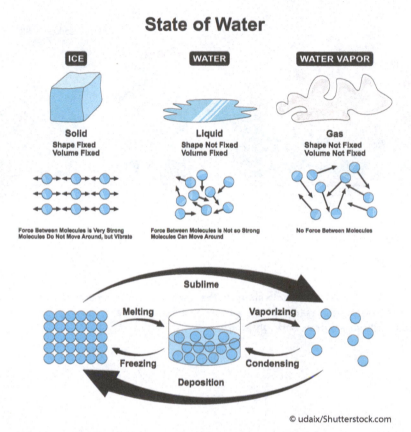

Figure 24.2b Phase transformations of water from solid (ice) to liquid (water) to gas (steam) reflect increasing energy and changes in the arrangement of water molecules in the material at each step.

Ocean water and that in lakes and other surface features cover about 75 percent of Earth's surface. Of all commonly occurring substances, water has one of the highest measures of heat capacity. This characteristic means that water requires a large amount of heat in order to increase its temperature. The equatorial regions of Earth receive the greatest amount of solar radiation, so this heat can be stored by the oceans. Evaporation is most effective at latitudes at or near the equator. These processes that absorb and release heat are key to driving the convection process in the atmosphere and the oceans.

Fronts and Mid-Latitude Cyclones

Whenever two different air masses collide with one another, they generally do not mix. A cold air mass with little moisture in it will not intermix with a warm, moisture-laden air mass. In North America cold air masses descend from the North Pole regions toward the equator. As these cold air masses pass through the mid-latitudes (30° to 45° north), they often collide with warm air masses that have moved northward from the Gulf of Mexico or off the western edge of the Atlantic Ocean. When a cold air mass collides with warm, moist air, the cold air pushes the warm air to higher altitudes. Cold air is more dense and so it is able to force warm air out of the way quickly. Condensation takes place as the moist air encounters colder temperature (**Figure 24.3**). A line of high rising, vertical clouds results.

When warm air collides with cold air, the warm air rises above the denser, cold air and pushes out along a long surface. The warmer air is at a higher altitude so condensation occurs. This elongated string of clouds produces cloudy conditions that extend over hundreds of kilometers (**Figure 24.3b**).

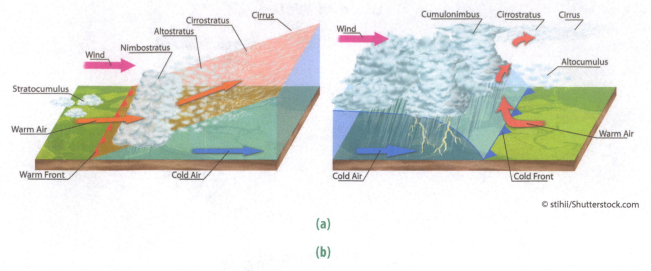

(a)

(b)

© stihii/Shutterstock.com

Figure 24.3 When a warm front overrides a cold air mass, the warmer air is spread out along a long distance, producing clouds. When a cold air mass encounters warm air, the warmer air is forced upward, causing condensation to produce thick clouds and the possibility of thunderstorms and lightning.

Latent Heat in Water

By Dean Fairbanks

Water or H_2O is the most abundant compound on Earth's surface. It is also one of the most important compounds for not only life, but as has been discussed for moving heat energy around. The Earth is covered 72% in liquid water that is roughly 1000 California's combined. The total volume of water on the surface and in the atmosphere combined is 1.4 billion cubic kilometers. However, before you find yourself in awe of all the water we have, its actually less water than you think, read **HERE**. The hypothesis is that the Earth's water was not manufactured on the planet when it was young but instead came from carbonaceous meteorites, which contain lots of water. This means that the early Earth's surface was struck by a substantial amount of these types of water-bearing meteorites.

The distribution of all water on Earth can be broken down into ocean 97.22% versus fresh 2.78%. The freshwater, which is important to our lives, can further be broken down into surface water 77.78%, groundwater 11.02%, deep groundwater 11.02%, and a miniscule amount of soil moisture 0.18%. On the whole it would seem that there is plenty of freely available surface water, however, of that 77.78% portion, if it is broken down 99.4% of it is locked up in ice and glaciers. Only 0.333% of the available water for us to consume on the surface is found in rivers, streams, and lakes. If we only look at water found in rivers and streams then the atmosphere contains 10× more by comparison. Useable water is truly precious.

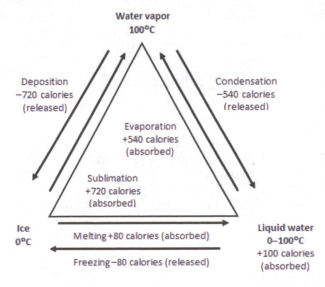

Figure 25.1   The states of water.

Here the concept of latent heat and the phase changes of water will be further defined. There are three states of water: solid (ice), liquid (water), and gas (vapor). As a basic rule heat energy is absorbed or lost at each phase change. The amount of heat lost or gained varies by phase change.

Changes in heat accompanying water phase changes are very important for atmospheric circulation. Latent heat energy transport is sometimes a little hard to visualize or understand because the energy is "hidden" in water vapor or water.

A solid to liquid phase change is melting, liquid to gas is evaporation, and sublimation is a solid to gas phase change. All require the addition of energy to occur. The needed energy may be consciously provided (putting a pot of water on the stove) or taken from surroundings (which causes surroundings to cool). Phase changes can also go the other direction. You can consciously remove energy from water vapor to make it condense. You take energy out of water to cause it to freeze (water in a freezer; energy would flow from the relatively warm water to the colder surroundings). If one of these phase changes occurs, without you playing a role, energy will be released into the surroundings (causing the surroundings to warm).

- Latent heat of **evaporation**—the energy required to convert liquid water to water vapor.
- Latent heat of **condensation**—the energy released when water vapor condenses to liquid water.
- Latent heat of **melting**—the energy required to convert solid to liquid water.
- Latent heat of **freezing**—the energy released when water goes from liquid to solid.
- Latent heat of **sublimation**—the energy required to convert solid water to water vapor. The best example of this is dry ice in a warm room changing to a gas.
- Latent heat of **deposition**—the energy released when water vapor condenses to solid. An example of this is the creation of frost, the process skips the liquid phase.

Shortwave insolation	Latent heat energy hidden	Energy released into atmosphere
Energy absorbed		
Liquid water evaporates	**Water vapor**	**Water condenses into clouds**
 Image © Shutterstock, Inc.	 Image © Shutterstock, Inc.	 Image © Shutterstock, Inc.
Ocean Energy added here ⇒	Flat/rolling landscape Moving across here ⇒	Mountains Energy reappears here to heat surrounding air

Figure 25.2 In the atmosphere, water and its latent heat energy-driven phase changes play a major role as a planetary heat pump.

Evaporation, Humidity and Lapse Rates
By Dean Fairbanks

The following sections' underlying motivation is to lead to the formation of clouds and precipitation. There are five conditions needed for the formation of major precipitation (rain, snow, and hail):

- Evaporation
- Air cooling
- Condensation and cloud formation
- An accumulation of moisture
- Growth of cloud droplets

Where there is liquid or solid water on the Earth's surface and energy, there is the potential for evaporation of that water into water vapor. The moisture we feel in air is termed humidity, which is the water vapor in air. Our atmosphere has a lot of water in it. Air near the ground or over a water body has more water vapor than higher in atmosphere. Scientists have devised several ways to measure and express the amount of water in the air. Water vapor in a given volume of air can always be converted to liquid water and weighed. **Specific humidity** is the mass of water vapor divided by the mass of air. While specific humidity will provide an exact quantity, it is difficult to make the measurements so relationships have been devised to relate air temperature to the amount of water it can hold. **Relative humidity** is the ratio of the water vapor content of air to the maximum water vapor content of air at a certain temperature. Imagine we have a volume of air at 21°C (70°F) and you are told that air has a relative humidity of 50%, then that means it is holding half of the amount of water a volume of air 21°C degrees could hold. The relative humidity can change if the moisture changes or if the temperature changes. Relative humidity can tell you how close a given volume of air is to saturation level, where maximum water vapor is possible at a given temperature. At saturation, the relative humidity is 100%, this is what is called the **dew point**, which is a much better indicator of moisture in the air and preferred by most weather scientists (meteorologists). The dew point temperature is that temperature where saturated air at 100% relative humidity condenses to liquid water. So, going back to the 21°C degree volume of air, if it starts to cool, it will eventually reach a temperature at which it can no longer hold the water vapor in it. Relative humidity varies in opposition to temperature (its anti- correlated), so higher temperatures have lower humidity and cool temperatures have higher humidity. Thus, during a 24-hr day in let's say the summer, starting at sunrise humidity is high from the cooler temperatures. As the air temperatures increase the relative humidity drops with the lowest percentages occurring in the late afternoon. By evening the relative humidity rises again as the air temperatures cool. In the early morning hours just before sunrise air can reach dew point and the water vapor in the air condenses on the surface—the cool morning dew, especially noticeable on tents when camping (**Figure 26.1**).

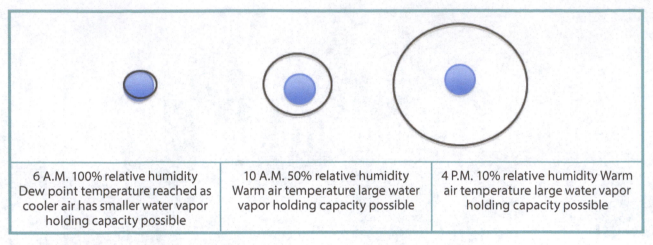

6 A.M. 100% relative humidity Dew point temperature reached as cooler air has smaller water vapor holding capacity possible	10 A.M. 50% relative humidity Warm air temperature large water vapor holding capacity possible	4 P.M. 10% relative humidity Warm air temperature large water vapor holding capacity possible

Figure 26.1 **The relationship between relative humidity and the dew point.**

For relative humidity, the water vapor holding capacity is set by air temperature (black open circle), and the actual vapor content is fairly constant (represented as the blue filled circle).

Finally, the last method is to measure the pressure of water vapor in a volume of air. The **saturation vapor pressure** (SVP) is the maximum amount of water that air can hold. The variation of SVP with temperature is nearly a pure exponential curve relationship. At any given temperature, air can hold a certain amount of water, for example, air at 30°C can hold just over 30 g of moisture, at 10°C it can hold 9.4 g. This is known as saturated vapor pressure (**Figure 26.2**).

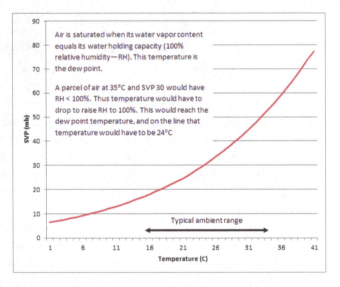

Figure 26.2 **As the temperature increases the saturation vapor pressure of water will also increase.**

As was learned earlier in this module as the air temperature is heated at the surface convectively it rises and cools. Imagine an air parcel shaped like a balloon. The air parcel cools internally as it expands under lower pressure with increase in altitude and becomes less dense. The balloon shaped air parcel will become larger, for example, child's balloon size at the surface to the size of a hot air balloon several thousand meters in altitude. The opposite occurs at higher altitude where the now cool dense air can sink with gravity. The air parcel heats internally as it is compressed by higher air pressure with decreasing altitude. Both cooling by expansion and heating by compression of an air parcel is the **adiabatic process**. Adiabatic means internal change. The adiabatic process describes the creation of low and high pressure areas. It is also how we can explain atmospheric stability and lapse rates. Air parcels are always being acted on by the two opposing forces: buoyancy-lifting and gravity-pulling down (**Figure 26.3**).

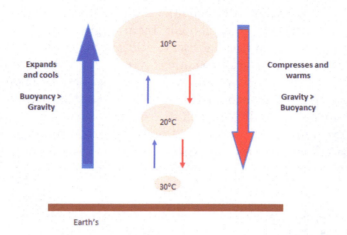

Figure 26.3   Adiabatic expansion/cooling of a rising parcel of air. Think of the convective movement in a lava lamp.

There are two types of adiabatic lapse rates, the rate at which atmospheric temperature decreases with increase in altitude: **dry adiabatic rate** (**DAR**) and **moist adiabatic rate** (**MAR**). The DAR is the rate at which unsaturated (dry) air cools, usually rounded up to 10°C/km. The MAR is the rate at which saturated air cools. Saturated air releases heat through condensation (cloud formation)—this offsets the cooling process. The MAR cools at a rate of between 4°C/km for warm air and 9°C/km for cold air (the average is 6°C/km). As discussed earlier, warm air contains more moisture, thus more heat is released during condensation, which reduces the impact of cooling. In synthesis, imagine an air parcel leaving the surface with a content of evaporated water making the relative humidity of the parcel 50%. The air parcel is not saturated so it cools as it rises at the DAR rate, as air cools the relative humidity of the parcel increases. When the air parcel reaches saturation or 100% relative humidity the water vapor condenses to suspended liquid—a cloud is formed. The saturated air continues to rise now at the MAR rate (**Figure 26.4**).

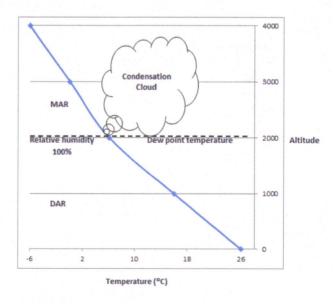

Figure 26.4   Cloud formation occurs when saturation is reached.

Lapse rates also lead to the concepts of stability and instability in air. **Stability** (stable conditions) occurs when a rising parcel of air cools more quickly than air surrounding it. The air surrounding the parcel is under the normal ELR condition. As it is colder than the ELR it is denser, and therefore sinks back to position. Stable air is associated with calm high pressure dry conditions. **Instability** is common, especially on hot days. Localized heating raises the temperature of air above it. The air begins to rise as it is warmer and less dense than the surrounding air it continues to rise. If it rises to sufficient height, condensation cloud development and rain may occur. Pilots handle unstable air conditions over airports in the afternoons, which makes for a bumpy takeoff. This is why it is always better to fly in the morning or evening hours. Finally there is conditional instability, when the ELR is lower than the DAR but higher than the MAR. Rising air parcels, cooking at the DAR become cooler than surrounding air, and should sink down to the ground. However, they may be forced to rise again. This may cause the air to cool to dew point. Once saturation occurs, condensation takes place. Therefore, the air parcel begins to cool at the MAR. If it becomes warmer than the surrounding air, it will continue to rise. The air is unstable on the condition that dew point is reached and it cools at the MAR.

Clouds, Fog and Lifting Mechanisms

By Dean Fairbanks

Clouds are aggregations of suspended microscopic water droplets and ice crystals. To form, the air must be saturated with water vapor (relative humidity = 100%) and there must be microscopic nuclei for the vapor to condense onto. These cloud-condensation nuclei tend to be aerosols representing dust or soot from combustion and their size is small at 2 µm diameter. As liquid water goes through a process of collision- coalescence in size to moisture droplets (20 µm diameter), these droplets build until they become at typical raindrop of 2 mm (2000 µm diameter) where now they are heavy enough for gravity force to be greater than the uplifting buoyancy force. During the winter air parcels can be cold enough with lifting and the surrounding colder air mass to develop snowflake formation. Cloud droplets break apart and freeze as water molecules, start to develop ice crystals, which enlarge further by absorbing more water molecules to large snowflakes.

When it comes to cloud types and identification there are four main groups based on their height above and appearance from the ground, with a fifth being the special cloud class formed at ground level known as fog (**Figure 27.1**). The cloud types (classifications) are based on Latin root descriptions such as the following: clouds can be flat (*stratus*), puffy (*cumulus*), or wispy (*cirrus*). Horizontal, layered clouds are *stratiform*; vertically developed clouds are *cumuliform*; and high, wispy clouds are *cirroform*.

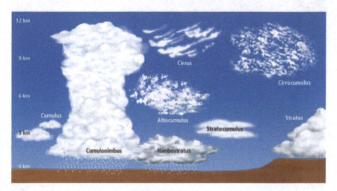

Figure 27.1   Cloud types. (NOAA)

The four main groups by altitude are as follows:

- **High clouds (6000–12000 m)**
  - Cirrus
  - Cirrocumulus
  - Cirrostratus

Figure 27.2 High clouds-Cirrus. (NOAA)

- **Middle clouds (2100–6000 m)**
  - Altocumulus
  - Altostratus

Figure 27.3 Middle Clouds-Altocumulus. (NOAA)

Figure 27.4 Middle clouds-Altostratus. (NOAA)

- **Clouds with vertical development (1500–9000 m)**
  - Cumulus

- **Low clouds (< 2100m)**
  - Stratocumulus
  - Nimbostratus
  - Stratus

Fog is a special type of cloud at ground level. It is confined to a restricted layer because it is "capped" by a thermal inversion. It mostly occurs in high pressure (calm) conditions, as winds tend to mix and disperse fog. There are different kinds of fog, but the basic mechanism of fog formation is the same as for clouds, just at ground level.

- **Advection fog**—common in spring and early summer along the coast and may penetrate in a few miles along coastal valleys. Formed by cooling of warm air by passage over a cold ocean. Advection is simply horizontal mixing of air in response to pressure gradients. High pressure over the ocean and low pressure over land (**Figure 27.5**).

Figure 27.5 Advection fog advancing across San Francisco, California. (photos.com)

- **Evaporation fog**—this fog forms when cold air overlies a warm body of water, like a lake. The warm lake is evaporating water which then comes in contact with the cold air and condensation occurs just over the water (**Figure 27.6**).

Figure 27.6 Evaporation fog in the early morning hours on a lake. (photos.com)

- **Radiation fog**—this kind of fog forms over land that undergoes nighttime radiative cooling (clear nights with little to no wind), bringing the overlying air to its dew point. This fog occurs in the winter months in inland areas, especially low-lying (valleys), moist ground. When air along the upper slopes of mountains begins to cool after sunset, the air becomes dense and heavy and begins to drain down (cold air drainage) into the valley floors below. As the air in the valley floor continues to cool due to radiational cooling, the air becomes saturated and fog forms. This is one of the thickest types of fog and very dangerous to drive in (**Figure 27.7**).

Figure 27.7 California's famous locally named tule fog is a radiation fog. (NASA)

Let's make it rain! We have been discussing lifting of air to dew point, forming a cloud, and cloud types but we have not discussed the vertical lifting mechanisms available to lift the air to dew point. There are four distinct mechanisms, with the same basic idea in all cases—move air to its dew point by expansion/cooling and then condensation occurs. All the lifting mechanisms develop types of rain (precipitation).

- **Convergence** rain—lifting associated with low pressure and general air convergence. This mechanism only happens at the intertropical *convergence* zone (ITCZ) where air converges from all points to low pressure zone and the displaced air is lofted.
- **Convection** rain—warm air rises. Unequal surface heating produces hotspots where "bubbles" of air will rise.
- **Orographic or relief** rain—air forced over a topographic barrier and cools as it ascends on the windward side then dry air warms as it descends on the leeward or rainshadow side.
- **Frontal** rain—lifting at frontal boundaries. Common at mid-latitudes where warm air rises over cold air, expands, and cools rapidly to form clouds and rains. On weather maps warm fronts denoted with a red line and filled semicircles in the direction of advance. When the cold front is moving its leading edge forces warm air aloft. On weather maps cold fronts are denoted with a blue line and triangles pointing in the direction of advance.

The patterns of precipitation (rainfall) tend to be geographically zonal across the planet with some slight modifications. There is an abundance of rain in the equatorial trough zones (ITCZ); moderate to large amounts in the mid-latitudes (polar frontal zone—jet stream and westerlies); relatively low rainfall in the sub-tropics (high pressure from Hadley cell circulation) and particularly at the poles. These general global precipitation patterns are deviated by orographic barriers (especially western mountain ranges running from Alaska to Chile) where very heavy orographic precipitation along the windward sides develops, for example, Olympic peninsula, Washington; Coastal Cascades, Oregon; Klamath Mountains, California. In addition ocean currents play a strong role, where cold ocean currents (left sides of continents) have less rain development and warm ocean currents (right sides of continents) have more rain development, generally.

Variations in Weather
By Dean Fairbanks

The difference between weather and climate is a measure of time. Weather is what conditions of the atmosphere (heat, pressure, moisture, winds) are over a short period of time, and climate is how the atmosphere operates over relatively long period of time. The time frame of weather is seconds to weeks, while climate is over years to thousands of years. An easy way to remember the difference is that climate is what you expect, like a very hot summer, and weather is what you get, like a hot day with thunderstorms and maybe a tornado in the Midwest.

Most people think of weather in terms of temperature, humidity, precipitation, cloudiness, brightness, visibility, wind, and atmospheric pressure, as in high and low pressure (NASA). There are really a lot of components to weather. Weather includes sunshine, rain, cloud cover, winds, hail, snow, sleet, freezing rain, flooding, blizzards, ice storms, thunderstorms, tornadoes, thunder and lightning, hurricanes, steady rains from a cold front or warm front, excessive heat, heat waves, and so on.

Air Masses

One of the keys to weather is the movement and clashing together of air masses, especially in the mid-latitudes (23.5°–66.5°). An air mass is an area of air which has similar properties of temperature and humidity. Air masses develop over areas of similar geographical character, like the polar ice caps, hot deserts, or oceans. Source regions for air masses tend to be flat terrain or ocean and areas of little wind. They are typically classified based on the moisture derivation (marine or continental) and temperature based on latitude. To qualify as an air mass their size is usually greater than 1600 km. The boundaries between air masses are called fronts.

Over North America there are five separate air masses that interact with the jet stream and each other over the seasons that drive the regions weather patterns (**Figure 28.1**). These include the following:

Table 28.1 Principal air masses over the world

Air Mass	Symbol Used	Description
Continental Arctic	**cA**	Form over sea ice/ocean or ice/land and descend toward equator. Summer = cool, dry; Winter = bitterly cold, extremely dry
Continental Antarctica	**cAA**	

Air Mass	Symbol Used	Description
Continental Polar	cP	Form over dry lands and descend toward equator. Summer = mild, dry; Winter = cold, dry
Continental Tropical	cT	Form over desert and plains and flow north. Summer = hot, dry; Winter = mild, dry
Martime Polar	mP	Form over cold oceans. North Pacific air mass moves westward with jet stream and westerlies. Summer = mild, moist; Winter = cool, moist
Martime Tropical	mT	Form over warm oceans and flow north mostly in summer. Summer = very warm, moist; Winter = warm, moist
Maritime Equatorial	mE	Form over very warm equatorial waters. Summer = very warm, very moist; Winter = very warm, very moist

When two air masses interact this is called a front. From the table above, cold and warm fronts from various source regions collide to develop weather systems. Many times these weather systems are very powerful and can lead to weather hazards (**Figure 28.2**).

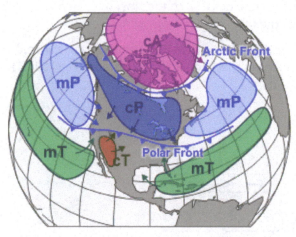

Figure 28.1 **Principal air masses over North America. (NOAA)**

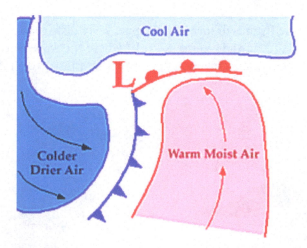

Figure 28.2 **Warm and cold front depiction on weather maps. (NOAA)**

Mid-Latitude Weather Systems and Fronts

Mid-latitude weather systems occur between the tropical and polar climates in both the northern and southern hemispheres. These areas are influenced by the meeting of warm air from the south and cold air from the north to form fronts. These give rise to low pressure systems (cyclones or depressions) and high pressure systems (anticyclones). Cold air on the move undercuts stationary warm air and warm air on the move rides over stationary cold air (**Figure 28.3**).

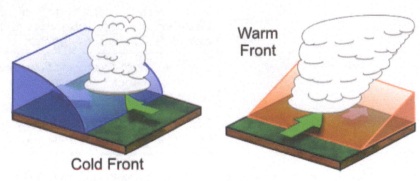

Figure 28.3   Cold front vs. warm front. (**NOAA**)

A **mid-latitude cyclone** (depression) is a migrating storm system that have low pressure cores, converging, ascending air and they rotate counterclockwise in the Northern hemisphere. They have a life cycle that has been documented consisting of the following:

- **Cyclogenesis**—There will initially be a boundary (front) separating cold air from warm air. The front is often stationary. The air masses are setting up for a collision (**Figure 28.4**).

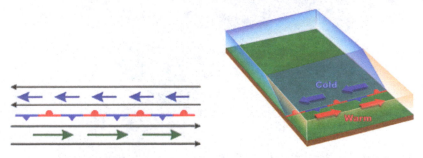

Figure 28.4   Cyclogenesis: view from weather map and perspective view. (NOAA, National Weather Service)

- **Open stage**—A wave on the front will form as an upper level disturbance embedded in the jet stream as it moves over the front. The front develops a "kink" where the wave is developing. Precipitation will begin to develop with the heaviest occurrence along the front (dark green) (**Figure 28.5**).

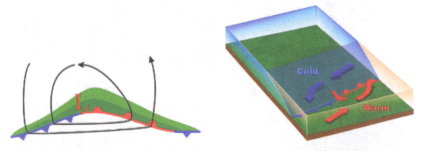

Figure 28.5   Open stage: view from weather map and perspective view. (NOAA, National Weather Service)

- **Occluded stage**—As the wave intensifies, both cold and warm fronts become better organized. Precipitation is heavy to moderate at this stage. During this stage of a cyclone's life cycle, the faster-moving cold front overtakes the warm front and wedges beneath it ("occluding" or closing it) (**Figure 28.6**).

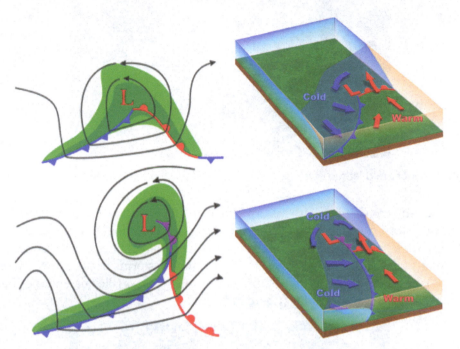

Figure 28.6 Occluded stage: view from weather map and perspective view. (NOAA, National Weather Service)

- **Dissolving stage**—As the cold front continues advancing on the warm front, the occlusion increases and eventually cuts off the supply of warm moist air, causing the low pressure system to gradually dissipate. There is still some counterclockwise air flow with light winds. During this stage, there is no more uplifting of air and the storm gradually fades (**Figure 28.7**).

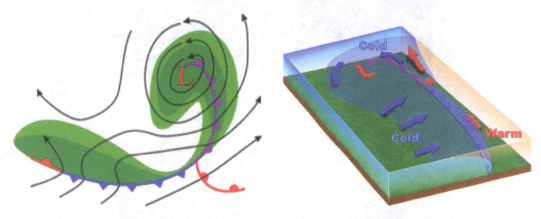

Figure 28.7 Dissolving stage: view from weather map and perspective view. (NOAA, National Weather Service)

Figure 28.8 Satellite view of a mid-latitude cyclone over the Midwest and Eastern US. (NOAA)

Mid-latitude cyclones and convectional storms can lead to violent weather (**Figure 28.8**). This would consist of thunder and lightning storms, including hail. Thunderstorms are commonly identified with cumulonimbus clouds, these clouds are characterized by rapidly rising bursts of air that condense and release latent heat, driving further uplift by increasing local buoyancy. Most thunderstorms occur under three conditions: (1) along cold fronts, (2) within moist air masses dominated by **mT** air mass (Florida and the Gulf coast), and (3) around mountains due to orographic uplift (**Figure 28.9–28.10**).

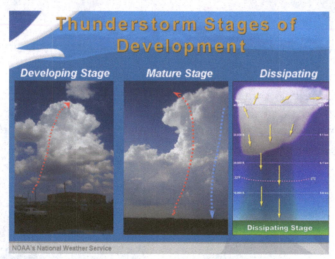

Figure 28.9 Development of a frontal storm. (**NOAA National Weather Service**).

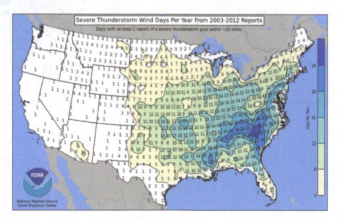

Figure 28.10 Map of thunderstorm activity in the US. (**NOAA**)

The following are usual components of thunderstorms, besides rainfall:

- **Lightning**—represents a flash of light from powerful electrical discharges. These discharges occur because interactions between the particles in a cloud leads to areas of positive and negative charge (**Figure 28.11–28.12**).

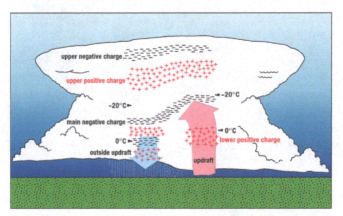

Figure 28.11 Ice and water particles separate; negative charges fall to Earth, positive electricity rise from the ground; and the two join and hit the ground. (NOAA)

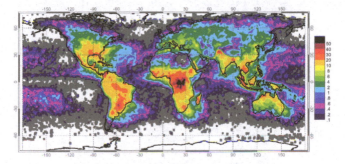

Figure 28.12 World map of lightning frequency. Units: flashes/km2/yr. (NASA, NSSTC Lightning Team)

- **Thunder**—violent heating and expansion of air surrounding a lightning discharge. The discharge can heat up air to 30,000°C—5× hotter than the surface of the sun! The noise you hear is a sonic shock wave.
- **Hail**—hailstones form inside cumulonimbus clouds when water droplets are lofted vertically to below freezing temperatures and then circulated downward to add on more water then back up again to freeze. They grow ice in layers like an onion (**Figure 28.13–28.14**).

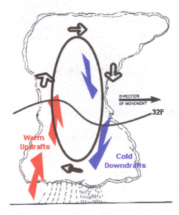

Figure 28.13 Creation of a hailstone. (NOAA)

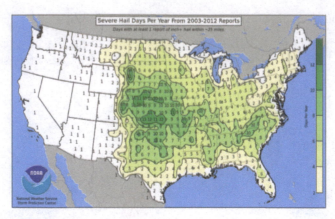

Figure 28.14 Severe hail days per year. (NOAA National Weather Service)

Tornadoes and Hurricanes/Tropical Cyclones

Tornadoes are some of the most violent short lived storms on the Earth (**Figure 28.15**). In the United States, a number of factors need to occur simultaneously for tornadoes to form:

Figure 28.15 Tornadoes can be very ominous from the ground. When they do not touch the ground they are called funnel clouds. (NOAA)

- A northernly flow of **mT** air from the Gulf of Mexico that is both very humid and has temperatures at the surface over 24 C.
- A cold, dry **cP** air mass flowing south from Canada or out of the Rocky Mountains at speeds greater than 80 km/h.

Then add in the jet stream combing from the west at speeds greater than 380 km/h.

All three of these flows of air produce shearing conditions, imparting spin to a thundercloud. Once the tornado has started and touched ground their core is usually less than 1 km wide and acts like a giant vacuum cleaner sucking up air and objects. The updrafting air can be rising at over 160 km/h. Most tornadoes last on average 10 minutes on the ground (**Figure 28.16**).

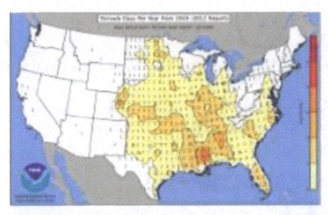

Figure 28.16 Map of tornado occurrence by month. (**NOAA**)

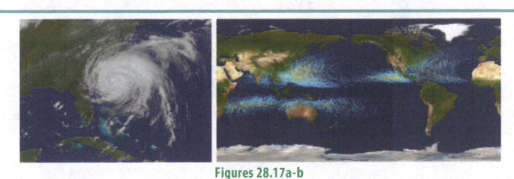

Figures 28.17a-b

| Hurricane Irene during 2011, notice the counterclockwise spin for the low in the Northern hemisphere. (**NASA**) | Map of main tracks across the planet. Tropical cyclones are prevalent across the tropical areas of the world. While in the United States we tend to focus on hurricanes along the Atlantic seaboard, the western Pacific sees the most activity —where they are called typhoons. (**NASA**) |

Tropical cyclones—**hurricanes** or typhoons are much more destructive than tornadoes. They are also much larger than tornadoes with a diameter of up to 800 km, although the very strong winds that cause the most damage are found in a narrower belt up to 300 km wide. They bring very intense hazards in the form of heavy rainfall, strong winds, high waves, and have direct impacts to flooding and mudslides (**Figure 28.17a-b**). Hurricanes typically develop the following:

As intense low pressure systems over tropical oceans.

A calm central area known as the eye, around which winds spiral rapidly (**Figure 28.18**).

From **mT** air masses, not from interacting air masses (like mid-latitude cyclones). Generated by evaporation of a warm tropical ocean in the presence of strong winds and lowered surface pressures.

To the entire height of the troposphere.

Under different locations around the world they have different names: Hurricanes in North America, Typhoons in the western Northern Pacific, and cyclones in Indian Ocean and western Southern Pacific.

West of 40° longitude will make land fall in the United States. Those that develop east of the 40° longitude will not and merely spin into the Atlantic. This is all due to the Coriolis force.

The factors affecting the development of hurricanes include the following:

Ocean water above 28° C

Winds need to be converging near the surface

The air needs to be unstable so it continues to rise

Air up to ~5500 m altitude needs to be humid as it is pulled into the storm (extra water vapor supplies more latent heat energy)

Pre-existing winds not associated with the storm help the storm if they are all coming from the same direction and speed at all altitudes so the storm does not rip apart

An upper atmosphere high pressure area helps pump away air rising in the storm.

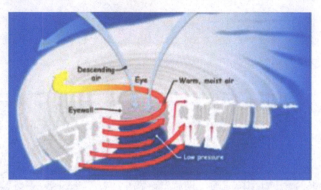

Figure 28.18   Cross section of a hurricane. Descending air in the eye causes high pressure and clear skies surrounded by very low pressures 880–970 mb. (**NASA**)

Monsoon

The **Asian monsoon** is the reversal of pressure and winds which gives rise to a marked seasonality of rainfall over south and southeast Asia. The monsoon is most noted over the Indian subcontinent. One way of explaining the monsoons is to see them as sea and land breezes over a very large region that change direction with the seasons—winter to summer. The influences that drive the monsoon include the effect of the Himalayas on the ITCZ through orographic lifting and the differential heating and cooling of land compared to the adjacent ocean.

- **Winter**: North-east monsoon—Temperatures over central Asia are low, leading to high pressure. An intense low pressure develops over the southern Indian ocean and northern Australia, where it is summer. Winds blow from Asian high pressure to the more intense Indian Ocean and Australian low pressures. These dry airstreams produce clear skies and sunny weather over most of India (November–May).
- **Summer**: South-west monsoon—In March and May the winds shift, and the upper westerly air currents begin to move north. Surface heating in central Asia leads to intense low pressure, separated by the Himalayas, from a smaller intense low pressure over northern India. The overhead sun migrates north to a position just over India and the ITCZ moves north bring the monsoon depression. High pressure develops over northern Australia and the Indian Ocean where it is winter; winds blow from Australia to the more intense Asian low pressure.

Monsoons are known for their intense rainfall and high levels of flooding.

Elements of Hazardous Weather

By Ingrid Ukstins and David Best

Elements of Hazardous Weather

The collision of air masses of different temperatures can create hazardous or severe weather. These weather events are often relatively short-lived in duration, but can be very destructive to property and can cause a significant loss of life if populated areas are affected. Most violent weather begins as picturesque cumulus clouds that eventually are pushed higher in the atmosphere, where their dynamics change (**Figure 29.1**).

Source: David M. Best.

Figure 29.1 Cumulus clouds are the early stage of thunderstorm development. As more updraft moves moisture to high altitudes, the storm enters the mature stage.

Thunderstorms

The conditions that must exist for a **thunderstorm** to develop include the heating and rising of moisture-laden air to higher altitudes where moisture will condense upon cooling, and the development of electrical charges that generate **lightning**. The generation of lightning through the atmosphere superheats the air and produces thunder, a tell-tale sign of such storms. One rule of thumb to remember is that once you see a bolt of lightning, for each three seconds you count, the lightning is one kilometer away—this is because light travels at about 300,000 km/second and sound travels at 332 m/second. Thunderstorms have three stages of formation: the developing or towering cumulus stage, the mature stage, and the dissipating stage.

> **thunderstorm:** A local storm produced by a cumulonimbus cloud and accompanied by lightning and thunder.
>
> **lightning:** A visible electrical discharge produced by a thunderstorm. The discharge may occur within or between clouds, between the cloud and air, between a cloud and the ground or between the ground and a cloud.

The developing stage involves the formation of cumulus clouds (puffy clouds that resemble large cotton balls or heads of cauliflower) that are pushed upward by a rapid **updraft** of rising warm air (**Figure 29.2**). Little or no rain forms as the clouds are coalescing at elevations of five to seven kilometers. The trigger for this can be ground warmed by solar radiation, wind blowing over high-elevation ground, or two winds converging and forcing air upwards.

> **updraft:** A small-scale current of rising air. If the air is sufficiently moist, then the moisture condenses to become a cumulus cloud or an individual tower of a towering cumulus.

A continuation of the updraft pushes moisture to higher altitudes where condensation takes place and precipitation begins to fall to the surface. Once the warm air cannot rise any further it begins to spread out and forms an anvil shape. The falling rain drags surrounding air with it and generates downdrafts. This combination of updraft and downdraft forms shear within the thunderstorm. The storm now has reached the mature stage, when it becomes its most violent. Strong downdrafts can generate high winds; hail and heavy rain can fall, and intense lightning can develop. If sufficient rotation exists within the storm, tornadoes can be spawned.

When the amount of downdraft exceeds the rising updraft, the storm reaches the dissipation stage. Descending cold air intercepts warm air near the base and prevents the warm air from rising to fuel the storm. Precipitation is in its final stages, although lightning can still be a threat.

Thunderstorms are of four main types. An isolated or single-cell storm is often a short-lived event, lasting maybe 20 or 30 minutes. They are also called air-mass thunderstorms and form from a single main updraft. Their isolated nature generally prevents them from becoming very severe as they do not have sufficient energy to become very large. Some storms can be classified as a severe thunderstorm if they have winds that are at least 93 km per hour, hail of 25 mm in diameter or greater, or if they have a funnel cloud or tornado. Severe thunderstorms can occur from any kind of storm but are most often produced from multicell clusters, multicell lines, and supercell thunderstorms.

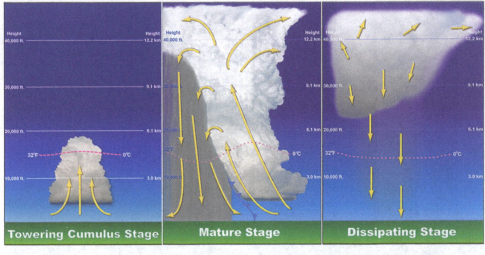

Source: NOAA.

Figure 29.2 **Stages of development of a thunderstorm. The developmental or towering cumulus stage begins as warm, moist air rises. The mature stage is characterized by updrafts and downdrafts. The dissipating stage occurs when the upward movement of air has ceased.**

The most common occurrence of thunderstorms is as a multi-cell cluster. Several storm cells move along as a unit, with each cell representing a different phase in the life cycle. This configuration typically has a mature cell near the center of the cluster with dissipating cells on the downwind edge of the cluster. Each cell in this group may only last 20 or 30 minutes but the entire group of cells could be active for hours.

A squall line or multicell line is an elongate line of thunderstorms that can be hundreds of kilometers in length and generally form along or ahead of a cold front. These lines are known for having strong downdrafts ahead of the line, with large hail, frequent lightning, and heavy rainfall accompanying the strong winds, along with possible tornadoes.

Supercell thunderstorms are large, last for 2 to 4 hours, and are characterized by the presence of a rotating updraft or mesocyclone (**Figure 29.3**). These kinds of storms are the least common but have the potential to be the most severe type of thunderstorm. Supercells are characterized by having extremely strong updrafts of about 112 km/hour and up to 280 km/hour in some cases and a significant amount of rotation. Hail can exceed 5 cm in diameter due to the extreme updrafts pushing the moisture repeatedly upward. Violent downdrafts are common, and strong to violent tornadoes are most commonly associated with supercell thunderstorms. The extreme updrafts prevent precipitation from falling through the center of the storm. The structure of a supercell is characterized by an overshooting top formed by the powerful updraft in the mesocyclone. As the rising warm air reaches the troposphere, it spreads out into an anvil-like shape once it reach about 15,000 to 21,000 meters, and juts out in front of the storm. Wall clouds form when humid air is pulled into the updraft and condenses, they appear to descend from the base of the supercell. The forward flank downdraft is the area of most intense precipitation.

Figure 29.3   **A supercell thunderstorm in Saskatchewan, Canada.**

The National Oceanic and Atmospheric Administration reports that the typical thunderstorm is 30 kilometers in diameter, and lasts about 30 minutes. Worldwide, about 2,000 thunderstorms are occurring at any given moment. In the United States approximately 100,000 thunderstorms occur each year with roughly ten percent of them being classified as severe. Thunderstorms can occur during any month, but the largest number occur during the summer months, due to the increased heating of the ocean and atmosphere, particularly in the Gulf of Mexico. All states experience these storms but the largest number occurs in Florida, where thunderstorms take place an average of 80 to 100 days per year (**Figure 29.4**). Unfortunately Florida also leads the nation in the number of deaths related to these storms. The high number of thunderstorms that occur in Colorado and northern New Mexico are due to the collision of warm and cold air masses along the Front Range of the Rocky Mountains. Note that California has very few thunderstorms because there is very little cold air in the atmosphere.

Hail

As thunderstorms move air vertically in the updrafts, moisture is being pushed to higher altitudes, where it freezes. Most hail takes on a spherical shape as the moisture droplets become larger as more water is frozen to its surface due to being repeatedly being moved to higher altitudes. At some point, the mass of the frozen particle is too much for the updraft to hold and the particle falls to Earth as **hail** (**Figure 29.5**). Although grapefruit-size hail has been reported in storms in the Midwest, hail seldom kills people (most people have sought refuge indoors). However, large hail can destroy agricultural crops and can damage roofs and automobiles.

hail: Solid, spherical ice precipitation that has resulted from repeated cycling through the freezing level within a cumulonimbus cloud.

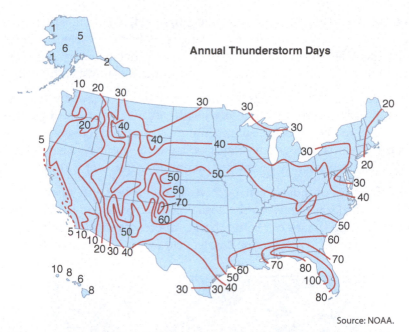

Annual Thunderstorm Days

Source: NOAA.

Figure 29.4 The average number of days that thunderstorms occur in the United States.

© Jack Dagley/Shutterstock.com

Figure 29.5 Hailstones the size of golf balls can produce a great deal of damage, especially to agricultural crops.

What Causes Lightning?

Lightning results from a sudden electrostatic discharge, and commonly occurs during thunderstorms. The charged areas temporarily re-equilibrate through the discharge of charged particles in a flash of bright light. This discharge can also be a lightning strike if it interacts with an object on the ground. The discharge is extremely hot and a lightning bolt can heat the surrounding air to temperatures five times hotter than the surface of the sun. As the lightning bolt passes through the air and heats it, it expands and vibrates rapidly, which is what generates the noise we hear as thunder.

Benjamin Franklin was fascinated by storms and lightning. In June 1752 he conducted experiments with his kite and discovered that lightning was a form of electricity. Soon after that he developed the lightning rod that was placed atop many buildings of the period. Lightning is a natural phenomenon that is present in large hurricanes, volcanic eruptions, extremely intense wildfires, heavy snowstorms, and (most commonly) thunderstorms.

The electrical imbalance that causes lightning is generated as a cloud grows, when water droplets in the bottom of the cloud are carried upwards in updrafts and interact with ice crystals from the top of the could that are being pushed down in downdrafts. As these particles bump into each other and interact, electrons are stripped off of them. The electrons move towards the bottom of the cloud and the positively charged particles move towards the top of the cloud, with the atmosphere between them acting as an insulator within the cloud itself (**Figure 29.6**).

On Earth's surface underneath a thunderstorm, positive charges collect and move along the surface with the storm. Eventually the electrical force between the negatively-charged cloud base and the positively-charged ground surface creates lightning (**Figure 29.7**). The charge differences within a storm tend to be much stronger, explaining why about 75 to 80 percent of lightning occurs within storm clouds.

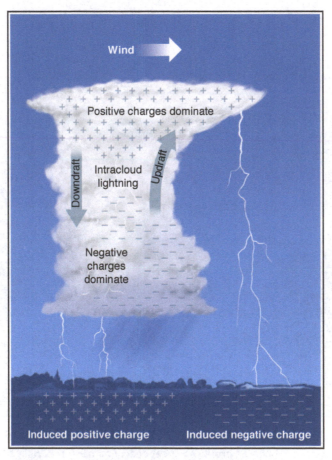

© Kendall Hunt Publishing Company.

Figure 29.6 Electrical charges in a thundercloud are separated with positive charges in the top and negative charges toward the bottom. Once the energy is strong enough, the opposite charge attract and connect to create lightning.

Types of Lightning

Ground flashes involve the ground or something attached to the ground being hit. Lightning can originate from the cloud, and strike the positively-charged surface of the Earth. Lightning can also start on the ground and move from the ground upward to the cloud. Lightning bolts travel about 100,000 kilometers per second and have an estimated width of 1 to 2 cm.

NOAA, Shane Lear.

Figure 29.7   **Lightning is a common feature of thunderstorms.**

Following a bold of lightning, thunder travels out at the speed of sound, roughly 1236 km per hour. The sound is the result of rapidly expanding gases that were superheated by the lightning. Air adjacent to a bolt is heated to 10,000°C. This superheated air expands and creates the rumbling sound that can be used to estimate the distance to the lightning strike. Thunder is generally heard within 25 km of a storm. Sound travels roughly 343 m per second. Begin counting when you see a lightning flash and count until you hear the thunder, then multiply your number of seconds by 350 m per second to get the approximate distance to the lightning.

Effects of Lightning on Humans

Lightning strikes have a mortality rate of from 10 to 30% and up to 80% of survivors have long-term injuries from high voltage induced nerve and muscle damage. Over the last 20 years an average of about 50 people were killed in the United States and about 300 more injured by lightning strikes. A study completed by the National Weather Service and the National Severe Storms Laboratory in the mid 1990s found that men accounted for 84 percent of the 3,239 deaths and 9,818 injuries caused by lightning between 1959 and 1994. Only flash floods and river floods caused more weather-related deaths during that period.

Property damage caused by lightning strikes increased substantially over the 35 years of the NWS/NSSL study. Most of the increase was due to population increases and new construction. The report, entitled "Demographic of United States Lightning Casualties and Damages from 1959 to 1994," by Holle and Lopez, described almost 20,000 property damage reports due to lightning strikes. Pennsylvania had the highest number of damage reports, while the highest rates of damage reports corrected for population differences were in North Dakota and Oklahoma.

In terms of casualties, Florida has twice as many lightning deaths and injuries as any other state, and ranks first among the states that have lightning casualties (**Figure 29.8**). Nationwide the greatest number of deaths and injuries occurred between noon and 4 p.m. (local standard time). Sunday had 24 percent more deaths than any other day; Wednesday was second (these are very popular golfing and fishing days). The worst month was July, when thunderstorms are most common across much of the country.

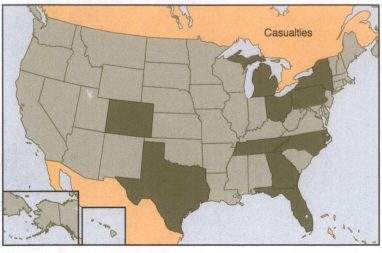

Source: NWS.

Figure 29.8 Of the ten states with the highest number of lightning casualties (deaths and injuries combined), Florida has the highest number.

In what location were most people struck?

- Walking in an open field
- Swimming
- Holding a metal object (golf club, fishing pole, umbrella)

If you are in the vicinity of a thunderstorm and lightning is being generated, seek shelter in order to reduce your exposure to the elements. Documented cases exist of lightning striking up to 15 km from a thunderstorm. If you can hear thunder, you are potentially vulnerable to being struck.

Tornadoes

The intense atmospheric dynamics of a thunderstorm can produce a **tornado**, a rapidly rotating column of air that ranks as the most violent type of naturally occurring weather condition (**Figure 29.9**). These funnel-shaped clouds, which extend down from the base of severe thunderstorms, may or may not make contact with the ground. The appearance of a tornado ranges from light gray in color (containing mainly moisture droplets) to pitch black (one with a high degree of dust, dirt, and debris). The siphoning effect of the updrafts carries material high into the atmosphere and spreads it across the landscape as the tornado moves along.

> **tornado:** A rotating column of air usually accompanied by a funnel-shaped downward extension of a cumulonimbus cloud and having a vortex several hundred yards in diameter whirling destructively at speeds of up to 600 kilometers per hour (350 miles per hour).

How Do Tornadoes Form?

The most common location for the formation of tornadoes is a supercell thunderstorm. These are highly organized, extremely intense storms that derive their energy from the strong updrafts that rotate and become tilted (**Figure 29.9**). Lasting more than one hour, these supercells contain vertical air currents that stretch 15 km in diameter and reach altitudes of 15,000 m. Rotation within the cloud system produces a mesocyclone which can be seen on Doppler radar (**Figure 29.10**). A tornado represents a very small extension of the larger roational cell.

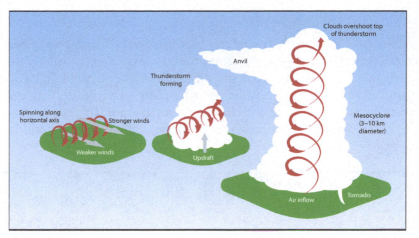

Figure 29.9   The rotational vortex of a tornado lifts dust and debris from the surface high into the funnel cloud.

Researchers do not fully understand how tornadoes form, in spite of years of lab research and observations in the field. A supercell must exist and have a well-defined rotating updraft. Winds traveling in two opposite directions or at different speeds at different altitudes can produce a shear force (similar to placing your flat palms together and sliding them in opposite directions). This shearing force produces a horizontal rotation that gets transformed into a vertical rotation by updrafts in the storm. Warm, moist air near the surface is lifted and cooled. The temperature differences in the air at altitude adds to the energy of the storm.

Fewer than 20 percent of supercells spawn tornadoes. Scientists are undertaking large-scale studies to learn more about these phenomena. Vortex2—the Verification of the Origins of Rotation in Tornadoes Experiment 2—a collaborative endeavor involving governmental agencies and universities, is examining the conditions that form tornadoes in the Midwest. This is the largest tornado research project in history to explore how, where and why tornadoes form. A fleet of 10 mobile radars and groups of scientists drove over 15,000 miles across the midwestern USA chasing thunderstorms that could spawn tornadoes for scientific research.

Once a tornado forms, the rotational vortex spins in a counterclockwise direction (less than 1 percent rotate clockwise). Most tornadoes form at the trailing end of a thunderstorm, stretching down from the cloud base. Tornadoes move along with the thunderstorm at velocities ranging from close to stationary to 100 km per hour. The rotational speed of winds can reach more than 500 km per hour. As tornadoes stretch down the base in contact with the ground can jump up so that they appear to skip along the surface.

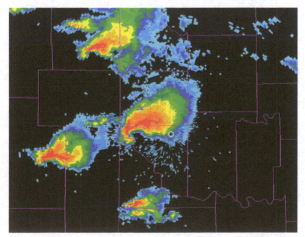

Source: NOAA.

Figure 29.10   Hook echo on Doppler radar image of supercell thunderstorm.

Tornado Intensity

Wind velocity values are used to classify the intensity of tornadoes. These velocities are either measured directly or extrapolated from the damage the storms produce. Twenty-eight different measures go into the determination and assignment of the value for a given tornado. The Enhanced Fujita Scale is used to assign a value to a given tornado (Table 29.1). The scale was originally set up in 1971 by Professor Ted Fujita, a world-renowned researcher who worked at the University of Chicago. His research into severe storms and weather led to the discovery of phenomena such as microbursts and downbursts, sudden violent blast of air that produces damaging results.

Table 29.1	The Enhanced Fujita Tornado Scale				
Fujita Scale			**Operational EF-Scale**		
F Number	**Fastest 1/4-mile (mph)**	**3 Second Gust (mph)**	**EF Number**	**3 Second Gust (mph)**	
0	40–72	45–78	0	65–85	
1	73–112	79–117	1	86–110	
2	113–157	118–161	2	111–135	
3	157–207	162–209	5	136–165	
4	208–260	210–261	4	166–200	
5	261–318	262–317	5	Over 200	

Source: http://www.ncdc.noaa.gov/oa/satellite/satelliteseye/educational/fujita.html

Frequency of Tornado Occurrences

Tornadoes occur everywhere across the world. However, the highest number of tornadoes occur in the United States, where all fifty states have experienced a tornado at some point in time. Because tornadoes are directly related to thunderstorms, they are more common in those regions struck by thunderstorms. Because clashes of cold and warm, moist air generate unstable atmospheric conditions, the Midwest has the greatest number of tornadoes (**Figure 29.11a**). Warm, moist air from the Gulf of Mexico drawn up into the Midwest collides with cold air from Canada (**Figure 29.11b**). The result creates severe thunderstorm and tornado conditions.

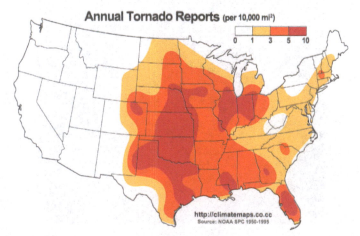

Source: NOAA. Mamatus/Wikimedia Commons.

Figure 29.11a Tornadoes have formed in every state, but are most common in the midwest.

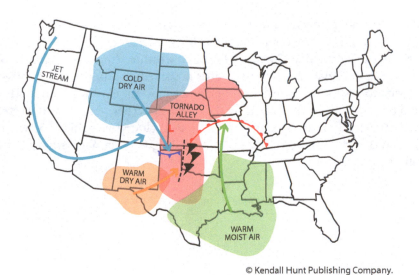

© Kendall Hunt Publishing Company.

Figure 29.11b Tornadoes form in the continental United States because of the confluence of warm, moist air and cold, dry air over the broad, flat central midwestern plains.

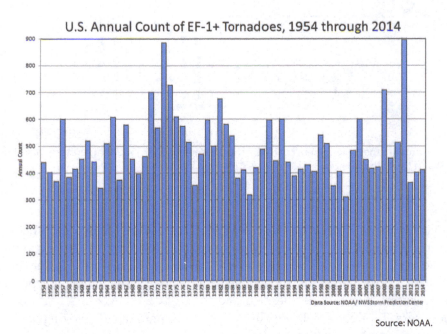

Source: NOAA.

Figure 29.12 The number of EF-0 tornadoes that have been reported in the United States has increased significantly since 1950.

Since 1950, the number of tornadoes that have been reported in the United States has increased about seven-fold (**Figure 29.12**). One possible reason for this dramatic increase is the introduction of new technology that can more readily identify tornadoes in the atmosphere, even though they might not be spotted by humans. Another reason for the increase could be that the population of the country has become more spread out in areas that once had few, if any, people. Also there could be an effect produced by **climate change** that has created more warm and cold air masses that collide with each other.

climate change: The long-term fluctuations in temperature, precipitation, wind, and other aspects of the Earth's climate.

The highest number of tornadoes occur during the spring and summer months, when clashing air masses of different temperatures and moisture content collide. Most tornadoes tend to occur in the afternoon, after temperatures have risen during the day (**Figure 29.13**). Long weather fronts often spawn swarms of tornadoes rather than isolated events. A megaswarm of tornadoes occurred on April 3 and 4, 1974 (**Figure 29.14**). Plots of the paths of the tornadoes showed the general northeast movement of tornadoes. This is normal for tornadoes in the United States, as they are driven by the overall weather pattern and westerlies. One-hundred forty-eight tornadoes struck 13 states in a 16-hour period. Included were six F-5 twisters. At the end 307 people lost their lives and more than 6,000 were injured. Property damage amounted to more than $600 million (in 1974 dollars).

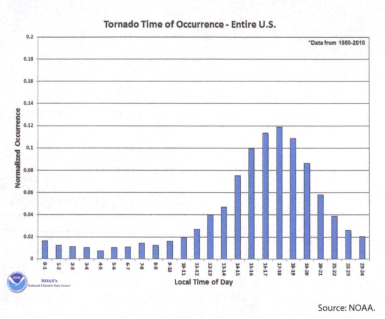

Source: NOAA.

Figure 29.13 Tornadoes are most likely to occur in the late afternoon, when land temperatures are at their highest and thunderstorms are most likely to develop.

Tornado Damage

Tornado damage is caused by high wind velocities and a large difference in atmospheric pressure between the tornado and its surroundings. The rotating winds can destroy weak structures, and the extremely low pressure inside the tornado generates strong pressure differences between the inside and outside of buildings. This pressure difference causes roofs to be lifted and removed. The high winds pick up smaller objects including small structures, animals, people, cars, and especially mobile homes, and can carry these objects up to several kilometers. The debris picked up by the winds becomes rapidly moving projectiles that could be lethal when hurled against a human body.

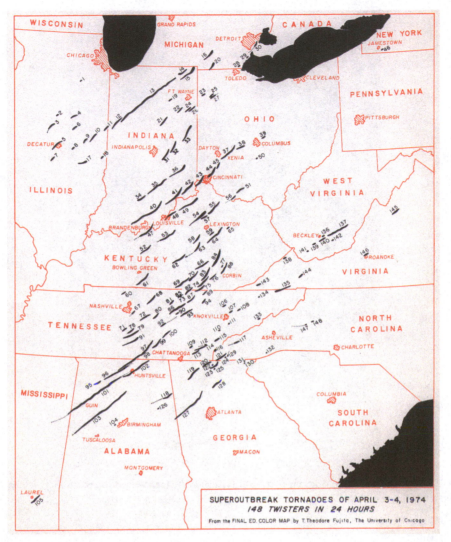

SUPEROUTBREAK TORNADOES OF APRIL 3-4, 1974
148 TWISTERS IN 24 HOURS
From the FINAL ED. COLOR MAP by T. Theodore Fujita, The University of Chicago

Figure 29.14 **The super outbreak of tornadoes across the Mississippi Valley region on April 3 and 4, 1974, generated 148 tornadoes.**

Homes in the Midwest have storm cellars that allow families to go underground to wait out the high winds. An EF5 tornado can turn a community into a flattened mass of rubble. Winds from a less violent EF3 storm can still have enough energy to strip trees of their leaves and throw building roofs into the cleared area (**Figure 29.15**).

FEMA, photo by Win Henderson.

Figure 29.15 An EF3 tornado hit Mena, Arkansas on April 11, 2009. The tree was stripped of all its leaves and caught an aluminum roof from a nearby building.

Tornado Prediction and Warning

Tornadoes cannot be predicted with precision. However, when strong thunderstorm activity is detected, a **tornado watch** is generally issued for all areas that may fall in the path of the thunderstorm. **Doppler radar** can detect rotating motion within a thunderstorm and when this is detected, or a tornado is actually observed, a **tornado warning** is issued for all areas that may fall in the path of the thunderstorm. Tornado safety (from the Federal Emergency Management Agency, FEMA) includes the following.

> **tornado watch:** This is issued by the National Weather Service when conditions are favorable for the development of tornadoes in and close to the watch area. Their size can vary depending on the weather situation.
>
> **Doppler radar:** Radar that can measure radial velocity, the instantaneous component of motion parallel to the radar beam (i.e., toward or away from the radar antenna).
>
> **tornado warning:** This is issued when a tornado is indicated by the WSR-88D radar or sighted by spotters; therefore, people in the affected area should seek safe shelter immediately.

If at home:

- Go at once to the basement, storm cellar, or the lowest level of the building.
- If there is no basement, go to an inner hallway or a smaller inner room without windows, such as a bathroom or closet.

- Get away from the windows.
- Go to the center of the room. Stay away from corners, because they tend to attract debris.
- Get under a piece of sturdy furniture such as a workbench or heavy table or desk and hold onto it. Use arms to protect head and neck.
- If in a mobile home, get out and find shelter elsewhere.

If at work or school:

- Go to the basement or to an inside hallway at the lowest level.
- Avoid places with wide-span roofs such as auditoriums, cafeterias, large hallways, or shopping malls.
- Get under a piece of sturdy furniture such as a workbench or heavy table or desk and hold onto it. Use arms to protect head and neck.

If outdoors:

- If possible, get inside a building.
- If shelter is not available or there is no time to get indoors, lie in a ditch or low-lying area or crouch near a strong building. Be aware of the potential for flooding. Use arms to protect head and neck.

If in a car:

- Never try to out-drive a tornado in a car or truck. Tornadoes can change direction quickly and can lift up a car or truck and toss it through the air.
- Get out of the car immediately and take shelter in a nearby building. If there is no time to get indoors, get out of the car and lie in a ditch or low-lying area, away from the vehicle. Be aware of the potential for flooding.

After the tornado:

- Help injured or trapped persons. Give first aid when appropriate. Don't try to move the seriously injured unless they are in immediate danger of further injury. Call for help.
- Turn on radio or television to get the latest emergency information.
- Stay out of damaged buildings. Return home only when authorities say it is safe.
- Use the telephone only for emergency calls.
- Clean up spilled medicines, chemicals, gasoline or other flammable liquids immediately.
- Leave the building if you smell gas or chemical fumes.
- Take pictures of the damage—both to the house and its contents—for insurance purposes.
- Remember to help your neighbors who may require special assistance—infants, the elderly, and people with disabilities.

Inspecting utilities in a damaged home:

- Check for gas leaks—If you smell gas or hear a blowing or hissing noise, open a window and quickly leave the building. Turn off the gas at the outside main valve if you can and call the gas company from a neighbor's home.
- If you turn off the gas for any reason, it must be turned back on by a professional.

Mitigation of Potential Tornado Damage

Tornadoes can occur wherever thunderstorms occur—basically anywhere. Therefore it is financially impossible to prepare every possible locality for a tornado. The best preparation is to ensure that people are aware of warning systems and have taken some precaution, such as building a storm cellar if they live in an area that is likely to experience tornadoes. Building codes can also require more sturdy construction. However, if a structure is hit directly by an EF 5 tornado it will certainly be severely damaged, if not totally destroyed. Warnings broadcast by governmental agencies must be timely, accurate, and heeded by those for whom they are sent.

Other Severe Weather Phenomena

By Ingrid Ukstins and David Best

Topic
30

Nor'easters

A **nor'easter** is an extratropical cyclonic storm that forms in a region away from the tropics. It has some of the characteristics of a tropical storm, winds that rotate in a counterclockwise direction, and a low pressure center. These storms that affect the east coast of the United States form off the south Atlantic coast. The counterclockwise winds come out of the northeast (hence the name). Because these usually occur in the fall and winter months, they are less intense than hurricanes because the colder ocean waters do not provide a large amount of thermal energy. However, damage can be extensive (**Figure 30.1**).

> **nor'easter:** A strong low pressure system with winds from the northeast that affects the mid-Atlantic and New England states between September and April. These weather events are notorious for producing heavy snow, copious rainfall, and tremendous waves that crash onto Atlantic beaches, often causing beach erosion and structural damage.

FEMA.

Figure 30.1   A nor'easter hit New England in April 2009. Damage in Saco, Maine, was extensive along the shoreline.

Winter Blizzards and Severe Weather

Severe winter weather is often accompanied by high winds, significant snowfall, and cold temperatures. These conditions can produce a **blizzard**, which the National Weather Service Service defines as having sustained 35 mph (56 kph) winds that lead to blowing snow, and causes visibilities of ¼ mile or less, lasting for at least 3 hours (**Figure 30.2**). Although no specific temperatures are associated with a blizzard, the high winds often produce sub-zero wind chill conditions. These conditions form when a ridge of high-pressure interacts with a low-pressure system; this results in the horizontal movement of air from the high-pressure zone into the low pressure area.

> **blizzard:** A severe winter storm that has the following conditions that are expected to prevail for a period of 3 hours or longer: sustained wind or frequent gusts to 35 miles an hour or greater and significant falling and/ or blowing snow (i.e., reducing visibility frequently to less than ¼ mile).

FEMA, photo by Michael Rieger.

Figure 30.2   A blizzard hits Denver, Colorado, in December 2006 with more than two feet of snow.

Other winter weather condition watches and warnings that are issued by the National Weather Service include:

- **Winter Storm Watch.** Conditions are favorable for hazardous winter weather conditions including heavy snow, blizzard conditions, or significant accumulations of freezing rain or sleet. These watches are issued by the Weather Service Forecast Office in Chicago and are usually issued 12 to 36 hours in advance of the event.
- **Winter Storm Warning.** Hazardous winter weather conditions that pose a threat to life and/or property are occurring, imminent, or likely. The generic term, winter storm warning, is used for a combination of two or more of the following winter weather events; heavy snow, freezing rain, sleet, and strong winds
- **Heavy Snow Warning.** Snowfall of 6 inches or more in 12 hours or less, or 8 inches or more in 24 hours or less.
- **Lake Effect Snow Warning.** Lake effect snowfall of 6 inches or more in 12 hours or less, or 8 inches or more in 24 hours or less. The source of moisture for these storms comes from large bodies of water, such as the Great Lakes. Michigan, northern Ohio and Pennsylvania, and western New York state are often hit with heavy snow that developed from coming from the Great Lakes.

Heat Waves

Several significant heat waves have occurred throughout the world over the past 30 years. A **heat wave** is a prolonged period of excessively hot weather, which may be accompanied by high humidity. There is no universal definition of a heat wave, as the term is relative to the usual weather in the area. Temperatures that people from a hotter climate consider normal can be termed a heat wave in a cooler area if they are outside the normal climate pattern for that area. Increased humidity adds to the effect of the elevated temperatures to create deadly conditions.

heat wave: A prolonged period of excessively hot weather and high humidity.

The term *heat wave* is applied both to routine weather variations and to extraordinary spells of heat which may occur only once a century. Severe, prolonged heat waves have caused catastrophic crop failures, thousands of deaths from hyperthermia, and widespread power outages due to increased use of air conditioning (**Figure 30.3**). In the 40-year period from 1936 through 1975, nearly 20,000 people were killed in the United States by the effects of heat and solar radiation. In the disastrous heat wave of 1980, more than 1,250 people died, mainly in the central and southern Plains. Extreme heat now causes more deaths in United States cities than all other weather events combined, and heat waves such as the one that hit the Northern Hemisphere in 2018 resulted in dozens of deaths across the US and Canada as well as Japan. In the USA, a high-pressure system locked in a dome of heat that resulted in temperatures over 90 degrees Farenheight in 44 of 50 states. Heat waves are especially deadly when nighttime temperatures remain high—the human body can't recover from the effects of extreme heat if the air temperature doesn't drop below 80 degrees F at night. The top 4 hottest years on record are 2016, 2017, 2015 and 2018. The continuing rise in global temperatures will result in more extreme heat in the future, and an estimated 150 Americans will die every summer day due to this extreme heat by 2040, totaling about 30,000 heat-related deaths each year.

© Jasper Suijten/Shutterstock.com.

Figure 30.3   Heat waves can result in extreme drought conditions and crop failure.

In the summer of 2003 Europe was hit by a heat wave that killed more than 37,000. Much of the heat was concentrated in France, where nearly 15,000 people died. Elderly people were the most affected. In July 2006, the United States experienced a massive heat wave, and almost all parts of the country recorded temperatures above the average temperature for that time of year. Temperatures in some parts of South Dakota exceeded 115°F (46°C), causing many problems for the residents. Also, California experienced temperatures that were extraordinarily high, with records ranging from 100 to 130°F (38 to 54°C). On July 22, the County of Los Angeles recorded its highest temperature ever at 119°F (48.33°C) (**Figure 30.3**).

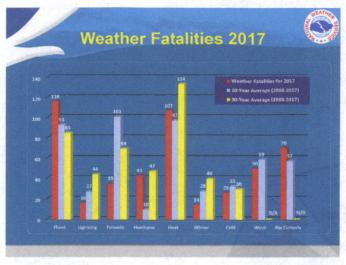

Source: NOAA.

Figure 30.4   Deaths attributable to weather related events show a wide range of values.

Drought and Famine

Drought is an extended period lasting months or years when a region receives much less than normal precipitation, resulting in a significant shortage of water. Both surface water and groundwater supplies are drastically reduced. This shortage rapidly reduces the ability of people to grow crops and to sustain life, and the ecosystem of the area is impacted to the point that it begins to deteriorate.

The onset of a drought is caused by changes in the upper-level air flow. In the United States it is associated with a prolonged ridge of high pressure that pushes dry air down to the surface. This sinking air is warmed by compression and begins to reduce the humidity even more. The dry air reduces moisture in the soil, reducing the ability to grow crops. In the heartland of the United States these conditions began to develop in the early 1930s and lasted for several years. The result was severe dust storms throughout the central part of the United States, giving the name "Dust Bowl" to the region (**Figure 30.4**). Widespread crop failures resulted in the malnutrition of hundreds of thousands of people, Numerous farms were abandoned and there was a mass migration to the West, especially California, where former farmers hoped to reestablish the way of life they had known previously. These events came on the heels of the Great Depression, making living conditions and survival very problematic for thousands of people.

© Everett Historical/Shutterstock.com.

Figure 30.5   A dust storm in Elkhart, Kansas, in May 1937.

Hurricanes, Cyclones and Typhoons

By Ingrid Ukstins and David Best

Why study cyclonic storms?

These events occur throughout the globe at all times of the year in both the northern and southern hemispheres. They generate destructive conditions that affect our communities, landscape, and daily lives. In the United States these storms, whether lower grade tropical storms or more intense hurricanes, affect more than half the population of the country. Of the ten most populated states, six of them (Texas, Florida, New York, Pennsylvania, Georgia, and North Carolina) lie in the paths of these events. Although not all of these states are impacted every year, when hurricanes do come ashore, their fury is extremely damaging and disruptive. In addition to property being destroyed, commerce and travel can come to a halt, producing economic upheaval. Since 2005, evidence has made us acutely aware of the forces of nature that alter our well-being. If you have not been affected by these events, perhaps you know people who have. As more people move to warmer regions of the country, including those mentioned above, they will come to understand the role that storms play in their lives.

Large-scale **cyclonic storms** occur in many areas of the globe, generated by warm, oceanic waters that create rising air circulations that lift moisture and heat from the water surface into rotating storms. Few regions on Earth are spared from these weather phenomena. Both the geologic and human impacts can be long lasting when these storms encounter continental regions.

cyclonic storm: A generic term that covers many types of weather disturbances that are typified by low atmospheric pressure and rotating, inwardly directed winds.

Hundreds of cyclonic storms have hit coastal and inland regions along the Pacific, Indian, and North Atlantic oceans, as well as the Caribbean Sea and Gulf of Mexico.

Areas of Cyclonic Storm Generation

Wind is the movement of air. It can be affected by changes in air pressure and the Earth's rotation. On a global scale the differences in heating near the equator and the cooling at the poles produce large circulation cells that drive large weather systems across Earth's surface. In some cases these are gentle breezes, but at other times winds can produce storms that become major natural disasters. In areas of low atmospheric pressure, which are associated with stormy conditions, rotational forces cause winds to create a swirling vortex.

As the rotation intensifies, a cyclonic disturbance forms and moves through the Earth's atmosphere, with the potential of becoming a major cyclonic storm (**Figure 31.1**).

National Geophysical Data Center, NOAA.

Figure 31.1   **A sequence of satellite images shows the position of Hurricane Andrew as it approaches and moves across Florida and the Gulf of Mexico. The storm image to the right was taken August 23, 1992; the center image is its position on August 24, and the image on the left was taken August 25, 1992. Notice the changes in the size of the eye of the storm.**

Three different terms are used to categorize these cyclonic storms, based on where they first develop. **Cyclones** are associated with the Indian Ocean and South Pacific Ocean and affect South Asia, Australia, and the east coast of Africa. **Typhoons** are generated in the western and southern Pacific, affecting islands in the Pacific Ocean as well as southeast Asia, China, and Japan. Cyclonic storms in the western portions of the North Atlantic Ocean, the Caribbean Sea, and the eastern Pacific off Mexico and Central America are termed **hurricanes** (**Figure 31.2**). Unless we refer to a specific storm in a portion of the globe other than the North Atlantic and adjoining regions, we will refer to cyclonic storms as hurricanes.

cyclone: A rotating mass of low pressure in the atmosphere that covers a large area; the warm air mass rotates counterclockwise in the northern hemisphere and clockwise in the southern hemisphere. The term is strictly applied to large low pressure storms in the Indian Ocean and South Asia.

typhoon: A cyclonic storm that forms in the central and western Pacific Ocean. Refer to cyclone and hurricane.

hurricane: Term applied to cyclonic storms that occur in the North Atlantic Ocean, Caribbean Sea, Gulf of Mexico, or eastern Pacific Ocean. Minimum wind velocity is 74 miles per hour. Refer to cyclone and typhoon.

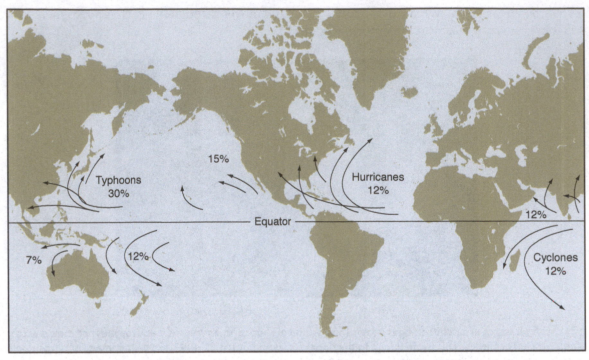

Figure 31.2 Global map showing where storms first form. The majority form in the Northern Hemisphere. Strong wind shear and lack of weather disturbances do not allow cyclonic storm formation in the areas east and west of South America.

Many variables are involved in the formation of these storms. They have a range of sizes and intensities as well as different forward velocities, so each storm is unique in terms of its characteristics and behavior. However, some traits are common to all cyclonic storms.

- They initially form in only a few areas over the eastern and southwest Pacific Ocean, eastern North Atlantic Ocean, and north Indian Ocean.
- Storms form between the latitudes of 5° and 20° N or S.
- They develop more frequently during the summer months of their respective hemispheres, but they can occur in any month.

The movement of cyclonic storms across Earth's surface is influenced by Earth's rotation and the jet stream. This rotation produces the Coriolis effect, which generates a force that causes storms in the northern hemisphere to curve to the right. The reverse movement is seen in the southern hemisphere. The physics of fluid dynamics also controls movement within the low-pressure zones. Areas at risk include the shorelines lying in the path of the storm in addition to the water surface over which the storm passes. Low-lying coastal areas are frequently subjected to major flooding produced by the landward movement of massive amounts of water being pushed ashore by the storm's winds. In general, east-facing coastal areas are affected in the Northern Hemisphere. An exception is the west coast of Mexico, which is often hit by storms coming off the eastern Pacific Ocean and the western side of India, which is affected by storms that form in the northern Indian Ocean.

Hurricane Formation and Movement Processes

In an average year, approximately 80 to 100 cyclonic storms develop in the tropical and subtropical regions on Earth, being born from the warm waters lying near the equator. Storms have a life span of several days up to a few weeks, during which time they can intensify from a small tropical depression into a major atmospheric event that can affect millions of people.

Hurricane Conditions

For a **tropical disturbance** to evolve into a **tropical depression** and eventually a hurricane, several conditions must be met:

- Warm, ocean surface water (exceeding 28°C), along with high humidity, and unidirectional, constant winds
- Rotating winds produced by thermal heating and from the Coriolis effect in the lower latitudes
- Low atmospheric pressure near the equator that creates an atmospheric, tropical depression (**Figure 31.3**) that can develop into more intense storms.

> **tropical disturbance:** The beginning stage of a cyclonic storm lasting at least 24 hours, originating in the tropics or subtropics; clouds and moisture become organized and a vertically rotating wind mass creates atmospheric instability.
>
> **tropical depression:** A slow-forming cyclonic storm with sustained surface winds of 38 miles per hour or less.

Cyclones form in tropical and subtropical latitudes, generally between 5° and 20° north or south of the equator. This region on Earth receives the highest amount of solar radiation, which warms ocean waters that evaporate into the atmosphere (**Figure 31.3**). Sea water must be a minimum temperature of 28°C to be warm enough for heat and moisture to contribute to the formation of a storm. The resulting low-pressure conditions begin to rotate as the heat and moisture rise. Circulation is counterclockwise in the northern hemisphere and clockwise in the southern hemisphere.

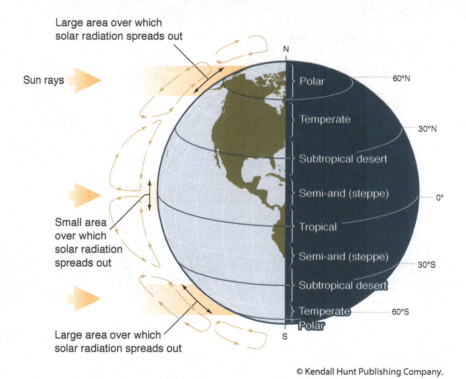

© Kendall Hunt Publishing Company.

Figure 31.3 **Global atmospheric circulation patterns. Note the formation of low pressure centered on the equator as heated air rises into the atmosphere and descends in the subtropical regions.**

The National Hurricane Center defines the progression of storm development as shown in Table 31.1. Over a period of days before the system reaches the threshold of a **tropical storm**, wind velocities increase and the storm becomes better defined; at this time it is assigned a name by the National Hurricane Center. Once hurricane conditions are reached, the high winds and torrential rainfall produce a significant threat to the environment and the communities the storm might impact. In the northern hemisphere, ocean conditions are at an optimum for storm development from August through October, when the largest number of tropical storms occur in the North Atlantic Ocean (Table 31.2). This period is near the end of summer as ocean waters need the summer months to collect enough solar radiation and heat to generate storms. Historically the greatest number of hurricanes that make landfall in the United States occurs in September, also the stormiest month.

tropical storm: A cyclonic storm with sustained surface winds between 39 miles per hour to 73 miles per hour. At this level of activity the system is assigned a name to identify and track it.

Thermal energy—energy that comes from heat drawn from surface waters—is what drives hurricanes. As this energy rises into the atmosphere, it rotates and produces unstable conditions aloft. Further rotation concentrates energy toward a central column and the velocity increases (the spinning ice skater effect). The ocean surface becomes more agitated, which increases the heat and moisture in the atmosphere.

thermal energy: The amount of energy in a system that is related to the temperature of its constituents; for hurricanes this is heat originally taken from the oceans.

Table 31.1 Definitions of Tropical, Low-Pressure Weather Systems (Storms begin as disturbances and can become hurricanes.)

System Designation	Conditions
Tropical disturbance	Organized mass of convectional air and thunderstorms with partial rotation present; generally 100 to 300 nautical miles in diameter; forms in the subtropics or tropics
Tropical depression	Closed circulation with sustained winds of 38 mph (33 kt or 62 kph) or less
Tropical storm	Sustained winds (1-minute measurement, 10 m above water) of 39 mph (63 kph or 34 kts) up to 64 kts. A name is now assigned to the storm.
Hurricane	Sustained winds of 74 mph (119 kph or 64 kts) or more (see **Box 31.1**)

Source: National Hurricane Center.

Table 31.2 Monthly Occurrence of Tropical Storms and Hurricanes in the North Atlantic Ocean, for the Period 1851 to 2017

January-April	6	*	2	*	0	*
May	24	0.1	4	*	0	*
June	87	0.5	30	0.2	19	0.11
July	107	0.6	50	0.3	27	0.16
August	355	2.1	238	1.4	78	0.47
September	469	2.8	337	2.0	110	0.66
October	282	1.7	170	1.0	52	0.31

(continued)

Table 31.2

Table 31.2 | **Monthly Occurrence of Tropical Storms and Hurricanes in the North Atlantic Ocean, for the Period 1851 to 2017 (*continued*)**

November	61	0.4	38	0.2	5	0.03
December	9	*	4	*	0	*
Year	1400	8.4	869	5.2	291	1.75

* Less than 0.05.
Source: National Oceanic and Atmospheric Administration.

BOX 31.1	Why Do Hurricane-Force Winds Start at 64 Knots?

In 1805–1806 Commander Francis Beaufort RN (later Admiral Sir Francis Beaufort) devised a descriptive wind scale in an effort to standardize wind reports in ship's logs. His scale divided wind speeds into 14 Forces (soon after pared down to thirteen) with each Force assigned a number, a common name, and a description of the effects such a wind would have on a sailing ship. And since the worst storm an Atlantic sailor was likely to run into was a hurricane, that name was applied to the top Force on the scale.

Beaufort Wind Scale

Force 0	Calm
Force 1	Light Air
Force 2	Light Breeze
Force 3	Gentle Breeze
Force 4	Moderate Breeze
Force 5	Fresh Breeze
Force 6	Strong Breeze
Force 7	Near Gale
Force 8	Gale
Force 9	Strong Gale
Force 10	Storm
Force 11	Violent Storm
Force 12	Hurricane

Contributed by Neal Dorst

During the nineteenth century, with the manufacture of accurate anemometers, actual numerical values were assigned to each Force level, but it wasn't until 1926 (with revisions in 1939 and 1946) that the International Meteorological Committee (predecessor of the World Meteorological Organization, an agency of the United Nations) adopted a universal scale of windspeed values. It was a progressive scale with the range of speed for Forces increasing as you go higher. Thus Force 1 is only 3 knots in range, while the Force 11 is eight knots in range. So Force 12 starts out at 64 knots (74 mph, 33 m/s).

There is nothing magical in this number, and since hurricane force winds are a rare experience, chances are the committee that decided on this number didn't do so because of any real observations during a hurricane. Indeed, the Smeaton-Rouse wind scale in 1759 pegged hurricane force at 70 knots (80 mph, 36 m/s). Just the same, when a **tropical cyclone** has maximum winds of approximately these speeds, we see the mature structure (eye, eyewall, spiral rainbands) begin to form, so there is some utility with setting hurricane force in this neighborhood.

Source: http://www.aoml.noaa.gov/hrd/tcfaq/.

tropical cyclone: A low-pressure system having a warm center that developed over tropical (sometimes subtropical) water and has an organized circulation pattern. The magnitude of its winds defines it as a disturbance, depression, storm or hurricane/typhoon.

Hurricane Eye

Eventually an **eye** forms at the center of the rotating vortex. The eye is a region of calm air and relatively clear conditions characterized by a column of descending cool, dry air (**Figure 31.4**). The column is bordered by the **eye wall**, a thick mass of moisture-laden clouds spinning at a high velocity, and represents the most violent portion of a hurricane. Upward rotational movement in the eye wall lifts warm, moist water to higher altitudes, where condensation occurs and heat is given off.

eye: The central core of a cyclonic storms, normally relatively small and lacking clouds, moisture, and wind.

eye wall: The boundary between the eye of a cyclonic storms and the inner most band of clouds.

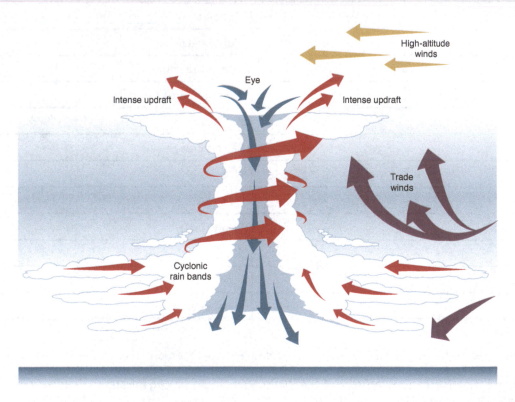

Source: National Geophysical Data Center, NOAA.

Figure 31.4   **Generation of the eye wall and circulation pattern of a hurricane.**

Rotational forces throw most of the air and moisture outward, but some air flow is directed toward the center of the eye, where it is heated by compression. This warmer air can now absorb additional moisture, so most water vapor in the eye will be absorbed into the air. Any clouds in the eye dissipate, producing a zone devoid of clouds. The rising air creates lower pressure along the storm-ocean water boundary, which causes more moisture and heat to be drawn upward and the process continues. In the early stages of storm development, the eye is not well-defined but has better definition as overall rotation of the entire storm system increases (**Figure 31.5**). Once formed, the eye ranges in diameter from 5 to 40 miles (8 to 65 km), while the entire storm can stretch across 250 to 400 or more miles (400 to > 650 km).

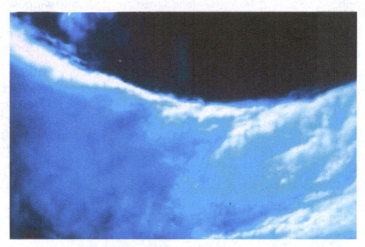

Source: The National Oceanic and Atmospheric Administration.

Figure 31.5   Photo taken in the eye of Hurricane Katrina, showing the eye wall. Note the clear conditions aloft and the rotating cloud pattern.

Measuring the Size and Intensity of a Hurricane

Hurricanes have a huge amount of thermal energy inside them. The National Oceanic and Atmospheric Administration (NOAA) reports that an "average" hurricane has at least 1.3×10^{17} Joules per day or 1.5×10^{12} watts of energy, which is equivalent to about half the world's entire electrical generating capacity. Because of the extensive size of a hurricane and its slow forward speed, winds blow for many hours in a given location. At first winds are blowing from one direction and then as the storm passes directly over an area, the winds diminish to almost none at all as the eye passes overhead. For the uninformed, this might signal the end of the storm. After the eye passes, winds increase again—but from the opposite direction—and will continue to blow until the storm moves out of the area.

Tropical storms, and those that grow into hurricanes, extend over very large areas. Gale-force winds (those exceeding 39 mph) can reach out from the center of the storm 300 to 400 miles. Hurricane-force winds, which exceed 74 mph, can extend out up to 100 miles or more from the center. Hurricane Sandy in 2012 is the hurricane that covered the largest area. Its gale-force winds were felt more than 1,000 miles from the storm center. Fortunately most of those winds were situated over open ocean to the east of the coastline of the mid-Atlantic states and New England. However, those winds created a massive storm surge that affected the entire region.

Hurricane forecasters assign storms to one of five categories, based on their wind speeds (Table 31.3). This designation provides an estimate of the potential property damage and flooding expected from the storm.

Table 31.3	The Saffir-Simpson Scale Used to Categorize Intensities of Hurricanes	
Category	**Winds**	**Effects**
1	74–95 mph	No real damage to building structures. Damage primarily to unanchored mobile homes, shrubbery, and trees. Also, some coastal road flooding and minor pier damage.
2	96–110 mph	Some roofing material, door, and window damage to buildings. Considerable damage to vegetation, mobile homes, and piers. Coastal and low-lying escape routes flood two to four hours before arrival of center. Small craft in unprotected anchorages break moorings.
3	111–130 mph	Some structural damage to small residences and utility buildings, with a minor amount of curtain wall failures. Mobile homes are destroyed. Flooding near the coast destroys smaller structures with larger structures damaged by floating debris. Terrain continuously lower than 5 feet above sea level may be flooded inland 8 miles or more.
4	131–155 mph	More extensive curtain wall failures, with some complete roof structure failure on small residences. Major erosion of beach. Major damage to lower floors of structures near the shore. Terrain continuously lower than 10 feet above sea level may be flooded requiring massive evacuation of residential areas inland as far as 6 miles.
5	Greater than 155 mph	Complete roof failure on many residences and industrial buildings. Some complete building failures with small utility buildings blown over or away. Major damage to lower floors of all structures located less than 15 feet above sea level and within 500 yards of the shoreline. Massive evacuation of residential areas on low ground within 5 to 10 miles of the shoreline may be required.

RECENT EXAMPLES

Category	Sustained Winds (MPH)	Description	Examples
1	74–95	Minimal	Hermine 2016 FL
2	96–110	Moderate	Arthur 2014 NC
3	111–130	Extensive	Irene 2011 CT
4	131–155	Extreme	Katrina 2005 LA | Harvey 2017 TX LA
5	>155	Catastrophic	Camille 1969 MS | Andrew 1992 FL

Source: National Hurricane Center, formulated in 1969 by Herbert Saffir, a consulting engineer, and Dr. Bob Simpson, director of the National Hurricane Center.

Causes of Damage

Wind Action

Forward motion of the entire mass of a tropical storm is created by global circulation patterns prevalent in the area of the storm. In **Figure 31.6a** we see several interesting features. As mentioned earlier, there are no tropical storms that occur near the equator and no storms exist on the oceanic sides of South America. The strongest storms are in the western Pacific Ocean. Very few of the storms in the southern hemisphere become major storms. In the North Atlantic Ocean, the prevailing westerlies push storms from their regions of origin in the east toward the west. The science of fluid mechanics also helps explain much of the movement of these storms. As a mass of fluid (in this case, water-laden, rotating air) moves from the equator toward the poles, it experiences increased rotation from a state of no rotation at the equator to a maximum value at the poles. In the northern hemisphere, there is a maximum rotation on the west side of a storm and a minimum value on the east side. This difference in the rotational strength, along with the Coriolis force, generates a deflection of the storm toward the north. Thus, hurricanes in the North Atlantic Ocean curve northward as they progress to the west (**Figure 31.6b**). In its earliest stages, the forward motion of a hurricane ranges from almost 0 mph to more than 10 or 12 mph (0–20 kph). When storms move out of the latitudes of the trade winds (about 30° north and south of the equator), they increase their velocity to about 15 to 20 mph (25 to 35 kph). This forward motion, when coupled with the counterclockwise circulation of winds makes the northeast side of a storm the most devastating, as the two wind velocities are added together (**Figure 31.7**).

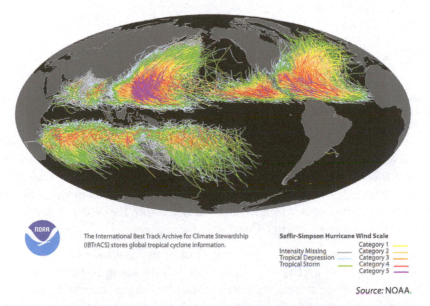

The International Best Track Archive for Climate Stewardship (IBTrACS) stores global tropical cyclone information.

Saffir-Simpson Hurricane Wind Scale

Intensity Missing
Tropical Depression
Tropical Storm

Category 1
Category 2
Category 3
Category 4
Category 5

Source: NOAA.

Figure 31.6a   Global tropical storm activity shows much stronger activity in the northern hemisphere.

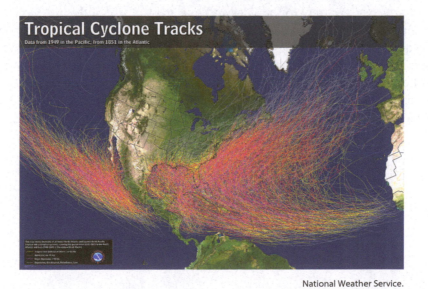

National Weather Service.

Figure 31.6b   Activity in the North Atlantic over the past 150 years shows the prevailing movement of storms as they travel westward and then to the northeast.

Topic 31 Hurricanes, Cyclones and Typhoons   **161**

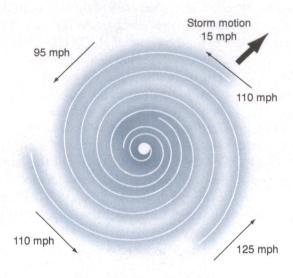

Storm motion
15 mph

95 mph

110 mph

110 mph

125 mph

© Kendall Hunt Publishing Company..

Figure 31.7 **The strongest winds are on the right side of the storm. This value considers both the forward motion of the storm as well as the wind speed around the eye.**

As storms get closer to the coast of the United States, they often encounter continental weather systems that have moved across the Midwest. These large weather fronts push the dissipating cyclonic low to the northeast and sometimes entrain the low pressure into its primary wind pattern, dragging it along and breaking it up. Tropical storms (which were previously hurricanes) have been recorded to move as fast as 50 mph (80 kph) across the mid-Atlantic and New England states northeastward into the North Atlantic, where colder waters fail to provide the energy needed to sustain the system.

Wind damage is perhaps the most obvious product of a hurricane. Photos and video images show buildings being torn apart, roofs flying through the air, and trees and limbs being bent. In many coastal areas that lie in the paths of hurricanes, building standards have improved to lessen the effect of wind damage, but major storms are capable of producing severe wind damage. Roofs are very susceptible as winds catch under shingles or corrugated metal and rip these off. Roofs on older buildings are frequently blown off as a single unit as they are often not attached securely to the sides and internal walls of buildings. Isolated or free-standing structures are especially affected if they lack walls (**Figure 31.8**). Roofs on new construction in coastal communities are now required to be tie-bracketed to interior and exterior walls to help maintain some structural integrity for the building when strong winds strike. Tall trees are often toppled because the ground holding their root systems becomes saturated and can't hold the trees in place (**Figure 31.9**). Occasionally trees are snapped off, but that is mainly the result of sudden gusts of higher velocity winds. Although the effects of wind are readily seen in an area hit by a hurricane, there is a more costly and devastating agent of destruction at work.

Figure 31.8 Hurricane-force winds can easily destroy weak, free-standing structures.

Figure 31.9 Tall pine trees, which have shallow roots systems, are among the first trees to be blown over due to saturated soils.

Storm Surge and Flood Hazards

Once a hurricane strikes land, it begins to lose its strength as it is no longer over the warm waters that feed it. Within a short period (usually less than one day), the hurricane is downgraded to a tropical storm or depression. However, these reduced winds, along with intense rainfall, will continue to affect areas in the storm's path. Extensive flooding often results from this stage of the storm. In June 1972, Hurricane Agnes developed off the Yucatan Peninsula of Mexico, gained strength as it passed over the warm waters of the Gulf of Mexico, and hit land in the panhandle of Florida. After it made landfall, it was downgraded to a tropical storm or depression and moved across Georgia and the Carolinas. It then joined another cyclonic low in the westerlies and regained strength as it went over the ocean to the east of Virginia. As it swung back inland, it was picking up moisture from the Atlantic Ocean and proceeded to drench the Mid-Atlantic states with intense rains (**Figure 31.10**). Portions of Maryland, northeastern Pennsylvania, upstate New York, and northern Virginia received more than 15 inches of rain. There were 122 recorded deaths in the United States, the majority caused by drowning and mudslides.

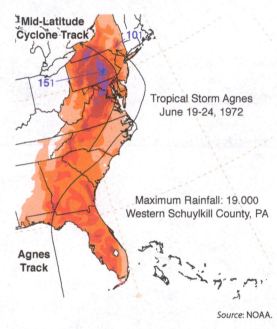

Source: NOAA.

Figure 31.10   **Rainfall map for tropical storm Agnes. The thin, solid line shows the location of the center of the storm as it moved northward.**

Sea water from storm surges produces the most extensive damage of a hurricane, especially when the surge strikes at high tide. The momentum of this water is difficult to recognize because the wind and driving rain catch everyone's attention. Within hours, rising waters can appear. The storm surge, which can be 10 to 15 ft (3 to 5 m) or higher, inundates the shore and adjacent inland areas. The surge rises as the hurricane makes landfall. The rise in water level is usually gradual but relentless (**Figures 31.11 and 31.12**). The greatest recorded storm surge was associated with Tropical Cyclone Mahina that struck Australia in 1899. Reports stated a surge in excess of 13 m (42 ft). Dolphins were reported to be lying on the ground 45 ft above sea level!

© Danny E. Hooks/Shutterstock.com

Figure 31.11 Onshore winds push water over low-lying areas and produce beach erosion.

© John Orsbun/Shutterstock.com

Figure 31.12 Large wind-driven waves produce significant erosion.

Because coastal areas are relatively flat and often have low spots, flood waters recede very slowly and the subsequent flooding can continue for several days or weeks. Flooding kills many more people than the wind. Hurricane Camille in 1969 was one of only three category 5 hurricanes to hit the continental United States mainland (the other two were the unnamed Florida hurricane of 1935 and Hurricane Andrew in 1992, since upgraded by a review of its activity). Most of the 256 deaths associated with Camille were a result of flooding along inland portions of the Gulf coast states. One hundred thirteen people died in Virginia due to flooding caused by extreme rainfall. From 1970 to 1994, 59 percent of the 589 deaths in the United States were by drowning in rain water that falls from an average hurricane at the rate of 2,300 cu meters per second. When measured against the average flow of the St. Lawrence River in New England and eastern Canada (a flow of 6,900 cu meters per second), an average hurricane produces that much water (in just three seconds). If rain falls in areas where streams funnel the runoff into developed areas, devastating conditions develop rapidly.

Hurricane Katrina, one of the strongest storms to hit the United States in the last 100 years, killed more than 1,800 people, most of whom drowned in flood waters that covered Alabama, Louisiana, and Mississippi. New Orleans, a city of almost 500,000 people before the onslaught of Katrina, is situated below sea level, surrounded by a system of levees that were constructed to hold back the Mississippi River and Lake Ponchartrain. Several major breaches in the levees resulted in rapid flooding that covered the city with more than 20 ft (7 m) of water (**Figure 31.13**).

NOAA. LCDR Mike Moran.

Figure 31.13 **New Orleans experienced severe flooding due to a breached levee system.**

Analysis of the region after Katrina passed led investigators to realize that the levees had not been sunk deep enough into the ground to withstand the enormous forces inflicted by the wind and water. Also, floodwaters undercut the bottom edges of the levees, thereby removing the material they rested on and they collapsed. Hurricane Katrina struck New Orleans as a category 3 storm, having lost some of its earlier punch when it was a category 5 storm over the central Gulf of Mexico. The relentless winds and storm surge battered the levees that were in place to protect the city. Once the levees broke, the entire city was flooded and remained in those conditions for several weeks until the water could be pumped out (**Figure 31.14**). Low-lying areas, such as the Ninth Ward, were inundated to depths exceeding 20 ft (6 m).

FEMA. Jocelyn Augustino.

Figure 31.14 Flooded New Orleans neighborhood near inundated roadway. Numerous rooftop rescues were carried out.

Any time there is a storm surge, there can be removal of sand from the beaches (**Figures 31.15 and 31.16**). This material is carried inland by the water and deposited once the surge abates. Thick sand deposits cover streets and yards and become a hazard (**Figure 31.17**). Beaches must be rebuilt to maintain the natural balance between the ocean and streams that normally supply the sand. One novel way to rebuild beaches is to use old Christmas trees to serve as wind breaks, allowing sand to reform coastal dunes (**Figure 31.18**).

© Steven J. Taylor/Shutterstock.com.

Figure 31.15 Beach erosion is rapid in areas without any grasses or other vegetation to hold sand in place.

Figure 31.16 This house once sat on sand dunes along the beach in the Outer Banks of North Carolina. Severe beach erosion associated with Hurricane Dennis in August 1999 removed large amounts of sand and isolated many structures along the shoreline.

Figure 31.17 Storm surge from Hurricane Frances in September 2004 pushed sand several hundred meters inland at Fort Pierce Beach, Florida.

Figure 31.18   Discarded Christmas trees are used to reestablish eroded beach dunes on the North Carolina coast.

The Historic and Recent Records and the Human Toll

Hurricanes have been reported by sailors since the first ships sailed the Pacific and Atlantic Oceans. Prior to that, the indigenous peoples of the Pacific Ocean and other regions all had stories of storms that were passed along through their oral histories. As countries took to the seas, sailors experienced the forces of nature when ships sailed through and around storms. Records of storms that have affected the United States have been kept for more than a century and detail some of the most destructive natural disasters to hit the U.S. mainland.

Often conflicting reports are given in terms of the number of deaths and the amount of property damaged by tropical storms and hurricanes. These data vary because of different reporting techniques that include a wide range of interpretations. We have attempted to provide reliable data that represent reports provided by U.S. government offices or by reputable research agencies, but differences will be encountered if you use a variety of sources.

The United States, certainly a country with a well-developed economic base and a means to deal with the ravages of hurricanes, is not exempt from the destruction that results when major storms hit the coastline. The National Oceanic and Atmospheric Administration (NOAA) reports that the average annual damage from hurricanes in the mainland United States is $4.9 billion. Damage increases exponentially with rising winds, so a category 4 hurricane can produce up to 250 times the damage of a category 1 storm. NOAA reported that in 2005 the country experienced a record 27 named storms, of which 15 reached hurricane status. Four hurricanes reached category 5, and five named storms developed in July—a record on both counts. This all-time record in terms of number and intensity of storms fortunately did not carry over into 2006, as some forecasters had feared. **Figure 31.19** shows that no hurricanes made landfall in 2006.

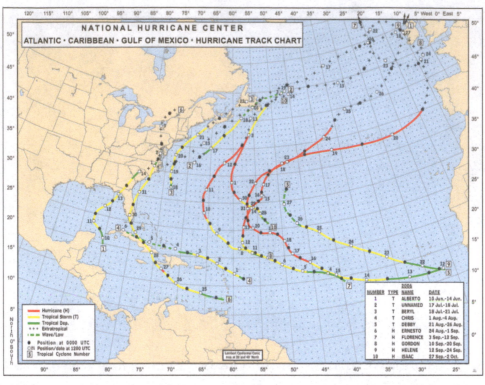

Source: NOAA.

Figure 31.19 Tropical storm tracks in the North Atlantic Ocean for the 2006 season. No hurricanes hit the United States and the season had only 10 named storms, a marked contrast to 2005, when 27 named storms formed.

Table 31.4 shows the ten most costly hurricanes in the United States through 2017. Texas and Florida have been very vulnerable to these storms. Note that in 2005 and 2017 these two states were hit by major events. Several factors contribute to the high loss of property. Resurgent construction and population growth frequently follows a devastating storm as people think another storm will not come along to inflict damage again, but obviously it happens. Data available from NOAA shows that in the period 1985 to 2015, almost 25 percent of catastrophic losses related to hurricanes occurred in Texas and Florida. These figures were calculated before the 2017 hurricane season which Hurricanes Harvey and Irma strike those regions with devastating results. As mentioned earlier, numerous hurricanes have struck the continental United States. Two major events that have done so recently are discussed in Boxes 31.2 and 31.3.

Table 31.4	Ten most costly hurricanes in the United States (not adjusted for inflation)			
Rank	**Hurricane**	**Year**	**Category**	**Damage (USD)**
1	Katrina (SE FL, LA, MS)	2005	3	$161,000,000,000
2	Harvey (TX, LA)	2017	4	125,000,000,000
3	Maria (PR, USVI)	2017	4	90,000,000,000
4	Sandy (Mid-Atlantic, NE US)	2012	1	71,000,000,000
5	Irma (FL)	2017	4	50,000,000,000
6	Ike (TX, LA)	2008	2	30,000,000,000

(continued)

Table 31.4 **Ten most costly hurricanes in the United States (not adjusted for inflation) (*continued*)**

Rank	Hurricane	Year	Category	Damage (USD)
7	Andrew (SE FL, LA)	1992	5	27,000,000,000
8	Ivan (AL, NW FL)	2004	3	20,500,000,000
9	Wilma (S FL)	2005	3	19,000,000,000
10	Rita (SW LA, N TX)	2005	3	18,500,000,000

Source: NHC

BOX 31.2   **Superstorm Sandy 2012**

Most major hurricanes that affect the Caribbean and continental United States occur between mid-August and mid-October each year. Late October 2012 saw the formation of a hurricane that was unusual due its late formation and its severity. Hurricane Sandy, the eighteenth named storm of 2012.

Within a six-hour period on October 19, a small tropical disturbance in the Caribbean developed into a tropical storm. Five days later it became a hurricane, the tenth of that year. Between October 24 and 27, Sandy's strength oscillated between being a Category 2 storm, a tropical storm, and back to a Category 1 hurricane as it moved northward. Its first encounter with the east coast of the United States was when it came ashore near Atlantic City, New Jersey, on the evening of October 29. Unfortunately, this landfall coincided with high tides associated with a full moon, driving the storm surge approximately 20 percent higher than normal. Extreme damage resulted from flooding and widespread power outages.

New York City was subjected to extensive flooding that filled tunnels and the subway system. Battery Park at the southern end of Manhattan experienced a surge of 4.2 meters. A major loss of electricity darkened Lower Manhattan and surrounding areas. Within a few days more than eight million people in the area were without electric power. The effects were far-reaching, as gas stations and other businesses closed, airlines canceled almost 15,000 flights, and normal operations throughout the immediate region ceased. In the end, the storm cost almost $20 billion in damaged property and an estimated $10 to $30 billion in lost business revenue and wages.

FEMA, Jocelyn Augustino.

Box Figure 31.2.1   **Damaged boardwalk at Rockaway, New York following Hurricane Sandy.**

As of mid-October 2018, the Atlantic hurricane season had experienced fifteen tropical depressions, including two major hurricanes (Category 3 or higher): Florence and Michael. As reported by the National Hurricane Center, Hurricane Michael was the third most intense hurricane to strike the mainland United States, behind the Labor Day hurricane of 1935 and Hurricane Camille in 1969.

As Hurricane Michael traveled across the warm waters off the west coast of Florida, it rapidly intensified into a Category 4 hurricane. Its extreme winds and storm surge devastated several cities in the Florida panhandle and then weakened as it moved northeastward toward the mid-Atlantic states. More than fifty people were killed. Estimated insurance losses totaled more than $8 billion.

Source: US Army Corps of Engineers.

Box Figure 31.3.1   Mexico City, Florida, experienced extreme damage due to the winds and storm surge of Hurricane Michael in October 2018.

The naming of North Atlantic and Caribbean storms began in 1953, and currently the World Meteorological Organization prepares lists of names to be used for potential storms each year (see **Box 31.4**). To date, 85 names have been retired, as explained in **Box 31.5**, as these names still evoke unfortunate memories for people affected by those storms' destructive power. Note that more than 40 percent of the retired names are for hurricanes that have occurred since 2000.

In the past century, hundreds of thousands of people have died worldwide in cyclones, typhoons, and hurricanes. The densely populated regions of South Asia and the Far East have been subjected to catastrophic storms that hit coastal areas where many people live because of their need to be near the sea for their livelihoods and the availability of easy, cheap transportation (Table 31.5). Unfortunately, it is the very sea these people rely upon that takes its toll and often wipes out entire communities and villages in the course of a few hours.

Topic 32

The Naming of Hurricanes
by K.L. Chandler

The naming of hurricanes is a recent phenomenon that has only become standard since the mid-20th century. Prior to this standardization, hurricanes (or tropical cyclones) were typically named for either when they made landfall (Saint's Days or holidays) or where they made landfall. Any preemptive radio coverage of these historic storms included descriptions that stated the latitude and longitude, but this was difficult due to the shifting and erratic movements, and proved confusing, especially during busy hurricane seasons when multiple cyclones were in the Atlantic Ocean concurrently.

In 1950 hurricanes were named for the year in which they occurred and alphabetically using the 1941 US Joint Army Navy Phonetic Alphabet (Table 32.1). As an example, the first hurricane in the 1950 season was named 1950-ABLE.

Table 32.1	1941 US Joint Army-Navy Phonetic Alphabet					
A – ABLE	E – EASY	I – ITEM	M – MIKE	Q – QUEEN	U – UNCLE	Y – YOKE
B – BAKER	F – FOX	J – JIG	N – NAN	R - ROGER	V – VICTOR	Z - ZEBRA
C – CHARLIE	G – GEORGE	K – KING	O – OBOE	S – SUGAR	W – WILLIAM	
D - DOG	H – HOW	L - LOVE	P - PETER	T – TARE	X – X-RAY	

Source: NOAA.

In 1953 the process changed to the use of short easy to remember names with a rotating list of women's names approved for the naming of the unpredictable and destructive weather systems. (The use of names hailed back to meteorologist Clement Wragge who in the late 1800s named Australian tropical storms after politicians he disliked.) In the 1970s the naming standardization was seen as defaming to women and upon petition the National Hurricane Center agreed to include men's names on the rotating list.

Today, the tropical cyclone lists are managed by the World Metrological Organization (WMO). The lists rotate and repeat every six years, but the names of especially destructive hurricanes can be retired from a list at a country's request. (Table 32.2) There are different oceanic lists based on hemispheric location, all maintained by the WMO. The most common lists for the United States use are the North Atlantic Ocean list with 21 names (Table 32.3), followed by the East North Pacific Ocean list with 24 names (Table 32.4).

Table 32.2
Retired Tropical Cyclone Names for the Atlantic Ocean and Gulf of Mexico

		1954 Carol Hazel Edna	**1955** Connie Diane Ione Janet	**1956**	**1957** Audrey	**1958**	**1959**	**1960** Donna .	**1961** Carla Hattie
1962	**1963** Flora	**1964** Cleo Dora Hilda	**1965** Betsy	**1966** Inez	**1967** Beulah	**1968**	**1969** Camille	**1970** Celia	**1971**
1972 Agnes	**1973**	**1974** Carmen Fifi	**1975** Eloise	**1976**	**1977** Anita	**1978** Greta	**1979** David Frederic	**1980** Allen	**1981**
1982	**1983** Alicia	**1984**	**1985** Elena Gloria	**1986**	**1987**	**1988** Gilbert Joan	**1989** Hugo	**1990** Diana Klaus	**1991** Bob
1992 Andrew	**1993**	**1994**	**1995** Luis Marilyn Opal Roxanne	**1996** Cesar Fran Hortense	**1997**	**1998** Georges Mitch	**1999** Floyd Lenny	**2000** Keith	**2001** Allison Iris Michelle
2002 Isidore Lili	**2003** Fabian Isabel Juan	**2004** Charley Frances Ivan Jeanne	**2005** Dennis Katrina Rita Stan Wilma	**2006**	**2007** Dean Felix Noel	**2008** Gustav Ike Paloma	**2009**	**2010** Igor Tomas	**2011** Irene
2012 Sandy	**2013** Ingrid	**2014**	**2015** Erika Joaquin	**2016** Matthew Otto	**2017** Harvey Irma Maria Nate	**2018** Florence Michael	**2019** Dorian	**2020** Laura Eta Iota	**2021** Ida

Source: NOAA.

Table 32.3	Tropical Cyclone Names for the North Atlantic Ocean and Gulf of Mexico				
2023	**2024**	**2025**	**2026**	**2027**	**2028**
Arlene	Alberto	Andrea	Arthur	Ana	Alex
Bret	Beryl	Barry	Bertha	Bill	Bonnie
Cindy	Chris	Chantal	Cristobal	Claudette	Colin
Don	Debby	Dexter	Dolly	Danny	Danielle
Emily	Ernesto	Erin	Edouard	Elsa	Earl
Franklin	Francine	Fernand	Fay	Fred	Farrah
Gert	Gordon	Gabrielle	Gonzalo	Grace	Gaston
Harold	Helene	Humberto	Hanna	Henri	Hermine
Idalia	Isaac	Imelda	Isaias	Imani	Idris
Jose	Joyce	Jerry	Josephine	Julian	Julia
Katia	Kirk	Karen	Kyle	Kate	Karl
Lee	Leslie	Lorenzo	Leah	Larry	Lisa
Margot	Milton	Melissa	Marco	Mindy	Martin
Nigel	Nadine	Nestor	Nana	Nicholas	Nicole
Ophelia	Oscar	Olga	Omar	Odette	Owen
Philippe	Patty	Pablo	Paulette	Peter	Paula
Rina	Rafael	Rebekah	Rene	Rose	Richard
Sean	Sara	Sebastien	Sally	Sam	Shary
Tammy	Tony	Tanya	Teddy	Teresa	Tobias
Vince	Valerie	Van	Vicky	Victor	Virginie
Whitney	William	Wendy	Wilfred	Wanda	Walter

Source: NOAA.

Table 32.4	Tropical Cyclone Names for the East North Pacific Ocean				
2023	**2024**	**2025**	**2026**	**2027**	**2028**
Adrian	Aletta	Alvin	Amanda	Andres	Agatha
Beatriz	Bud	Barbara	Boris	Blanca	Blas
Calvin	Carlotta	Cosme	Cristina	Carlos	Celia
Dora	Daniel	Dalila	Douglas	Dolores	Darby
Eugene	Emilia	Erick	Elida	Enrique	Estelle
Fernanda	Fabio	Flossie	Fausto	Felicia	Frank
Greg	Gilma	Gil	Genevieve	Guillermo	Georgette
Hilary	Hector	Henriette	Hernan	Hilda	Howard
Irwin	Ileana	Ivo	Iselle	Ignacio	Ivette
Jova	John	Juliette	Julio	Jimena	Javier
Kenneth	Kristy	Kiko	Karina	Kevin	Kay
Lidia	Lane	Lorena	Lowell	Linda	Lester
Max	Miriam	Mario	Marie	Marty	Madeline
Norma	Norman	Narda	Norbert	Nora	Newton
Otis	Olivia	Octave	Odalys	Olaf	Orlene
Pilar	Paul	Priscilla	Polo	Pamela	Paine
Ramon	Rosa	Raymond	Rachel	Rick	Roslyn
Selma	Sergio	Sonia	Simon	Sandra	Seymour
Todd	Tara	Tico	Trudy	Terry	Tina
Veronica	Vicente	Velma	Vance	Vivian	Virgil
Wiley	Willa	Wallis	Winnie	Waldo	Winifred
Xina	Xavier	Xina	Xavier	Xina	Xavier
York	Yolanda	York	Yolanda	York	Yolanda
Zelda	Zeke	Zelda	Zeke	Zelda	Zeke

Source: NOAA.

Climate and Climate Change

By Dean Fairbanks

Climate is weather over time; climate is not weather. Climate is what we should expect in a location any season of the year and weather is what we get. Climate is statistical averaging of weather to identify characteristics of locations and to study trends in weather variables.

Climate Measurement and Classification: World Climatic Regions

Climate scientists have deduced that a minimum of 30 years of weather station data is required to truly characterize an area's climate. From weather station data and statistical analysis (in space and time) climatic regions are developed which define areas with similar weather statistics. The factors influencing the world climatic regions include insolation, pressure, air masses, heat exchange from ocean currents (gyres), distribution of mountain barriers, pattern of prevailing winds, distribution of land and sea, and altitude. All of these have been covered to this point, and while there are further details not listed they are enough to drive the two main climatic components that need to be measured at every weather station, these two parameters are the minimum required: temperature and precipitation. Using these measurements, classification of climatic regions can be processed by grouping the data into related categories. These types of classifications are empirical classifications since they are based on statistical data analysis of the long-term measurements of average (and the ranges) temperature and precipitation. The most famous and widely used is the **Köppen-Geiger climatic classification**. It was developed by German botanist-climatologist Wladimir Köppen in the late 19th century and later revised in 1940 with German climatologist Rudolf Geiger, and finally a last revision in 2007. It is based on using dominant vegetation to best define the climate as the surrogate along with temperature and seasonality of rainfall. Other empirical systems also include evaporation as a parameter for refinement.

Another type of climate classification that has been developed is called the **genetic classification** that is based on air masses. This modern approach to climate classification are designed to make up for the shortcomings of the empirically classification schemes that produce distinct boundaries between climate zones rather than represent their true gradual transition nature. In the genetic classification system, the definition of air masses leads to a reduced number of categories: Dry Polar (cP), Dry Moderate (Hadley cell descending dry and warm compressing air), Dry Tropical (cT), Moist Polar (mP), Moist Moderate (a hybrid between mP and mT), and Moist Tropical (mT, mE and maritime monsoon).

Climographs: Reading the Climate

Climatic graphs or climographs describe the seasonal pattern of rainfall and temperature (**Figure 33.1**). These graphs can be based from the 30 year averages and ranges of precipitation and temperature or from a year's worth of data. The diagram created for the 12 months typically show the mean monthly average of temperature as a point-line graph and total monthly precipitation as a bar chart all on the same diagram. There are two Y-axis scales to depict the temperature scale on the left-hand side and the rainfall scale on the right-hand side. Extensions to these diagrams include the mean monthly minimum and maximum temperature. These are the average of all the minimum or maximum temperatures for each day of the month. Sometimes an added point-line graph of potential evaporation is added for refinement.

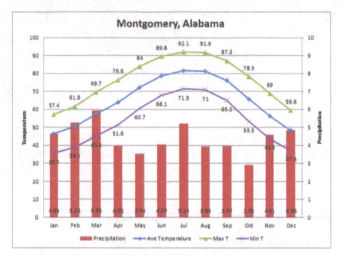

Figure 33.1 Here is an example of a climograph for Montgomery, Alabama. Scales are in Fahrenheit and inches. (**NOAA**)

When reading a climograph the following should be looked for to understand the climate of a location:

- Total rainfall
- Rainfall seasonality—when does it most and least occur
- Maximum and minimum temperature
- Range of temperature
- Length of time below freezing (0°C)
- If potential evaporation is provided then look at the relationship with rainfall (precipitation > potential evaporation or precipitation < potential evaporation)

El Niño and La Niña

El Nino—the "Christ Child"—is a warming of the eastern Pacific that occurs at intervals of 3 to 7 years and may last for many months up to one year. The early Spanish explorers and colonists on the West coast of South America near Peru and Ecuador noticed changes in the climate and warming water temperatures during the Christmas holidays and named the phenomenon. This warm water current is part of a much larger system. It has the ability to disrupt the world's climates and cause economic hardship. In the last 50 years the 1997–1998 El Nino was the worst ever recorded.

During normal conditions in the tropical Pacific Ocean, trade winds generally drive the surface waters westward. Walker circulation is present as an east-west circulation that occurs in low latitudes. Near South America winds blow offshore, causing upwelling of the cold, nutrient rich waters. By contrast, warm surface water is pushed into the western Pacific by the Trade winds. Normally, sea surface temperatures in the western Pacific are over 28°C, causing an area of low pressure and producing high rainfall.

By contrast, over coastal western South America sea surface temperatures are lower, high pressure exists and conditions are dry.

During **El Nino** episodes, the pattern is reverse. As warm water from the western Pacific flows into the eastern Pacific, water temperatures rise. This is due to the weakening of the eastern Trade winds, which normally blow west across the equatorial Pacific. As they nearly cease to blow, the warm waters piled up in the western Pacific (up to one-half meter difference in sea level height) rush back to the east, hitting equatorial South America and deflecting north and south of the equator. During El Nino events, sea surface temperatures of over 28°C extend much further across to the eastern Pacific. Low pressure develops over the eastern Pacific and high pressure over the western Pacific. Consequently, heavy rainfall occurs over coastal South America as far south as southern Chile and as far north as California, whereas the western Pacific island nationals experience warm, dry conditions. Since the height of an El Nino occurs during the winter season in the northern hemisphere, US West coast is already receiving its cooler mid-latitude storms generated in the Gulf of Alaska's low pressure region via the polar jet stream, but now warmer sub-tropical storms are also flowing northward generated by the warm waters off the coast (flowing against the north Pacific gyre generated cold California current). Other global effects of a strong El Nino condition include heavy rain in the south of the United States; drought in Australia, Indonesia, the Philippines, southern Africa, and north-east Brazil; increased risk of malaria in South America; lower rainfall in northern Europe; and higher rainfall in southern Europe (**Figure 33.2**).

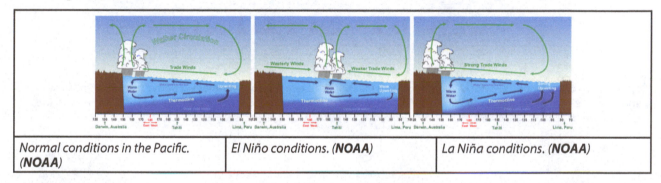

Normal conditions in the Pacific. (**NOAA**)	El Niño conditions. (**NOAA**)	La Niña conditions. (**NOAA**)

Figure 33.2   El Niño and La Niña offer two unique warm and cool climate patterns that influence the world.

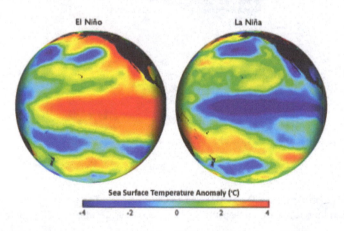

Figure 33.3   Differences in Pacific ocean temperatures during El Nino and La Nina. (**NOAA**)

The exact opposite of an El Nino condition is the **La Nina** condition, which is a resumption of the eastern Trade winds pushing the surface flow of warm waters back to the western Pacific. However, it is much stronger than during normal conditions. This leads to a very strong upwelling of cold waters off South America (cooler than usual ocean temperatures occur on the equator between South America and the International Date Line, 180° longitude) and a much larger high pressure air mass region in the sub-tropical latitudes of the eastern Pacific. Consequently, a La Nina event leads to a shift of the polar jet stream farther north, bringing warmer and drier winters to the western United States and therefore increased drought conditions, but leaving extra-wet conditions in the Pacific Northwest where the polar jet is largely contained (**Figure 33.3**).

The El Nino-La Nina cycle is known by scientists as the Southern Oscillation, which scientists have noticed that global temperature has a strong interannual variability tied to the Southern Oscillation. They have shown that the correlation of 12-month running mean global temperature and Niño 3.4 index is maximum with global temperature lagging the Niño index by 4 months. Thus the 1997–1998 very strong El Nino had a timing that maximized 1998 global temperature. In contrast, the 2011 global temperature was dragged down by a strong La Niña. The Southern Oscillation is shown to be strongly tied to global warming—climate change (**Figure 33.4-33.5**).

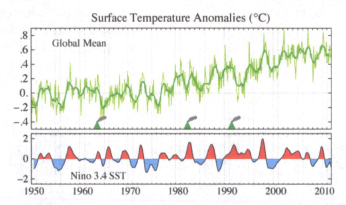

Figure 33.4 Global monthly and 12-month running mean surface temperature anomalies relative to 1951–1980 base period, and 12-month running mean of the Niño 3.4 sea surface temperature (SST) index. Red peaks are El Nino years and blue dips are La Nina years. (**NASA**)

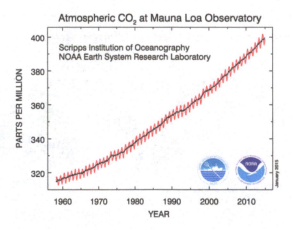

Figure 33.5 Measurement of atmospheric CO2 at Mauna Loa Observatory. The peaks in the seasonal line are in winter and spring, while the troughs are in late summer corresponding to seasonal vegetation cycling. (**NOAA**)

Enhanced Greenhouse Effect: Climate Change

A synthesis must cover the role of the greenhouse effect and climate change in general and global warming specifically. The greenhouse effect is the process by the gases of carbon dioxide (CO_2), methane (CH_4), nitrous oxides (NO_2), and water vapor absorb outgoing long wave thermal radiation from the Earth, and return some of it back to Earth. Greenhouse gases vary in their abundance and contribution to global warming. While biotic life on Earth needs the greenhouse effect and also affects two of the main gases (CO_2 and

CH4) connected with the greenhouse effect, there are consequences to an overabundance of these gases over time. This has led to an enhanced greenhouse effect called global warming.

The concern about global warming is the build-up of greenhouse gases from human sources. CO2 levels have risen globally from about 315 ppm (parts per million) in 1950 to 399 ppm in 2014 and are expected to reach 530 ppm by 2050. The increase is due to human activities (burning fossil fuels and deforestation). At the beginning of the industrial revolution in 1750, CO2 was at 280 ppm. Up until that point in the last 1.3 million years the CO2 ppm ranged between 175 and 280 ppm according to long-term ice core analysis from both Greenland and Antarctica. The observed warming has been greatest at the higher latitudes, and in the Northern hemisphere. Over the last 40 years average winter temperatures have increased by 6°C in central Siberia for example. The Earth's average temperature has increased by more than 0.62°C in the last century (**Figure 33.6-33.7**).

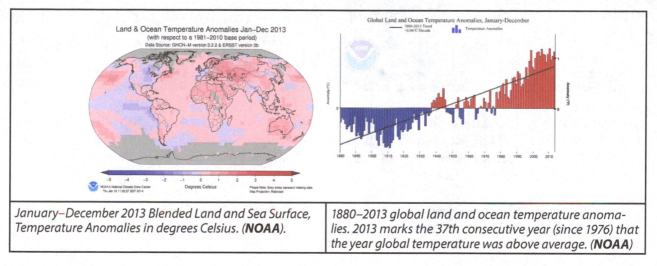

| January–December 2013 Blended Land and Sea Surface, Temperature Anomalies in degrees Celsius. (**NOAA**). | 1880–2013 global land and ocean temperature anomalies. 2013 marks the 37th consecutive year (since 1976) that the year global temperature was above average. (**NOAA**) |

Figure 33.6-33.7

Is Global Warming Occurring?
By Ingrid Ukstins and David Best

For centuries people have recorded temperatures in cities and other locations throughout the world. Annual data for the past 125 years show that average global temperatures have been increasing significantly (**Figure 34.1**), with an increase of more than 1°C in the past century. When viewed over the past 15,000 years, temperatures have risen more than 4°C.

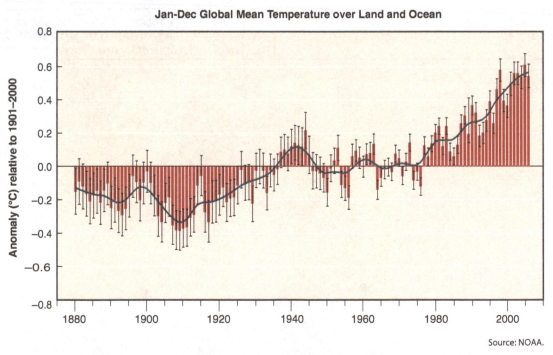

Jan-Dec Global Mean Temperature over Land and Ocean

Source: NOAA.

Figure 34.1 Annual global temperatures have been increasing over the past 125 years, although most of the increase has occurred since 1980.

Effects of Global Warming

Increased warming causes sea level to rise as glaciers and ice sheets melt. Because water has a high capacity to retain heat, the oceans absorb more heat, causing the molecules to expand, thereby raising sea level further. Large bodies of water are slow to heat up, but they are also slow to cool down.

Warmer ocean waters have increased thermal energy, which leads to more tropical storms. Higher ocean temperatures allow cyclonic storms to intensify as they pass over these heat sources. Such was the case with Hurricane Katrina as it moved into the Gulf of Mexico before striking the Louisiana-Mississippi coastline in August 2005 (**Figure 34.2**). Temperatures in the Gulf of Mexico were well above average, thus providing more thermal energy to intensify the hurricane.

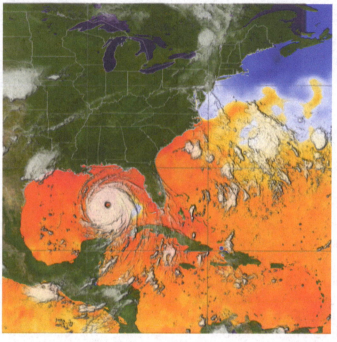

Source: NASA.

Figure 34.2 Hurricane Katrina is situated in the Gulf of Mexico several days before striking the Gulf Coast region of the southern United States. Orange to red colors in the Gulf of Mexico reflect the warmer seawater temperatures providing energy to the storm.

Continental areas are affected by global warming as changes in the hydrologic cycle alter the distribution of precipitation, which can affect vegetation patterns. Droughts can destroy once productive agricultural areas, causing famine and the possible need to relocate people.

Increased temperatures in high latitudes melt permafrost, surface material that is normally permanently frozen below a certain depth. Because these ecosystems contain large amounts of organic material, decay processes are accelerated, which release trapped methane into the atmosphere. This methane then becomes part of the greenhouse gases. This is an example of a positive feedback cycle—warming causes methane release that causes more warming.

Long-Term Climate Changes

Variations in climate have occurred since the formation of the atmosphere several billion years ago. In addition to the role the oceans play in these variations, continents also contribute in less obvious ways. The positions of the continents have changed drastically through geologic time. The location of these large land masses has a direct effect on large-scale circulation patterns in the oceans. Today, many ocean currents flow near the edges of continents, moving massive amounts of water and heat across Earth's surface. Near the end of the Permian Period about 250 million years ago, one large land mass existed (called Pangaea). As it began to break up, the earliest shapes of the present-day continents began to form. This break-up altered oceanic circulation. Today's configuration of the continents allows water to move readily between the poles and the open oceans (**Figure 34.3**).

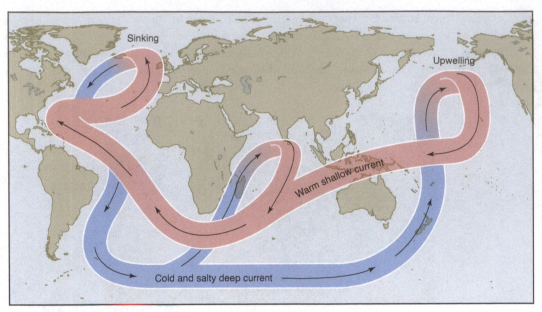

Sinking

Upwelling

Warm shallow current

Cold and salty deep current

© Kendall Hunt Publishing Company.

Figure 34.3   Movement of waters in the ocean can be thought of as a conveyor belt transporting deep, cold water to the surface where the warm, shallow water moves near the surface.

The rearrangement of the continents also puts the land masses in different locations. Positioned near the equator, an area would experience warmer temperatures and more precipitation than a land mass situated at a high latitude. Think of portions of central Africa and southern and central Alaska today. The warm equatorial regions of Africa have a wide range of life forms, while the harsh, colder climate of Alaska limits the varieties of plants and animals that can survive there (**Figure 34.4**).

National Park Service.

Figure 34.4   Terminus of Hubbard Glacier, Alaska.

Short-Term Climate Changes

The cause of short-term changes in Earth's climate is generally controlled by the amount of incoming solar radiation. Both atmospheric and astronomical factors contribute to these changes. Atmospheric factors include the amount of greenhouse gases and the amount of dust and other aerosols in the air. Researchers have examined ice cores from the polar regions and found that periods of increased glaciation correspond to increased particulates in the ice core. The additional atmospheric dust reflected solar energy back into space, thereby producing a drop in near-surface temperatures. As discussed previously, short-lived decreases in global temperatures can also result from large volcanic eruptions throwing aerosols into the air.

Astronomical influences are mainly related to Earth's position in space relative to the Sun. Variations in Earth's orbit around the Sun produce changes in the distance between the two bodies. The Earth is approximately 150 million kilometers away from the Sun, which is defined as one Astronomical Unit. The Earth's orbit is not perfectly circular, so this distance can vary from about 147.5 million kilometers when it is closes to the Sun—called perhelion—to 152.6 million kilometers when it is furthest away—called aphelion. Eccentricity is a measure of how circular versus elliptical the Earth's orbital path is. As Earth revolves around the Sun, its path changes due to the gravitation attraction of the Sun, Moon, and other more distance objects. When the orbit become more circular, the distance between the Earth and the Sun is less than when the path is more elliptical (**Figure 34.5a**). These variations occur over periods of about 100,000 years and the ellipticity changes about 5%.

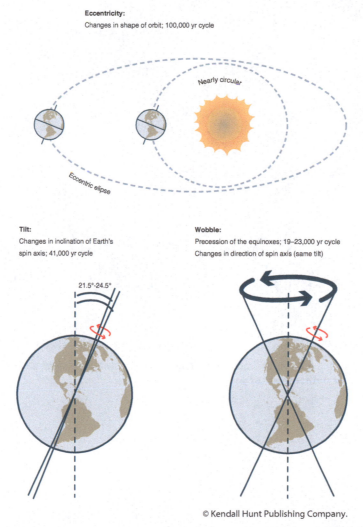

© Kendall Hunt Publishing Company.

Figure 34.5a   The eccentricity, tilt, and wobble of the Earth as it moves around the sun contribute to long-term changes in the climate.

As the Earth rotates on its axis, there are times when the angle of tilt of the axis ranges from 22.1 degrees to 24.5 degrees. Currently Earth is tilted 23.5° from the vertical. The period of change is roughly 41,000 years. The planet's tilt reached a maximum in 8,700 BCE, we are currently about halfway between the maximum and minimum, and will reach a minimum in 11,800 CE. Increased tilt means that each hemisphere's summer will get more incoming solar radiation as the pole is tipped towards the sun, and less solar radiation in winter as the pole is tipped away from the sun. Our current trend of decreasing tilt and the changes in solar energy reaching the poles promotes warmer winters and cooler summers. As Earth rotates, it also tends to wobble on its axis, in the same way a spinning top begins to wobble as its rotation rate decreases. This wobble, also termed **axial precession**, has a period of about 26,000 years. It is caused by the tidal forces from the Sun and the Moon on the solid Earth. The net effect of all these astronomic factors occurring together is that their period, together with their maximum influences, corresponds to times when active glaciations happened, due to major changes in the amount of solar radiation striking Earth over time (**Figure 34.5b**).

axial precession: The wobble that occurs when a spinning object slows down.

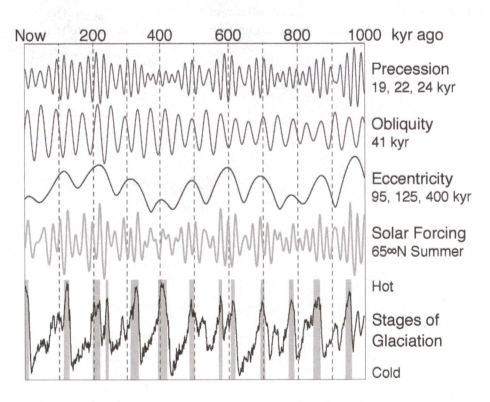

Source: NOAA National Centers for Environmental Information.

Figure 34.5b   Milankovitch cycles show how changes in the Earth's orbit, its tilt and distance from the Sun affect the climate on the scale of thousands to tens of thousands of years.

Hydrological Cycle and Water Resources
By Dean Fairbanks

Earth is a closed material system; the only additive component is energy from the sun. The oxygen and hydrogen atoms that make up the molecule H_2O have been on the planet since it formed billions of years ago. The water cycle moves molecules of H_2O around the planet as liquid water, frozen water, and water vapor. Water is both a solvent and a transporter. The services of water support all life. The flow of water shapes the landscape through erosion and in doing so, supports the formation of soils. The services of water help to regulate climate. Water provides humans with cultural services such as beautiful views, a sense of place, or a religious foundation (**Figure 35.1**).

Scientists estimate that the total volume of water on Earth distributed throughout ocean, atmosphere, and terrestrial stocks is 1.4 billion cubic kilometers. The particular volume in any of these stocks varies over the long-term with climate. Significant volumes of water are tied up in glaciers and ice caps during glacial periods and then move back the ocean during interglacial periods. Higher temperatures also increase *evaporation* and *transpiration*. Evaporation is the flow of water vapor from liquid sources in the ocean, lakes, rivers, and soils to the atmosphere. Transpiration is the flow of water vapor from plants to the atmosphere. Evaporation and transpiration are typically measured as a combined rate called *evapotranspiration* (ET) (**Figures 35.2–35.3**).

Energy from the sun drives the hydrological (water) cycle. The **hydrological cycle** is a model of the flow of water from place to place. Water flows through the atmosphere and across the land, where it is also stored as ice, standing water, and within soils and deeper as groundwater. Thus, the hydrological cycle is the total movement, exchange, and storage in vapor, liquid, and solid states of water. The components of the hydrological cycle include transport, condensation, precipitation, ET, interception, infiltration, percolation, vadose zone, groundwater (aquifer), and surface runoff. Each of these components will be explained in turn.

Illustrates the flow of water out of the stock in the ocean (*evaporation*) into the atmospheric stock. It then flows into terrestrial stocks that include snow and glaciers, lakes, rivers, aquifers, soils, plants, and animals (*condensation, precipitation, infiltration, photosynthesis, and consumption*). From there water flows either back to the atmosphere (*ET and sublimation*) or to the ocean (*runoff and discharge*).

One or more of the following determines the rates that water **transports** between these stocks: energy from the sun and temperature, photosynthesis and respiration, topography and gravity, composition of soils and aquifers, and human demands. From the point of view of a water balance model, evaporation and precipitation are in net balance with each other. Evaporation from the oceans and ET from the land accounts for 86% and 14%, respectively. This water vapor condenses in the atmosphere and moves advectively (horizontally) until such point that precipitation is initiated. Precipitation occurs 78% over the oceans and 22% over the land surface.

Figure 35.1 Standard US National Weather Service weather station focusing on measuring temperature, precipitation and potential ET. (**NOAA**)

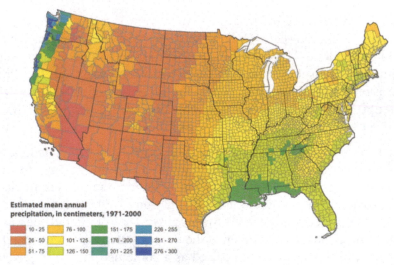

Estimated mean annual
precipitation, in centimeters, 1971-2000

10 - 25	76 - 100	151 - 175	226 - 255
26 - 50	101 - 125	176 - 200	251 - 270
51 - 75	126 - 150	201 - 225	276 - 300

Figure 35.2 Estimated mean annual precipitation for the conterminous US for the period 1971-2000. (**USGS**)

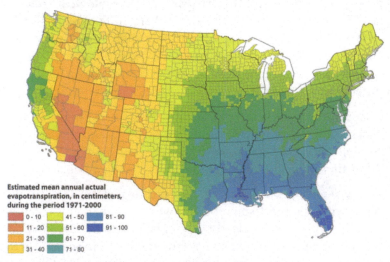

Estimated mean annual actual
evapotranspiration, in centimeters,
during the period 1971-2000

0 - 10	41 - 50	81 - 90
11 - 20	51 - 60	91 - 100
21 - 30	61 - 70	
31 - 40	71 - 80	

Figure 35.3 Estimated mean annual actual evapotranspiration (ET) for the conterminous US for the Period 1971-2000. (**USGS**)

If the precipitation is greater than the infiltration rate then overland flow can be initiated as *runoff* into streams, lakes, wetlands, marshes, or ocean for the cycle to begin again. The length of time that any water molecule resides in ocean, atmospheric or terrestrial stocks can vary from seconds to millions of years in the deep ocean, glacial ice, or aquifers (**Figure 35.4**).

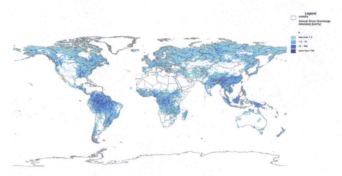

Figure 35.4   Global annual river runoff. (**Oak Ridge National Labs**)

Water that infiltrates surface soils and percolate deeper will come in contact with the **vadose zone** then recharge geologic formations called **aquifers**. The vadose zone is a zone above the ground water table (saturated rock) and the upper soil-moisture zone that is unsaturated, and is mainly open pore spaces that water primarily moves downward through (**Figure 35.5**). Water beneath the vadose zone that is beyond the soil-root zone; fills the pores in the rock is known as an aquifer—a permeable layer of rock that conducts (allows flow of) groundwater. Depending upon the geologic formation in which an aquifer is found, the composition can range from sand and gravel to various types of rock including sandstone, shale, and basalt (**Figure 35.6**). The *water table* is considered the surface of the aquifer that is saturated with water. The flow of water into an aquifer is called **recharge**. Rates of recharge vary from slow (100–1000 years) to fast (days) dependent on the system (**Figure 35.7**).

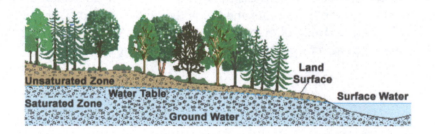

Figure 35.5   The vadose zone. (**USGS**)

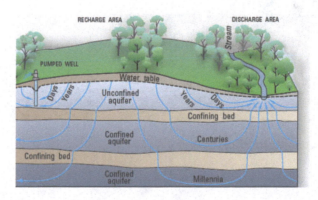

Figure 35.6   Aquifer model. (**USGS**)

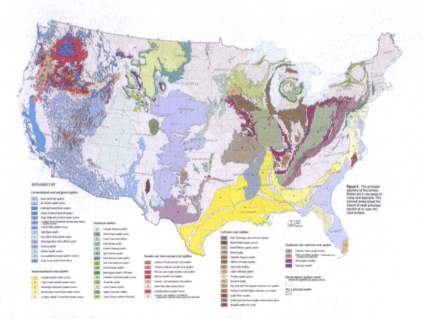

Figure 35.7 Major aquifers in the US. **(USGS)**

Aquifers are described as either *confined or unconfined*. In unconfined systems groundwater and surface water interact. Imagine a streambed that is made up of large rocks, water can easily flow into and out of the streambed. Some streams in certain areas leak to the underlying aquifer (influent conditions) and in others receive water from an aquifer (effluent conditions). A confined aquifer is one that is isolated by an impermeable layer (**aquiclude**) of material such as clay, hardpan, or layer of basalt laid down by lava flows. Water may have originally flowed into the aquifer from an edge or the surface but at some point weathering or geologic changes effectively sealed it from additional recharge (or reduced the rate of recharge to the point the aquifer is considered non-renewable on human time scales). Because unconfined aquifers interact with surface water, there is the potential for the inflow of pollutants and the potential for "flushing" or outflow of nutrients or pollutants. If natural or human-derived pollutants contaminate a confined through a well or other disturbance of the confining layer the aquifer it will remain contaminated (**Figures 35.8–35.9**).

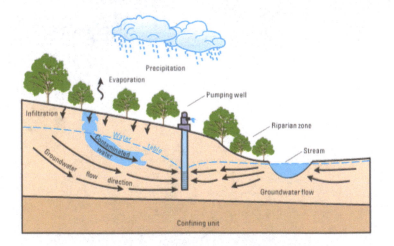

Figure 35.8 Groundwater pollution potential. **(USGS)**

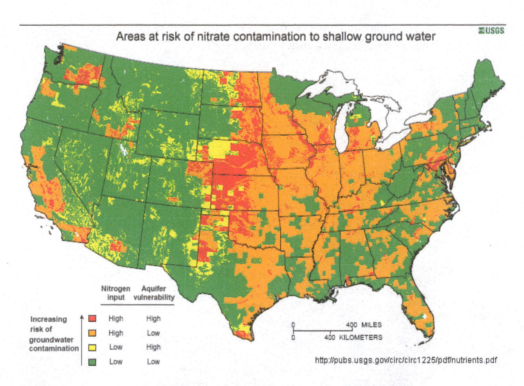

Areas at risk of nitrate contamination to shallow ground water

Increasing risk of groundwater contamination	Nitrogen input	Aquifer vulnerability
(red)	High	High
(orange)	High	Low
(yellow)	Low	High
(green)	Low	Low

400 MILES
400 KILOMETERS

http://pubs.usgs.gov/circ/circ1225/pdf/nutrients.pdf

Figure 35.9 US vulnerability to groundwater pollution from nitrates from farming. (USGS)

The water cycle is impacted by human demands on freshwater resources resulting from the increase in population and accompanying demand for food, energy resources, sanitation, and an equitable quality of life. Furthermore, human utilization of water and other resources including fossil fuels has a direct impact on water quality. Climate scientists, hydrologists, chemists, biologists, ecologists, oceanographers, and others are working to better understand the feedbacks between climate change, human utilization of water, and ecosystem function.

Groundwater is utilized throughout the world to supply agriculture, municipalities, and industry. Some of this water is consumed, some infiltrates back into the aquifer. Many regions of the world are utilizing ground water at a rate that is higher than recharge (**Figure 35.12**). This may result in ground subsidence (structural collapse of the aquifer), reduced water in streams and lakes, wells drying up, and saltwater intrusion (**Figures 35.10–35.11**).

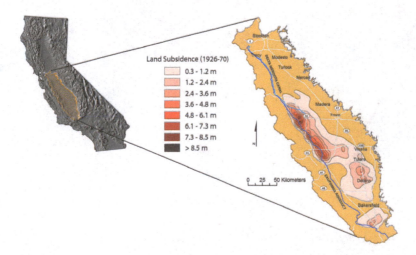

Land Subsidence (1926-70)	
	0.3 - 1.2 m
	1.2 - 2.4 m
	2.4 - 3.6 m
	3.6 - 4.8 m
	4.8 - 6.1 m
	6.1 - 7.3 m
	7.3 - 8.5 m
	> 8.5 m

0   25   50 Kilometers

Figure 35.10 Land subsidence is a common problem in dry climates with high populations. (USGS)

Problems with over pumping wells can lead to subsidence of the ground as the pore spaces collapse. This has been happening in California's San Joaquin Valley since the 1920s.

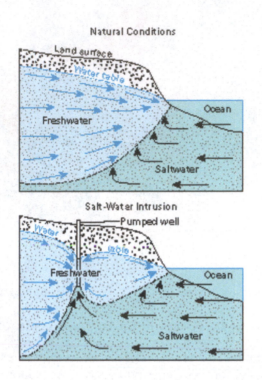

Figure 35.11 Salt water intrusion into groundwater from over pumping. (**USGS**)

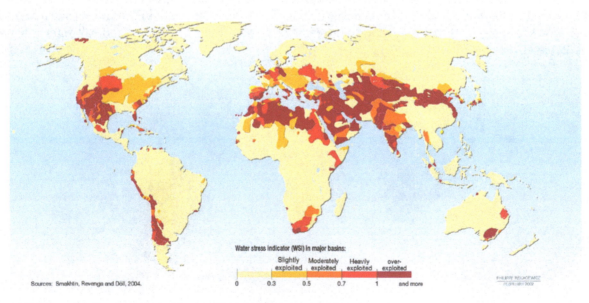

Figure 35.12 Global water scarcity. (**United Nations Environmental Program**)

Changes to the water cycle impact all regulating and supporting services, many of which feed directly back into the water cycle increasing the dynamics of change. Changes in precipitation directly impact all aspects of biological life and human economic systems. Changes to the water cycle directly impact local water quality and the health of the humans and other organisms utilizing that water. Lower flows may mean less dilution of pollutants. The world is increasingly challenged with water scarcity as more surface water and groundwater are used for agricultural, industrial, and human consumption. On top of this use there is the very large concern of the freshwater that is available becoming polluted. The pollution can either lead to complete non-use of the water resource or a very expensive clean- up process. Fresh water is precious, despite the Earth being covered 72% in water. The hydrological cycle is intimately tied to the atmosphere, and as will be discussed next the development of weather systems and long-term climate change.

Hydrologic Cycle and Water Crisis
by K.L. Chandler

The **hydrosphere** is a closed system comprised of all water on Earth, whether liquid, solid, or gas. It includes the water in the biosphere, on or beneath the Earth's surface, and also includes the water in the atmosphere. Water is the most common substance on the planet and covers 71% of Earth's surface. Without water, Earth could not sustain life of any kind.

Hydrologic Cycle

Water moves continuously within the Earth and its atmosphere in a closed system known as the hydrologic or water cycle. The hydrologic cycle is a biogeochemical process in which water is transferred and stored across the Earth. (**Figure 36.1**)

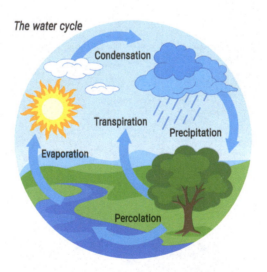

© 3xy/Shutterstock.com

Figure 36.1   The hydrologic cycle, also known as the water cycle, is a closed system of fresh water.

Collected water warmed by radiation from the sun's solar rays changes from a liquid to a gas, and evaporates and rises into the atmosphere. Water as a gas, called water vapor, cools as it rises, condensing into microdroplets in the air and creating clouds and fog where it can be stored and moved.

The amount of condensed moisture the air can hold is temperature dependent and once the air has reached its maximum (saturation vapor pressure), precipitation will fall to the Earth's surface in the form of rain, snow, hail, fog drip, graupel (snow pellets), or sleet. The precipitation that falls is always fresh water.

Once precipitated water falls one of several things can occur. Often plants and trees intercept the falling water - collecting and storing it until it is evaporated back into the atmosphere (transpiration). Precipitation in a liquid state might also be absorbed into the surface of the Earth where it percolates through the layers of the soils and rocks until it is stored for later usage in aquifers as groundwater. Water that is not absorbed by plants, trees, or the ground, called overland runoff, is transferred across the landscape by gravity until it merges and recharges the surface water (streams, rivers, lakes, oceans.) While water falling as snow may accumulate to form glaciers in high elevations or may be stored seasonally until it melts becoming either groundwater or surface water.

It is here that the process begins again with radiation and evaporation.

Water Crisis

One would think a planet that is nearly three-quarters under water would have no problem with scarcity, but the truth is that only about 3% of it is the freshwater humans need to survive, and just 1.2% of that amount is usable. In addition to lakes, rivers, and streams, fresh water can be found in glaciers, ice caps, and the deep recesses of soils – difficult, if not impossible, locations to access. UNICEF estimates that 844 million people around the world lack access to clean water. Sometimes, as in India, and Pakistan, this shortage is a matter of **physical scarcity**: the region does not have enough fresh water supply to meet the demand for agriculture or humans' use. It may have been diverted, the water table could be lowering from drought or seasonal influences, the supply could be overallocated, or water resources could be overdeveloped. When groundwater levels become threatened, aquifers, become fragile, causing possible ground collapse, and sinking buildings.

In other cases, the challenge is **economic scarcity**. Water may be available, but it is inaccessible, polluted, or toxic from a lack of infrastructure, poor resources, mismanagement, or underdevelopment. On average, countries with lower economic indicators are more likely to have economic water scarcity problems. Other issues, such as conflict over water usage within a region or poor practices of sanitation and waste disposal, can compound these situations. (**Figure 36.2**)

© Sk Hasan Ali/Shutterstock.com

Figure 36.2 KHULNA, BANGLADESH – Bangladeshi people carry drinking water after collecting it from a freshwater source. In less developed areas of Africa and Asia, women and girls walk an average of 3.5 miles each day to provide water for their households.

Studies report that over 3.5 million people die from water-related diseases and pollution every year, and 2.2 million of those people are children. Water-borne disease is the number one killer of children under the age of five. The effect of this crisis is not limited to humans; it also jeopardizes the **flora** (plant life) and **fauna** (animal life) of the biosphere.

Water Cycle and Water Resources

By Ingrid Ukstins and David Best

The water cycle, which is also referred to as the hydrologic cycle, shows where water is located and how it moves through the atmosphere, hydrosphere, lithosphere, and biosphere (**Figure 37.1**). The movement of water is related to the three states or conditions in which water can exist—as a liquid, solid, or a gas. More than 75 percent of Earth's surface is covered by water in either a liquid or solid form. Initially, water formed on the surface through its being expelled by volcanic eruptions from water-bearing magmas and also by comets from outer space bringing frozen ice that later melted.

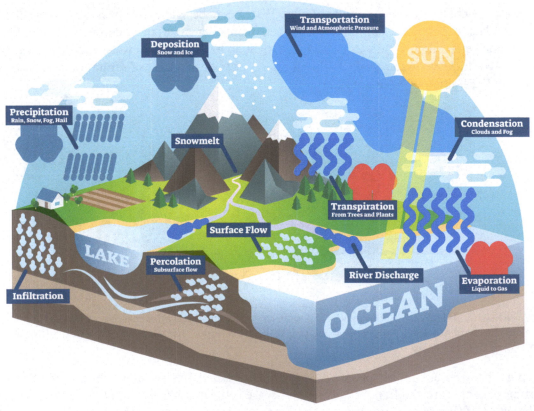

© VectorMine/Shutterstock.com

Figure 37.1   The water cycle (also referred to as the hydrological cycle) shows how water in its various phases moves or rests on Earth.

The United States Environmental Protection Agency (EPA) was formed in 1970 to "protect human health and to safeguard the natural environment—air, water and land—upon which life depends (part of the EPA Mission statement). The EPA sets the standards for the amounts of potentially harmful substances that can affect human life and the natural environment.

Federal legislation has addressed the need for guidelines to establish and maintain water and air standards. The Clean Water Act is a U.S. federal law that regulates the discharge of pollutants into the nation's surface waters, including lakes, rivers, streams, wetlands, and coastal areas. Passed in 1972 and amended in 1977 and 1987, the Clean Water Act was originally known as the Federal Water Pollution Control Act. The Clean Water Act is administered by the EPA, which sets water quality standards, handles enforcement, and helps state and local governments develop their own pollution control plans.

Water Pollution

Water is essential for plant and animal growth. Water must be clean enough for its consumer to use it without adversely affecting the health and well-being of the plant or animal.

Most individuals do not think about where the water they drink comes from or where the water that goes down the drains winds up. If asked, their responses range from—"the city provides my water" to "I don't know, it just flows when I open the faucet." Only when we are without water do we consider where it is and ask if it is drinkable.

"Is it drinkable?" becomes an important question. Because water is the universal solvent, most chemicals will eventually become incorporated. None of us would intentionally drink water with harmful chemicals such as arsenic, cadmium, or lead, in it. Thus we trust that our water suppliers, be it a city or a bottler of water, have some standards in what can and cannot be in the water we drink. The EPA sets water standards for the nation, and the 50 states can set the standards to be the same or be even more strict.

When you flush anything down the drain, for example, some expired prescription drugs, the water treatment plant may or may not remove those drugs before the treated water is discharged into a watercourse. Downstream another city takes in that water and treats it to destroy harmful bacteria and viruses, but what about the other substances in the water? Where would you like to live? Upstream or downstream of such a community?

Much of the freshwater we have on the planet is used in agricultural endeavors to feed and nourish our ever-increasing population. Pumps run 24/7/365 to move water from deep aquifers to the surface for irrigation (**Figure 37.2**). These aquifers are not recharging at anywhere near their depletion rate. Thus the water is being mined in a permanent sense. In some areas of our country where there is a preponderance of agricultural activities and associated irrigation, saltwater encroachment is occurring in the aquifers.

© alexmisu/Shutterstock.com

Figure 37.2 Irrigation of fields uses large amounts of ground and surface water. The water must be of good quality to provide the moisture needed to raise crops and other agricultural products.

The pollution of water in aquifers beneath cities and military installations occurs as a result of leaks and accidental spills into the ground, which eventually enters the water table. For example, petroleum pipelines connecting refineries and distribution centers around the country occasionally break. Certainly if the leak is detected mediation (clean up) measures are instigated. But what about undetected leaks, or leaks from years ago when regulations were lacking or ignored, many times out of ignorance. Today we have monitoring wells around sites of known hazardous leaks. These wells are regularly sampled to ensure that the chemical plume's vicinity to ground water is known. But again these are the sites we know about.

When the Cuyahoga River in Cleveland, Ohio caught on fire in June 1969, a wakeup call was sounded to the nation and the world. The call was simple—we cannot keep using our water supplies as a place to dump. The mentality of "out of sight, out of mind" had to stop. We are still working on this one.

Runoff from agricultural and livestock feeding facilities represents another source of water pollution. Fertilizers, insecticides, fungicides, and herbicides that are sprayed on fields to improve crop yields find their way into streams, rivers, lakes, and eventually oceans. Fertilizers, while enriching the soil, can end up in the water supply and alter the biota by accelerating the growth of certain algal components. Bacteria feed on the decaying algae, depleting the dissolved oxygen levels, causing fish kills with their associated ramification to the environment. It takes one domino falling into another to cause the remaining ones in line to fall.

Usually not mentioned but certainly on the list of water pollution is thermal pollution. Downstream outflows from power generating stations, which use river water to cool the machinery of power production, alter temperature regimes, which alter habitat and thus the biota. Some organisms flourish while others die off in the heated water.

As stated above, oceans are the final receiver of continental runoff (**Figure 37.3**). Oceans and lakes seem vast and bottomless, but in some areas, usually at the mouths of rivers, they may be virtual cauldrons of toxins. These toxins may be suspended in the water column or accumulate in the sediments. Ocean food chains are linked to the water column and the sediments.

© Gary Webber/Shutterstock.com

Figure 37.3 Surface drainage and sewage can enter the ocean without any monitoring or treatment for water quality.

Organisms feeding in these areas accumulate toxic compounds that are then later accumulated at higher levels in larger organisms (biomagnification). Biomagnification was introduced by Rachel Carson in her book, *Silent Spring*. Considering that many species of ocean life are consumed by humans around the world, and depending on where in the world we are, ocean life is sometimes a staple in the diet and sometimes a delicacy. Either way, the toxins placed by us, either accidently or by intent, return to us in an often harmful way.

Coastal Systems: Waves
By Dean Fairbanks

The landscapes of coastal areas are important to human populations. The estimated length of coastlines of the world is 1.6 million kilometers. Ocean waves and currents have created them over time. In the United States, 50% of the population lives within 50 miles off the coast; about 80% within 200 miles.

"The vulnerability of coastal areas to disasters and the need to take this into account in planning and coastal zone management is well known. High population densities, infrastructure and property development in coastal areas all contribute to higher social, economic and environmental consequences in case of disasters. Approximately three billion people, half the world's population, live within 200 km of a coastline. The average population density in coastal areas is about 80 persons/km2, twice the global average (UNEP 2005). Of the world's 17 largest cities, 14 are located along coasts - and 11 of these are in Asia (**Figure 38.1**). In addition, two-fifths of cities with populations of one million to 10 million people are located near coastlines". *United Nations Environmental Program, 2005*

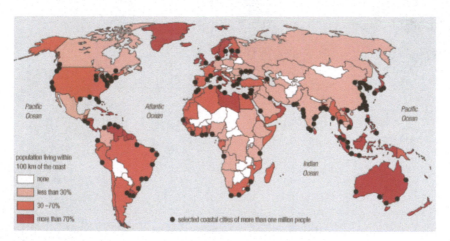

Figure 38.1 World Population and coastlines (**UN Environmental Program, 2002**)

We humans are creatures of the coastal areas of the planet. Thus understanding their natural processes of development and how our infrastructure can change their character is important to lead to more resilient management measures of the coastal zone.

The principal processes of coastal areas are **wave** and **current** motions. Similar to fluvial, wind, and ice processes, waves and currents involve entrainment, erosion, transport, and deposition of sediment. The erosion of the coastline and the deposition in particular parts of a coastline leads to associated landforms. The region between the high water line during a storm and a depth at which storm waves are unable to move sea-floor sediments is called the **littoral zone or intertidal** (also known as the nearshore). Wave energy determines their capacity for erosion and sediment motion.

Ocean and nearshore waves are the main driving process behind coastal evolution (**Figure 38.2**). Waves represent regular oscillations in the water surface of large water bodies (ocean or freshwater lakes). Waves are created by the wind, just like they create the large oceanic gyre currents. They are created through the frictional drag across the water. As the waves form in the offshore environment, the surface becomes rougher and it is easier for the wind to grip the water surface and intensify the waves. Waves are composed of the following components, which determine the energy in them:

- **Wavelength** (amplitude)—the distance between two successive crests
- **Wave period**—the time in seconds between two successive crests
- **Wave frequency**—the number of waves per minute
- **Wave height**—the distance between the trough and the crest

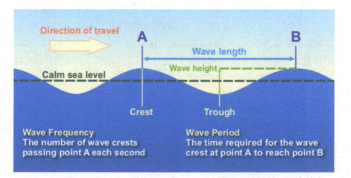

Figure 38.2 **Anatomy of a wave. (NOAA)**

Waves on the open water (called swells) appear to be going up and down in motion, but in fact the motion is circular (**Figure 38.3-38.4**). The orbital motion causes an object on the surface to bob up and down, forward and backward as the wave energy passes under it. The speed at which a wave moves depends on the wave's length and the depth of the ocean. The longer the wave length the faster it moves.

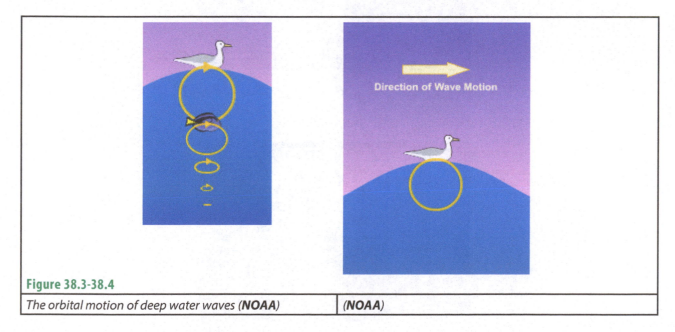

Figure 38.3-38.4

*The orbital motion of deep water waves (**NOAA**)* | *(**NOAA**)*

Deep water waves are unaffected by the sea floor. The energy is transmitted cyclically through the orbital oscillation of the water. The water itself does not actually travel in the direction of the energy that is passing through it. The characteristics of deep water waves, that is, height, wavelength, and period—thus power—are determined by:

- **Wind speed**—wave height generally proportional to square of wind speed
- **Wind duration**—time during which wind blows in one direction
- **Fetch**—distance over which the wind blows in one direction

As a deep water wave reaches the nearshore, there is a sequence of transitions that the wave goes through. These include:

- **Transitional wave**—the oscillations are beginning to be influenced by the ocean floor.
- **Shallow water wave**—oscillations of the wave are strongly interfered with by the ocean floor. Wave speed is reduced by friction from shoaling. This is the point where the depth of the water is one-half of the wave's length. The wave will slow down, grow taller and become shaped like peaks.
- **Breaking wave**—As the depth rapidly decreases, the wave frequency remains *constant*, the wave length *decreases*, the wave speed *decreases*, and the wave height and steepness *increases*. During this transition to a breaking wave, the energy in the wave remains constant (conservation of energy). The wave peaks reach a height of instability and gravity takes over and breaks the wave over and the energy and water move toward the shore in the form of **swash** up the beach, then **backwash** (retreating wave), water moving down the beach. Depending on the slope gradient of the beach a wave can be classified as **plunging** (destructive) or **spilling** (constructive). Plunging waves are erosional waves that occur on steep beach gradients and can be augmented by storms. Their backwash is greater than swash. Spilling waves are depositional waves that occur on shallow beach gradients. Their swash is greater than their backwash (**Figure 38.5-38.6**).

Figure 38.5   Plunging waves (NOAA)

Figure 38.6   Spilling waves (NOAA)

Coastal Systems: Refraction, Longshore Drift, and Tides

By Dean Fairbanks

Topic
39

The spatial concentration of energy in waves is affected by the shape of a shoreline in terms of **wave refraction** (bending process). The refraction occurs when waves approach an irregular shoreline, usually at an oblique angle. Refraction reduces wave velocity causing wave fronts to break parallel to shore (**Figure 39.1**). Energy is concentrated (converges) on the flanks of headlands and dissipates (diverges) energy in bays. However, since wave refraction is rarely complete, **longshore drift** occurs. Longshore drift is the movement of sediment by swash and backswash in the general direction of the prevailing winds. The long-term result of the concentration of the energy on headlands is coastal straightening since the erosional energy of the waves is concentrated on any coastal geology that sticks out from the main coastline.

Figure 39.1   Waves in Shallow water will break and release energy along the shoreline. (**USGS**)

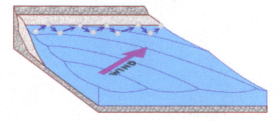

Figure 39.2   Wind blowing across the water surface generates waves. (**USGS**)

Waves move sediment toward the shore at an oblique angle. Backwash carry sediment away from beach at 90°, perpendicular to the shore. The effect of these processes is longshore transport (drift) of sediment parallel to shore in the general direction of the prevailing winds (**Figure 39.2**).

Waves also have the capability of generating currents, including longshore currents that run in shallow water parallel to the shoreline and rip currents that are created from variable wave heights. Rip currents are strong swift currents that run perpendicular to shore and flow out to the ocean (**Figures 39.3–39.4**).

Figure 39.3 "Diagram of a rip current." (**NOAA**)

Figure 39.4 Rip current. (**NOAA**)

Tides and the tidal cycle come into play with the movement of sediment and the erosional ability of waves. Tides are regular oscillations in the sea's surface height, caused by the gravitational attraction of the moon and sun on the oceans. The moon accounts for the large share of the pull. Spring tides occur just after new and full moons and are very low and very high, respectively. Neap tides occur one week later when the moon is at right angles to the sun, producing moderate tides. Each month are two sets of each type of tides (**Figure 39.5**).

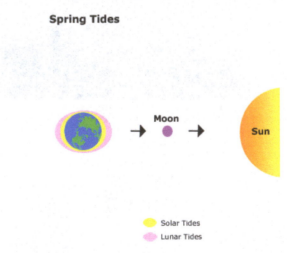

Figure 39.5 Animation of spring and neap tides. (**NOAA**)

Coastal Systems: Erosion and Deposition

By Dean Fairbanks

Coastal areas are worked on by similar processes we have seen earlier for fluvial and wind: *entrainment, erosion, transport,* and *deposition*. All of these processes are carried out by waves and currents. The modifications to the basic theory relate to the special nearshore processes of oscillating motion (swash and backwash) provided by the power of waves, which were provided their energy via wind. The specific erosion processes found in coastal environments include:

Figure 40.1 A large slump is an impressive landscape feature at Palomarin Beach, Marin County. At low tide, a plunging anticline is exposed in a wave-cut platform below the toe of the great slump. **(USGS)**

Figure 40.2 Sea stack and arches. Point Reyes National Seashore, California. **(USGS)**

- **Abrasion**—the wearing away of coastlines by material carried away by waves
- **Hydraulic impact**—the force of water on rocks, especially during storm surges
- **Solution**—the wearing away of rocks by acidic water. Limestone and chalk are especially affected.
- **Attrition**—the rounding and reduction of particles carried by waves.

The evolution of a coast with cliffs is one of a steep cliff being replaced by a lengthening platform and lower angle cliff. **Wave-cut platforms** (terraces) develop sometimes with multilevels as sea levels change or due to the rising of a coastal landscape under tectonic development (**Figure 40.1**). Other erosional features found along coasts are landslides along the **sea cliffs, notched cliffs, sea caves, sea arches**, and **sea stacks** (**Figure 40.2**).

The material that is eroded is transported either by direct wave action as swash or by longshore drift. The source of material comes from eroded cliffs, beach sand deposits, and river sediments. Depositional features include:

- **Tombolo**—a sandbar that links the mainland to an island by wavedeposited material (**Figure 40.3**)
- **Barrier spit**—a sand beach linked at one end to land. Spits are found on indented coastlines (Cape Cod) or at river mouths.
- **Bay barrier (bar)**—a ridge of sand that blocks off a bay or river mouth. Creates lagoons
- **Barrier islands**—lumpy sand bars that appear parallel to the coastline, not blocking bays or river mouths. Found in chains of a few long islands with tidal inlets between them.

Image © Shutterstock, Inc.

Figure 40.3   **Tombolo at Point Sur, California.**

Coastal Systems: Coastal Management

By Dean Fairbanks

People love living on the coast, but lack of understanding and engineering solutions to make coasts static has been destructive. **Coastal management** is an attempt to keep the sediment mass balance in balance. There are a number of ways in which people affect coastal systems, for example, dredging sediment and building wave-protection schemes. The problem is that attempting to protect the coast in one place may increase pressure elsewhere. This is often seen in the development of groins. **Groins** are rock or concrete barriers that run perpendicular from a beach into the area of breaker waves (**Figure 41.1**). They are an attempt to force the deposition of sand from longshore drift on the one side of the groin. They can prevent longshore drift from removing a beach. However, by capturing the sediment they deprive another area of the coast of that sediment for beach replenishment. Finally, the most destructive engineering solution is the seawall. Their design leads to increased scouring of the beach and the undercutting of the wall, which then leads to its toppling and failure. What is the answer to coastal management?

The best protection for any coast is a beach. Without a thick wide layer of beach sand or even pebbles/cobbles, a coast is increasingly vulnerable to erosion. Try not to cut off access to sediment inputs from rivers, which mean also rethinking dams on rivers. One thing is very clear: Fluvial landscapes and littoral landscapes are highly connected and should be managed with that connective understanding (**Figure 41.2**).

Figure 41.1 Groins, seawall and riprap (large boulders dumped on beach). (**USGS**)

Figure 41.2 US national assessment of coastal vulnerability to sea-level. (**USGS**)

Drainage Basin: Rivers
By Dean Fairbanks

River Discharge and Storm Hydrographs

The measurement of water flowing through a river/stream is a measurement of its average velocity and **discharge** (Q). This is the volume of flow passing through a cross section of a river during a given period of time. A storm hydrograph measures the speed at which rainfall falling on a watershed/drainage basin reaches the river channel. It is a graph on which river discharge during a storm of runoff even is plotted against time (**Figure 42.1**).

A stream's discharge and its pattern is made up of a minimum *base flow level* (provided by seepage of groundwater into channel), a *cresting storm flow*, and the *interflow* which is the range in between the rising limb to the storm crest and the falling limb back to the base flow level. The discharge is higher in larger basins or the higher the rainfall the greater the discharge. In relation to river/stream profile, steep watersheds will have lower infiltration rates so high storm flow peaks, while flat watersheds will have high infiltration so lower peaks (**Figure 42.2**).

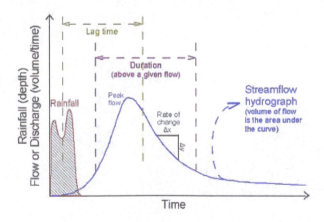

Figure 42.1   A single-event flow hydrograph. (EPA)

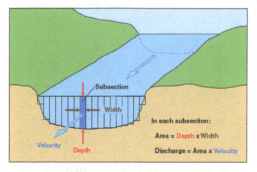

Current-meter discharge measurements are made
by determining the discharge in each subsection of a channel
cross section and summing the subsection discharges to obtain
a total discharge.

Figure 42.2 Discharge is the volume of water moving down a stream or river per unit of time, commonly expressed in cubic feet per second (cfs) or cubic meters per second (cms). **(USGS)**

To work out discharge (Q), the average channel velocity needs to be measured. The velocity accounts for the effects of the gradient slope and channel walls and allowing for variation in the velocity due to width and depth. Therefore the total discharge (Q) of a stream is the total volume per unit time flowing through a stream channel at a given location as a function of:

$$Q = W*D*V$$

- W = width
- D = flow depth
- V = average stream velocity

The discharge is able to determine the ability to transport sediment, the rates of erosion and deposition, and the form of the river channels.

The river is a subsystem of a large unit—the watershed or drainage basin. The sediment system within a river is a further subsystem depending on many variables: discharge, climate, relief, and the rock type. Keep in mind that the key processes of denudation by the flow of water over a landscape include entrainment, erosion, transport, and deposition of material. The flowing water acts on weathered geologic material, often in granular form.

In upland areas of a watershed, discharge is low (but velocity is high from larger slope gradients) and the bed material is large (e.g., cobbles and boulders), so within the channel (endogenetic) sediment yield is low (**Figure 42.3**). In the lowland areas of a watershed discharge is high (but velocity is low from smaller slope gradients) and bed and bank material is not resistant to the larger erosion power of the river, so sediment yield will be high. The erosive processes are summarized in the **Hjulmstrom curve**. The Hjulmstrom diagram is a good approximation of particle sizes transported by a river or stream based on velocity, where higher

Figure 42.3   Higher energy mountain stream—boulder and cobble bed. **(USGS)**

Figure 42.4 Lower energy lower course stream—sand and silt bed. **(USGS)**

velocities can carry gravel, cobbles, and boulders (representing higher energy) and lower velocities can carry clay, silt, and sand (representing lower energy) (**Figure 42.4**). The equilibrium of a river/stream is as such:

- Steeper gradient = increase in velocity → beginning of erosion and increase in load (degrading the river/stream bed)
- Gentler gradient = decrease in velocity → decrease in erosion and beginning of deposition (aggrading the river/stream bed)

The role of **fluvial transport** is to move the eroded weathered material downstream to be deposited. The moving of sediments and dissolved materials as a result of water flow within (and at times beyond) a stream channel. The *capacity* of a river/stream refers to the total weight of material carried by a river. The *competence* of a river/stream refers to the diameter of the largest particle that the river can carry. A river/streams load is represented by:

- **Dissolved load**—dissolved ions carried in solution derived from precipitation, chemical weathering, erosion, and pollution (e.g., sodium, calcium)
- **Suspended load**—carried with the body of the current (held in suspension by turbulence) representing small silt clay particles. A faster velocity can bring larger fine-medium sands derived from the channel bed.
- **Bed load**—large-grained sand to cobbles that move by sliding, traction (rolling), or saltating (bouncing along the bottom).

The processes that derive the Hjullmstrom curve and thus fluvial sediment transport are based on discharge of water per unit width of a river/stream and the slope of the landscape surface. Water can transport 4x more sediment if the water discharge doubles. It helps explain why stream channels are formed (**Figure 42.5**).

River Patterns: Channel Morphology

When considering the river channel, we must focus on the types of flow along the entire longitudinal profile broken into its main components: Source à Intermediate sections, which include bedrock and alluvial channel sections and the floodplain à mouth (**Figures 42.6–42.9**).

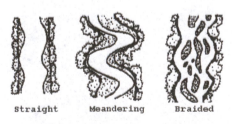

Figure 42.5 Various river channels diagrammed. **(USGS)**

Figure 42.6   Bedrock channel (solid rock). Image © Shutterstock, Inc.

Figure 42.7   Alluvial channel (sediments). Image © Shutterstock, Inc.

Figure 42.8   Meandering river (**USGS**)

Figure 42.9   Braided river (**USGS**)

Stream flow is very complex. The velocity and therefore the energy is controlled by the:

- Slope gradient
- Volume of water
- Shape of the channel
- Channel roughness which indicates friction

The efficiency of a channel is measured by the cross-sectional area, which is affected by river level and channel shape. The rule of thumb with shape: deep and narrow is very efficient (low relative friction), while wide and shallow is inefficient (high friction). The channel shape can be linked back to the sections they are found in the longitudinal profile. In the upper courses (steeper gradient) the channel is very efficient, while in the middle to lower courses (gentle gradient) the channel is inefficient. The main river/stream patterns can be linked back to the slope gradient in the following way: Straight shape—high slope angle Meandering shape—moderate to low slope angle. Occurs most frequently on floodplains Braided shape—very low slope angle (almost none—flat) The increasing slope is equal to the power of a river/stream to move water and the sediment load. The channel forms minimize the work of the stream in moving sediment.

The development of meandering is linked to the development of the floodplain and its widening over time (**Figure 42.10**). While meanders are linked to a slope threshold, they also have strong links to pools and riffles. They are caused by turbulence in the flow. Eddies cause disposition of coarse sediment (riffles) at high-velocity points and fine sediment (pools) at points of low velocity. Riffles have a steeper gradient than do pools, which leads to sinuousity. The sinuosity (meandering) increases with the **thalweg**, a line tracing the deepest and fastest water. The thalweg moves from side to side within the channel and also corkscrews in cross section. In a meander, the faster thalweg flow starts to erode the outside bank in a bend, while the slower water deposits on the inside bank forming a point bar. A **flood plain** valley is then widened by lateral erosion especially during **bankfull** periods. Bankfull stage is the elevation at which a stream first begins to overflow its natural banks onto the active flood plain. In time, the meander migrates down its floodplain. Over time, strongly looped meanders can have two cutbanks eroding toward each, which then cuts off the meander resulting in deposition and blocking the former meander from the main river. This creates an **oxbow lake** on the flood plain (**Figure 42.11**). This occurs because river channel forms always work towards forms that minimize the work (the flow) of the stream in moving sediment—straighter path rather than a large meander loop.

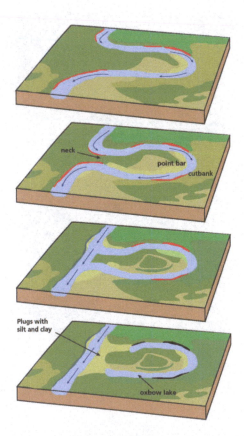

Figure 42.10 Evolution of a meandering stream includes the following: (1) stream channel within meander belt; (2) development of a nearly closed meander loop; (3) high water flowing across the neck of loop, making a cutoff; (4) deposition of sediment sealing the loop and creating an oxbow lake. (**NPS Geologic Resources Division**)

Figure 42.11   Oxbow off the Buffalo Fork River in the Northern Rockies. (**USGS**)

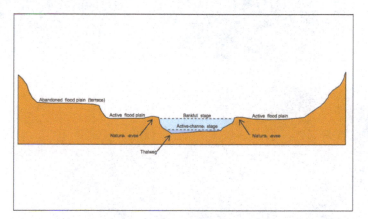

Figure 42.12   Schematic diagram showing geomorphic features of a stream channel. (**USGS**)

At the end of the river/stream longitudinal profile where base level is reached, deposition of the load occurs. This is at the river mouth (**Figure 42.12**). Forward motion of the discharge stops and the depositional fallout of sediment create deltas and estuaries. **Deltas** are formed when river sediments are deposited as a river enters a standing body of water such as a lake, lagoon, or ocean. Deposition occurs because velocity is checked. Deltas formation will be more likely on gentle coastline where wave or tidal energy is low. **Estuaries** occur where a coastal area has recently subsided or the ocean level has risen, causing the lower part of the river to be drowned.

Flooding and Stream Processes

By Ingrid Ukstins and David Best

Figure 43.1 Intense rainfall can produce significant runoff, The park in Scottsdale, Arizona, was flooded by an intense thunderstrom more than 20 km away.

The Role of Water

Water, considered by many to be the "staff of life", is an essential part of processes that are found on Earth. More than 70 percent of Earth's surface is covered by water, whether in liquid or solid form, as well as being in the atmosphere. The continental regions on which people live have lakes, rivers and streams, and a small amount of surface covered by ice. When these areas are affected by water-laden storms, the additional water can create flooding that can be very disruptive, dangerous and costly. Oftentimes this occurs without warning. People who live in some regions, such as low-lying areas along coastlines, are aware of the potential for problems, while those who live in arid climates often do not think about the potential for disastrous events. Everyone should recognize that water has the ability to generate a natural disaster with potentially catastrophic consequences.

All climates, especially arid ones, are prone to flooding. Every year, numerous floods affect millions of people throughout the world (**Figure 43.1**). Weather systems can stall over an area and generate copious amounts of rainfall, causing streams and rivers to overflow their banks when too much water enters a drainage basin.

Rapid melting of snow and ice also create abnormal water discharge. Areas that are most affected tend to lie toward the lower portions of a river system, as the topography there is less steep and the increased downstream flow cannot readily move water away from the affected region.

Floods can also be the result of poor urban planning or dam construction that causes water to flow unexpectedly in places where it was not meant to go. Coastal regions become flooded when maritime storms land ashore. The strong winds of hurricanes, cyclones, and typhoons generate storm surges that bring massive amounts of water inland, flooding low-lying areas. Tsunami, although relatively rare, rapidly push sea water past beach zones, inflicting severe damage to communities in the path of the waves. Floods can also result from volcanic eruptions that melt snow and ice atop a volcano, the condition that contributed to the lahars of Mount St. Helens.

The National Weather Service reports that flooding in the United States annually causes an average of 82 deaths and almost $8 billion in damages. In many low-lying areas, home and landowners cannot purchase flood insurance and the losses they might sustain are not covered. Often the lower socioeconomic classes reside in these topographically lower locations and suffer immense losses in terms of property and human life. Although floods can be caused by storms, tsunamis, and volcanoes, this topic will address flooding associated with streams.

Stream Processes

Earth's hydrosphere includes all water at or near the surface in addition to what is contained in the atmosphere. Water, which can be a liquid, a gas (water vapor), or a solid (ice), moves around in the **hydrologic cycle**—a continuous circulation of water around the globe. Solar energy drives this cycle; it stretches from the equator to the poles, where water movement is obviously slower but nevertheless part of the cycle.

> **hydrologic cycle:** The cycle that moves water and water vapor among the oceans, land, and atmosphere through evaporation, condensation, precipitation, transpiration, and respiration.

Geologists define a **stream** as a body of water that flows within a confined channel on the surface. When several streams or **tributaries** join to produce a larger flowing body, the term **river** is then applied (**Figure 43.2**). Water that falls onto Earth's surface can produce **runoff**, which travels along the surface and ends up in a channel, or it can infiltrate into the ground, where it temporarily resides in underground collection areas

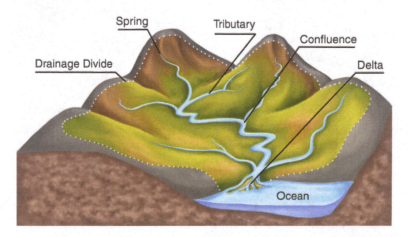

© stihii/Shutterstock.com

Figure 43.2 Streams flow from high to low elevations. In the upper portions small streams form the tributary system. Water flows downslope to feed other larger streams or it flows into lake, sea, or ocean.

before eventually finding its way into a stream system. In arid climates, infiltrated water often evaporates and reenters the hydrologic cycle as water vapor.

> **stream:** A body of water that flows downhill under the influence of gravity and lies within a defined channel.
>
> **tributary:** A stream that flows into another larger stream.
>
> **river:** A large stream.
>
> **runoff:** Water that flows across a surface and into a stream or other body of water.

Drainage Basins

A **drainage basin** or **watershed** is an area on the surface that collects water that flows into a stream and forms a drainage pattern. Drainage divides, which are often ridges or a series of high points that divide downhill slopes, separate one watershed from its adjacent neighbors (**Figure 43.3**). The amount of water collected by a drainage system is dependent on the amount of precipitation, the size of the area, and the subsurface characteristics that control infiltration and runoff. As long as the amount of water flowing into streams and falling on the watershed is carried away by the existing system, water remains in the channels and does not present a problem.

> **drainage basin:** An area that drains water to a given point or feature, such as a lake.
>
> **watershed:** See drainage basin.

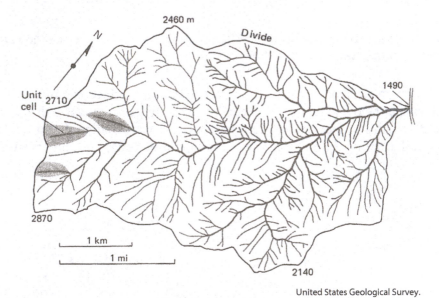

United States Geological Survey.

Figure 43.3   A watershed is outlined by its divide, which separates it from adjacent watersheds. Elevations are in meters. Notice how the stream patterns collect water and channel it into a larger stream downhill.

Think of a stream configuration as the transportation system that is moving water to some end point, either a temporary one, such as a lake, or the ultimate end point—a sea or an ocean. A stream begins to cut a channel into the landscape, thereby creating the "highway" that is allowing water to move through an area. Each channel has a cross sectional view, in which we see the width and depth of the channel. This area, coupled with the length of a particular stream segment and its drop in elevation, defines the volume or how much water is contained (and moved) by the stream and at what velocity (**Figure 43.4**). **Discharge** is the volume

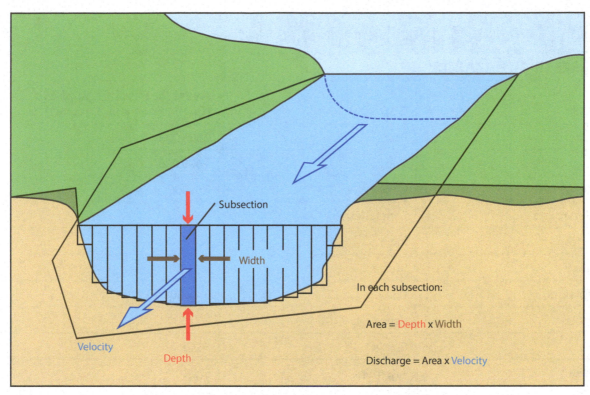

In each subsection:

Area = Depth x Width

Discharge = Area x Velocity

United States Geological Survey.

Figure 43.4 Discharge measurements are made by determining the discharge in each subsection of the channel cross section and t hen by summing these discharge amounts, the total discharge is calculated.

of water that flows downstream past a given point in a given period of time. We measure this volume in cubic feet per second (cfs) or cubic meters per second (cms), and the amount can vary widely depending on weather, surface conditions, and stream characteristics. The world's largest rivers move enormous volumes of water each second (Table 43.1). All of these rivers have very large watersheds and many of the rivers flow through regions that have wet, temperate climates (Table 43.1).

discharge: The volume of water flowing through a stream channel in a given period of time, usually measured as cubic feet per second or cubic meters per second.

Floodplain

A stream channel is bordered by its banks and the area to either side, which is termed the **floodplain**. When water spills over the banks, it is no longer moving in its channel. As it spreads out, the velocity of the water decreases rapidly, causing any sediment carried by the stream to be deposited. This process is repeated every time the stream floods. The continual buildup of sediment along the banks creates natural levees that increase in height, thereby deepening the stream channel. Repetition of this process permits the stream to carry more water than before, because its cross-sectional area has increased. When water spills over onto the floodplain, it does not tend to drain back into the main river channel because the natural levees now act as a dam. This water slowly moves downhill on the floodplain, creating a yazoo tributary. Eventually this water finds a site where it can flow back into the main stream or off the floodplain (**Figure 43.5**).

floodplain: The flat area alongside a stream that becomes flooded when water exceeds the banks of the stream.

Table 43.1 World's 10 Largest Rivers by Discharge

TERRESTRIAL HAZARDS

River	Country	Average Discharge at Mouth (Cubic Feet per Second)
Amazon	Brazil	7,500,000
Congo	Congo	1,400,000
Yangtze	China	770,000
Brahmaputra	Bangladesh	700,000
Ganges	India	660,000
Yenisey	Russia	614,000
Mississippi	USA	611,000
Orinoco	Venezuela	600,000
Lena	Russia	547,000
Parana	Argentina	526,000

Source: http://www.waterencyclopedia.com/Re-St/Rivers-Major-World.html.

Source: NASA.

Figure 43.5 This image of the Mississippi River at Burlington, Iowa, shows the floodplain on either side of the river. Periodic flooding of the rivers covers the floodplain and impacts the cities of Burlington and West Burlington.

Streams provide water, transportation, food, irrigation, and soil to people living nearby. The finer silt carried by streams is deposited to create arable farmland, such as in the delta region of the Nile River in Egypt. Before the High Aswan Dam was completed in 1970, these rich farmlands were frequently flooded, destroying crops, homes, and killing people, but also depositing valuable silt that enriched farming areas near the mouth of the river. Construction of dams lessens the occurrence of major floods, but the dams hold back valuable silt, rendering many downstream regions unsuitable for growing crops.

FACT BOX	Equivalent Measures and Weights of Water at 4°C.
1 gallon = 0.134 cubic foot = 8.35 lb (3.79 kg)	
1 cubic foot per second (cfs) = 7.48 gal/sec = 62.4 lbs/sec (28.3 kg/sec) = 449 gals/min	

Topic 44

Types of Floods
By Ingrid Ukstins and David Best

Regional River Floods

Flooding that is related to seasonal rains or snow melts, or a combination of rain falling on snow, produces a large volume of water that cannot be handled by existing stream systems. Such flooding is common in wetter climates that have generally larger, more established rivers. Floods can occur anywhere along the length of a stream. In the upstream regions, such floods are caused by intense rainfall or snow melt over a watershed that flows into smaller streams and tributaries. When several watersheds feed into a larger stream, the volume of water can be immense and produce widespread flooding.

These vast amounts of water can have an effect on the landscape. One foot of water covering one acre is termed an **acre foot**. This amounts to 325,851 gallons of water, so for every inch of rain falling on an acre, there are 27,154 gallons that can flow across the surface. If one acre foot of water moves into a river, it will contribute 43,560 cubic feet (1233 cu meters) of flow to the stream's volume. Generally, these intense rainfalls are so rapid that very little water percolates into the subsurface, producing stream flows that become rushing torrents (**Figure 44.1a–b**). As the gradient of the stream flattens out and the downstream river channels become wider, the velocity drops and the water becomes calmer. However, the volume of water continues to increase, especially if more streams are present in the system.

> **acre foot:** The amount of water that covers one acre to a depth of one foot; equivalent to 325,851 gallons of water.

Many watersheds consist of hundreds or thousands of acres, resulting in massive amounts of water moving downslope and downstream. The lateral and down cutting erosive power of the water causes dimensions of the stream channel to increase. Upstream regions generally have fairly steep slopes and the longitudinal profile of the stream shows a rapid drop in the elevation of the stream (**Figure 9.1**). Therefore water will tend to move downslope quickly and generate large-scale flooding.

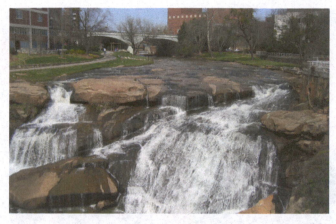

Photo by James S. Best, by permission.

(a)

Photo by James S. Best, by permission.

(b)

Figure 44.1   The Reedy River in Greenville, South Carolina. (a) normal flow conditions in March; (b) following a June thunderstorm.

As higher elevation streams move water downstream, they join with other streams and increase the size of the trunk stream (in a fashion similar to the trunk of a deciduous tree having many branches that feed into the main trunk). If a widespread rainstorm covers several different watersheds and feeds water into the trunk stream, the downstream region can experience flooding. Widespread saturation of the ground prevents water from percolating into soils, so the water must flow under the force of gravity to lower elevations. This downslope flow is not instantaneous, so there is often a lag that gives some warning to communities at the lower end of a drainage area.

Downstream communities generally have some preventive measures in place in preparation for recurring floods. Increasing the height of levees and deepening the river channel help the river to handle a greater flow, thus preventing flooding. However, these measures are not always successful, particularly if there is a flow that greatly exceeds the capabilities of the infrastructure to handle the runoff. The result can be flooding so severe and widespread that the governor of the state requests presidential declaration of the region as a disaster area so that it can receive federal help (**Figure 44.2**). As we see from the map in **Figure 44.2**, very few regions in the United States are exempt from flooding over a period of several decades.

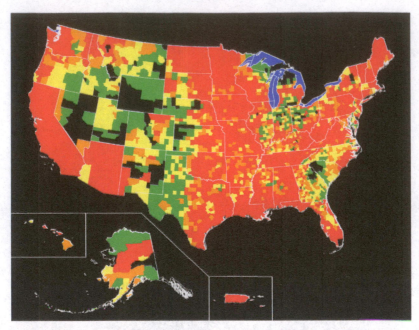

Figure 44.2 Presidential disaster declarations related to flooding in the United States, shown by county: Green areas represent one declaration; yellow areas represent two declarations; orange areas represent three declarations; red areas represent four or more declarations between June 1, 1965, and June 1, 2003. Map not to scale.

Other natural hazards play a role in changing the drainage regime of an area. The 1991 eruption of Mount Pinatubo, a volcano in the Philippine Islands, deposited massive amounts of ash across the countryside. Within a few months, copious rainfalls struck the region and moved the ash downslope as sediment. When flooding occurred in the streams, deposits were created along the banks and formed natural levees that increased the depth of the channels. After several episodes of this redistribution of the ash, streambeds were flowing at a level higher than the original surface. When the streams experienced later flooding, the water easily flowed into the lower lying areas adjoining the raised stream channels. Removal of ground cover increases surface runoff that causes sediment to affect stream flow.

Flash Floods

Flash floods occur with little or no warning. They are usually caused by torrential rainfall that takes place over a very short time, often associated with severe, localized thunderstorms or from a series of storms continually soaking an area. Although flash floods can occur anywhere, they tend to be more devastating in two areas: (1) in mountainous areas, where steep slopes funnel water into narrow streams, and (2) in desert regions, where normally dry or low flow streambeds are quickly transformed into raging torrents. Rainfall does not have any chance of infiltrating into the subsurface and is often flowing in a rapid, sheet-wash manner across impervious surfaces to collect in dry stream beds. The intense nature of flash floods makes them capable of causing extensive damage and loss of life. When people drive their vehicles through normally dry washes and streambeds that are filled with fast-moving water, their vehicles begin to float, are pushed along by the flow, or are overturned, trapping the victims inside. The force with which fast-moving, sediment-laden water hits a surface is great (Box 44.1)

BOX 44.1	How One State Deals with Not-So-Smart Drivers

The State of Arizona has enacted the Stupid Motorist Law (Arizona Revised Statutes 28-910), which imposes a fine of up to $2,000 on drivers who have to be rescued from a flooded area. In spite of this law, people continue to drive around barricaded crossings and attempt to get through flooded roadways. A lack of adequate storm culverts causes water to flow across low points on streets and highways. The depth of water, even when it is flowing across a roadway that might be familiar to the driver, is very uncertain and the force of the water is much greater than one realizes.

DO NOT DRIVE THROUGH FLOODWATERS!

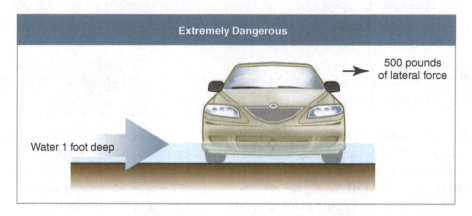

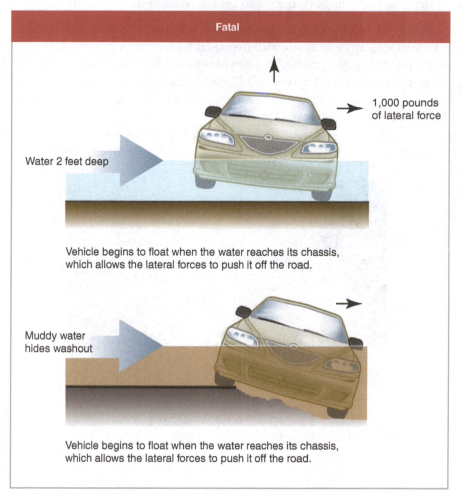

Vehicle begins to float when the water reaches its chassis, which allows the lateral forces to push it off the road.

Muddy water hides washout

Vehicle begins to float when the water reaches its chassis, which allows the lateral forces to push it off the road.

United States Geological Survey.

Box Figure 44.1.1a–44.1.1b

Winter Climate-Driven Floods

Ice-jam floods are a problem in regions where rivers freeze and then begin to thaw or receive surface water from rainfall or nearby melting. In the winter, some regions receive excessive snowfall, which rests on frozen ground. Rivers that normally drain the snowmelt freeze and cannot move any water downstream. Ice builds up whenever a small period of melting occurs and then refreezing takes place, thus making the drainage situation worse.

> **ice-jam flood:** A flood, usually in the spring, that results from broken pieces of river ice blocking the flow of a river, thereby flooding areas adjacent to the river.

During the winter of 1996 and 1997, the watershed of the Red River of the North, which forms the state boundary of North Dakota and Minnesota, was besieged by excessive rainfall and a series of blizzards. Precipitation totals were more than three times normal and cold weather early in the winter froze the soil, preventing any percolation of water. In early April the region experienced record low temperatures along with another 10 to 12 inches (25 to 30 cm) of snow. Within a 10-day period, daytime temperatures swung from single digits to highs in the upper 50s. Rapid melting occurred that produced extensive flooding.

Because the Red River of the North flows north—one of few rivers in the United States to do so—the water was draining into an area where the river was still frozen. The surrounding farmland had a very low gradient (less than a few inches per mile), so the water had nowhere to go. The result was that almost 4.5 million acres were covered in water. Grand Forks, a city of about 48,000 on the banks of the river, was flooded (**Figure 44.3**) and more than 24,000 homes were destroyed. The downtown area was under several meters of water. News photos recorded fires burning that destroyed several key buildings in the downtown area. Farms were under water, major crops were lost, and more than 120,000 cattle died. Losses amounted to several billion dollars. The damage continued into Canada, where more than $800 million worth of property was destroyed.

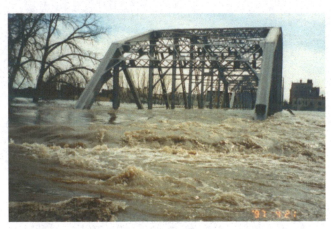

United States Geological Survey.

Figure 44.3 Sorlie Bridge, which connects Grand Forks, North Dakota, with East Grand Forks, Minnesota, lies under water in the April 1997 floods. Losses exceeded $3.5 billion in these cities.

Dam-Related Floods

Vaiont Dam, Italy, 1963

Completed in 1961, the Vaiont Dam, the world's sixth highest dam, was built to generate hydroelectric power for northern Italy (**Figure 44.4**). The region behind the 262 m high dam included steep hillsides that were underlain by sedimentary rocks, including shales, in the Dolomite region, about 100 km north of Venice. In October 1963, after the reservoir filled naturally, abnormal rainfall caused a block of approximately

270 million cubic meters (9.4 billion cu ft) to detach itself from one wall and slide into the lake at velocities of up to 30 m/sec or 65 mph. This generated a wave 100 m high that went over the top of the dam and blasted down the valley below. More than 50 million cubic meters (13.2 billion gallons) of water shot downstream. More than 2,500 people died in the disaster, which was very preventable. During planning and construction of the dam, geologists and engineers failed to note the potential of rock slippage. Fortunately the dam itself did not fail.

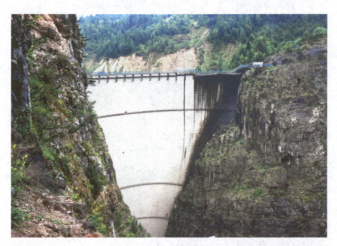

© Migel/Shutterstock.com

Figure 44.4 Vaiont Dam in northern Italy was topped by a massive wave generated by a landslide that went into the impounded reservoir. The resulting flood killed more than 2,500 people in villages along the downstream reach of the Vaiont River, in northern Italy.

Buffalo Creek, 1972

During a five-year period (1972 to 1977), several major dams failed in the United States. In 1972, a privately controlled, slag-heap dam on Buffalo Creek in West Virginia gave way following excessive rainfall, resulting in 125 deaths and more than $50 million in damage. After the main dam broke, water rushed downstream and destroyed two more dams. In a matter of minutes more than 1,000 people were injured and 4,000 were left homeless. Interestingly, the U.S. Department of the Interior had warned officials in the state in 1967 that the potential for a disaster existed, but nothing was done to address the issues.

Teton Dam, 1976

On June 5, 1976, the Teton Dam on the Teton River in eastern Idaho, near the city of Rexburg, collapsed. The earthen dam was originally designed to serve as a multipurpose structure, providing irrigation water to agricultural land in the area, hydroelectric capabilities, recreational opportunities, and flood control. The dam was the subject of lawsuits in federal court during its planning and early construction, but it was completed and put into service in January 1976. It failed on the first filling of the impounded reservoir. In the end, the federal government paid out almost $400 million in damage claims, and the dam was not rebuilt. The final toll was 11 lives lost and 13,000 head of livestock drowned. Flaws in the design of the dam were considered the source of the failure.

Human Interactions and Flooding

By Ingrid Ukstins and David Best

Streams will naturally flood at some point during their existence. Whether flooding occurs regularly or on a very periodic time scale, water will flow over the banks and onto a floodplain. However, mankind has introduced new parameters into the flooding process that create flooding conditions more frequently than the natural processes.

Construction of Levees and Channels

Construction of natural levees, although meant to allow a stream to flood in a natural way less often, upsets the normal balance of a stream system. Increased height of levees causes more water to flow in the deeper channel, so when it does occasionally break through, major problems develop. Water cannot easily flow back into the main channel.

Construction of concrete channels to contain flow through urban areas only speeds up any excess water from heavy rainfall or snowmelt that enters the channels. The Los Angeles River has been channelized as the river passes through the city of Los Angeles. More than 60 percent of its watershed is covered by impervious material (asphalt and concrete), so infiltration is greatly reduced. Water is sent to lower elevations in the channel system. There are times when too much water accumulates at the lower reaches of these channels, and flooding results in extensive, unexpected damage.

When flooding occurs, sandbags are often used to contain the flow of water. Bags are placed along the banks of a river as temporary, "quick-fix" levees that hold most of the water in a channel. Bags are also used in low-lying areas to slow the flow of water into buildings and their doorways or low windows. Sandbags work well in the short term for localized flooding but do not help with major overbank conditions. Sometimes large sheets of plastic are placed over the bags but these sheets are not continuous and will allow leaks. All of these techniques are insufficient in solving the larger problem of streams overflowing their banks.

Flood Control

When water is out of control, it creates major problems. Several techniques have been devised to help control and channel potential flood waters. Channelization can involve the clearing or dredging of an existing channel to speed up the flow. Another method is to construct artificial cutoffs that shorten the length of a stream, increasing its velocity and gradient. The result is that more water flows through an area with higher velocities. The goal is to reduce the chances of flooding, but channelization has produced mixed results.

Recall that natural levees form along rivers and deepen the channel. Piling extra earth atop natural levees or putting earth alongside a river creates artificial levees that deepen the channel and increase the volume of water in the stream channel. However, these are only temporary solutions that eventually prove unsuccessful in controlling floods.

Dams help to reduce flooding by storing water and releasing it in a controlled manner. In addition to the impounded water, however, the dams collect sediment that normally would have been transported downstream to enrich floodplain farmland or to create riparian habitats along the stream. Dams create an imbalance in the ecology of the river environment, a fact that was not recognized until recently. On the other hand, the thousands of dams in the United States also generate hydroelectric power, provide water for agricultural irrigation, and serve as recreation areas.

Topic 46

Lithosphere
By K.L. Chandler

Lithosphere

Earth's **lithosphere** is comprised of its crust and upper mantle. The crust's thicker continental segments and thinner oceanic areas account for the lithosphere's varying 40- to 100-mile thickness. Until the twentieth century, scientists assumed Earth to be fixed and rigid, with the continents locked in place. Enter Alfred Wegener (1880-1930), a German meteorologist and geophysicist, who presented a radical theory of supercontinents and continental drift in 1912. He proposed that an original supercontinent, **Pangaea**, included nearly all of Earth's land masses before it broke off over time and drifted into the land mass arrangement seen today, citing the apparent ability of some of the continents, such as South America and Africa, to fit together like puzzle pieces. (**Figure 46.1**)

Despite categorical criticism from the scientific community as a whole, Wegener remained dedicated to this theory until his death. Geologic evidence suggests North America and Europe were once a single land mass. Similar fossils appear on far-flung continents, and mountain chains that are currently separated by oceans look as if they may have been part of the same ranges in ages past. Technological advances allowed scientists to study seafloor spreading in the 1960s and revisiting Wegener's research proved much validity in 1978, thanks to the newly developed global positioning system.

Wegener's work formed the basis of modern plate tectonics. This theory studies the combination of the eight major and thirteen minor tectonic plates, in addition to nearly 60 microplates.

EIGHT MAJOR TECTONIC PLATES

- African Plate
- Antarctic Plate
- Australian Plate
- Eurasian Plate
- Indian Plate
- North American Plate
- Pacific Plate
- South American Plate

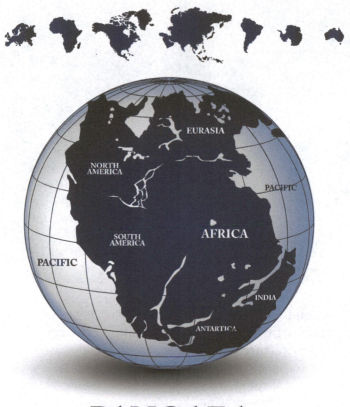

PANGAEA

© robin2/Shutterstock.com

Figure 46.1   A rendering of how Pangaea may have appeared before continental drift.

THIRTEEN MINOR TECTONIC PLATES

- Amur Plate
- Arabian Plate
- Burma Plate
- Caribbean Plate
- Caroline Plate
- Cocos Plate
- Nazca Plate
- Okhotsk Plate
- Philippine Sea Plate
- Scotia Plate
- Somalian Plate
- Sunda Plate
- Yangtze Plate

Tectonic plates, massive pieces of Earth's continental and oceanic crust, float atop the **asthenosphere**, a layer of magma that is directly under the lithosphere and acts like a conveyor belt. Propelled along the asthenosphere by convection currents of heat, the plates move slowly toward each other at a rate of 0-100 millimeters per year. As a result of this movement, North America is approximately three centimeters further away from Europe than it was a year ago.

Geological Structure and Processes

By Dean Fairbanks

What is a natural landscape and how does one form? These are the questions that pertain to the Earth's crust—the **lithosphere**. These are the lands that we as humans reside and go about our business with the other animals and plants. The structure and processes on landscapes are both informed not only by the atmosphere and hydrosphere (**exogenic** processes) but also by deep geologic processes and deep crustal Earth energy that drive dynamic shaping and creation of new geological materials (**endogenic** processes) (**Figures 47.1–47.4**). Natural landscapes are formed by underlying geologic "raw" materials, processes leading to landform development, and present their own nature of resulting landforms. Physical geographers and geoscientists study natural landscapes to understand and solve various environmental problems as they pertain to human settlements and enterprise.

Figure 47.1 Iceland lava field. Image © Shutterstock, Inc.

Figure 47.2 Colorado River. Image © Shutterstock, Inc.

Figure 47.3   The Mississippi River. Image © Shutterstock, Inc.

Figure 47.4   Yosemite National Park. Image © Shutterstock, Inc.

The "raw" materials from which landforms develop are those formed and transformed by various surface and subsurface (including deep) geologic processes. We can only understand landforms in terms of geologic materials and the processes acting on them over both short and long time frames. The understanding must start with the fundamentals of matter—substance or substances of which any physical object consists or is composed. In Module 1, the fundamentals of matter were discussed in terms of atoms, elements, and molecules. In landforms and the geologic material from which they are comprised, minerals are the focus.

A **mineral** is a naturally occurring substance that is solid and stable at room temperature, represented by a chemical formula and has an ordered elemental structure, often crystalline. It is different from a rock, which can be an aggregate of minerals or nonminerals (basic elements) and does not have a specific chemical composition. There are over 4,900 known mineral species. The silicate minerals compose over 90% of the Earth's lithosphere (such as SiO_2—quartz crystal), since oxygen and silicon constitute approximately 75% of the lithosphere (**Figure 47.5**). The diversity and abundance of mineral species is controlled by the Earth's chemistry. Minerals are distinguished by various chemical and physical properties. Differences in chemical composition and crystal structure distinguish various species, and these properties in turn are influenced by the mineral's geological environment of formation. Changes in the temperature and pressure of a rock mass can cause changes in its mineralogy.

Table 47.1

Element	Percentage of Earth's Crust
Oxygen (O)	46.6
Silicon (Si)	27.7
Aluminum (Al)	8.1
Iron (Fe)	5.0
Calcium (Ca)	3.6

(continued)

Element	Percentage of Earth's Crust
Sodium (Na)	2.8
Potassium (K)	2.6
Magnesium (Mg)	2.1
All others	1.5

Figure 47.5 Quartz crystal (SiO_2) represents the dominant mineral in the continental crust. **(USGS)**

Structure of the Earth
By Dean Fairbanks

Taking Earth in a cross section reveals a layered structure and internal energy in the layering. The following pattern is revealed starting from the planet's center and working out to the surface:

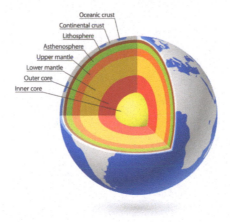

Image © Shutterstock, Inc.

Figure 48.1 The structure of the Earth.

- **Inner Core**—solid, consists of iron and nickel. It is approximately five times denser than the surface rocks and is 1,230 km thick.
- **Outer core**—liquid, consists mainly of iron. It is believed that the Earth's magnetic field is generated by movements in the liquid outer core, which is 2,250 km thick.
- **Mantle**—solid lower and semiliquid upper subdivisions. Consists of lower density material known as peridotite, a material composed of silicate minerals and olivine in the upper mantle. Lower mantle consists of olivine (iron/magnesium silicates). This layer is 2,590 km thick.
- **Asthenosphere**—plastic-like qualities. It is 180 km thick.
- **Crust**—solid, divided into two different types, continental and oceanic crust. Depth varies between 6 and 70 km.

During the Earth's early formation, the heavy elements went to the core via gravity and the lighter elements ended up in the crust. While there are many scientific limitations to knowing the interior of the planet, geoscientists have been able to utilize seismic waves that are created from earthquakes and satellite gravity mapping to provide evidence relating to the internal structure.

The final concept to consider in order to understand the Earth's surface connection to the internal structure is to address **isostatic adjustment**. This concept is one that looks at the equilibrium between the crust (lithosphere) and the asthenosphere such that the crust is broken up into plates that float at an elevation that depends on their thickness and density. When continental mountains are building, they displace downward pressure into the plastic like asthenosphere. This is like when one lies on a water bed, water is displaced due to the weight of the body. Over time, mountains have their elevations reduced due to weathering and erosion, they grow lighter and the asthenosphere rebounds against the crust causing uplift at the surface. This adjustment of thickening crust displacement and then crustal thinning upliftment is the major broad-scale process of topographic development on the planet's surface. Isostatic adjustment is the interplay between endogenic and exogenic processes.

Plate Tectonics and the Cycling of the Lithosphere

By Ingrid Ukstins and David Best

Plate tectonic theory provides us with a framework for interpreting the composition, structure, and internal processes of Earth on a global scale and how they contribute to producing natural events such as earthquakes and volcanoes. **Plate tectonics** (*tekton*, "to build") is a theory developed in the late 1960s to explain how the lithospheric outer layer of the Earth moves and deforms on the ductile asthenosphere. The theory revolutionized the way we think about the Earth and has proven useful in predicting geologic hazard events and explaining many aspects of the natural processes we see on Earth.

> **plate tectonics:** The theory that the Earth's lithosphere consists of large, rigid plates that move horizontally in response to the flow of the asthenosphere beneath them, and that interactions among the plates at their borders (boundaries) cause most major geologic activity, including the creation of oceans, continents, mountains, volcanoes, and earthquakes.

The lithosphere is divided into segments called tectonic plates that slide continuously over the asthenosphere. They are also referred to as **lithospheric plates**, or simply plates. The plates can encompass both continental and oceanic crust (note the North American plate), but some are composed largely of oceanic crust (e.g., the Pacific plate for example). The continuous movement of the plates is powered by internal heat forming convection currents within the mantle. The slow movements of different plates (usually a few centimeters per year, or about as fast as your fingernail grows) create natural hazard zones of volcanic activity and earthquakes where their margins meet.

> **lithospheric plate:** A series of rigid slabs of lithosphere that make up the Earth's outer shell. These plates float on top of a softer, more plastic layer in the Earth's mantle known as the asthenosphere.

Plate Boundaries

Along these margins, or **plate boundaries**, the plates diverge, converge, or slide horizontally past each other (**Figure 49.1**). Again, these plate boundaries are important because the plates interact where they meet and are zones where deformation of the Earth's lithosphere is taking place and generating potential hazard areas.

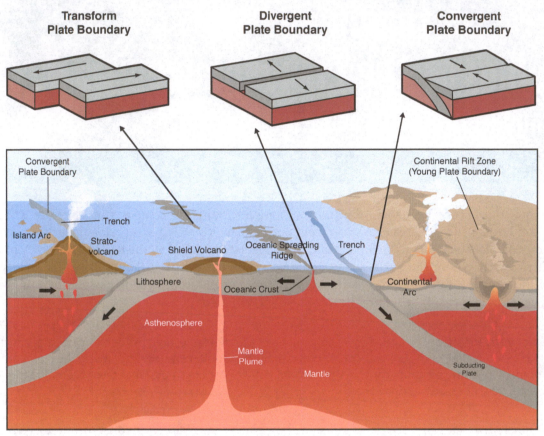

United States Geological Survey.

Figure 49.1 Plate boundaries consist of three types: divergent, where plates move apart; convergent, where plates come together; and transform, where the plates slide past one another.

Divergent Plate Boundaries

At **divergent plate boundaries**, the tectonic plates move away from each other due to tensional stresses. Divergent plate boundaries in oceanic crust generate new oceanic crust as magmas rises from the underlying asthenosphere (**Figure 49.2**). The magmas intrude and erupt to form basalt rock along the newly created edge of the plate. The margins of divergent plate boundaries occur mostly in the oceanic plates and they are marked by **mid-ocean ridges** (e.g., the Mid-Atlantic Ridge) that are essentially linear underwater mountain ranges uplifted by the hot underlying mantle. These boundaries are also known as *oceanic spreading centers*, because oceanic lithosphere spreads away on each side of the margin at the ridge as a result of underlying *convection cells* in the mantle. While most diverging plate boundaries occur at the ocean ridges, sometimes continents are split apart along zones called *rift zones*, such as at the East African Rift in eastern Africa. Volcanism and earthquakes are common along diverging plate boundaries but are of relatively low intensity.

divergent plate boundary: A boundary in which two lithospheric (tectonic) plates move apart.

mid-ocean ridge: An uplifting of the ocean floor that occurs when mantle convection currents beneath the ocean force magma up where two tectonic plates meet at a divergent boundary. The ocean ridges of the world are connected and form a global ridge system that is part of every ocean and form the longest mountain range on Earth.

© Kendall Hunt Publishing Company.

Figure 49.2 Iceland lies along a divergent plate boundary as two separate plates move away from each other. The island consists of volcanic material extruded upward from greater depths. Active volcanoes are shown by triangles.

Convergent Plate Boundaries

Convergent plate boundaries are where two tectonic plates move toward each other by compressional stresses and where most lithosphere is destroyed. In this manner, the Earth maintains a global balance between the creation of new lithosphere and the destruction of old lithosphere. At such boundaries, when one of the plates is colder and denser than the other, it sinks down into the mantle in a process called **subduction**. Convergent boundaries are complex because the margins can involve convergence between two plates carrying oceanic crust, between a plate carrying oceanic crust and another carrying continental crust, or between two plates carrying continental crust. Each type of boundary behaves differently based on the density of the two different plates colliding—continental crust is light and buoyant and resists subduction whereas oceanic crust can be cold and dense and subduct more readily.

> **convergent plate boundary:** A boundary in which two plates collide. The collision can be between two continents (continental collision), a relatively dense oceanic plate and a more buoyant continental plate (subduction zone) or two oceanic plates (subduction zone).
>
> **subduction:** Process of one crustal plate sliding down and below another crustal plate as the two converge.

At **subduction zones**, the oceanic plate subducts (sinks) beneath another oceanic plate, *ocean-ocean convergence*, or the oceanic plate subducts beneath a continental plate, *ocean-continent convergence*. Where subduction occurs, an **oceanic trench** is formed on the seafloor that marks the plate boundary. The subducting plate is heated by the surrounding mantle and as it warms up it releases water and other volatiles from the ocean crust rocks, overlying sediments, and from water-rich minerals in the rocks. This fluxes the overlying mantle with water as it descends into the hotter asthenosphere and generates magma that rises to the surface through the overriding upper plate, and if it reaches the surface the erupting lava forms a chain of volcanoes (known as a **volcanic arc**). The oldest oceanic crust on Earth is about 180 million years old because the denser oceanic lithosphere continuously recycles back into the mantle at subduction zones. Rocks found in continents are as old as 3.96 billion years because subduction consumes very little of the lighter buoyant continental crust.

> **subduction zone:** Also called a convergent plate boundary. An area where two plates meet and one is pulled beneath the other.
>
> **oceanic trench:** Deep, linear, steep-sided depression on the ocean floor caused by the subduction of oceanic crustal plate beneath either other oceanic or continental crustal plates.
>
> **volcanic arc:** Arcuate chain of volcanoes formed above a subducting plate. The arc forms where the downgoing descending plate becomes hot enough to release water and gases that rise into the overlying mantle and cause it to melt.

In ocean-ocean convergence, two oceanic crustal plates collide and the older, colder, denser crust subducts beneath the younger less dense crust. The subducting plate releases volatiles into the overlying mantle and generates magmas from mantle melting, and the magmas rise and erupt to form a chain of volcanoes built on the overlying plate of ocean crust, called **island arcs**. Well-known examples of island arcs include the islands of the Caribbean, the Aleutian Islands, and the island chains of Japan, Indonesia, and the Philippines. In ocean-continent convergence, the magma rises through the continental plate to the surface, where it erupts to form a volcanic mountain chain along the edge of the continent. Well-known examples of this type of volcanic arc are the Cascade Volcanoes (including Mount Saint Helens) of the northwestern United States and the Andes of South America. The hazards associated with these two boundaries include some of the largest earthquakes ever recorded, frequent volcanic eruptions, and tsunami.

> **island arc:** An arc-shaped chain of volcanic islands produced where an oceanic plate is sinking (subducting) beneath another.

When two continental plates converge, both plates are made of light, buoyant continental material and neither one wants to be subducted. Instead, they smash together to form a collision boundary. The force of collision forms a mountain range from the compression of the two continents at their convergent boundary. The Himalayan Mountains between India and China are being formed in this way today, as were the ancient Appalachian Mountains about 450 million years ago. This boundary also produces frequent and powerful earthquakes.

Transform Plate Boundaries

Transform plate boundaries occur when two plates slide past one another horizontally along faults known as **transform faults**. These are also zones of frequent and powerful earthquakes, but generally not zones of volcanism since material is not transferred between the asthenosphere and lithosphere. A well-known example of a continental transform fault is the San Andreas fault of California, which forms one part of the boundary between the Pacific plate and the North American plate (**Figure 49.3**).

transform plate boundary: An area where two plates meet and are moving side to side past each other.

transform fault: A strike-slip fault with side to side horizontal movement that offsets segments of an a continental or oceanic plate.

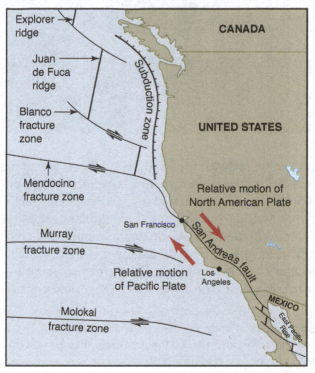

Figure 49.3   The Pacific Plate is sliding along against the western edge of North America. This contact has formed the San Andreas fault, a right-lateral strike slip fault. Notice that California actually lies in two different plates.

The Rock Cycle
By Ingrid Ukstins and David Best

A **rock** is an aggregate of one or more **minerals** which are naturally occurring inorganic solids, each having a definite chemical composition and a crystalline structure. Chemical composition and crystalline structure are the two most important properties of a mineral and give them their unique physical properties, such as hardness. Geologists classify minerals into groups that share similar compositions and structures. The most important and common minerals are the silicates, which are composed of combinations of oxygen and silicon with (or without) metallic elements. Geologists classify rocks into three categories based on how they were formed: *igneous*, *sedimentary*, and *metamorphic*. Each group contains a variety of different rock types that differ from one another in composition or texture (size, shape, and arrangement of minerals).

rock: A naturally occurring aggregate of minerals. Rocks are classified by mineral and chemical composition; the texture of the constituent particles; and also by the processes that formed them. Rocks are thus separated into igneous, sedimentary, and metamorphic rocks.

mineral: Any naturally occurring inorganic substance found in the earth's crust as a crystalline solid.

Rocks may seem permanent and unchanging over a human lifetime. However, over geologic time rocks are constantly exposed to different physical and chemical conditions (or environments) that change them. The processes that change rocks from one type to another are illustrated by the rock cycle (**Figure 50.1**). The interactions of energy, Earth materials, and geologic processes act to form and destroy rocks and minerals. The rock cycle is the slowest of Earth's cyclic processes with rocks recycled on a scale of hundreds of millions of years. This slow process is responsible for concentrating the planet's nonrenewable resources that humans depend upon. Plate tectonics is responsible for recycling rock materials and drives the rock cycle, as well as determining to a large degree how and where different rock types will form. The rock cycle provides a way of viewing the interrelationship of internal and external processes and how the three rock groups relate to each other.

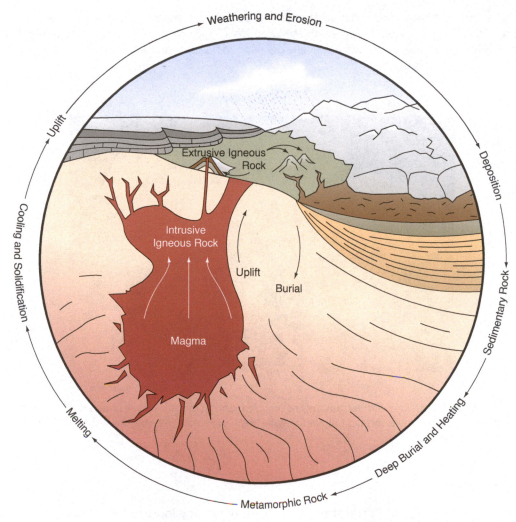

Figure 50.1   In the rock cycle material is continually reworked to form sedimentary, igneous, and metamorphic rocks.

© Kendall Hunt Publishing Company.

Igneous Rocks

Igneous rocks result when molten rock, called magma, cools and solidifies or volcanic ejecta such as ash accumulate and consolidate (**Figure 50.2**). Magma can be formed from partial melting of the mantle or melting of continental crust. Magma that cools and solidifies (a process called crystallization) slowly beneath the surface produces *intrusive igneous rocks* whereas magma that erupts and cools on the surface produces *extrusive igneous rocks*. Igneous rocks make up approximately 95 percent of the Earth's crust, but their great abundance is hidden on the Earth's surface by a relatively thin but widespread layer of sedimentary and metamorphic rocks.

> **igneous rock:** A rock formed when molten rock (magma) has cooled and solidified (crystallized). Igneous rocks can be intrusive (plutonic) or extrusive (volcanic).

Rocks exposed at the Earth's surface break down physically and chemically by **weathering** caused by exposure to gases and water in the atmosphere and hydrosphere. Weathering processes result in particles and dissolved ions that are reworked and redeposited as sediment and by erosional agents such as streams, glaciers, wind, or waves.

© Linda_K/Shutterstock.com.

Figure 50.2 Igneous rocks can form either from the cooling of magma in the subsurface or by solidifying after being erupted onto the surface. These rocks formed following a catastrophic eruption of a volcano in the southwestern United States more than 27 million years ago.

Sedimentary Rocks

Sedimentary rocks result from the lithification (the process of turning loose sediment into stone) of unconsolidated sediment (**Figure 50.3**). Sediments are usually lithified into sedimentary rock when compacted by the weight of overlying sediment layers, or when cemented by minerals as subsurface water containing dissolved ions moves through the pore spaces between sediment particles and those ions are precipitated as secondary minerals filling the open spaces. Sedimentary rocks can form from the consolidation of rock fragments and mineral grains, precipitation from solution, or compaction of plant and animal remains. Since these rocks form from sediment at the Earth's surface, geologists can determine the type of environment in which sediments were deposited and the transporting agent. Sedimentary rocks also contain evidence of past life-forms preserved as fossils and thus are useful in interpreting Earth's history.

© Leene/Shutterstock.com

Figure 50.3 Sedimentary rocks are typically layered, having been formed by the deposition of sediments derived from weathering of the continents.

Metamorphic Rocks

Metamorphic rocks result from the transformation of other rocks when deeply buried and subjected to high heat and pressure as well as chemical activity with fluids (**Figure 50.4**). Rocks recrystallize under high heat and pressure and the new minerals align themselves in a similar orientation to become foliated; if there is no directional pressure as the new minerals crystallize, then the mineral grains form with a random orientation and are nonfoliated. Foliation is the parallel alignment of minerals due to pressure, which gives the rock a layered or banded appearance. With the addition of higher pressures or increased temperatures, the metamorphic rocks will melt, creating magma, beginning the cycle again.

> **metamorphic rock:** A rock that has been altered physically, chemically, and mineralogically in response to strong changes in temperature, pressure, shearing stress, or by chemical action of fluids.

The rock cycle can follow many different paths. For example, weathering may turn uplifted metamorphic rocks into sediment, which becomes lithified into a sedimentary rock. An igneous rock may become metamorphosed into a metamorphic rock. The rock cycle simply shows that rocks are not permanent but change over geologic time through internal and external processes. The rock cycle illustrates several types of interactions between Earth system's spheres. Interactions occur among rocks, the atmosphere, the hydrosphere, the cryosphere, and the biosphere. Water and air, aided by natural acids and other chemicals secreted by plants and animals, weather solid rocks to form large amounts of *clay* (an important silicate mineral) and other sediments. During these processes, water and atmospheric gases react chemically to become incorporated into the clay minerals and transfer matter from the atmosphere and hydrosphere to the solid material of the geosphere. Rain and gravity then wash the sediment into streams, which is later deposited (mostly in the oceans at the edge of continents). Energy from the Sun powers the hydrologic cycle to evaporate water to form rain, which in turn feeds flowing streams. The interaction of the hydrologic cycle is also part of the rock cycle, illustrating again that all of Earth's processes are interrelated.

Figure 50.4 Metamorphic rocks form when rocks are heated, subjected to pressure, or have chemically active fluids introduced into their composition. They do not melt, these chemical reactions take place in a solid state and the final composition of the metamorphic rock is usually similar to the starting composition of the protolith—the material that was metamorphosed.

Topographic Relief
By Dean Fairbanks

The deep Earth processes that drive plate tectonics have created a varied crustal surface below and above the waters that cover the planet (**Figure 51.1**). These variations in **topography** are what are called surface relief features.

Figure 51.1   Earth's surface relief features. **(NOAA)**

The relief features are scaled based on crustal orders of relief. We are concerned here with the relative relief based on elevation variations on the terrestrial landsurface, i.e., above mean sea level. The order of relief concerns the spatial scale of features on a terrestrial landsurface.

Table 51.1		
Order of Relief	**Scale (detail)**	**Example**
First order	Coarse	Continents and ocean basins
Second order	Moderate	Mountain ranges, plains, mid-ocean ridges, and abyssal plains
Third order	High	Hills, valleys, and individual mountains

Hypsometry is the measurement of land elevation relative to sea level, whereas **bathymetry** is the underwater equivalent.

Tectonic Processes

The processes of crustal formation in particular and tectonic activity in general are very slow (**Figure 51.2**). The rates of plate motion range between 2.5 and 15 cm a year. Despite the slow movement, endogenic processes result in gradual uplift and new landforms, with major mountain building occurring along plate boundaries. There are three categories:

- Residual mountains and continental **cratons,** which represent inactive remnants of ancient tectonic activity. Cratons are ancient crystalline rock (i.e., granites), which are called **shields** when they are exposed at the surface.

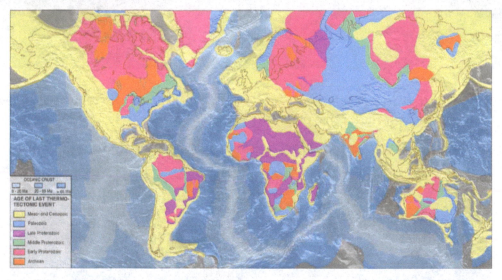

Figure 51.2 **Age of last major tectonic process. (USGS)**

- Tectonic mountains and landforms derived from active folding and faulting
- Volcanic features that are either inactive or active

There are three major types of active plate boundaries: constructive interaction—spreading zones; destructive interaction—converging and subducting plates; and conservative interaction—sliding plate boundaries (**Figure 51.3**).

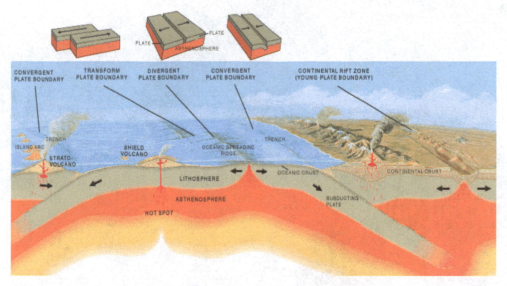

Figure 51.3   Cross section illustrating the main types of plate boundaries. **(USGS)**

Constructive interaction motion occurs when two plates are spreading away from each other, thinning the crust between them and causing a gap where new material can emerge in the form of magma to fill the gap. In the oceanic basins, this situation is known to create **mid-ocean ridges**, like the most famous one snaking down through the center of the Atlantic Ocean. In East Africa, the spreading on the continental crust is called a **rift zone** (Great Rift Valley), where the crust is thinning as it is stretched thin.

Topic 52

Tectonic Processess
By Dean Fairbanks

Tectonic Processes

Destructive Interaction motion consists of two types of converging zones: one that subducts and the other that compresses. In the case of subduction, one plate is usually heavier and the other lighter plate rides up over the heavier plate which forces the heaviest plate to plunge back into the asthenosphere. This subduction causes destruction of the plate in the form of melting back into magma; it is one way of viewing our planet's materials recycling program. There are two different ways this can occur with two different landscape outcomes created. One is when the subduction happens between two oceanic plates (**Figure 52.1**). The magma melt from one plate erupts under the other oceanic plate to form volcanic island arcs, Japan, New Zealand, and Alaska's Aleutian islands being the most famous forms of this interaction. In the other case, the denser oceanic crust subducts under the lighter continental crust, thus melting and allowing magma to erupt on the surface of the continental plate creating a volcanic chain; the Northwestern US Cascade Range of volcanoes represents this type of interaction (**Figure 52.2**). The southern end of the Cascade Range occurs in California with the two most prominent volcanoes of Mt. Lassen and Mt. Shasta being featured. In the north, Mt. Hood (Oregon), Mt. St. Helens (Washington), and Mt. Rainer (Washington) are prominent features from this subduction zone.

Finally, there is the case of **destructive collision** or compression between two plates. This is not unlike having two cars having a front-in collision with each other. In the case of plate activity, it is usually seen between two continental plates smashing into each other and instead of subduction there is the creation of folded/faulted mountains because the crustal material becomes compressed upward (**Figure 52.3**). The most powerful example of this is the compression of the Indian plate into the Eurasian plate forming the Himalayan Mountains.

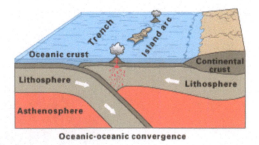

Oceanic-oceanic convergence

Figure 52.1   Oceanic–continental destructive convergence illustrating subduction. (USGS)

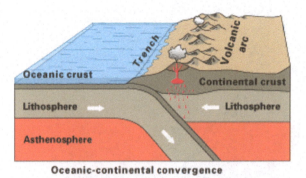

Oceanic-continental convergence

Figure 52.2 Oceanic–continental destructive convergence illustrating subduction. **(USGS)**

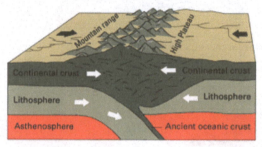

Continental-continental convergence

Figure 52.3 Continental–continental convergence zones. **(USGS)**

Conservative interaction represents the interaction between two plates that slide past one another. The transverse motion that this type of boundary represents on continents is called strike-slip faulting. In California, the San Andreas Fault is the best representative of this action, with the Pacific plate sliding northwest past the much slower North American plate at the rate of 2 inches per year (**Figure 52.4**). While the North American plate is also moving northwest, it is moving slower so it appears as if the plates are going in opposite directions along their slip boundary. This is not unlike the relative motion one witnesses when passing another car on the freeway, even though both cars are going in the same direction. The slipping along oceanic plates is called a transform fault.

In faulting, rocks on either side of a fracture displace relative to the other side (**Figure 52.5**). The main types of fault include:

- **Normal fault**—This fault is the result of tension (stretching). Rock strata are pulled apart and one side is thrown down. Movement is down the dip of the fault plain. Land area is increased at the surface and a fault scarp is produced.
- **Reverse or thrust fault**—This is the result of compression (shortening). Beds on one side of the fault plain are thrust over the other, i.e., overthrusting up the dip. This causes overlapping of the rock strata and the surface is decreased. An overhanging fault scarp (hanging wall) is formed.
- **Strike-slip fault**—This is formed where the shift is horizontal (shear) although the fracture is vertical.

The landforms produced by faults can be quite dramatic, such as the **Horst and Graben** landscapes (**Figure 52.6**). Upward-faulted and downward-faulted blocks are produced by a sequence of faults. Horsts stand out above relatively low land on either side, bounded by fault scarps on either side. This type of landscape is most notable as the Basin and Range topography of Nevada (**Figure 52.8**). The Great Rift Valley of Eastern Africa is a faulted Graben trough let down by parallel faults with stretching tension dropping the block over time (**Figure 52.7**).

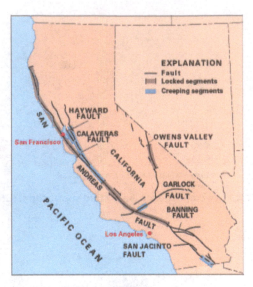

Figure 52.4   Generalized map of major faults in California. (**USGS**)

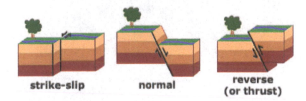

Figure 52.5   The three main types of fault structures. (**USGS**)

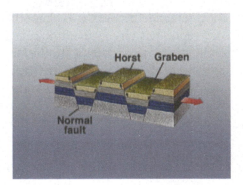

Figure 52.6   Horst (upfaulted block) and Graben (downfaulted block) landscapes. (**USGS**)

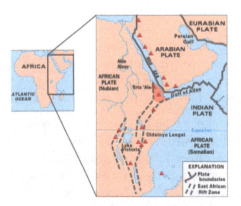

Figure 52.7   Great Rift Valley of East Africa. (**USGS**)

Thelin, G.P., and Pike, R.J., 1990, Digital shaded relief map of the conterminous United States: Menlo Park, California, U.S. Geological Survey digital image processing, scale 1:3,500,000

Figure 52.8 Basin and Range landscape of Nevada. Where the alternating valleys and low mountain ranges of Nevada contrast sharply with the high, rugged California mountain ranges that surrounded the low-lying Central Valley. **(USGS)**

Earthquakes

Along a fault, the moment of fracture releases a sharp jot of energy, which is an earthquake. The release of energy sends waves traveling through the Earth's crust at the moment of rupture. The aftershocks occur later as stresses are redistributed. The general sequence is as follows:

Rocks under stress → Rocks deform → Breaking point is reached, rocks shear → Stored energy is released as seismic waves

The energy is released underground, where the rocks break from the shearing. This location is called the **focus**. Directly vertical from the focus at the surface is called the **epicenter** of the earthquake. This is the geographic location (latitude, longitude) that is mapped and described by news agencies. However, the actual earthquake of energy release is underground (**Figure 52.9**). Geologists are still unable to predict *when* an earthquake is likely to occur, but they are becoming better at predicting *where* earthquakes are likely to occur (**Figure 52.12**).

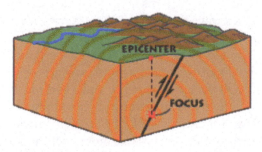

Figure 52.9 Epicenter versus focus of an earthquake. **(USGS)**

Scientists can measure the size of an earthquake in two ways:

- **Moment magnitude scale** - The measurement of magnitude (sometimes still referred by major news media as the "Richter scale") is based on a seismograph's record of the amount of shaking (energy released) during an earthquake. Using a diagnostic evidence of the Earth's internal structure. The waves move away from the focus in all directions in time proportional to distance. Seismic waves come in two forms: body waves and surface waves. Body waves such as **P** (primary) and **S** (secondary) **waves** move through the Earth's interior and travel much more quickly than surface waves. The P wave (~8 km/sec) is comparable to sound waves, which compresses and dilates the rock as it travels forward through the Earth. The S wave shakes the rock vertically as it advances at half (~4 km/sec) the P-wave speed. Surface waves such as L (Love) and Rayleigh waves move over the surface of the Earth and cause much of the destruction during earthquakes by shaking sideways.
- **Intensity** (Mercalli) **scale** - The intensity of an earthquake measures its effect on people, objects, and structures. Reports by people who experience an earthquake determine the intensity. It is based on a scale of I–X+ (nothing felt to extreme shaking/heavy damage) (**Figure 52.10**).

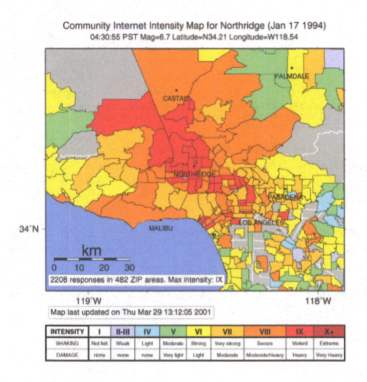

Figure 52.10 **Intensity map of the 1994 Northridge, California earthquake. Magnitude was 6.7. (USGS)**

In all earthquakes, energy is released as the two sides of the fault slide past one another. Seismic waves carry this energy through rock, generating the ground shaking that causes much of the damage during earthquakes. The transmission of the shock wave through the planet varies according to temperature and the density of various layers within the planet. It is these seismic waves that are used as indirect diagnostic evidence of the Earth's internal structure. The waves move away from the focus in all directions in time proportional to distance. Seismic waves come in two forms: body waves and surface waves. Body waves such as P (primary) and S (secondary) waves move through the Earth's interior and travel much more quickly than surface waves. The P wave (~8 km/sec) is comparable to sound waves, which compresses and dilates the rock as it travels forward through the Earth. The S wave shakes the rock vertically as it advances at half (~4 km/sec) the P-wave speed. Surface waves such as L (Love) and Rayleigh waves move over the surface of the Earth and cause much of the destruction during earthquakes by shaking sideways (**Figure 52.11**).

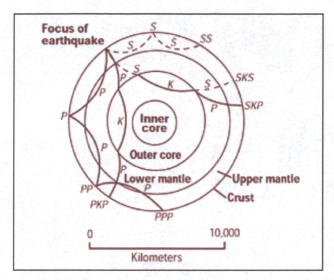

Figure 52.11 Cross section of the whole Earth, showing the complexity of paths of earthquake waves. The paths curve because the different rock types found at different depths change the speed at which the waves travel. Solid lines marked P are compressional waves; dashed lines marked S are shear waves. S waves do not travel through the core but may be converted to compressional waves (marked K) on entering the core (PKP, SKS). Waves may be reflected at the surface (PP, PPP, SS). (**USGS**)

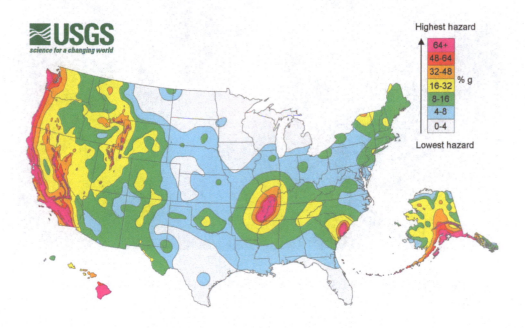

Figure 52.12 Potential earthquake hazard map of the US. (**USGS**)

Volcanic Eruptions
By Ingrid Ukstins and David Best

© fboudrias/Shutterstock.com

Figure 53.1 An eruption of Volcán de Fuego (Spanish for Volcano of Fire) in Guatemala. This is one of Central America's most active volcanoes.

Living with Volcanic Risk

More than 800 million people live within 100 kilometers of a volcano, but few of the potential hazards associate with all of the active volcanoes on Earth are monitored regularly. Forecasting the hazards and impact of eruptions accurately is an ongoing issue—scientists are combining satellite measurements of ground movement and artificial intelligence to more accurately monitor volcanic activity, and may one day be able to predict volcanic eruptions. Detailed research looking at the internal 'plumbing' of volcanoes like Etna in Italy also helps us understand eruptions and how quickly magma moves to the surface. The United States Geological Survey predicts that California's next big hazard could be a volcanic eruption rather than an earthquake. Auckland, the largest city in New Zealand, has 1.5 million people who live on a volcanic field made up of 53 different volcanoes—the youngest of which is only 600 years old. Residents may have as little as a few hours to several days of warning before the next eruption.

© Steven Bostock/Shutterstock.com

Box Figure 53.1.1   Auckland, New Zealand, city skyline with Mount Victoria volcano—now a public park and greenspace—in the front center of the picture.

The Plate Tectonic Connection

Earlier discussions of plate tectonics showed us that volcanic activity has played a major role in the development of Earth's outer crust. Volcanoes are the result of molten rock (called **magma**) that has worked its way to the surface and has been extruded onto either the continents or the ocean floor as **lava**. Lava can form volcanic mountains (**Figure 53.2**) and flows on land (**Figure 53.3**).

© Ecuadorpostales/Shutterstock.com

Figure 53.2   Tungurahua Volcano in Ecuador was quiet for many years before it began erupting in the year 2000. The last activity of this current eruptive episode ended in March of 2016.

> **magma:** Molten rock that lies below the surface.
>
> **lava:** Molten rock that flows onto the surface and cools.

Magma rises under the mid-oceanic ridges as plates move away from each other in a divergent manner (**Figure 53.4**). The extruded lava is cooled and quenched relatively rapidly and forms lava flows and pillow **basalts**, which are tubes of lava similar to toothpaste squeezed from a tube, and form because the outer shell

© Wead/Shutterstock.com

Figure 53.3 **Lava flowing down Mount Etna, Italy.**

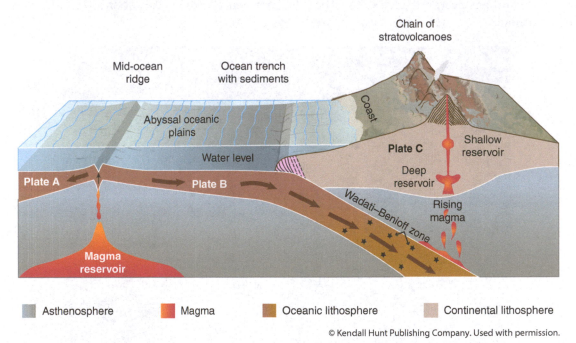

© Kendall Hunt Publishing Company. Used with permission.

Figure 53.4 **Cross section showing the divergent movement of oceanic lithospheric Plates A and B away from the mid-ocean ridge; Plate B is subducted underneath continental Plate C at a convergent plate boundary.**

of the lava quenches extremely fast underwater. As these tubes stack up, they drape over each other and form what looks like pillows or blobs in cross-section (**Figure 53.5**).

> **basalt:** An extrusive, fine-grained, dark volcanic rock that contains less than 50 percent silica by weight and a relatively high amount of iron and magnesium.

As tectonic plates continue to move away from the ridge, more magma flows out onto the ocean floor and hardens, forming the new ocean crust of basaltic oceanic lithosphere. This activity is not readily observed

Photo by A. Malahoff, USGS.

Figure 53.5 Underwater lava flows add material to the ocean floor. Notice the spherical shape of the pillow basalts.

as it occurs under several kilometers of water in most places except Iceland, where the mid-ocean ridge is exposed above sea level.

Oceanic plates that move away from the mid-oceanic ridge in one area of the Earth eventually collide with other plates. When this collision occurs between two oceanic plates or between an oceanic plate and a continental plate, basaltic oceanic lithosphere and sediments that have accumulated on the oceanic plate are subducted and driven downward into the mantle where they heat up. The trapped water in these subducted rocks is driven out of the sediments, rocks and minerals by the increased heat at depth. This dehydration process triggers melting of the overlying mantle rocks. At depths of roughly 100 kilometers there is sufficient heat to begin melting and produce new magma, which begins to rise toward the surface.

Near the continental margin the subducted oceanic **lithosphere** is relatively cold as it is usually old and far removed from its heat source at the mid-oceanic ridge. The descending oceanic lithosphere is being pulled down by the effect of gravity tugging on the sinking, colder slab. In **Figure 53.4** we see the Wadati-Benioff zone, a region where earthquakes occur along the cold, brittle, subducting slab. As the subducted oceanic plate descends, sometimes it breaks into pieces. Seismic evidence shows that there are regions in the subsurface of subduction zones where no earthquakes are occurring. This produces a seismic gap on the surface.

lithosphere: The solid outer layer of the Earth that rests on the mobile, ductile asthenosphere. Continental lithosphere, having a granitic or granodioritic composition, ranges in thickness from about 30 to 60 km; basaltic oceanic lithosphere ranges between 2 and 8 km thick.

Earth's surface consists of seven major lithospheric plates and including the many smaller ones about 20 total. Notice that continental plates such as the African plate and the North American plate also include oceanic areas as the plates extend to the mid-oceanic ridges and continues to form new crust from that boundary. The continents themselves consist of less dense, granitic material that "floats" atop the denser, underlying **asthenosphere**.

asthenosphere: The semi-plastic to molten zone beneath the rigid lithosphere.

There are about 1,500 known active volcanoes, and of these about 500 have erupted during historical times. More than 90 percent of these are associated with plate boundaries and approximately two-thirds of all active volcanoes are located around the Ring of Fire that surrounds the Pacific Ocean (**Figure 53.6**). Those areas along the Ring of Fire that display an almost continuous line of active volcanoes, such as in the Aleutian Islands southwest of Alaska (**Figure 53.7**) or the area around Japan and the Kurile Islands, are regions where very active subduction is taking place. Both volcanic eruptions and significant earthquake activity occur in these areas as the Pacific Plate is subducted under the adjacent continental and other oceanic plates.

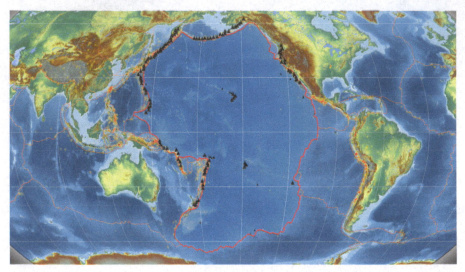

© Yarr65/Shutterstock.com

Figure 53.6   Major volcanoes found in the circum-Pacific Ring of Fire and adjoining regions.

Source: NASA.

Figure 53.7   Mount Cleveland in the Aleutian Islands last erupted on June 28, 2018, which was identified by satellite images that showed a lava flow in the summit crater of the volcano. Cleveland Volcano is on the uninhabited island of Chuginadak, which makes it difficult to monitor. Cleveland volcano is responsible for the only known fatality from an Aleutian Island volcano, when it erupted as a small group of soldiers were stationed there during June of 1944.

Of the world's 15 most populous countries, portions or all of five of the countries (the United States, Indonesia, Japan, Mexico, and the Philippines) lie on the Ring of Fire and have areas that could experience the direct effects of major volcanic eruptions. The hazard potential for large portions of Indonesia, Japan, and the Philippines is high. Volcanic eruptions during the past two decades have been common in these regions.

Volcanoes also occur in **rift valleys**—regions on the Earth that are undergoing stretching due to large scale, extensional tectonic forces. These are areas where the lithosphere is being pulled apart. The East African Rift System, which passes through Uganda, Kenya, Tanzania, and Ethiopia, is home to many volcanic peaks, including Mount Kilimanjaro (**Figure 53.8**). At an elevation of 5,895 m (19,340 ft), it is the world's highest free-standing mountain. The East African Rift results from the African plate stretching and splitting apart, which is attributed to elevated heat flow from the mantle forming thermal uplifts that stretch and fracture the outer brittle crust. This process is often associated with volcanic eruptions.

rift valley: A depression formed on the surface caused by the extension of two adjacent blocks or masses of rock.

© Volodymyr Burdiak/Shutterstock.com

Figure 53.8 Mt. Kilimanjaro, the world's highest free-standing mountain, is the result of volcanism related to the East African Rift System.

Types of Volcanoes
By Ingrid Ukstins and David Best

Topic

54

Types of Volcanoes

Shield volcanoes are the largest volcanoes in size and are so named because they have the broad appearance of an inverted shield. They are formed when low viscosity magmas are slowly extruded onto the surface and flow as effusive lavas out over large areas. These flows build up over time and produce very large structures such as those found in the Hawaiian Islands and Iceland. Mauna Loa, the largest volcano on Earth, has an elevation of 4,175 m (13,697 ft) above sea level and rests on the ocean floor in 5,800 m (19,000 ft) of water (**Figure 54.1**). Generally eruptions associated with shield volcanoes are quiet and move across the surface at from 1 to 10 meters an hour. Fluid lava flows travel large distances and shield volcanoes have very flat slopes of about 5 degrees.

shield volcano: A broad volcano that has gentle slopes consisting of low viscosity basaltic lava flows.

© dirkr/Shutterstock.com

Figure 54.1   Mauna Loa, on the island of Hawaii, appears along the horizon.

Cinder Cones

Cinder cones build up through many explosions of gas-rich mafic lava that form small to medium sized **pyroclastic** particles of ash and lapilli that build up around the vent of the volcano (**Figure 54.2a**). As they are cast out of the volcano, they pile up and create a conical structure called a cinder cone. Because little or no lava flows out the top of the volcano, the cinders remain as loose grains that are susceptible to erosion. The unconsolidated material forms a pile based on how well the particles interlock, and the slope of the surface is called the angle of repose. Angular particles interlock well and cinder cones have steep slopes of about 30 degrees (**Figure 54.2b**).

cinder cone: A conical-shaped hill created by the build up of cinders (lapilli) and other pyroclastic material around a vent.

pyroclastic: Related to material that is thrown out by a volcanic eruption.

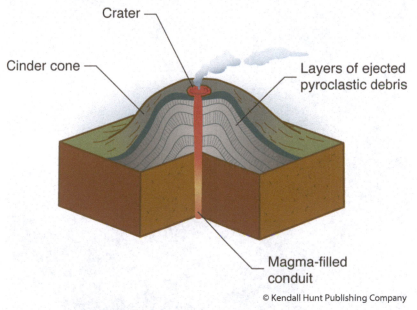

© Kendall Hunt Publishing Company

Figure 54.2a   Block diagram showing how pyroclastic debris builds up to form a cinder cone.

© ollirg/Shutterstock.com

Figure 54.2b   Cinder cone on Mount Etna. Sicily, Italy.

Cones are generally no more than 300 m (1,000 ft) in height and pose a threat only to areas in the immediate vicinity of the eruption. Wind can transport some of the smaller ejecta over considerable distances (**Figure 54.3**).

© Alexander Piragis/Shutterstock.com

Figure 54.3   An aerial view of actively erupting Tolbachik Volcano cinder cone, Kamchatka Peninsula, Russia.

Composite Volcanoes

Perhaps the most widely recognized volcano shape is a **composite volcano**, or stratovolcano, which has a symmetrical, generally conical shape. The volcanic peaks of Japan, the Philippines, Alaska, and the Pacific Northwest in the United States are excellent examples of composite volcanoes. They form from deposits of pyroclastic debris that build up near the vent and then are covered by occasional lava flows that move down the flanks to "glue" the looser debris in place. This repeated combination of loose material and lava produces stratified layers that create the awe-inspiring, cone-shaped mountains we associate with volcanoes (**Figure 54.4**). These are made mostly of andesite but usually include some rock types that are both higher and lower in silica. Composite volcanoes can be up to 2500 meters from base to summit, and are second in size behind shield volcanoes. They have slopes of about 10 degrees on their flanks and as much as 30 degrees near the summit, reflecting the more fluid lavas that flow far from the vent and the pyroclastic material that builds up around the summit area. These are in many ways like a combination of shield volcanoes with lava flows and cinder cones with explosive eruptions.

composite volcano: A volcano that forms from alternating layers of lava and pyroclastic debris; also known as a stratovolcano.

© WorldStockStudio/Shutterstock.com

Figure 54.4 Mayon volcano, rising to an elevation of 2462 m in southeast Luzon, is the most active volcano in the Philippines. It has steep upper slopes averaging 35–40 degrees that are capped by a small summit crater. A single central conduit system has allowed the volcano to build up into a classic, symmetrical profile.

Lava Domes

During the late stage of many volcanic eruptions, as the magma in the vent cools down and begins to crystallize, the viscosity increases to the point where the lava can't be erupted but can only be pushed out as a semi-solid or solid mass, like a blob of toothpaste. Magma extruded at the vent thickens and creates a **lava dome** either just under or on the surface. Within a year or two of the major eruption of Mount St. Helens in May 1980, a dome formed in the inner crater and volcanic activity slowed down (**Figure 54.5**). A dome located at the top of a volcano can act like a cork in a bottle and hold in gases and new rising magma until enough pressure builds up in the system that will blow the dome out of the way. Following the main eruption of Mount St. Helens in May 1980 there have been additional eruptions that have added almost 150 million cubic yards of material to the dome and surrounding area (Box 54.1).

> **lava dome:** A dome-shaped mountain formed by very viscous lava flows.

© Alexander Piragis/Shutterstock.com

Figure 54.5 View of the dome inside the crater of Mount St. Helens, Washington.

Calderas

In terms of the destructive potential of various types of volcanoes, **calderas** are by far the biggest threat. However, they are also the least likely to occur, so we are relatively safe from the mega-hazard their explosions would produce. Mega-eruptions are termed ultra-Plinian and are characterized by extreme volumes of ejected material exceeding 10^9 or more cubic meters (1 cubic km) of material. The largest of these result in subsidence bowls or tectono-volcanic depressions (calderas) over the magma chamber, which form toward the end of the eruption process after magma has been erupted and the overlying crust or lid to the magma chamber collapses in on the empty space created from the eruption. Yellowstone National Park, located in the northwestern corner of Wyoming and stretching into portions of Idaho and Montana, is centered on a massive caldera covering almost 3,500 sq km (1,350 sq mi). This area has experienced three major eruptions. The first occurred approximately 2.1 million years ago, followed by another about 1.3 million years ago. The most recent major eruption took place about 630,000 years ago (**Figure 54.6**). The result of these eruptions was the ejection of almost 3,800 cu km (900 cu miles) of volcanic ash that covered much of the continental United States. The area continues to be geologically active today with a well-developed geyser system emitting steam into the atmosphere (**Figure 54.7**). Earthquake activity is ongoing in the park and adjoining regions.

caldera: A large basin-like depression that is many times larger than a volcanic vent.

© Cromagnon/Shutterstock.com

Figure 54.6 Ashfall from Yellowstone eruptions 2 million years ago (in yellow) and 0.63 million years ago (in orange) compared to ash from the 1980 eruption of Mount St. Helens (in red) and Long Valley Caldera in California at 0.76 million years ago (in purple).

© Jakub Barzycki/Shutterstock.com

Figure 54.7 Stokkur geyser in Iceland.

| BOX 54.1 | The Eruption of Mount St. Helens May 18, 1980 |

FACTS ABOUT THE LATERAL BLAST

Summit elevation

Before	9,677 ft (2,950 m)
After	8,363 ft (2,549 m)
Removed	1,314 ft (401 m)

Crater dimensions

East-west	1.2 miles (1.93 km)
North-south	1.8 miles (2.9 km)
Depth	2,084 ft (635)
Volume of material removed	3.7 billion cu yds (2.8 billion cu m)

FACTS ABOUT FATALITIES

Human loss of life	57
Wildlife	At least 7,000 big game animals and 12 million salmon fingerlings in hatcheries; countless nonburrowing wildlife in blast zone

FACTS ABOUT THE ERUPTION COLUMN AND CLOUD

Height	Reached 80,000 ft (24,384 m) in 15 minutes
Downwind extent	Spread across the United States in 3 days; circled the globe in 15 days
Volume of ash	1.4 billion cu yds (1.07 billion cu m)
Ash fall area	Detectable amounts covered 22,000 square miles (56,980 sq km)
Ash fall depth	10 in (25.4 cm) at 10 miles (16 km) downwind (ash and pumice); 1 in (2.5 cm) at 60 miles (100 km) downwind; fractions at 300 miles (500 km) downwind

FACTS ABOUT PYROCLASTIC FLOWS

Area covered	6 sq mi (15.5 sq km); reached 5 mi (8 km) north of crater
Volume and depth	155 million cu yds (199 million cu m); many flows 3 to 30 ft thick; up to 120 ft thick (37 m) in some locations
Velocity	Estimated at 50 to 80 mph (80 to 130 kph)
Temperature	At least 1,300°F (700°C)

FACTS ABOUT THE LANDSLIDE

Area and volume removed	23 sq miles; 3.7 billion cu yds (60 sq km; 2.83 billion cu m)
Deposit depths	Buried 14 miles (22.5 km) of North Fork of Toutle River Valley to an average depth of 150 ft (46 m) (deepest deposits 600 ft or 183 m)
Velocity	70 to 150 mph (113 to 241 kph)

FACTS ABOUT THE LATERAL BLAST

Area covered	230 sq mi (596 sq km); extended to 17 mi (27 km) NW of crater

Volume of deposit	250 million cu yds (190 million cu m)
Velocity	At least 300 mph (480 kph)
Temperature	Up to 660°F (350°C)
Energy released	24 megatons (MT) of thermal energy (blast generated 7 MT; remainder was released heat)
Trees flattened	4 billion board ft (enough to build 300,000 2BR homes)

FACTS ABOUT LAHARS

Velocity	Between 10 and 25 mph (16 to 40 kph) (50 mph or 80 kph) on steep slopes on side of volcano)
Destruction caused	27 bridges; almost 200 homes
Effects on Cowlitz River	Reduced flood stage capacity at Castle Rock from 76,000 cfs (2150 cms) to less than 15,000 cfs (425 cms)
Effects on Columbia River	Reduced channel depth from 40 ft (12 m) to 14 ft (4.3 m); 31 ships stranded in ports upstream

Source: http://pubs.usgs.gov/fs/2000/fs036-00/

The U.S. Geological Survey (USGS) reported that the bulge and surrounding area slid away in a gigantic rockslide and debris avalanche, releasing pressure and triggering a major pumice and ash eruption of the volcano. Thirteen-hundred feet (400 meters) of the peak collapsed or blew outward. As a result, 24 square miles (62 square kilometers) of valley were filled by a debris avalanche; 250 square miles (650 square kilometers) of recreation, timber, and private lands were damaged by a lateral blast; and an estimated 200 million cubic yards (150 million cubic meters) of material were deposited directly by lahars (volcanic mudflows) into the river channels. Fifty-seven people were killed.

United States Geological Survey. Photo by Austin Post.

Mount St. Helens, May 18, 1980.

Economic losses resulting from the eruption of Mount St. Helens included almost $900 million in revenues lost for timber, agriculture, and fisheries. Damage to the surrounding infrastructure, including roads and bridges, along with dredging costs of rivers, amounted to almost $460 million (Blong, 1984).

Source: http://pubs.usgs.gov/fs/2000/fs036-00/-United States Geological Survey.

Topic 55

Volcanic Hazards
By Ingrid Ukstins and David Best

Volcanic Hazards

Gas Emission

In addition to gas being emitted from the volcano itself during an eruption or while it is restless, on rare occasions, an eruption of only gas can occur. A documented case happened in August 1986 in Cameroon, West Africa. Lake Nyos, situated in the crater of a volcano at an elevation of 1,091 m, is relatively deep given its small size (about 200 m deep, 1 to 1.5 km wide). Little intermixing of its cold, CO_2-saturated bottom waters with shallow, warm water normally takes place. However, the carbon dioxide built up to critical levels and the lake suddenly overturned, releasing dense suffocating carbon dioxide gas and killing, more than 1,700 people along with several thousand head of cattle. The denser carbon dioxide cloud moved down several stream valleys extending almost 100 meters above the ground surface and killing people up to 25 km away from the lake. Some people, who were unconscious for more than 36 hours, revived once the cloud lifted, only to find all of their family, neighbors and livestock were dead.

Lava Flows

Lava flows are produced when molten rock extrudes out onto Earth's surface. The ease with which it flows is controlled by the composition and temperature of the magma. Magmas that are rich in silica, such as those with a rhyolitic composition, have a high viscosity and barely flow at all. Basaltic magmas, those containing less silica, flow readily onto and across the surface. The island of Hawaii has been constructed by many episodes of eruption of basaltic lavas that continue today (**Figure 55.1**). As the lava cools and hardens, it forms new land on which future flows will travel and subsequently harden. Given the relatively slow movement of lava flows, it is easy to avoid them. However, infrastructure such as homes, roads and other buildings that cannot be moved can be destroyed as it is inundated by lava.

United States Geological Survey.

Figure 55.1 Lava flowing from Puʻu ʻŌʻō, Kilauea Volcano, Hawaii, through the Royal Gardens housing subdivision in February of 2008. The lava here is about 3 meters across.

Lahars

A **lahar** is a hot or cold mixture of water and volcanic material that forms a mudflow or debris flow. Lahars generally occur on or near composite volcanoes because of their height and steep topography and the lahar flows down the side of the volcano, usually following river valleys. Lahars can reach speeds of 200 km/hr (120 mph) and look like a rolling, churning slurry of wet concrete which can crush, bury or carry away almost anything in their path. Lahars can happen with or without volcanic eruptions—eruptions of hot material can melt snow and ice or mix with water from a crater lake and trigger lahars, but they can also form when a large amount or a long duration of rain occur during or after a volcanic eruption. The fine-grained, loose volcanic material that makes up the steep slopes of composite volcanoes is easily eroded. The initial flow may be small but lahars incorporate everything in their path and accumulate additional water through addition of melting snow or ice, or river or lake water. As they move downslope, lahars increase in size and can grow to 10 times larger than their initial size. As they move downslope and flow onto lowlands surrounding volcanoes, they slow down, spread out, and come to rest like a sheet of concrete (**Figure 55.2**). Lahars are one of the greatest volcanic hazards because they can occur during eruptions and also long after an eruption is over, they can occur without warning and can travel long distances beyond a volcano, and can inundate areas in distal, low-lying terrain.

lahar: A volcanic mudflow or landslide that contains unconsolidated pyroclastic material.

United States Geological Survey.

Figure 55.2 One year later, cleaning up from lahar deposits after the eruption of Mount Pinatubo, Philippines, on June 15, 1991. The cataclysmic eruption was the second largest of the 20th century.

The far-reaching effects of lahars became evident during the eruption of Nevado del Ruiz in Colombia in November 1985, when the resulting lahar traveled more than 75 km from the volcanic vent, covering the city of Armero and surrounding villages (**Figure 55.3**). More than 23,000 people died. The eruption of Mt. Pinatubo in the Philippines in 1991 was followed by more than 200 lahars produced by the rainy season in the spring and summer. These have continued for years after the eruption. These lahars erode and transport loose volcanic material far downstream from the volcano, and this deposition can lead to severe and chronic flooding in the river system. They can travel onto floodplains and bury entire towns and valuable agricultural land.

United States Geological Survey.

Figure 55.3 The town of Armero was destroyed and more than 20,000 people were killed by lahars from the eruption of Nevado del Ruiz, Colombia in November 1985. This was the second-deadliest volcanic disaster of the 20th century.

Ash Fall

Erupting volcanoes eject a wide variety of material into the air, ranging in size from ash to blocks and bombs. The small particles of ash are carried up in the eruption column and fall out in a downwind plume to form ash fall deposits. These fine particles accumulate like snow, the largest pieces fall out close to the volcano and the thickest deposits occur closest to the volcano. They are glassy fragmented bubble shards and commonly have sharp edges, which makes them very abrasive. They can act as irritants to the eyes and lungs, damage airplanes vehicles and houses, contaminate water supplies and damage or kill crops and livestock, and if the ash layer is thick and heavy enough, can cause roof collapse (**Figure 55.4**).

United States Geological Survey.

Figure 55.4   Volcanic ash is highly abrasive and can be remobilized after eruptions and obstruct visibility as well as damage motor vehicles.

Pyroclastic Density Currents

A pyroclastic flow is a hot mixture of gas and volcanic particles that flows away from the volcano as a density current with speeds from 100 to 700 km/hour and temperatures up to 1000°C. They form during explosive eruptions and usually result from collapse of an eruption column. This provides their extreme heat and speed. Pyroclastic density currents that are very dilute and have a high proportion of gas to volcanic material are called pyroclastic surges. Pyroclastic flows and surges represent two end-members on a continuum of pyroclastic density currents (**Figure 55.5**). Deposit volumes can be small, but they are also some of the largest eruptions on Earth with volumes of thousands of cubic kilometers in individual eruptions. The largest flows can travel for hundreds of kilometers and have thicknesses of 100 meters or more. These flows have tremendous energy and can flatten everything in their path (**Figure 55.6**). They also have enough speed that they can flow up and over topography (**Figure 55.7**). Their high temperatures and hot speed make them nearly impossible to escape from and they incinerate any object in their path. The residents of Pompeii and Herculaneum were killed by pyroclastic density currents from Vesuvius in 79 AD (**Figure 55.8**).

United States Geological Survey.

Figure 55.5 Pyroclastic density current from the 1980 eruption of Mount St. Helens.

Source: NASA.

Figure 55.6 The blast from the eruption of Mount St. Helens in 1980 flattened or damaged enough trees to build 150,000 homes.

© Photovolcanica.com/Shutterstock.com

Figure 55.7 A pyroclastic flow from Soufriere Hills Volcano on the Caribbean island of Montserrat speeds down the Tar River Valley in January of 2010.

© Andrea De la Parra/Shutterstock.com

Figure 55.8 Residents of Herculaneum took refuge in the boat storage sheds by the beach at the start of the 79 ad eruption of Vesuvius but were killed there by a pyroclastic density current that flowed through their town.

Directed Blasts

In many volcanic eruptions the vent is perpendicular to the surface and the eruption column is directed vertically, transporting a large amount of ash and volcanic particles up into the atmosphere. Large eruptions can eject massive amounts of ash that can remain in the atmosphere for months or longer causing global cooling events (**Figure 55.9a**). Evidence of this is seen in spectacular sunsets caused by this ash. The eruption of Tambora in Indonesia in April 1815 had a worldwide effect the following year on weather to the degree that in the northeastern United States there was a significant decrease in solar radiation and temperature. The year 1816 became known as the "year without a summer". Lower temperatures had a direct effect on crops. This climatic effect was caused by dust particles and sulfur dioxide reflecting and absorbing solar radiation. Atmospheric gases, particularly sulfur derivatives such as sulfur dioxide and sulfur trioxide, are capable of absorbing large amounts of infrared radiation energy, thus producing a longer term effect (**Figure 55.9b**).

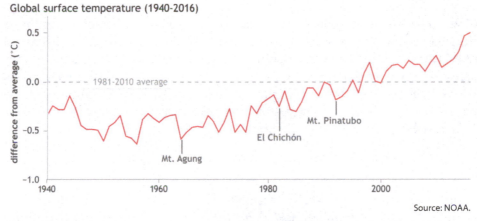

Source: NOAA.

Figure 55.9a Three volcanic eruptions in the tropics had climate-cooling effects in the second half of the 20th century—Mt. Agung (Indonesia, 1963), El Chichon (Mexico, 1982), and Mt. Pinatubo (Philippies, 1991). Volcanic gases like sulfur dioxide have to reach the stratosphere in order to cause this temporary global cooling effect.

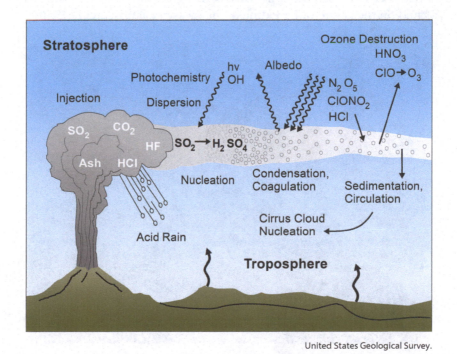

United States Geological Survey.

Figure 55.9b Volcanic gases react with the atmosphere and sulfur dioxide (SO_2) is converted to sulfuric acid (H2SO$_4$). This sulfuric aerosol can cause global cooling of up to 0.5 degrees C (ca. 1 degree F) that lasts for several years after an eruption, such as the 1816 eruption of Mount Tambora in Indonesia.

If the blast is directed laterally out the side of the volcano, the energy is oriented along the ground and can have a devastating effect on everything in its path (**Figure 55.6**). Such was the case with the May 1980 eruption of Mount St. Helens in southwestern Washington. The initial blast was directed to the northeast and thousands of acres of trees were flattened and large amounts of debris clogged the rivers and streams. The volcano sent ash and debris vertically to a height of 25 km (80,000 ft) in 15 minutes. The U.S. Geological Survey reported that ash clouds stretched across the United States in a period of three days and ash circled the Earth in 15 days. Ash remained aloft for several years.

Landslides and Tsunami

The steep slopes of a volcano are very unstable because of the interlayered lavas and clastic deposits, as well as the flux of hydrothermal fluids and sulfuric acid altering the interior of the structure. In addition to lava and other pyroclastic debris, large sectors of the volcanic edifice itself can collapse and race down to form large landslides with velocities up to 800 kilometers per hour. Volcanic eruptions may further destabilize already unstable volcanic flanks. Researchers believe that a catastrophic landslide could result if there is an eruption of Cumbre Vieja Volcano on the island of La Palma in the Canary Islands (**Figure 55.10**).

Source: NASA.

Figure 55.10   Cumbre Vieja Volcano in La Palma Island in the Canary Islands.

A review of geological evidence on the island and offshore indicates that the western edge of the volcano could slide into the Atlantic Ocean. Were this to happen, an estimated 150 to 500 cubic kilometers of material entering the ocean will generate a tsunami that could inundate the east coast of Florida with waves exceeding 50 m in height. If this did occur, coastal residents would only have a few hours warning. This would be an insufficient amount of time to allow any significant evacuation of the millions of people who would be affected.

Submarine and near-shore volcanic eruptions can generate tsunamis or seismic sea waves that travel at very high speeds across the open ocean. Reaching heights of several tens or even hundreds of meters, these waves wreak havoc on coastal regions. Large landslides are associated with volcanic activity near the shore on the island of Hawaii. Massive blocks measuring hundreds of cubic kilometers are moving into the sea at a relatively slow rate (several cms per year). However, the recurrence rate for large volcanic-induced tsunamis is about every 100,000 years—we shouldn't experience that any time soon (Decker and Decker, 2006). In 1975 a magnitude 7.5 earthquake on the island of Hawaii created a landslide that moved large blocks of the island into the sea, which resulted in a tsunami that killed two people.

Predicting Volcanic Eruptions
By Ingrid Ukstins and David Best

Predicting Volcanic Eruptions

The terms **active**, **dormant**, and **extinct** have been used to indicate the eruptive potential of volcanoes. Different definitions exist for these terms, but generally they include the following guidelines. An active volcano is one that has shown some activity in the past several thousand years (historic time), but it could be currently inactive. A dormant volcano is one we think of as sleeping or inactive but could awaken at any time. An extinct volcano has shown no activity in historic time and is not considered likely to do so.

> **active:** A term applied to a volcano that has erupted in recorded history or is currently erupting.
>
> **dormant:** A term applied to a volcano that is not currently erupting but has the likelihood to do so in the future.
>
> **extinct:** A term applied to a volcano that no longer is expected to erupt.

Blong (1984) points out that the terms active, dormant, and extinct are unsatisfactory because some extinct volcanoes become active. A review of past eruptions shows that many volcanoes have periods of no activity much longer than their historic record. Also the historic record differs for various parts of the world. The Mediterranean Sea region has a longer historic record than does the western United States. Hence, more volcanism during historical record-keeping time has been documented in Italy and Greece than in the Pacific Northwest of the United States.

Precursors to Volcanic Eruptions

Scientists can make several observations to forecast the possible eruption of a volcano. Among these are changes in seismic activity near the volcano, an increase in measured heat on the surface, and the bulging of the volcano's surface.

Seismic Activity

The rise of magma is a relatively slow process that can be monitored by scientists who can detect precursors to an eruption. The convective overturn of molten material within the magma chamber produces a series of small earthquakes (usually with magnitudes <4) that have a rather specific pattern in terms of

Earth movement. This is caused by the reverberation of energy within the closed magma chamber as energy bounces off the walls of the chamber. Seismic detection equipment records harmonic tremors that produce a unique pattern (**Figure 56.1**). The occurrence of these tremors is a sign that magma is rising and that an eruption could happen; an increase in seismic activity often signals that an eruption is on the verge of occurring.

© Andrey VP/Shutterstock.com

Figure 56.1 Seismic stations can detect, monitor and record earthquakes locally or, if they are large enough, from anywhere in the world.

Surface Heat

Ascending magma brings increased heat closer to the surface. This heat can be directly measured by heat flow instrumentation. Should the magma be rising beneath a volcano that is covered with snow and ice, the heat will melt the frozen water, which will then begin to flow downhill. Such was the case with the imminent eruption of Mount St. Helens in the spring of 1980. Usually the top of Mount St. Helens was covered by a solid snow pack, but scientists began to notice in March of that year the snow was becoming much thinner. This was an obvious signal to everyone that magma was moving closer to the surface. Monitoring heat flow and increased seismic activity provided evidence that magma was approaching the surface. The first eruption of Mount St. Helens in almost 125 years occurred in late March 1980 and continued with increased frequency until the main eruption on May 18 (Box 3.3).

Surface Bulge

Ascending magma also causes the Earth's surface to bulge (**Figure 56.2**). Measurements of the bulging can be taken using tiltmeters capable of measuring one unit of elevation change over a distance 1 million times greater than the rise. This is equivalent to detecting the increase in elevation produced by a dime coin over a distance of 1 kilometer. The detected bulges show scientists where the upward forces are the greatest and give an indication of where magma could be forced out. Laser beams reflected off targets placed on a volcano will show changes in travel times of the beams and hence a change in the distance as the volcano expands or contracts.

Other Techniques

Other techniques such as recording gas compositions and small changes in the magnetic and electrical fields give an indication of stress and pressure variations in the ground. Changes in gas compositions appear to be effective indicators of changes in the magma chamber. Emissions of sulfur dioxide and hydrogen have been indicators of eruptions at Mount St. Helens and at several Hawaiian volcanoes.

Research into past events allows geologists to produce hazard maps for areas that are likely to experience volcanic eruptions in the future. Analysis of historical records along with detailed mapping of volcanic deposits permits scientists to generate maps that show the likelihood of volcanic deposits covering areas in the vicinity of potential eruptions. Many people living in the region south of Seattle and in the suburbs of Tacoma are located in areas that could be affected by an eruption of Mount Rainier.

Earthquakes and Regions at Risk

By Ingrid Ukstins and David Best

Regions at Risk: The Plate Tectonic Connection

Major lithospheric plates move about on Earth's surface and eventually collide with one another, generating forces that result in earthquakes. Depending on the actual collision mechanism, which creates movement and fracturing in rocks, the earthquakes occur at varying depths, ranging from near the surface to depths approaching 700 km. The majority of earthquakes occur at depths less than 100 km in the lithosphere.

Types of Plate Boundaries

There are three types of plate boundaries that produce earthquakes. **Divergent boundaries** are associated with areas where plates are moving apart—for example, along mid-oceanic ridges where plates are being pushed apart. As the newly formed oceanic crust moves away from its heat source at the mid-oceanic ridge, it becomes colder. The leading edge of the oceanic plate tends to sink because it is far removed from the heat source. This sinking action creates tensional forces that pull the plate away from the ridge. Because magma is hot, stresses are unable to build up and the rocks at depth do not normally break. Seismic activity along mid-oceanic ridges is caused by rising magma as it bubbles its way to the surface. As magma ascends, the overlying material is pushed up and outward. Earthquake activity along mid-oceanic ridges is typically confined to relatively shallow depths, taking place in the upper 20 km of Earth's crust (**Figure 57.1**). Divergent motion can also occur in continental regions where the land surface is being split apart by extensional stresses. Rising heat in the mantle creates a bulge that can fracture the brittle continental crust. An example of this is in the eastern part of Africa along the East African Rift Zone.

divergent boundary: An area where two or more lithospheric plates move apart from each other, such as along a mid-oceanic ridge.

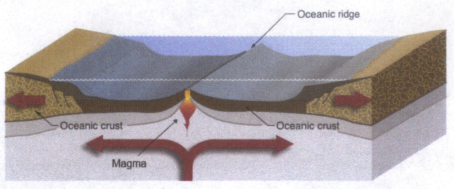

© Kendall Hunt Publishing Company.

Figure 57.1 A divergent plate boundary occurs when two plates are moving apart at the oceanic ridge. Tensional forces pull the two plates away from each other and new oceanic crust is being formed.

Convergent boundaries are regions where plates are colliding as a result of compressional forces. These conditions produce the largest earthquakes. Three types of convergent configurations exist. A collision that involves two oceanic plates is an ocean-ocean plate convergence. In this instance, one of the two plates is pushed under the other one, resulting in subduction of the downgoing slab. Once the slab starts its downward motion, gravitational forces pull the slab to greater depths. An example of this type of boundary is where the Caribbean Plate collides with the westward moving North American plate. The result is the string of volcanic islands that form the Lesser Antilles.

> **convergent boundary:** An area where two or more lithospheric plates are coming together, such as along the west coast of South America.

A convergence setting in which an oceanic plate collides with a continental plate is an ocean-continent plate boundary. The oceanic plate, consisting of predominantly basalt, is denser than the continental granite so the oceanic plate is subducted. This subducted material can maintain its integrity as a solid slab to depths of several hundred kilometers.

The third example of convergent plate collisions involves the convergence of two continental plates. Each plate consists of the same general rock type—predominantly granite or granodiorite. As we have seen, continents rest atop the asthenosphere and denser mantle, so this type of plate convergence will not result in any significant subduction. An example is the collision of the Indian subcontinent with the lower edge of the Eurasian plate. Because no subduction takes place, the two colliding land masses basically rumple themselves up, producing the Himalayas, Earth's highest mountain range, which is located in South Asia. The regions surrounding this mountain range have experienced some of the most destructive earthquakes recorded over the past century; these have been especially catastrophic because of the poor construction standards and high population densities in that region of the world. (See Box 57.1, which describes the 2005 Pakistan earthquake.)

BOX 57.1	The Pakistan Earthquake of October 2005

The Mediterranean-Trans-Asiatic belt runs in an east-west direction above the Mediterranean Sea stretching eastward through the upper Middle East into the lower portion of Asia. Stress in this region results from the African plate moving northward and colliding with the western portion of the Eurasian plate. There are also major strike-slip faults that run through Turkey into Iran and Iraq. The entire region experiences major earthquakes on a regular basis and the results are often devastating. Buildings and infrastructure here are old and poorly constructed, so a moderate amount of ground motion destroys villages and kills many people. Major earthquakes in this region of the world kill tens of thousands of people every decade or two.

The October 8, 2005, a magnitude 7.6 earthquake in Pakistan was related to major fault systems that are part of the Mediterranean-Trans Asiatic belt. The U.S. Geological Survey reported 80,361 people killed, more than 69,000 injured, and extensive damage throughout northern Pakistan. In some areas entire villages were destroyed and more than 32,000 buildings collapsed. The maximum intensity was VIII. An estimated 4 million people in the area were left homeless. Landslides and rock falls damaged or destroyed several mountain villages, along with roads and highways, thereby cutting off access to the region for several weeks. In the western part of the country there were reports of liquefaction and sand blows. Seiches, which are waves generated in closed bodies of water that produce a sloshing motion, were observed in West Bengal, India, and many places in Bangladesh.

Transform boundaries are another type of large-scale features associated with plate collisions. They form when sections of two plates slide against each other. This occurs in oceanic regions when portions of oceanic plates move at different rates. Earthquakes occur along these transform faults, but the magnitudes of these events are usually small.

transform boundary: An area where two plates move past each other in a horizontal, sliding motion.

When transform plate boundaries occur in continental areas, the resulting motion can produce very large earthquakes. In addition to the San Andreas Fault System is California, an excellent example of this process is seen in the North Anatolian Fault, which runs through Turkey and adjoining countries (**Figure 57.2**). The plate motion occurs along the boundary of the Eurasian Plate which lies north of the fault and the smaller Anatolian Plate to the south.

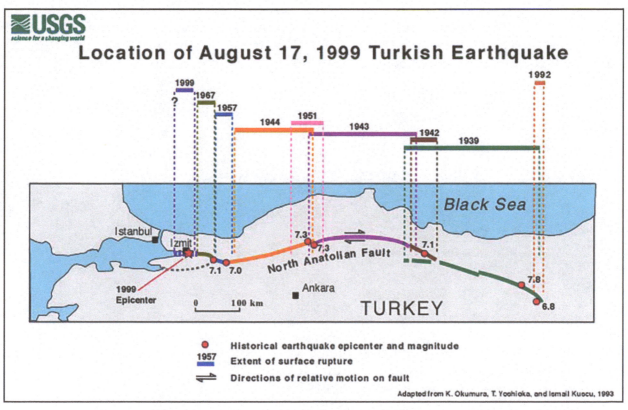

United States Geological Survey.

Figure 57.2   The North Anatolian Fault is an very active feature. It is a major transform plate boundary displaying right-lateral strike-slip motion.

Extending for 1500 km, the North Anatolian Fault passes about 20 km south of Istanbul, a city of more than 14 million people. Since 1939, eight earthquakes of magnitude 7.0 or larger have taken place along the fault. In 1999 a magnitude 7.6 event killed more than 17,000 people near the city of Izmit. Scientists believe that the likelihood of a major earthquake occurring in the vicinity of Istanbul in the near future is high.

Seismic activity along the North Anatolian Fault has shown a migration of epicenters since 1939, except for the 1992 events. The right-lateral strike-slip motion is the result of the Anatolian Plate slipping westward relative to the Eurasian Plate.

Case Study: Cascadia Subduction Zone

By Ingrid Ukstins and David Best

Cascadia Subduction Zone—What Could Possibly Happen?

The Cascadia Subduction Zone is part of the Ring of Fire, a region that encircles more than three-quarters of the Pacific Ocean (**Figure 58.1**). The Ring of Fire stretches from New Zealand through Indonesia, the Philippines, Japan, and along the Aleutian Islands and the coast of Alaska. It continues southward off the coast of British Columbia, along the western edge of the United States, Mexico, and Central America, and encompasses the entire western edge of South America. This Circum-Pacific belt is extremely active seismically and has many volcanoes due to continuous collisions of oceanic plates with continental plates.

In recent years there have been major earthquakes occurring in Chile (February 2010, magnitude 8.8), New Zealand (February 2011, magnitude 6.1), and Japan (March 2011 magnitude 9.1). These events have relieved tectonic stresses in three sections of the Ring of Fire; only portions of the northeast edge, near the Shumagin Gap section of the Alaskan coast and along coastal portions of the Pacific Northwest of the United States, have been spared.

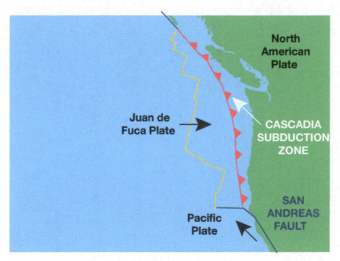

© Kendall Hunt Publishing Company.

Intro Figure 58.1   The small Juan de Fuca Plate is colliding with the North American Plate. This activity generates many earthquakes in the region and has produced a string of volcanoes on the continent.

The Juan de Fuca plate lies off the coast of the Pacific Northwest (**Intro Figure 58.1**). As it moves to the northeast, it is converging with the westward moving North American plate at a rate of approximately three centimeters per year and is being subducted under the North American plate. The zone of subduction is situated about

283

100 kilometers to the west of the shoreline of southwestern British Columbia, Washington, Oregon, and northern California. As the Juan de Fuca sinks into the subsurface, it generates rising magma, which has produced a line of volcanoes stretching from Lassen Peak in northern California northward to Mount Meager in British Columbia. This zone includes many well-known volcanoes, including Mount Rainier, Mount St. Helens, Crater Lake, and Mount Shasta. The subducted area is termed the Cascadia Subduction Zone, which approximately parallels the shoreline for about 1,400 kilometers.

The North America-Juan de Fuca collision has generated many significant earthquakes in historic time, the largest of which was the magnitude 9.0 event that occurred on January 26, 1700. In the past century, numerous earthquakes of magnitude 6.0 or greater have rocked the region, including more than ten magnitude significant events, such as the Puget Sound earthquake of 2001 (magnitude 6.8).

The absence of a great earthquake of magnitude 8.0 or larger in the region has produced growing concerns about the likelihood of such an event taking place in the foreseeable future. Several articles have brought this possibility to the attention of both disaster management personnel and the general public. Kathryn Schutz, in her 2015 Pulitzer Prize winning article, "The Really Big One," generated both recognition and concern to the people of the Pacific Northwest. Research by USGS and academic scientists has resulted in a significant amount of data that demonstrate that the region must be prepared for the potential of a catastrophic event occurring in the near future.

Unfortunately, it is virtually impossible to predict exactly when a mega earthquake will happen along the Cascadia Subduction Zone. As Stein and others (2017) pointed out in their study of the Cascadia Subduction Zone, such an earthquake could be relatively soon or in several centuries. What seems certain is that the Cascadia Subduction Zone will experience many earthquakes in the future, some of which will be quite large.

Thousands of earthquakes occur every day, but fortunately very few of them are large enough to create problems. Most major earthquake activity is associated with regions where tectonic plates are colliding with one other to generate stresses that cause rocks to snap, sending seismic energy through the Earth. As we saw earlier in the discussion of global plate tectonics, earthquakes and volcanoes—which can cause two of nature's greatest types of catastrophes—are associated with one another and will often occur together. Most earthquakes occur along plate boundaries, those areas where tectonic plates are moving apart, colliding, or sliding past one another.

Mechanics and Types of Earthquakes

By Ingrid Ukstins and David Best

When a force is applied to a surface, the result is termed **stress** (measured as force per unit area). Movement within the Earth occurs when the exerted stress exceeds the frictional force preventing movement from taking place. When blocks of rock move, slippage occurs along a surface or fault plane. An object that is stressed will experience **strain**, which can be elastic, thereby returning the object to its original shape. An example of this is a rubber band or piece of elastic, when pulled (extensional stress), returns to its original shape. If either the rubber band or piece of elastic is overly stressed, it breaks and the original single object is now two pieces, not the original one. If the object breaks, the strain is inelastic, such as what occurs in brittle rock.

> **stress:** The force being applied to a surface; forces can be compressional, extensional, or shearing.
>
> **strain:** The response of an object which is being stressed; strain can be elastic, in which case the object returns to its original shape or inelastic, when the object does not recover its shape, thereby being deformed.

Faults

Orientation of the fault plane can range from nearly horizontal to vertical. The fault plane separates two adjacent blocks called the hanging wall and the footwall. It is easy to assign names to each of these blocks. Picture yourself standing on the fault plane. The block underneath your feet is the footwall; the block over your head is the hanging wall.

Figure 59.1 shows three kinds of faults and the stresses related to each one. During the movement along a fault, if the hanging wall, which is the block lying above the fault plane, moves down relative to the footwall, the fault is classified as a **normal fault**. This type of fault is sometimes called a gravity fault because the downward force of gravity causes the hanging wall to move downward. Normal faults form when extensional or tensional forces are pulling on two blocks. This type of fault occurs at divergent plate boundaries, such as those found at mid-ocean ridges where oceanic plates are moving away from each other.

> **normal fault:** A plane along which movement has occurred such that the upper block overlying the fault has moved down relative to the lower block.

Different kinds of faults... Different stress
Brittle materials change a little and then break suddenly.

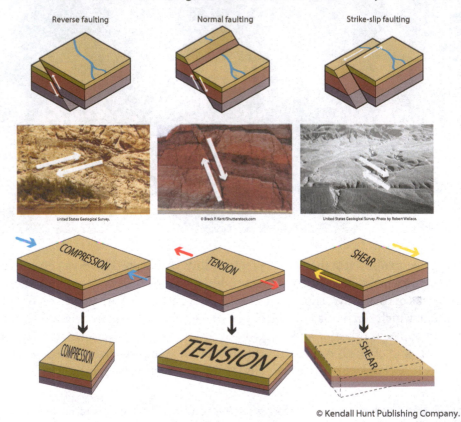

© Kendall Hunt Publishing Company.

Figure 59.1 The type of faulting depends on the type of stress applied to the rock. In the diagrams for reverse and normal faulting, the footwall and hanging wall are labeled. The terms do not apply to strike-slips faults.

When the hanging wall moves up relative to the footwall, a **reverse fault** is produced. Compressional forces squeezing two blocks together will generate a reverse fault. Reverse faults are commonly found at convergent plate boundaries where the plates are colliding.

> **reverse fault:** A plane along which movement has occurred such that the upper block overlying the fault has moved up relative to the lower block.

A **strike-slip fault** is formed when two blocks slide one past the other with very little or no vertical displacement. In this instance a pair of parallel forces is acting one against the other, producing a shear couple. A well-known example of a strike-slip fault is the San Andreas fault in southern California.

> **strike-slip fault:** A fault in which the motion of the two adjacent blocks is horizontal, with little if any vertical movement.

When rocks do fracture, the initial point at which breakage occurs in the subsurface is called the **focus** (or hypocenter) of the earthquake. Energy radiates outward in a spherical pattern from this point and travels in all directions. The point on the surface directly above the focus is termed the **epicenter**, a location that can be assigned latitude and longitude coordinates indicating where the earthquake occurred.

focus: The point in the subsurface where an earthquake first originates due to breaking and movement along a fault plane; sometimes referred to as a hypocenter.

epicenter: The point on Earth's surface directly above the focus. This is the position that is reported for the occurrence of an earthquake as a latitude and longitude value can be assigned to the point.

Earthquake foci have a wide range of depths. **Shallow-focus earthquakes**, those occurring between the surface and a depth of 70 kilometers, can form as upper-plate, thrust, or intraslab events. **Intermediate-focus earthquakes** (70 to 300 km) and **deep-focus earthquakes**, those reaching depths of 700 km, take place in brittle, subducted rock. As the subducted material sinks to depths of about 700 km, fewer events occur due to the heating of the rock, which make it more ductile and hence less likely to break.

shallow-focus earthquake: An earthquake that has its focus located between the surface and a depth of 70 kilometers.

intermediate-focus earthquake: An earthquake that has its focus located between a depth of 70 kilometers and 300 kilometers.

deep-focus earthquake: An earthquake that has its focus located between a depth of 300 kilometers and roughly 700 kilometers.

Seismic Waves
By Ingrid Ukstins and David Best

Energy radiates out from the focus of an earthquake (**Figure 60.1**) and is transmitted as seismic waves (sound and light waves are other examples of energy being moved through space). This energy produces two types of seismic waves: **body waves** and **surface waves**.

> **body wave:** A seismic wave that is transmitted through the Earth. P-waves and S-waves are body waves.
>
> **surface wave:** A seismic wave that moves along a surface or boundary.

Body Waves

Body waves are transmitted through the body or interior of the Earth. Body waves consist of two types. **P-waves** are the primary waves that result from a push-pull action. They are also referred to as compressional or longitudinal waves, because the back and forth motion moves in the direction of wave propagation (**Figure 60.2a**). A compressional wave moves along by squeezing the material; the compressed particles then rebound or expand, which compresses the next particles adjacent to the ones that had been compressed. P-waves are the fastest moving of all the seismic waves, and they can travel through any type of material (solid, liquid, or gas). Their velocities average approximately 6 km/sec, with some variations depending on rock type.

> **P-wave:** The primary wave, which is a compressional wave; this is the fastest moving of all the seismic waves, and arrives first at a recording station. P-waves are a type of body wave.

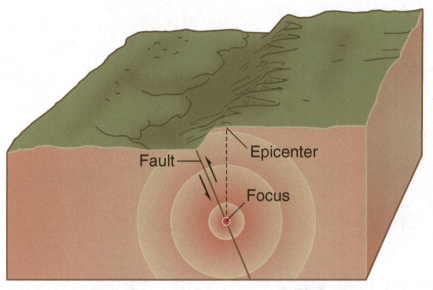

Figure 60.1 Location of the focus, epicenter, and fault plane. Notice how seismic energy radiates out from the focus in spherical paths.

S-waves, also known as secondary or shear waves, are so named because they are the second ones to arrive at a detection point. They travel at about 3 km/sec and have a lower frequency and higher amplitude. S-waves are also referred to as shear or transverse waves because their actual motion causes particles to move at right angles to the direction of wave propagation (**Figure 60.2b**). An important characteristic of S-waves is that they do *not* pass through liquids. This fact has allowed seismologists to determine that certain parts of Earth's interior are liquid because S-waves are not detected at all recording stations. Seismic recording stations that are located between distances of 103° and 143° from an earthquake focus will not record S-waves because the waves hit the liquid outer core and do not get transmitted any farther.

S-wave: The secondary wave, which is a shear wave that moves material back and forth in a plane perpendicular to the direction the wave is traveling. S waves do not travel through liquids. S-waves are a type of body wave.

Surface Waves

(**Figure 60.2c**) shows surface waves that travel along a surface or boundary between layers. This group of waves consists of **Rayleigh** (R) and **Love** (L) waves, two types that are named after mathematical physicists who each did research into their respective characteristics. For shallow and some intermediate-focus earthquakes, surface waves have the greatest amplitude and the most obvious signal (**Figure 60.3**). Vertically directed R-waves make the ground ripple up and down, similar to the wave motion created when you jump into a swimming pool. The lateral motion caused by faster moving L-waves is similar to that seen when a sidewinder snake moves across the desert floor. As the snake moves forward, its body moves side to side (**Figure 60.4**). The key point about each of these surface waves is that they involve a rolling motion of particles and they move much slower than either type of body wave. Surface waves are the most destructive because they produce the most ground motion and take longer to move through an area. Rayleigh waves have a velocity of approximately 12,500 kilometers per hour, while Love waves move at 16,000 kilometers per hour.

P-wave motion
Compressional wave

Wave direction

Adapted from usgs.gov.

(a)

S-wave motion
Shear-wave crest

Wave direction

Adapted from usgs.gov.

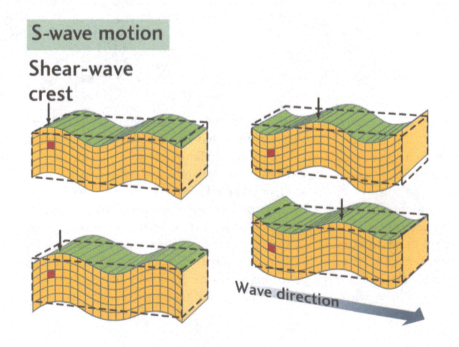

(b)

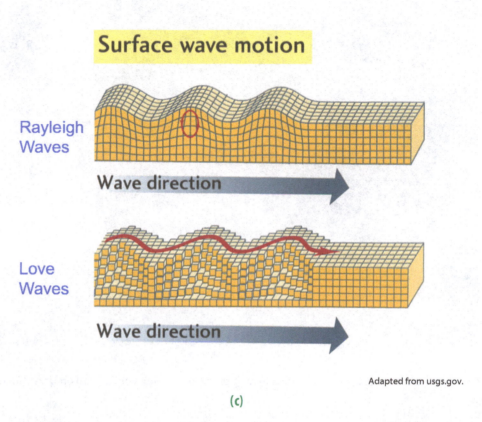

Surface wave motion

Rayleigh
Waves

Wave direction

Love
Waves

Wave direction

Adapted from usgs.gov.

(c)

Figure 60.2a–c Seismic motion travels in different ways through material. P-waves ave push-pull movement, S-waves wiggle back and forth (either vertically or horizontally) and surface waves tend to roll or slide.

When an earthquake occurs, all these types of motion (P, S, R, and L) are generated so the energy is passed through an area at different velocities, different times, and varying amplitudes. The velocity relationship is $V_P > V_S > V_L > V_R$, but the amplitudes of these waves are essentially in reverse order. The rocks and buildings that are being affected undergo a great deal of mixed stresses that produce the damage we observed after the shock waves pass. Engineers can design buildings to withstand most of these forces, but the underlying geology (bedrock, loose soil, or sediment, for example) must also be considered in the design process.

Rayleigh waves: These move particles in a rolling, up and down sense that travel in the direction of propagation of the energy.

Love waves: These move particles with a twisting, side to side motion.

Surface waves

P-wave   S-wave

© Kendall Hunt Publishing Company.

Figure 60.3 Seismogram showing the arrivals of the P-, S-, and surface waves. Notice the changes in amplitudes of the various waves as they arrive at the detection station.

Topic 60 Seismic Waves   **291**

Source: Bill Caid. Used with permission.

Figure 60.4 A sidewinder rattlesnake displays a horizontal ripple motion along with forward progress similar to that of an L-wave.

Detecting, Measuring and Locating Earthquakes
By Ingrid Ukstins and David Best

As seismic energy moves through rocks, it abruptly displaces them, sometimes on a large scale. As the ground moves, the motion is detected by a **seismograph**, a very delicate instrument that responds to small changes in the vertical and horizontal positions at a point. This movement is amplified before being sent to a recording device that produces a record of the ground motion as a **seismogram** (**Figure 61.1. a–b**). Careful analysis of a seismogram can provide information about the strength or size of the earthquake and its distance from the recording station. An important parameter is to know the precise time related to the recorded signal so the source region of an event can be located.

> **seismograph:** A device that records the ground motion of an earthquake.
>
> **seismogram:** The written or electronic record of ground motion detected by a seismograph.

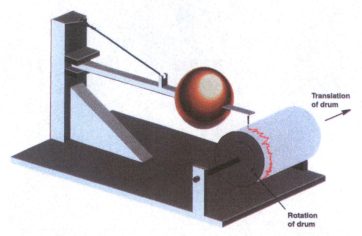

© Kendall Hunt Publishing Company.

Figure 61.1a   A basic design of a seismograph shows a pendulum at the end of a horizontal bar with an attached pen that records ground motion.

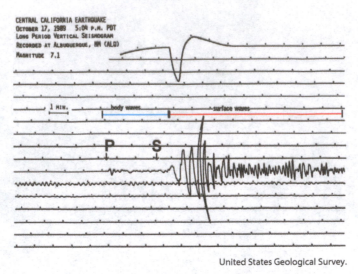

United States Geological Survey.

Figure 61.1b A recording station in Albuquerque, New Mexico, generated this seismogram of the arriving body and surface waves produced by the Loma Prieta, California, Earthquake of October 17, 1989. Given the distance of this recording station from the epicenter of the event, the P- and S-wave arrivals are small, but the surface waves are much larger.

There are numerous recording stations located throughout the world. Within the United States there are more than 100, with a significant number located in earthquake-prone regions including California and Alaska (**Figure 61.2**). These stations continually record all ground motion and report any unusual activity to the National Earthquake Information Center (NEIC) in Golden, Colorado, which issues reports and warnings of earthquake activity. Rapid computer analysis of earthquake data enables the NEIC to send information and warnings to all parts of the globe.

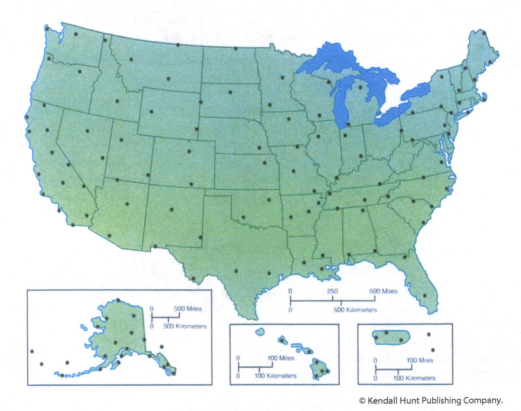

© Kendall Hunt Publishing Company.

Figure 61.2 A national network of more than 100 seismic stations throughout the continental United States provides data for detailed earthquake analysis.

Measuring the Size of an Earthquake

Several factors control the recorded size of an earthquake. The amplitude of the seismogram is related to the recording station's distance from the epicenter. The greater the distance, the less pronounced is the record. Also the magnitudes of the recorded motion depend on where the seismographs are placed. Loose soil moves more rapidly than solid rock and can give a false record of the true scale of the event. Seismographs are usually located on a solid base to minimize extraneous ground movement.

Earthquakes located near a recording station will generate a sudden jolt that will be detected by the seismograph as a spiked signal. Earthquakes occurring at a greater distance tend to produce longer period signals which have a lower amplitude (**Figure 61.3**). Records of the ground motion of the Northridge, California, earthquake are shown in **Figure 61.3** and the event is discussed in Box 61.1.

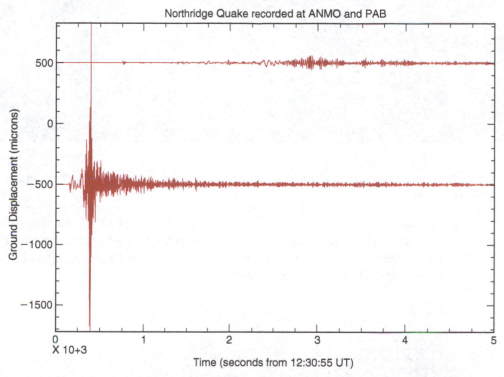

Source: http://www.quaketrackers.org.nz/curr/seismic_waves.htm.

Figure 61.3   Recordings of the Northridge, California, earthquake of January 17, 1994. The upper record is from a recording station in Pablo, Spain; the lower record is from Albuquerque, New Mexico. Notice the greater amplitude for the signal recorded at the station that was much closer to the event.

BOX 61.1	Northridge Earthquake of January 1994

Many of the residents of the Los Angeles, California, area had planned to take the Martin Luther King Day holiday on Monday, January 17, 1994. Unfortunately, Mother Nature had decided it was a day of work as the city of Northridge, a suburb on the northern edge of the city of Los Angeles, was rocked by a magnitude 6.7 earthquake at 4:31 a.m. The United States Geological Survey reported that 57 people were killed and more than 8,700 injured. More than 114,000 buildings were destroyed or severely damaged; numerous gas lines ruptured and several key freeway overpasses collapsed, blocking major traffic arteries in the Los Angeles area. The Northridge earthquake still ranks among the most costly natural disasters in U.S. history with damage estimates ranging from $13 to $44 billion. Following this earthquake major retrofitting of freeways, bridges, and overpasses in the Los Angeles area took place to minimize similar failures when the next large earthquake strikes the area.

FEMA by Robert Eplett.

Box Figure 61.1.1 Damage was very widespread in Northridge and surrounding communities. The interstate highway system was disrupted for weeks.

Figure 61.4 is a seismic record of a minor earthquake in central California. The horizontal lines are recorded by a pen that marks paper on a drum that turns once every 15 minutes, hence four lines per hour. Notice the sudden onset of this signal, which tells us this event was close to the recording seismograph. The recorded ground motion for this event lasted about two and one-half minutes.

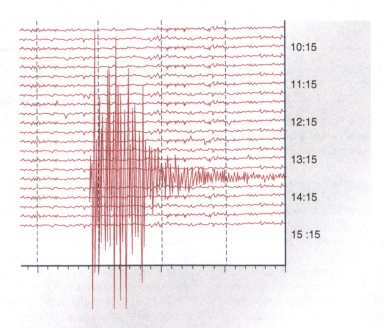

United States Geological Survey.

Figure 61.4 Magnitude 4.0 earthquake near Cloverdale, California, on January 10, 2000. The area lies about 150 km northwest of San Francisco and faults in the area are part of the San Andreas Fault System.

Measuring the Magnitude of an Earthquake

The magnitude of an earthquake is a number that is used to characterize the relative severity of an earthquake. Several different magnitude scales are used to describe this value. These differences are based on which of the recorded seismic waves are analyzed, but research has shown that minimal differences exist in assigned values for events of magnitude 6 or less. The majority of earthquakes that occur each year are in this smaller range. Larger events are more complex and thus require more data to determine their magnitude. We will use two scales to describe magnitudes: (1) local magnitude (M) for local or nearby earthquakes and those with values of 6.5 or less, and (2) the moment magnitude (Mw) scale for those with larger values.

The **local magnitude**, M, is used for moderate-sized earthquakes measured close to the epicenter. This method uses the **Richter scale**, which was first developed in 1935 by Charles F. Richter of the California Institute of Technology based on data from local earthquakes in California. Richter used the maximum amplitude of the S-wave and took into account the distance from the recording station to the epicenter to establish the magnitude of an event. It is an open-ended scale, so it can include the very smallest events or the largest possible ones that occur. The U.S. Geological Survey (USGS) reports that each year more than 10,000 earthquakes occur in the southern California area. Most of them are very low magnitude with only a few hundred measuring greater than magnitude 3.0, and only about 15 to 20 being greater than magnitude 4.0.

> **local magnitude:** A term that describes the size of an earthquake in an area near the epicenter; see Richter scale.
>
> **Richter scale:** A measurement scale developed by Charles Richter to determine the size of earthquakes in California. This is commonly used to describe many types of earthquakes but it is more correctly used for small, localized events.

Earthquake magnitude is a logarithmic scale signifying the size of the event. If we were the same distance from the epicenters of two different events, ground motion would be 10 times greater during an M = 6 earthquake than that of an M = 5 earthquake. This 10-fold increase in ground motion translates to approximately a 32-fold increase in the amount of energy produced by the larger event. This larger amount of energy helps explain why higher magnitude earthquakes are generally more destructive. A magnitude 6 earthquake would have more than 1,000 times the energy of an ML = 4 event.

The **moment magnitude** (Mw) is used to measure the size of large earthquakes. It is dependent on the shear strength of the rock, the area affected by the fault rupture, and the average displacement along the fault. This scheme is used for large earthquakes having magnitudes greater than an Mw value of 8. Seismic moment is considered to be the best gauge of earthquake as it describes the amount of strain energy released by an event.

> **moment magnitude:** A measure of an earthquake that is based on the area affected, the strength of the rocks involved, and the amount of movement along the primary fault.

A review of the 10 largest earthquakes in recorded history (Table 61.1) shows that all of them are related to active convergent plate boundaries, with nine of them occurring in the circum-Pacific Ring of Fire. The Assam, Tibet, event of August 1950 (along the India-China border) was caused by the collision of India with the Eurasian plate (**Figure 61.4**), which is a continent-continent plate collision.

When we examine the 10 most deadly earthquakes in history, only one of them is also listed among the 10 largest events—the December 2004 earthquake and tsunami off the coast of Sumatra. In terms of events producing the most fatalities, we notice that four of the occurred in China, a country that has always had a very large population. One other note to consider is that four of the most deadly earthquakes occurred before 1500. The fatality count for those has been determined by historians who have researched those ancient civilizations and based the estimates on historical records.

Table 61.1 Ten Largest and Ten Most Deadly Earthquakes

Ten Largest Recorded Earthquakes

Date	Location	Magnitude
May 1960	Off the coast of Chile	9.5
March 1964	Prince William Sound, Alaska	9.2
December 2004	Off the west coast of Sumatra	9.1
March 2011	Near the east coast of Honshu, Japan	9.1
November 1952	Off the east coast of Kamchatka, Russia	9.0
February 2010	Off the coast of Ecuador	8.8
January 1906	Off the coast of Ecuador	8.8
February 1965	Rat Islands, Aleutian Islands	8.7
August 1950	Tibet, near the China-India border	8.6
March 2005	Off the west coast of Northern Sumatra	8.6

Ten Most Deadly Earthquakes

Date	Location	Deaths
January 1556	Shaanxi, China	830,000
July 1976	Tangshan, China	255,000
August 1138	Aleppo, Syria	230,000
December 2004	Sumatra	228,000 The majority of deaths were due to the tsunami
December 856	Iran	200,000
December 1920	Haiyuan, China	200,000 Modified Mercalli Intensity XII
March 893	Iran	150,000
September 1923	Kanto, Japan	142,800 Extreme destruction of Tokyo and Yokohama
October 1948	Turkmenistan, Russia	110,000
September 1290	Chihli, China	100,000

Source: United States Geological Survey.

Determining the Intensity of an Earthquake

The intensity of an earthquake is a measure of the damage caused by ground shaking and how buildings and other structures respond to seismic energy. The observed damage is a subjective assessment because different observers might assign a variety of levels of severity to a given area depending on their interpretation and evaluation of the damage.

In 1902 Italian scientist Giuseppe Mercalli established a scale that assigned values to damaged areas based on how different materials responded to an earthquake. The scale was modified in 1931 by American seismologists Frank Neumann and Harry Wood, who refined the scale to better address building standards in the United States. The scale is based on 12 different levels of damage (values are assigned Roman numerals), with I being the lowest level of damage and XII being the highest (Table 61.2). The Modified Mercalli scale differs from magnitude scales because there is no mathematical basis for the values given to the damage observed in an area. The Modified Mercalli Intensity Scale is more relevant for engineers and homeowners as they can relate to the impact of the earthquake.

Table 61.2 Abbreviated Descriptions for Levels of the Modified Mercalli Intensity Scale

I. Not felt except by a very few under especially favorable conditions.

II. Felt only by a few persons at rest, especially on upper floors of buildings.

III. Felt quite noticeably by persons indoors, especially on upper floors of buildings. Many people do not recognize it as an earthquake. Standing motor cars may rock slightly. Vibrations similar to the passing of a truck. Duration estimated.

IV. Felt indoors by many, outdoors by few during the day. At night, some awakened. Dishes, windows, doors disturbed; walls make cracking sound. Sensation like heavy truck striking building. Standing motor cars rocked noticeably.

V. Felt by nearly everyone; many awakened. Some dishes, windows broken. Unstable objects overturned. Pendulum clocks may stop.

VI. Felt by all, many frightened. Some heavy furniture moved; a few instances of fallen plaster. Damage slight.

VII. Damage negligible in buildings of good design and construction; slight to moderate in well-built ordinary structures; considerable damage in poorly built or badly designed structures; some chimneys broken.

VIII. Damage slight in specially designed structures; considerable damage in ordinary substantial buildings with partial collapse. Damage great in poorly built structures. Fall of chimneys, factory stacks, columns, monuments, walls. Heavy furniture overturned.

IX. Damage considerable in specially designed structures; well-designed frame structures thrown out of plumb. Damage great in substantial buildings, with partial collapse. Buildings shifted off foundations.

X. Some well-built wooden structures destroyed; most masonry and frame structures destroyed with foundations. Rails bent.

XI. Few, if any (masonry) structures remain standing. Bridges destroyed. Rails bent greatly.

XII. Damage total. Lines of sight and level are distorted. Objects thrown into the air.

Source: United States Geological Survey.

Figure 61.5 shows the results of the Loma Prieta earthquake of October 17, 1989 (see Box 61.2). The simplified geologic map shows those areas that consist of soft mud, sand and gravel, and bedrock. Portions of the Nimitz Freeway (I-880) were built on mud deposits, very unstable material. The roadway collapsed in those areas because the ground was severely jolted by the seismic energy. This is evident in the seismograms for the mud, sand and gravel and bedrock outcrops. Clearly, it was faulty engineering that produced this deadly outcome.

United States Geological Survey.

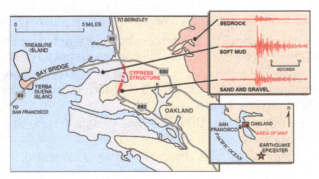

United States Geological Survey.

Figure 61.5   The Loma Prieta earthquake had a profound effect on the Nimitz Freeway located on the east side of San Francisco Bay.

<table>
<tr><td>BOX 61.2</td><td>The Loma Prieta Earthquake of October 1989</td></tr>
</table>

On October 17, 1989, at 5:04 p.m. (PDT), a magnitude 6.9 earthquake severely shook the San Francisco and Monterey Bay regions. The U.S. Geological Survey reported that the epicenter was located near Loma Prieta Peak in the Santa Cruz Mountains, approximately 14 km (9 mi) northeast of Santa Cruz and 96 km (60 mi) south-southeast of San Francisco.

As is true in many disasters, timing means a lot. The Great San Francisco Earthquake of 1906, the San Fernando quake of 1971, and the Northridge earthquake of January 1994, all struck in the early morning hours, sparing numerous people from imminent disasters. The Loma Prieta earthquake was no different in terms of timing. Baseball fans from the Bay Area were gathered in Candlestick Park in South San Francisco to watch Game 3 of the World Series between the San Francisco Giants and the Oakland Athletics.

More than 60,000 fans were in the relative safety of the stadium and off the highways. Many more were at home or in buildings watching the game so they were spared the disaster that struck the area. A portion of I-880, the Nimitz Freeway, collapsed, crushing cars and killing 47 people (this figure would have been much higher, given the rush hour timing of the earthquake, had not so many people been watching the World Series game). One section of the freeway was constructed on loose soil and the poor design of the vertical columns holding up the highest set of traffic lanes caused them to experience increased vertical resonance in the support columns. They all disintegrated under the downward force. Refer to Levy and Salvadori (1995) for details on the engineering issues.

The earthquake obviously caused the baseball game to be cancelled and, although the stadium was only slightly damaged by the seismic waves passing underneath it, the games were moved to Oakland after a 10-day postponement. The A's swept the Giants in four games.

United States Geological Survey. Photo by C. E. Meyer.

Box Figure 61.2.1 **A section of the San Francisco-Oakland Bay Bridge collapsed during the Loma Prieta Earthquake of October 1989.**

United States Geological Survey. Photo by H. G. Wilshire.

Box Figure 61.2.2 **A section of I-880, the Cypress viaduct and Nimitz Freeway that disintegrated during the Loma Prieta earthquake. This road is usually heavily traveled at rush hour, and thousands were spared from death because of the scheduled World Series game that kept drivers off the freeways.**

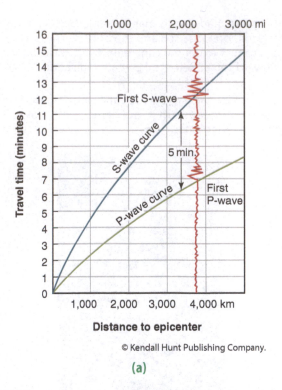

Distance to epicenter

© Kendall Hunt Publishing Company.

(a)

Figure 61.6a Travel-time graph showing the P-wave and S-wave travel times as measured against the distance they have moved from the earthquake focus; the S-P time separation is 5 minutes.

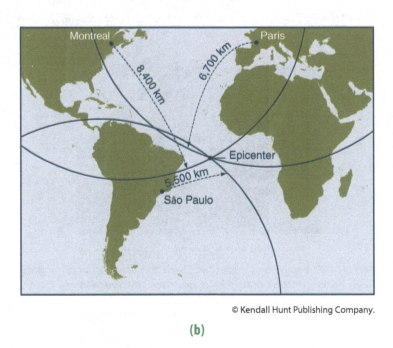

© Kendall Hunt Publishing Company.

(b)

Figure 61.6b Using arrival data from three different stations, it is possible to determine the location of the earthquake epicenter.

Locating an Earthquake

Whenever an earthquake occurs, often several seismic stations record the event. In the case of a major earthquake, hundreds of stations produce seismic records. Today many stations are set up to digitally record earthquakes and then transmit the information to some central office where a determination is made regarding the location and magnitude of the earthquake. In the United States the primary reporting station is the National Earthquake Information Center in Golden, Colorado. If sufficient data are available at a single recording station, it is possible for scientists to make a quick determination of the distance and magnitude of a single earthquake from that station.

To get a quick fix on the location of an earthquake, arrival data are needed from a minimum of three recording stations. Each station will have a seismogram that shows the P- and S-wave arrivals of the event. The S-P time is noted and then converted to a distance using a travel-time curve (**Figure 61.6a**). When this distance is determined for a single station, the locus of all points located the same distance from the station is a circle. To be more precise in assigning a location for the epicenter, a similar procedure must be completed for two more stations. When this is done, the three circles should intersect at a single point, but because of minor differences in estimated rock velocities used to construct the travel-time graph, the lines usually establish a region of intersection where the earthquake occurred. As more data are analyzed from other stations, this region is eventually refined to more precisely define the epicenter of the earthquake (**Figure 61.6b**).

Evidence from the Geologic Record

Earthquakes have occurred on Earth ever since crustal material was cold and brittle enough to break. Numerous examples exist of faults being preserved in rocks that are billions of years old. If a rock was faulted and later covered by sedimentary or volcanic material, the fault is preserved and we can reason that movement occurred prior to the later deposition of the overlying sedimentary material. Undoubtedly, major earthquakes occurred throughout the world in the past, breaking up rock in many locations. This fracturing process allowed once solid material to become more easily eroded and removed, thereby erasing the record of earthquake activity. Whenever the broken material was rapidly covered by sediment and other debris, evidence of the fault was preserved.

Along the San Andreas Fault in California, geologists have dug a series of trenches across a section of the fault to detect areas where soil and sediment layers have been offset. By age-dating the layers using radiocarbon techniques, it is possible to determine when earthquakes occurred and thus generate a recurrence interval for a given area. With sufficient data they can extrapolate back in time to determine what happened and to make some educated predictions for future activity. However, because data are often not very continuous in terms of time, the recurrence intervals carry a significant amount of error.

Mexico City experienced a major earthquake in 1985 that literally flattened scores of buildings. The city is constructed on ancient lake beds that provide a very weak geologic foundation. When the seismic energy passed through the area, it shook the ground as if the city rested on gelatin. Numerous buildings collapsed in a pancake fashion and many others crumbled (**Figure 61.7**). More than 9,500 people were killed and an estimated 30,000 were injured.

United States Geological Survey.

Figure 61.7 Damage was extensive as the General Hospital in Mexico City was destroyed following the September 19, 1985, earthquake that was centered off the west coast near the subduction zone, where the Cocos plate dives under the North American plate.

The central region of Mexico experienced a magnitude Mw 7.1 earthquake at 1:14 p. m. on September 19, 2017. Occurring at a depth of 51 km (32 miles), the epicenter was located approximately 120 km (75 miles) southeast of Mexico City. Severe ground shaking lasted for more than 20 seconds and produced major damage in the epicentral area. The surrounding area, which included Mexico City, experienced significant damage, as many structurally weak buildings collapsed. More than 360 people died throughout the region, with more than 240 deaths occurring in Mexico City. The devastation and loss of life in Mexico City is attributable to the city having been built on an ancient lake bed. When seismic energy moves through the underlying sand and clay, ground motion is greatly amplified. The USGS reported that there have been 19 earthquakes of magnitude 6.5 or greater within 250 km of the September 19, 2017 event.

Two interesting facts are related to this event. The date, September 19, is the anniversary of the 1985 Mexico City earthquake which killed approximately 10,000 people. As a reminder of the event, annual earthquake drills are held throughout the country on that day. September 19, 2017, was no exception as the drill began at 11 a.m., a little more than two hours before the earthquake struck. The city has a warning system in place but many people failed to heed the true warning just prior to the onset of the earthquake because they thought it was a continuation of the scheduled drill.

Regions that experience earthquakes must strengthen their building codes and attempt to retrofit existing structures that tend not to withstand seismic events well. Following the 1994 Northridge, California, earthquake, the state has spent billions of dollars to reinforce bridges and other structures to try to preserve the critical infrastructures that are often devastated by major earthquakes.

Earthquakes of equal or very similar magnitudes can produce extremely different effects, depending on where they occur. In December 2003 an earthquake of M 5 6.6 killed 26,200 people in southeastern Iran, while a magnitude 6.7 event in Hawaii in October 2006 did not kill anyone. The primary cause of the large death toll in Iran was the poor construction standards used in many of the older buildings. The lack of reinforced concrete or other building material allows buildings to collapse very easily. Careful planning and the enforcement of strict building codes has reduced the effects of earthquakes in many developed countries.

BOX 61.3 Haiti Earthquake of January 12, 2010

A major earthquake (M 7.0) struck the country of Haiti on January 12, 2010, at 4:53 p.m. local time. The region lies south of the boundary of the westward moving North American plate and the eastward moving Caribbean plate. This left-lateral strike slip motion has produced numerous earthquakes to the east in the Dominican Republic, the country that lies on the eastern side of the island of Hispaniola.

Two major east-west trending, strike-slip fault systems occur in the area. The Septentrional fault system is found in northern Haiti and the Enriquillo-Plantain Garden fault system traverses the southern part of the country in a similar manner. It was the latter fault system that ruptured, producing the earthquake. The last major earthquake in the vicinity of Port-au-Prince occurred in 1770. Thus stresses in the rock had been accumulating for 240 years, even though the average movement along these faults is about 21 mm per year.

More than four dozen significant aftershocks rocked the area, creating very unsafe conditions for rescue workers struggling to free trapped people from the rubble. Eight days after the main shock an aftershock of M 5.9 occurred within 10 km of the main shock.

The capital city of Port-au-Prince, with a population of more than two million people, suffered severe damage. Thousands of buildings were destroyed and an estimated 250,000 to 300,000 people died with projections reaching as high as 230,000 dead. More than 300,000 people were injured and more than 1.5 million people are homeless. The shallow depth of the event, about 13 km below the surface, added to the large amount of destruction.

Haiti, which is among the poorest on Earth, has experienced other major disasters in recent years including four hurricanes in 2008. The country lacks the resources and infrastructure to cope with such events and has been the recipient of worldwide aid to begin rebuilding.

The geological setting for this area is Haiti is very similar to that in southern California, where the right-lateral strike slip motion of the San Andreas fault system passes under heavily populated areas. In the next few decades possible movement along the southern San Andreas fault could possibly generate a similar catastrophe.

Source: Department of Defense.

Box Figure 61.3.1 **Thousands of buildings were destroyed or severely damaged, including many poorly constructed homes.**

© 21th Design/Shutterstock.com.

Box Figure 61.3.2 Epicenter of the Haiti earthquake of January 12, 2010 was located at a distance about 25 km west southwest of Port-au-Prince.

BOX 61.4	The Great Chilean Earthquake of February 2010

On February 27, 2010 at 3:34 a.m. local time, the earth shook in the vicinity of Maule, Chile, located about 115 km north-northeast of the city of Conception. This event measured M 8.8, making it the fifth largest earthquake in recorded history. The region lies at the boundary between the eastward-moving Nazca plate and the westward-moving South American plate. Over the past forty years, the region has experience 30 earthquakes of magnitude 7 or greater. The epicenter for this event is located about 275 km north of the location of the May 1960 magnitude 9.5 earthquake, the largest earthquake in recorded history.

© Yai/Shutterstock.com.

Box Figure 61.4.1 Buildings and homes were severely damaged or destroyed by the February 2010 earthquake.

Given the proximity of the earthquake to the Pacific Ocean, the Pacific Tsunami Warning Center in Hawaii issued several tsunami warnings to alert more than fifty nations and territories bordering the Pacific Ocean. However, the warnings was canceled after it was determined that these areas would be spared from the destructive waves. Countries as far away as Japan and Russia were expected to receive tsunami about 21 hours following the main shock, but only a slight wave arrived in those areas. One contributing factor to the warning was the relatively shallow depth of the focus—only 35 km below the surface. More than 150 significant aftershocks were generated in the first two days following the main shock. These, of course, can be destructive from the standpoint of destroyed buildings that were weakened by the main shock. The aftershock region extended for more than 700 km along the Chilean coastline. By the end of the third day the number of fatalities reached more than 700 people.

Table 61.3 Two Largest Earthquakes in the History of Chile

Date	Mw	Location and Depth	Death toll	Notes
May 22, 1960	9.5	38.29S 73.05W Depth = 33 km (est.)	1,655	Largest earthquake in recorded history; epicenter located in the southern portion of the Nazca-South American plate boundary
February 27, 2010	8.8	35.846S 72.719W Depth = 35 km Conception 115 km Santiago 325 km	525	Fifth largest earthquake in recorded history; epicenter located in the southcentral portion of the collision of the Nazca and South American plates

Two major earthquakes in early 2010 produced major damage in the Western Hemisphere. The earthquakes in Haiti (Box 61.3) and Chile (Box 61.4) are among the most deadly and strongest in decades. Notice that these two earthquakes were unrelated as they occurred on different tectonic plates. In 2011 Japan was hit with a great earthquake that produced an extremely deadly tsunami. The tsunami killed far more people than did the actual earthquake.

Mitigation involves reducing the risk that earthquakes create. Revising building codes to strengthen existing buildings, freeway bridges, and other structures lessens the damage they incur when they undergo shaking by seismic energy. Buildings experience damage because of the direct waves that pass through an area; if these do not destroy buildings, they can weaken them to the point that energy from aftershocks can finish the job of leveling the edifice.

Earthquakes will always occur because of the dynamics associated with plate tectonics, whether the events are the result of direct collision of plates or the result of volcanic activity that is also attributable to plate motion. Being aware of regions on Earth that are most likely to generate earthquakes will make people understand why they occur, although the actual timing of the events Is not predictable.

Tsunamis

By Ingrid Ukstins and David Best

So, you think we are safe?

Lurking off the coast of the Pacific Northwest, the Juan de Fuca Plate is inching its way to the east. In doing so it is colliding with a portion of the North American Plate which is moving westward. There is a real potential for a mega-earthquake occurring in the region, triggering a catastrophic tsunami.

This event will cause utter destruction and loss of life, as did the March 2011 earthquake and tsunami that occurred in Japan. In that event 98 per cent of the more than one million destroyed or damaged homes were the result of the tsunami, not ground shaking caused by the earthquake. Ninety-two per cent of the more than 20,000 fatalities were caused by the tsunami, not the earthquake itself. The low lying coastal regions of northern California, Oregon, and Washington are heavily developed and densely populated. Given the close proximity to the potential epicenter, there will be very little time to alert communities that are in the path of the tsunami. In addition, the highway systems will be virtually impassable due to debris from the event and countless vehicles on the roadways attempting to flee the area.

Tsunami have existed on Earth for millions of years, but it took the events of December 26, 2004, to remind the world of their existence and destructive power. The earthquake and resulting tsunami stunned the world when the destructive power of a mega-tsunami struck the coastal regions of Southeast Asia. Off the west coast of the island of Sumatra in Indonesia, two oceanic plates converged. The resulting earthquake was a M 9.1 event, the fourth largest in recorded history. The vertical movement of the ocean floor was more than 16 meters along a length of 1200 kilometers, the distance between Chicago and New York. This displaced thousands of cubic kilometers of water, generating a tsunami that raced across the adjoining oceans, killing an estimated 230,000 people. Those people who lived near the source had very little time to react to the event, as there is no tsunami warning system in place in the Indian Ocean, unlike the Pacific Ocean. Once officials learned what had happened, countries farther away were notified and thus were able to warn residents of the impending disaster. Coastal regions in East Africa did not experience the large loss of life that occurred in Sumatra, Sri Lanka, India, and Thailand because of the warnings. Obviously if a warning network had been in place throughout the region around the Indian Ocean, more inhabitants would have been informed in time to save many lives.

What Is a Tsunami?

The word **tsunami** (soo-NAH-mee) is derived from two Japanese characters: *tsu* meaning harbor, and *nami* which means wave (the word is both singular and plural in Japanese, as is the word *deer* in English). A tsunami (also termed **seismic sea wave**) is generated by a large displacement of sea water. A tsunami can be generated by an earthquake, a volcanic eruption, a landslide, or a meteorite impact striking the ocean. The energy moves out from the source, producing a series of waves that have long wavelengths. The time between the wave pulses can be as much as an hour or more.

> **tsunami:** A series of giant, long wavelength waves produced by the displacement of large amounts of ocean water by an earthquake, volcanic eruption, landslide, or meteorite impact.
>
> **seismic sea wave:** A large ocean wave that is produced by a major disturbance in the ocean, such as an earthquake, volcanic eruption, or a landslide.

Throughout the long documented history of Japan, these waves have occurred rather frequently. Almost 200 tsunami have been recorded hitting the islands. The location of Japan on the northwest edge of the Pacific Ocean is a focal point for these destructive waves as they travel across the Pacific. The shoreline of the Japanese islands consists of many harbors and inlets. When a seismic sea wave hits these indentations, the energy is rapidly focused into a small area. Depending on slope and elevation of the ground along the coast, water can move inland up to several kilometers. Other areas bordering the Pacific Ocean have also been struck by these waves, including Alaska, Hawaii, Guam, and California.

Tsunami commonly occur several times a year throughout the world as active tectonic plates produce large magnitude earthquakes that displace massive amounts of sea water. Although the majority of tsunami result from the vertical displacement of the sea floor or adjoining areas, strike-slip motion is thought to produce about 15 percent of all tsunami. Energy from in collisional subduction zones strike-slip mechanisms tends to remain close to the source, thus not producing large-scale disasters. The lateral slip motion produces little, if any, vertical offset that could displace the sea water. As populations have increased in the areas surrounding the Pacific, more people have settled in low lying, coastal areas that are in harm's way when these waves strike the shoreline. Coastal inhabitants are aware of the potential disasters and have developed escape routes to allow them to evacuate low areas (**Figure 62.1**).

© Michael Ledray/Shutterstock.com

Figure 62.1 Evacuation routes are set up in areas that can experience tsunami. These routes direct people to higher ground and a safer environment.

Characteristics of Tsunami

Tsunami are most often generated in deep water when vertical movement occurs along a fault or water that is displaced by an earthquake, a volcanic eruption, or a landslide. Plate boundaries are areas where movement is often very slow. The interface along the boundary edges can be thought of as being stuck in place. However, the buildup of force can cause the rocks to suddenly break. When this occurs, there is often some degree of vertical displacement. The initial vertical motion projects water upward, which then collapses, creating a splash that generates waves in all directions. This upward motion is not very large, perhaps on the order of several meters. However, a massive volume of water is moved. Successive waves move out from this point, separated by a distance measured in tens or hundreds of kilometers. A similar condition is produced when you drop a stone into a pond. The waves radiate from the point where the water was displaced. If you were on a ship in the open ocean when a tsunami passed underneath, it would be difficult to detect it. In addition to the long **wavelengths** (the distances between the crests of adjacent waves), the amplitude of the wave is very small, on the order of a meter or two. However, ocean buoys can be set in deep water to detect these waves and transmit the information to warning centers.

wavelength: The distance separating two adjacent crests (or troughs) on a wave form.

Energy in the oceans is transmitted by waves that have the characteristics. Tsunami differ from wind-generated waves in that tsunami have a much longer period, the time separating adjacent crests of the waves, and they also have much larger wavelengths. The periods of tsunami range from about 5 minutes to as many as 60 or more minutes. Normal, wind-generated waves typically have periods of between 5 and 20 seconds. Wavelengths for tsunami range between tens and hundreds of kilometers, while wind-generated waves average around 100 to 200 meters.

Tsunami have been referred to in the past as tidal waves, although their genesis is not related to tides at all. Perhaps the misnomer came about because, as tsunami approach the shore, they appear similar to breaking waves. Tsunami come ashore and continue moving inland; they do not recede in an ebb-and-flow sense that is typical of normal tides.

Frequency of Tsunamis
By Ingrid Ukstins and David Best

The National Centers for Environmental Information (NCEI), a division of the National Oceanic and Atmospheric Administration (NOAA), maintains a database of tsunami activity since 2000 B.C.E. to the present. A key part of the destruction of tsunami is the runup of water on land adjacent to the ocean. Numerous sources of information have been examined to develop a better understanding of where tsunami and runup events occur (Table 63.1). Eighty-three percent of all tsunami have occurred in the Pacific Ocean and the Mediterranean Sea. Major tsunami that affect a large area occur about once per decade.

Extensive documentation exists, especially in Japan and the Mediterranean areas, where inhabitants have recorded the occurrence of natural disasters for centuries. The first recorded tsunami occurred off the coast of Syria in 2000 B.C.E. Since 1900 (the beginning of instrumentally located earthquakes), most tsunami have been generated in Japan, Peru, Chile, New Guinea, and the Solomon Islands. However, the only regions that have generated remote-source tsunami affecting the entire Pacific Basin are the Kamchatka Peninsula, the Aleutian Islands, the Gulf of Alaska, and the coast of South America. All of these are very active areas of tectonic activity. Hawaii, because of its location in the center of the Pacific Basin, has experienced tsunami generated in all parts of the Pacific.

Table 63.1 Comparative Amounts of Tsunami and Runup Events, by Region

Region	Percentage of Tsunami	Percentage of Runup Events
Pacific Ocean	61	82
Mediterranean Sea	22	4
Indian Ocean	6	9
Caribbean Sea	4	2
Atlantic Ocean	7	3
Black Sea	1	0
Red Sea	<1	0

Source: National Centers for Environmental Information, NOAA

Regions around the Mediterranean and the Caribbean have experienced numerous, locally destructive tsunami. A small subduction zone exists where the African Plate is colliding with the Eurasian Plate. The Caribbean Plate is surrounded by the North American and South American Plates on its eastern edge and the North American and Cocos Plates on its western edge.

Only a few tsunami have been generated in the Atlantic and Indian Oceans. In the Atlantic Ocean, there are no subduction zones at the edges of plate boundaries to generate such waves, except small subduction zones under the Caribbean and Scotia Arc, which is located at the southern end of South America. Subduction is active along the eastern border of the Indian Ocean, where the Indo-Australian plate is being driven beneath the Eurasian plate at its eastern margin. Thus, most tsunami generated in this region are propagated toward the southwest shores of Java and Sumatra, rather than into the Indian Ocean. The December 2004 Indonesian tsunami was anomalous in that it affected all the countries in the immediate and distant areas because of the size of the earthquake that generated the seismic sea wave (**Figure 63.1**).

© capturefoto/Shutterstock.com

Figure 63.1 Damage caused by the December 2004 tsunami was far-reaching. Areas such as this one near Banda Aceh, Indonesia, were overrun by water that stripped the landscape clean.

Not all major earthquakes in the oceans produce tsunami. In a two-week period in January 2018 two such events did not generate any significant wave action. A magnitude 7.6 earthquake off the coast of Honduras was determined to be related to strike-slip motion along a fault bordering the contact of the Caribbean Plate and the North American Plate.

On January 23, 2018 a magnitude 7.9 event, located 220 kilometers southeast of Kodiak Island, Alaska, occurred along a strike-slip portion of the Pacific Plate–North American Plate boundary. Although a warning was issued, no tsunami materialized. Two previous large events (magnitudes 7.8 and 7.9) occurred in the same region in the late 1980s, producing no known damage.

Sumatra, Indonesia, December 26, 2004

December 26, 2004, began very quietly for many vacationers who were visiting the beaches of Sumatra, Thailand, India, and nearby countries. Unknown to many people, one of the largest earthquakes in history occurred, displacing trillions of gallons of water in a tsunami wave that ended up going all the way around the globe.

At 7:58 in the morning local time, a tsunami generated by a huge (M 9.1) earthquake began its journey across the Indian Ocean. Within about 15 minutes, it struck Sumatra, killing over 130,000 people. Damage in Sumatra was complete—trees were ripped out of the ground by their roots, entire towns were leveled, and water flooded the land to a depth of up to 25 m (**Figure 63.2**). Within two hours, the tsunami had struck Sri Lanka and Thailand, claiming another 40,000 lives.

© Frans Delian/Shutterstock.com

Figure 63.2 Banda Aceh, Indonesia, was one of the most devastated cities along the Sumatra coast, following the December 26, 2004, tsunami.

In all, at least 230,000 people died in one of the worst natural disasters in human history. The waves dramatically pointed out the vulnerability of coastlines to tsunamis, and the devastation caused when little warning is available.

Tsunami Prediction, Warning and Mitigation
By Ingrid Ukstins and David Best

Tsunami are often associated with large-magnitude earthquakes that occur in or near oceans. When a large earthquake occurs in a region near the ocean, instruments alert scientists about the potential for tsunami to be formed and warnings can be issued.

In the Pacific Ocean the Pacific Tsunami Warning Center located at Ewa Beach, near Honolulu, Hawaii, becomes the focal point of activity when an earthquake could potentially generate a tsunami. This facility is operated by the National Weather Service (NWS) and serves as the headquarters for the Operational Tsunami Warning System, a group that coordinates the monitoring and reporting of seismic activity around the Pacific Ocean, the Indian Ocean, and the Caribbean Sea. The West Coast/Alaska Tsunami Warning Center, located in Palmer, Alaska, monitors the northern Pacific Ocean, including Alaska and the Aleutian Islands. It also oversees tsunami activity along all coastal regions of the United States (except Hawaii), the coastal provinces of Canada, Puerto Rico, and the Virgin Islands.

Warnings are only as good as the people who heed them. If people decide to rush to the shore to watch waves come in, they will obviously be placing their lives in danger; such was the case in Crescent City, California. In March 1964 people went to the shore to watch tsunami generated by the great earthquake in Alaska, resulting in 12 deaths.

BOX 64.1	Tohoku, Japan Earthquake and Tsunami, March 11, 2011

As the Pacific Plate moves to the northwest and collides with a portion of the North American Plate, the resulting subduction has produced the volcanic islands that form Japan. The region has historically experienced many earthquakes including a major historic event in 1923.

On March 11, 2011, the earth ruptured 130 kilometers east of Sendai, a city of one million people on the eastern coast of Honshu. The shallow event had a focal depth of twenty-four kilometers and produced a magnitude 9.1 earthquake, which ranks as the worst in Japan's long history of seismicity and the world's fourth largest recorded since 1900. As often occurs with great earthquakes (magnitude 8.0 or larger), there were several major foreshocks, including a magnitude 7.2 event two days earlier. More than 11,000 aftershocks were recorded. The vast majority of the 20,000 or more deaths and missing persons resulted from the tsunami that was formed by the sudden fault movement offshore. Damage estimates were more than $360 billion US dollars.

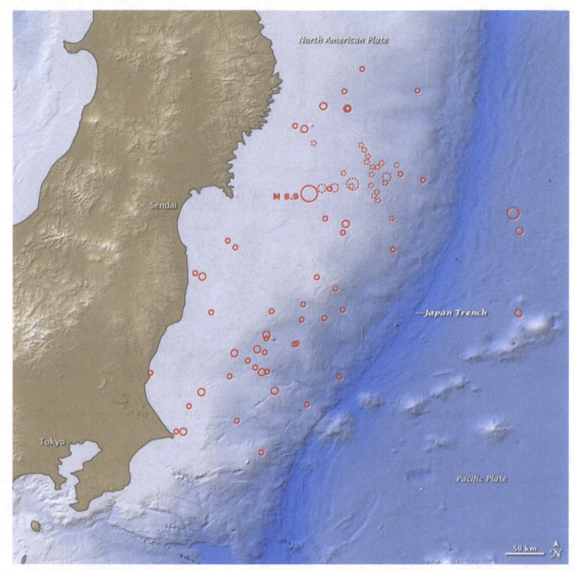

Source: NASA.

Box Figure 64.1.1 Location of the epicenter of the March 11, 2011, magnitude 9.1 earthquake. The Pacific Plate is being subducted under the North American Plate.

The resulting tsunami raced across the Pacific Ocean at the speed of a jet airplane. Within an hour of the earthquake, a wave 2.8 meters high struck the city of Hanasaki on the island of Hokkaido. Numerous other reporting stations throughout the Pacific reported waves, including Midway Island (five hours later, wave height 1.3 meters), Maui, Hawaii (eight hours later, wave height 1.7 meters), and Crescent City, California (eleven hours later, wave height 2 meters). Areas that have arcuate coasts and harbors experienced large surges due to the focusing of the wave energy. The Japanese government estimated that more than 1.5 million tons of debris floated out into the Pacific Ocean, some of which made its way to the west coast of the United States.

Seawater driven ashore by the tsunami damaged the cooling systems of the Fukushima Daiichi Nuclear Power Station. Three nuclear reactors overheated, producing a meltdown of the one reactor and releasing radioactive material into the atmosphere and surrounding water bodies. Nuclear power plants throughout the country were shut until inspections and maintenance were completed.

© Smallcreative/Shutterstock.com.

Box Figure 64.1.2 Extensive damage was caused by the March 11, 2011, tsunami as it came ashore at Sendai, Japan.

In order to lessen the loss of life and property damage, fewer people must live near the shoreline, especially in areas where there is very little relief above sea level. Oncoming waves that are as much as 15 m or higher will move rapidly onshore and flood flat-lying areas that are only slightly above sea level. All property will be inundated and likely totally destroyed from the onslaught of water.

Tsunami Detection Devices

The National Oceanic and Atmospheric Administration (NOAA) operates a network of stationary tsunami detection buoys in several oceans throughout the world. The system that employs the recording technology is the Deep-ocean Assessment and Reporting of Tsunamis (DART). These buoys are primarily deployed in the Pacific Ocean, but there are several in the western Atlantic Ocean and in the Gulf of Mexico (**Figure 64.1**). Only six buoys were operational in the Pacific prior to the 2004 Indonesian event, and none were located in the Indian Ocean. Therefore there was no means to notify nations around the Indian Ocean of the impending disaster coming their way.

The buoys float on the water surface and receive a signal from a source located on the seafloor. As a tsunami passes the sensor, there is a rapid change in pressure which is sent to the hydrophone underneath the buoy. The instruments are programmed to recognize this change and then transmit a signal to a warning center, which then sends the appropriate information to areas in the possible path of the tsunami.

If a tsunami is created, the first information available is related to the earthquake and its source area. This information gives scientists a general idea of the area that could be affected. As the tsunami travels across the ocean, its movement is tracked by data furnished from the recording buoys. As more data become available, the path of the tsunami can be projected more accurately, thus allowing steps to be taken to reduce the destruction ahead of the waves.

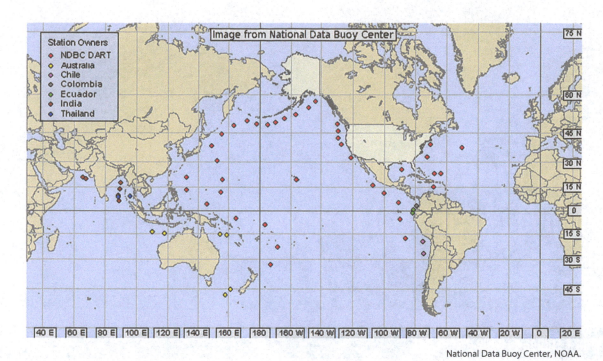

National Data Buoy Center, NOAA.

Figure 64.1 Detection buoys provide early warnings around the Pacific Ocean, the Caribbean and the east coast of the United States for areas prone to tsunami.

Mass Movement
By Ingrid Ukstins and David Best

It Won't Happen to Me—We'll See

The small town of Union Gap, Washington, (population 6,110) is located about six kilometers south of Yakima. In early October 2017 townspeople noticed a crack forming on the surface of Rattlesnake Hills, located just east of town. At first the hill was moving to the south at the rate of approximately thirty centimeters per week. Geologists and engineers set up monitoring equipment to better observe the activity on the twenty-acre tract of land. The hill consists of basalt overlying a weaker sedimentary layer. Within a few months the movement had increased to almost seventy centimeters a week. The potential is for an estimated three million cubic meters of material to slide downslope. Steps have been taken to slow the movement with large barricades at the base of the hill, which is slowing movement toward Interstate 82. A quarry at the foot of the hill has closed, and nearby homeowners have been evacuated. Although water does not appear to be a factor, the potential certainly exists for a heavy rainfall to work its way into the surface crack and flow downhill along the interface between the basalt and the sedimentary rock. Only time will tell if this disaster will occur.

Although the ground we live on may appear to be stable and unchanging, it is in fact a dynamic place where the force of gravity is constantly acting on Earth's materials, over time causing the downward movement of rock and loose surface material. **Mass movement** (also called *mass wasting*) is the downward movement of Earth's surficial material from one place to another under the direct influence of gravity. Earth's surface consists mainly of slopes ranging from gentle to steep and often covered with several components of the surface. **Regolith** consists of all material lying above unaltered bedrock, including unconsolidated and fragmental material which breaks down to form soil. Although gravity acts constantly on all slopes, the strength and resistance of materials usually hold the slope in place. However, natural processes or human activity may destabilize a slope, causing failure and mass movement of material. This movement may be so slow that it is almost undetectable (known as *creep*), or be sudden and swift, as in devastating **landslides**. Mass movements do not always have to involve failure on a slope, however. **Subsidence** is the vertical (downward) motion of earth materials, and includes vertical movements such as the gradual sinking of the land surface when fluids are withdrawn, or the sudden collapse into subterranean voids creating **sinkholes**.

mass movement (mass wasting): A term used more by geologists to describe the down-slope movement of material.

regolith: The layer of varied material that overlies unaltered bedrock and includes unconsolidated and fragmental particles.

landslide: A general term used to describe the down-slope movement of material under the force of gravity.

Mass movement is an important natural erosional process but can be a serious hazard where expanding populations are building homes on or near steep hillsides or over abandoned mine areas. Rapid downslope movements mostly involve landslides, which every year cause loss of life, property damage, or economic hardship to people around the world.

Landslides have occurred in all 50 States (**Figure 65.1**), and damage from these landslides exceeds $1 billion every year, while claiming 25 to 50 lives. In 1985, a massive landslide in Puerto Rico killed 129 people, making it the greatest loss of life from a single landslide in U.S. history. The Thistle, Utah landslide of 1983 caused an estimated $400 million in losses, making it one of the most expensive single landslide in U.S. history, while the 1997–98 El Niño rainstorms in the San Francisco Bay area produced thousands of landslides, causing over $150 million in losses. Landslides can also be associated with other hazards, such as earthquakes, floods, storm surges, severe storms, and volcanic activity. **Slope failures** are a secondary effect of wildfires that remove the vegetation holding soil and surface materials in place. Often they are more damaging and deadly than the associated hazard event.

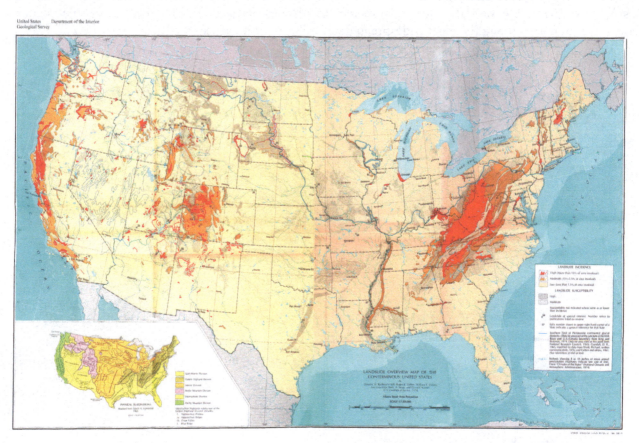

United States Geological Survey.

Figure 65.1 Landslides occur in all 50 states (Alaska and Hawaii not shown). Moist, unstable terrain in the eastern United States is especially affected by landslides. Darker shading shows areas that experience the highest likelihood of landslides.

What makes many mass movements, such as landslides and collapse sinkholes, so hazardous is that they can occur at anytime and almost any place with little or no warning, and the event is over very quickly. Knowledge about the relationships between the local geology and mass movement processes can aid in better development planning and a reduction in vulnerability to such hazards. Therefore, we will look at the various types of mass movements, their underlying causes, factors that affect ground and slope stability, and what we can do to reduce the vulnerability and risk of living on unstable ground.

Landslides: Mass Movements on Slopes

Earth's surface materials that make up slopes comprise a complex physical, chemical, and biological system. These materials are a mixture of rock, regolith, and soil, with a variable amount of water and organic material. The amount of water and organic material (mostly vegetation) may vary on a slope from season to season and year to year. This variability depends upon such factors as the nature and extent of precipitation, rates of materials added or removed to the slope by deposition or erosion, and activities that affect the vegetation, such as human impact or fire. These complex interactions involving the strength and composition of earth materials are responsible for the different shapes of slopes, and how slopes may fail under the influence of gravity. The resulting mass movement plays an important part in the erosional process, moving material downslope from higher to lower elevations.

The term *landslide* is used in the general sense to describe a wide variety of mass movement landforms and processes involving the downslope transport of earth material under gravitational influence. Landslides have a great variety of shapes (morphologies), rates of motion, and types of movement, and can range in size from a small area to a region of many square kilometers. Landslides play a significant role in the evolution of the landscape, and are among the most widespread natural hazards on Earth.

Topic 66

Weathering
By Dean Fairbanks

Weathering is the decomposition and disintegration of rocks *in situ*. Decomposition refers to the **chemical weathering** process and creates altered rock substances whereas disintegration or **mechanical weathering** produces smaller, angular fragments of the same rock. Weathering is a key process for landscape evolution as it breaks down rock and facilitates erosion and transport of material down slope.

Figure 66.1   Southwestern Colombia mountainous terrain riddled with slides and debris avalanches. (**USGS**)

Weathering is part of **landmass denudation**, where denudation refers to all processes that cause degradation of landforms. The principal landscape denudation processes include *weathering, mass movement, erosion, transportation,* and *deposition* all produced by water, air, waves, and ice (**Figure 66.1**). The weathering process is to *transform* rock into materials that are less resistant to *gravity, erosion,* and *transport.* Taking this a bit further, when considering the significance of the weathering processes and its products in understanding landscapes, we can note breaking down rock, preparing rock for transportation, encompassing the basis for soil development (basis for plant life), and ultimately weathering as the basis for the chemical composition of the world's oceans.

Since weathering simply prepares materials for erosion and transport, there are several levels of weathering within landscape material. **Bedrock** is the parent rock material from which weathered material is derived. On top of bedrock but below the soil is broken-up rock called **regolith**. Loose sediment and soil is derived from transported and deposited regolith.

Table 66.1

Weathering Grade	Description
Fresh	No visible sign of material weathering
Slightly weathered	Discoloration
Moderately weathered	Less than half material decomposed
Highly weathered	More than half of rock is decomposed
Completely weathered	All material decomposed to soil

Mechanical vs. Chemical Weathering

Mechanical (physical) weathering represents the breaking and disintegrating of rock without any chemical alteration. It relies on the granular disintegration, which tends to work along lines of weakness in rocks. Some lines of weakness are created as unloading of rocks, which occurs as a result of erosion of overlying materials. There are four main types:

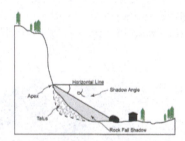

Figure 66.2   The distance a tumbling boulder would travel. (**USGS**)

- **Freeze–thaw** (congelifraction)—Also known as **joint-block separation** or **frost wedging**, it occurs when water in joints and cracks freezes and expands, thus exerting pressure on the two sides of rock. This further opens the crack and eventually breaking the rock in two. It is most effective in environments where moisture is plentiful and there are frequent fluctuations above and below freezing point, e.g., mountain environments and periglacial (permafrost) regions (**Figure 66.2**). A common landscape feature that is developed from frost wedging is the development of scree or talus slopes of broken rock (the former being smaller rock broken rock (the former being smaller rock and the latter being larger rock to boulders). These slopes of broken rocks are very common below cliff faces (**Figure 66.3**).
- **Disintegration**—There are two recognized types: **thermal expansion** and **exfoliation**. The first is found in hot desert areas where there is a large diurnal temperature range. The rock heats up and expands by day and cool and contract by night causing stress along lines of weakness. The other type, exfoliation, occurs particularly with granites. The granite unloads in large sheets as the stresses occur only in the outer layers of the rock. This tends to give granite landscapes a rounded appearance as most noted in Yosemite National Park (**Figure 66.5**). **Pressure release**—a process whereby overlying rocks are removed by erosion thereby causing underlying ones to expand and fracture parallel to the surface. Common with tectonic uplift, as erosion strips away overlying material the pressure from the uplift causes joint to develop in the rock.
- **Root wedging**—a process whereby plant roots, especially trees, develop into the joints and fractures of rocks and act like crowbars over time widening the cracks as the roots grow. Eventually the rock becomes completely broken up and weak. We can see this in urban areas as tree roots uplift concrete sidewalks.

Figure 66.3 Talus slopes displaying broken rock slopes. (**USGS**)

Figure 66.4 An aerial view of a large sinkhole in Florida. (**USGS**)

Figure 66.5 Northeast side of Half dome in Yosemite National Park illustrating granite exfoliation. (**USGS**)

Figure 66.6 Carlsbad cavern in New Mexico. (**USGS**)

Chemical weathering represents the decomposition and decay of the constituent minerals in rock through chemical alteration of those minerals at their surfaces. The susceptibility to chemical weathering depends on *mineral stability*. The rate of weathering depends strongly on *surface area*, i.e., more surface area increases weathering. *Water is essential*, with decomposition rates keyed to temperature and precipitation values. There are four main types:

Carbonation solution—Rainfall and dissolved carbon dioxide forms a weak carbonic acid. Any rocks containing calcium carbonate, e.g., limestone and chalk, will react with the acid water and form calcium bicarbonate, which is soluble and will thus be removed by the water. Limestone is a very abundant landscape type on Earth (**Figure 66.9**). They are called karst landscapes, which form a distinctive topography from chemically weathering with poorly developed surface drainage and solution features, many underground, that appear pitted and bumpy (**Figure 66.8**). Key features of this landscape type are sinkholes, tower karst, and caverns (**Figure 66.4** and **Figure 66.6**).

Figure 66.7   Oxidation of iron at the Iron Mountain Mine site, California. **(USGS)**

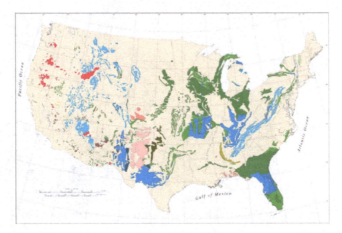

Figure 66.8   Map of karst and other karst like landscapes in the lower US. The red areas indicate lava tubes. **(USGS)**

Figure 66.9   Landscape with limestone towers and rice fields in Vietnam.

- **Hydrolysis**—a process where water (carbonic acid) is added to the softer minerals in granites, i.e., feldspar. The reaction forms new compounds of clay, like kaolinite (china clay).
- **Hydration**—a process whereby certain minerals absorb water, expand, and change their structure (become hydrates).
- **Oxidation**—a process that occurs when iron compounds react with oxygen to produce rust, as a crust-like coating (**Figure 66.7**).

Controls on Weathering

The major determinants that effect rates of weathering follow.

- The *type of rock* (igneous, sedimentary, and metamorphic), which is based on its
  - chemical composition;
  - nature of cements in sedimentary rock; rock structure-joints, cracks, faults, and bedding planes.
- *Climate* based on moisture availability and average annual temperature (see Peltier's classification below (1950). Frost wedging increases as the number of freeze–thaw cycles increases. Chemical weathering increases with moisture and heat. **Van't Hoff's law** states that for every 10° C increase in temperature, the rate of chemical weathering increases by 2-3 times (up to 60° C).
- Availability of *moving water*, which acts more as an erosional agent
- *Vegetation*, which acts as a protective barrier.

Table 66.2

Climatic Region	Mean Annual Temperature (°C)	Mean Annual Precipitation (mm)	Weathering Processes
Glacial Periglacial	−20 to −5 −15 to 0	0–1,100 125–1,300	Moderate chemical
Boreal	−10 to 5	250–1,500	Moderate frost action
Maritime	5 to 20	1,250–1,500	Moderate chemical
Tropical Rainforest Temperate	15 to 30 5 to 30	1,400–2,250 800–1,500	High chemical Moderate frost action
Savanna Semi-arid/ Mediterranean	10 to 30 5 to 30	600–1,250 250–600	Moderate chemical Moderate chemical
Arid	15 to 30	0–350	Moderate frost action Moderate chemical High thermal action

Slope Processes and Stability
By Ingrid Ukstins and David Best

Two types of forces are involved in mass movements on slopes: **driving forces** and **resisting forces**. Driving forces are those that promote movement, while resisting forces are those that tend to prevent movement. The main *driving force* responsible for mass movement is gravity. Gravity acts everywhere on the Earth's surface, pulling everything in a vertical direction toward the center of the Earth. On a flat surface the force of gravity acts directly downward, so, as long as the material remains on the flat surface, it will not move. On a slope, the force of gravity can be divided into two components: one acting perpendicular to the slope (the **normal force**) and one acting parallel to the slope (the **shear force**) (**Figure 67.1**). The normal force holds the material in place on the slope, whereas the shear force moves the material down the slope.

driving force: A force that produces down-slope movement caused by gravity.

resisting force: A force that tends to prevent downhill movement.

normal force: A force that is acting perpendicular to a surface.

shear force: A force that acts parallel to a surface.

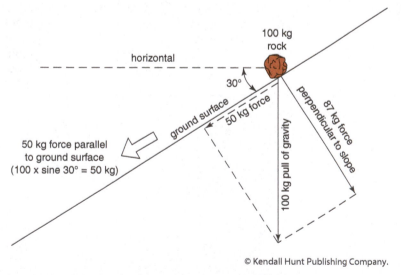

© Kendall Hunt Publishing Company.

Figure 67.1   Forces acting on a 100 kilogram boulder resting on a 30° hillslope.

On and below the ground surface, resisting forces, such as friction and particle cohesion, act in opposition to the driving forces. When the driving forces acting downhill exceed the resisting forces, mass movement results.

The **Factor of Safety (FoS)** can be defined for a slope as:

$$FoS = resisting\ forces/driving\ forces.$$

> **Factor of Safety (FoS):** The ratio of resisting force to driving force.

If the resisting forces are greater than the driving forces, the FoS is greater than 1 and the slope is stable. If the driving forces are greater than the resisting forces, the FoS is less than 1 and slope failure occurs. The magnitudes of the resisting and driving forces can change as a result of weather conditions or slope configuration.

Slopes are in a continuous state of dynamic equilibrium by constantly adjusting to new conditions caused by changes in the driving and resisting forces. Although we view mass wasting events as disruptive and often destructive processes, it is simply a way for a slope to adjust to new and changing conditions. For example, when a building or a road is constructed on a hillslope, the equilibrium of the slope is upset and the slope may become unstable. This could lead to slope failure as the slope adjusts to the new set of conditions (additional mass) placed upon it. The role of gravity is to level out all slopes and create a horizontal landscape.

Causes and Triggers That Influence Slope Stability

Rock, regolith, and soil remain on a slope only when the driving forces are unable to overcome the resisting forces keeping the material in place. Long before a landslide occurs, slopes undergo various processes that weaken rock material and gradually make the surface more susceptible to the pull of gravity. Eventually, the strength of the slope is weakened to the point that some event (known as a *trigger*) allows it to suddenly cross from a state of stability to one of instability.

Following a landslide, geologists and engineers often look for clues (the contributing factors) to explain why the slope failed. The reasons for slope failure can be broken down into two categories: *causes* and *triggers* (Table 67.1a–b). The difference between causes and triggers of landslides is subtle, but important to understand. **Causes** are considered to be the controlling factors involved in making the slope vulnerable to failure in the first place; in other words, they contribute to the slope becoming weak and unstable. A **trigger** is the single event that temporarily disturbs the equilibrium of the slope and initiates mass movement, resulting in slope failure. Triggers that initiate movement include earthquakes, heavy rainfall, and volcanic activity. The causes are the explanation for why a landslide occurred in a certain location, and include geological factors, morphological factors, and factors associated with human activity. Landslides most often have multiple causes that combine to make a slope vulnerable to failure, and then a trigger finally initiates the movement.

> **cause:** A controlling factor that makes a slope vulnerable to failure.
>
> **trigger:** An event that disturbs the equilibrium of a slope, causing movement to occur.

In most cases it is relatively easy to determine the trigger after the landslide has occurred, but the causes leading up to slope failure can be much more difficult to identify, because once a landslide occurs, any pre-existing evidence is usually destroyed.

Table 67.1a Common Landslide Causes

1. Natural Geological Causes

a. Weak or sensitive materials
b. Weathered materials
c. Sheared, jointed, or fissured materials
d. Adversely oriented discontinuity (bedding, schistosity, fault, unconformity, contact, and so forth)
e. Contrast in permeability and/or stiffness of materials

2. Natural Morphological Causes

a. Tectonic or volcanic uplift
b. Glacial rebound
c. Fluvial, wave, or glacial erosion of slope toe or lateral margins
d. Subterranean erosion (solution, piping)
e. Deposition loading slope or its crest
f. Vegetation removal (by fire, drought)
g. Thawing
h. Freeze-and-thaw weathering
i. Shrink-and-swell weathering

3. Human Causes

a. Excavation of slope or its toe
b. Loading of slope or its crest
c. Drawdown (of reservoirs)
d. Deforestation
e. Irrigation
f. Mining
g. Artificial vibration
h. Water leakage from utilities

Table 67.1b Common Landslide Triggers

1. Physical Triggers

a. Intense rainfall
b. Rapid snowmelt
c. Prolonged intense precipitation
d. Rapid drawdown (of floods and tides) or filling
e. Earthquake
f. Volcanic eruption
g. Thawing
h. Freeze-and-thaw weathering
i. Shrink-and-swell weathering
j. Flooding

Source: USGS Fact Sheet 2004-3072 and Circular 1325-508.

Common Controlling Factors (Causes) in Slope Stability

The following is a discussion of the more common controlling factors that influence slope stabilities, and the role they play in making them vulnerable to failure. Although they are discussed separately, most of them are interrelated and can collectively affect a slope's stability.

The Role of Slope Materials

Some rocks are inherently more stable than others. Massive, uniformly-textured rocks such as granite, basalt, and quartzite, have interlocking crystals and grains that give them nearly equal strength in all directions. Such rocks will hold their position against gravity, and form steep slopes only when they become fractured and jointed. Over time, chemical and physical weathering processes produce unconsolidated regolith, and change its resistance to gravity. Freezing and thawing of water in cracks, or shrinking and swelling of clay minerals decrease friction and cohesiveness and promote slope failure.

Unconsolidated earth materials such as soil, clay, sand, and gravel are more susceptible to movement. Clay becomes plastic and weak when it absorbs water, causing it to act as a lubricant for mass movements. For sediment coarser than clay, stability on slopes is influenced by grain shape (whether angular or round) and grain roughness. The presence of plant root networks and mineral are very effective in holding material in place, thus promoting stability.

The Role of Water

The water content in slope material strongly influences its stability in several ways, and is a critical factor in many causes of mass movements, as well as being a trigger. As a driving force, slope material can become saturated with water (after heavy rains, melting snow, or rising groundwater) by filling pores and fractures. This increases the weight (load) of material on the slope and increases the stress acting parallel to the surface (water weighs 1 kg per liter or 8.35 lbs per gallon). However, this extra weight of water is probably less important than the reduction of the resisting forces of the material. Water percolating through slope material helps decrease friction between grains and contributes to a loss of cohesion among particles.

Another aspect of water that affects slope stability is fluid pressure, which in some cases can build in such a way that water in the pores can actually support the weight of the overlying material. When this occurs, friction is greatly reduced, and thus the resisting forces holding the material on the slope is also reduced, causing slope failure. Water can also reduce rock strength by circulating through the pores of some rocks and dissolving soluble cementing materials, such as calcium carbonate. This process reduces cohesion as well. Another effect of water is that it can soften layers of shale, and even cause some types of clay minerals to expand, reducing friction between rock layers. Clay consists of platy particles that easily slide over each other when wet. This lubricating effect of water with clay is the reason why clay beds are frequently the slippery layer along which overlying material slide down-slope.

The Role of Slope Angle

Slope angle is a major factor that influences mass movement. Commonly, the steeper the slope, the less stable it is, and therefore, steep slopes are more likely to experience mass movement than gentle slopes. The steepest angle that a slope can maintain without collapsing is called its **angle of repose** (**Figure 67.2**). For dry, unconsolidated materials, the angle of repose increases with increasing grain size, but usually lies between about 33° and 37° on naturally formed slopes. At this angle, the resisting forces of the slopes material counterbalance the force of gravity. Stronger material such as massive bedrock can maintain a steeper slope. If the slope angle is increased, the rock and debris will adjust by moving downslope.

> **angle of repose:** The natural angle of a slope that forms in a pile of unconsolidated material.

© Djordje Zoric/Shutterstock.com

Figure 67.2   The angle of repose is the natural angle of a slope formed by a pile of material.

A number of processes can make a slope too steep and cause it to become unstable. **Undercutting** by stream and wave action is a common process that removes the base of a slope, increasing the slope angle, and increasing the driving force acting parallel to the slope. Waves pounding against the base of a cliff, especially during storms, often result in mass movements along the shores of oceans and lakes. Human activities also often create unstable slopes that become prime sites for mass movement. Mass movement occurs when grading and cutting into the slope too steeply increases the downhill forces and the slope is then no longer strong enough to maintain the steep angle. Such actions by humans is analogous to undercutting by streams and has the same result, explaining why so many mountain roads and building sites are plagued by frequent mass movements.

undercutting: A process whereby a slope or hillside has supporting material removed by erosion.

The Role of Vegetation

Plants protect regolith and soil against erosion, and contribute to the stability of slopes (**Figure 67.3**). Vegetation helps absorb rainfall and leads to a decrease in water saturation of the slope material that would otherwise lead to a loss of shear strength. The root systems of plants also help stabilize the slope by binding soil particles together and holding the soil to bedrock. Where vegetation is lacking, mass movement is enhanced. The removal of vegetation by natural events (fires, drought) or human processes (timber, farming, or development) frequently results in unconsolidated materials moving down-slope.

Figure 67.3 Vegetation on slopes is effective in holding the soil and regolith in place as root systems help stabilize the material. Whenever grasses and trees are removed, soil and regolith can readily move downslope.

The Role of Overloading

Overloading occurs when additional weight is added to a slope. As mentioned earlier, water can become a driving force by increasing the weight of slope material after prolonged periods of rain. However, overloading is most often the result of human activity, and typically results from the dumping, filling, or piling up of material on a slope (**Figure 67.4**). The additional weight created by the overloading increases the water pressure within the material, which in turn decreases the stability of the slope. If enough material is added to the slope, it will eventually fail, sometimes with tragic consequences. Water can become a driving force by increasing the weight of slope material after prolonged periods of rain by completely filling pore spaces.

The Role of Geologic Structures

Plate tectonic forces can reorient rocks after their formation, causing them to be rotated and tilted. If the rocks underlying a slope dip in the same direction as the slope, mass movement is more likely to occur than if the rocks are horizontal or dip in the opposite direction of the slope. When rocks dip in the same direction as the slope, water can percolate along the bedding planes and decrease the cohesiveness and friction between adjacent rock layers (**Figure 67.5**). This is particularly important when clay layers are present, because they become slippery when wet.

Even if the rocks are horizontal or dip into the opposite direction of the slope, other structural weaknesses (such as joints, foliations, and faults) may dip in the same direction as the slope. The water percolating through the rock structures weathers and expands the openings further until the weight of the overlying rock causes it to fail.

© Harvey Hessler/Shutterstock.com

Figure 67.4 Unstable slopes are created when tailings and other material from mining operations are piled up. These slopes are unable to develop soil layers that can support vegetation, so in addition to being unsightly they are likely to move if a triggering mechanism occurs nearby.

© Maksimilian/Shutterstock.com

Figure 67.5 Bedding planes dipping in slope direction lead to slope failure.

The Role of Human Activity

There are many ways in which human activities can affect slope stability and promote landslides. One is by the clearing of stabilizing vegetation during logging operations, exposing sloping soil to rain. Many types of construction projects can lead to over-steepening of slopes. Highway roadcuts, quarrying, or open-pit mining, and construction of homes on benches cut into hillsides are among the most common activities that can cause problems (**Figure 67.6**). Where dipping layers of rock are present, removal of material at the lower end of the layers may leave large masses of rock unsupported.

© Baxtar/Shutterstock.com

Figure 67.6   Terraces are cut in an open pit mine in order to help stabilize the steep walls. Notice the trucks on the roadway.

In the 1870s throughout Europe there was a large demand for slate, which was used to make blackboards. To meet this demand, slate quarry miners near Elm, Switzerland dug slate quarry at the base of a steep cliff containing excellent planar foliation that dipped toward the quarry. By September 1881, the quarry was 180 m long and 60 m into the hill below the cliff, and a fissure above the cliff had opened to 30 m wide. Falling rocks were frequent in the quarry and almost continuous noises were heard coming from the overhang above the quarry. On September 11, 1881, a 10 million cubic meters mass of rock above the quarry suddenly fell to the quarry floor, and produced an avalanche moving at 180 km/hr that traveled over 2 km, burying the village of Elm and killing 115 people.

Slopes cut into unconsolidated material at angles higher than the angle of repose without planting stabilizing vegetation becomes extremely unstable. The very act of building a house above a naturally unstable or artificially steepened slope adds weight to the slope, thereby increasing the stress acting parallel to the slope. Other activities connected with the presence of housing developments on hillsides include watering the lawn, use of septic tanks, or an in-ground swimming pool from which water can seep slowly. These are all activities that increase the water content of the soil and can render the slope more susceptible to movement. Even homes built on apparently solid rock could be prone to destruction (**Figure 67.7**).

Topic 67 Slope Processes and Stability   **333**

© Alexander Chaikin/Shutterstock.com

Figure 67.7 This home in Valparaiso, Chile, is situated on solid rock. However, earthquakes are a common occurrence in Chile, and this home could easily collapse down the hill.

Common Triggers of Slope Failure

The factors discussed thus far all contribute to slope instability and are considered causes that make the slope vulnerable to failure. A sudden triggering event may then happen that initiates movement on the unstable hillside. Movement could eventually occur without a trigger if slope conditions became more unstable over time. However, sudden, rapidly moving landslides usually involve a triggering mechanism. Water, volcanic activity, and seismic activity are three common landslide trigger mechanisms that can occur either singly or in combination (refer to Table 6.1a–b).

Water as a Trigger

Rainfall. When a slope becomes saturated with water, the increased mass of the water, and its lubricating effects, create instability. Intense or prolonged rainstorms act as a trigger of the mass movement. Porous soil and layers of underlying rock allow the pore pressure to increase, thereby decreasing the adhesion of the grains. Failure results. Heavy rainfall occurs during tropical storms and hurricanes, short-lived, severe thunderstorms, and prolonged rainfall related to slow-moving weather systems. Flooding and landslides are commonly the result (**Figure 67.8**).

© PHOTO999.Shutterstock.com

Figure 67.8 These houses were destroyed by landslides associated with Typhoon Morakat, which hit Taiwan in August 2009.

Snowmelt. In regions that experience prolonged cold weather that allows snow packs to increase in size, seasonal melting will produce large amounts of water in relatively short time periods. Melting can be accelerated by short periods of unseasonably warm weather. In addition to direct runoff of melt waters, water can percolate into the subsurface and saturate the soil above the zone of permanently frozen ground (permafrost). If there is any slope to the saturated layer, the fluids and solid material will rapidly move downhill as increased pore pressure reduces the shear strength of the soil. Depending on the amount of water present and the angle of the slope, rapid or slow mass movement can occur.

River and Wave Undercutting. Erosional processes caused by moving water in streams, or wave action along a shoreline, can remove material at the base of a slope. If sufficient removal occurs, the slope is no longer supported and slope failure occurs. A large-scale example of this is seen at Niagara Falls along the border of the United States and Canada (**Figure 67.9**). Water cascading over the falls erodes the relatively soft underlying shale layers. The upper rocks are no longer supported and collapse. Erosion there is occurring at the rate of about one meter per year, causing the falls to retreat upstream. The falls are retreating about one meter per year.

Source: David M. Best.

Figure 67.9 Undercutting occurs as water flows over Niagara Falls. The softer, underlying rock (shale) is eroded, thereby removing support for the more resistant overlying dolomite, which subsequently falls into the area below.

Volcanic Activity as a Trigger

Some of the largest and most destructive landslides known have been associated with large volcanic eruptions. These occur either in association with a violent eruption of the volcano itself, or as a result of mobilization of its weak deposits that are formed as a consequence of volcanic activity. Volcanic eruptions produce seismic shaking similar to high-energy explosions and earthquakes, and can trigger flank collapses that create fast-moving debris avalanches. Hot volcanic lava or ash eruptions can melt snow and ice on a volcano summit at a rapid rate, causing a flow of rock, soil, ash, and water that accelerates rapidly on the steep slopes of volcanoes. These volcanic debris flows (known as *lahars*) can reach great distances, once they leave the flanks of the volcano, destroying anything in its path.

The 1980 eruption of Mount St. Helens triggered a massive debris avalanche on the north flank of the volcano (**Figure 67.10**). The 1985 Nevado del Ruiz eruption in Colombia created pyroclastic flows that melted ice and snow at the summit, forming lahars up to 50 meters thick. These massive flows traveled more than 100 kilometers down several river valleys, destroying many houses and towns. The town of Armero was completely covered by the lahar, killing approximately 23,000 people, making it the second-deadliest volcanic disaster in the 20th century (after the 1902 eruption of Mount Pelée, which killed 29,000 people).

Figure 67.10 **A massive debris avalanche in the Coldwater Lake area near Mount St. Helens, Washington.**

Seismic Activity as a Trigger

Earthquake activity can cause potentially unstable slopes to fail as seismic energy passing through rock shakes the material. Consolidated, solid rock, such as granite is not affected by the energy passing through it. However, loose, unconsolidated material or mud can be turned into a liquid if sufficient water is present. A combination of vertical and horizontal stresses creates failure in slopes when the earthquake trigger occurs. In regions where water is common, pore pressure is increased by the seismic waves, and liquefaction occurs.

Most of the monetary losses due to the 1964 Great Alaska Earthquake were caused by widespread slope failures and other ground movement (**Figure 67.11**). In other areas of the United States, such as California, Oregon, and the Puget Sound region of Washington, slides, lateral spreading, and other types of ground failure due to moderate to large earthquakes have been experienced in recent years. Worldwide, landslides caused by earthquakes kill more people and damage more structures than in the United States.

United States Geological Survey.

Figure 67.11 The ground in Anchorage, Alaska, was affected greatly by the M 9.2 earthquake of March 1964. Up to 3 m of subsidence occurred on the left side of the photo.

Slopes and Relief
By Dean Fairbanks

Weathered material is ready to be moved down slope via erosion due to gravity. **Slopes** (rise/run) can be defined as any part of the solid land surface that are curved, inclined surfaces that form the boundaries of landforms. They can be **aggradational** (depositional), **degradational** (eroded), or **transportational**, or any mixture of these. Physical geographers study the **hillslope,** which is the area between a ridge top (or drainage basin divide) and the base of a slope that may or may not contain a stream channel. The form of a slope refers to the shape in cross-section (e.g., uniform-straight, convex, concave, convex-concave) (**Figure 68.1**). Both endogenic and exogenic events provide new sets of relationships on a landscape, and therefore the dynamic activities of slope processes which in turn drives the evolution of a slope over time (**Figure 68.2**). A slope is a dynamic conflict between forces that act to balance into an incline (slope degree) that is optimum. These forces include *gravity, friction* (varies with weight of particle and slope angle; water can overcome friction), *degree of cohesion* (binding of the particles) of the material, and *pore-water pressure* (degrees of particle saturation). The *shear stress* measures the component of gravitational force that act in the downslope direction. When any of these items change, then the *angle of repose* (optimum slope from balance of forces) is altered as weathered material is moved down slope via gravity and erosion. All the forces on the slope then compensate by adjusting to a new dynamic equilibrium.

Slope Mechanics and Form

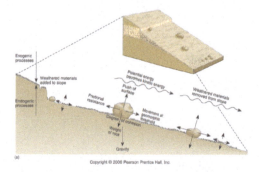

Copyright © 2006 Pearson Prentice Hall, Inc

Figure 68.1 Directional forces (noted by arrows) act on material along an inclined slope.

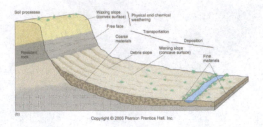

Figure 68.2 The principal elements of a slope. (Robert Christopherson, *Geosystems*, 7th Edition, pg. 404, figure 13.3, ISBN: 978-0-13-600598-8)

In simple terms, erosion downslope or mass movement increases with slope angle. However, we need to expand this simplification to the factors controlling slope development:

- *Climate*—For example, humid slopes are rounder due to chemical weathering, whereas arid slopes are jagged or straight from mechanical weathering.
- *Geology*—Rock type affects slopes through mineral strength, angle of emplacement (dip), and orientation of joints and bedding planes.
- *Vegetation*—offers protection to slopes by roots holding the material together.
- *Aspect*—linked to insolation loads, NE-facing receives less than SW-facing aspects. A SW-facing aspect is subjected to many cycles of extreme heating and cooling, compared with the nearly stable cooler shade conditions on the NE-facing aspect. Thus weathering rates are higher on SW-facing aspects.
- *Soil cover*—Thinner soil covers are related to the waxing and free face slopes, whereas thicker soils are related to the waning slope.
- *Human activity*—Humans can inadvertently accelerate processes of slope development through destabilizing hillslope materials: overloading a slope with additional material (e.g., mine tailings); removing material supporting a slope (e.g., road cuts), or de-vegetating a slope (e.g., forestry clear-cuts).

Classifications and Types of Mass Movement

By Ingrid Ukstins and David Best

Often, the term *landslide* is used in the general sense to describe many types of slow to rapidly descending masses of material down a slope. However, the term does not tell us anything about the processes involved in the movement. Most geologists prefer the term *mass movement*, because slope movements can occur on the ocean floor and not just on land. Also, many slope movements occur by mechanisms such as falling, flowing, and creeping rather than by sliding and more descriptive terms such as *rockfall, debris flow*, and *rockslide* are used. Geologists and civil engineers distinguish different types of mass movement based on three characteristics:

- **Type of Material.** The type of material involved in a mass movement depends upon whether the descending mass began as unconsolidated material or as bedrock. If regolith dominates, terms such as *debris, mud*, or *earth* are used in the description. The term *earth* refers to material that is composed mainly of sand-sized or finer particles, and debris is composed of coarser fragments. If the massive bedrock breaks loose and moves downslope, the term *rock* is used as part of the description.
- **Type of Motion.** The type of motion describes the way material moves down the slope. The most common motions described are *fall, slide*, or *flow*. Falls move unimpeded through the air and land at the base of a slope. Slides move in contact with the underlying surface. Flows are plastic or liquid movements in which the mass breaks up during movement. Other movements include topples and lateral spreads.
- **Rate of Movement.** The movement of slope material occurs at a wide range of speeds. Some mass movement occurs so slowly that it may take years before the movement is noticeable from the downslope displacement of trees, fences, and walls. Other mass movements occur very rapidly and reach speeds in excess of 200 kilometers per hour. Water also affects the rate of movement. Commonly, the higher the water content, the faster the rate of movement.

Landslide Hazards
By Ingrid Ukstins and David Best

BOX 70.1	Landslides in Washington State

Each year the Pacific Northwest is subjected to a large amount of moisture that comes off the Gulf of Alaska and the Pacific Ocean. The result of these weather systems is the formation of numerous landslides, usually due to road cuts and naturally occurring slopes along highways. Early in the morning of October 11, 2009, a massive landslide cascaded down a hill in southwest Washington, about 35 km northwest of Yakima. Covering approximately 80 acres, the slide moved more than one million cubic meters of material, damming the Naches River. State Highway 410 was covered in debris that was more than 12 meters deep, cutting off more than 600 people from nearby towns.

Source: Washington State Department of Natural Resources.

Box Figure 70.1.1 More than 1 million cubic meters of material slid down this hillside, destroying about 800 m of highway and temporarily damming the Naches River.

Less than five years later, a much larger landslide occurred on March 22, 2014. After almost six weeks of heavy rainfall (about 200 percent of normal), a hillside gave way near the small community of Oso, 80 km northeast of Seattle. The slide was 460 m long, 1,300 m wide and ranged in depth from 7 to 21 m. The Oso mudslide, which killed 43 people, ranks as the deadliest mass movement event in the United States, excluding those associated with earthquakes or volcanic eruptions.

United States Geological Survey, Mark Reid.

Box Figure 70.1.2   The scar of the Oso landslide shows the large amount of material that destroyed the forest and blocked the North Fork of the Stillquamish River. Notice the highway in the lower left side of the image.

Landslide Hazards

Since the hazards associated with landslides are highly variable and depend on the material involved and mechanism of movement, the common types of landslides are discussed below, along with their associated hazards. These definitions are based mainly on the terminology discussed in USGS Fact Sheet 2004-3072, which explains landslide types and processes. The type of movement describes the actual internal mechanics of how the landslide mass is displaced: fall, topple, slide, spread, or flow. Thus, landslides are generally described using two terms that refer respectively to material and movement, for example, rockfall, debris flow, rock avalanche, and earthflow.

Slides

Slides refer to mass movements where there is a distinct zone or surface of weakness that separates the overlying slide material from more stable underlying material. The two major types of slides are rotational slides and translational slides, which are characterized by the surface of rupture.

slide: The movement of material along a curved or flat plane.

Rotational Slide. This is a slide in which the surface of rupture is curved concavely upward and movement is roughly rotational about an axis that is parallel to the ground surface and perpendicular across the slide (**Figure 70.1a**). Rotational slides occur most frequently in homogeneous materials and are the most common landslides occurring in loose or unconsolidated material. They are usually associated with slopes ranging from 20 to 40 degrees, and travel extremely slowly (less than 0.3 meter every 5 years) to moderately fast

(1.5 meters per month) to rapid. Most are triggered by intense and/or sustained rainfall or rapid snowmelt that can lead to the saturation of material resting on a slope.

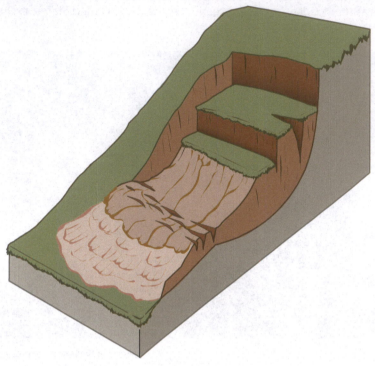

Source: United States Geological Survey.

Figure 70.1a   A rotational slide has a curved plane along its base, giving the sense of rotation of the material.

A **slump** is an example of a small rotational landslide and commonly occurs in unconsolidated sediments and in some weaker rocks (**Figure 70.1b**). Slumping is commonly caused by erosion at the base of a slope, which removes support for the slope material. The erosion may be natural, such as cutting away of the base of a coastal cliff by waves, undermining of a river bank by stream flow, or the result of human activity, such as road construction. When slope failure occurs, the slump block rotates downward, a *scarp* (cliff) is formed at the head of the slope and the toe moves outward over the slope below. The leading edge or toe provides the primary resisting force, so when it is upset by erosion or removal, the material generally moves downhill.

slump: A mass movement in which generally unconsolidated material moves down a hillside along a curve, rotational subsurface plane.

Source: NOAA.

Figure 70.1b   Slumping along the Pacific Palisades of Southern California.

Slumping is an especially serious hazard where structures are built on cliffs above a shoreline when wave energy erodes the base of the cliffs, causing slope failure (especially during storms). In many cases, homes built along the cliffs are destroyed and the scarp progresses inland, posing a hazard for the next line of houses on the cliff.

The Great Alaska Earthquake of March 27, 1964 (M 9.2) caused serious land displacement in the city of Anchorage, which is mostly underlain by the Bootlegger Cove Clay. This former marine clay layer originally held saltwater in its pores that was later replaced by freshwater, making this area an unstable *quick-clay*. The ground shaking that accompanied the earthquake caused the clay to liquefy and the ground fail, producing a series of slump scars. Some of the worst damage was in the subdivision of Turnagain Heights built on a flat-topped bluff some 30 m above the level of Cook Inlet (**Figure 70.2**). Slumping began in the overlying glacial deposits as soon as the underlying clay lost its cohesion. Within minutes, the flat bluff area was changed into a mass of rotated slump blocks, covered with twisted trees and destroyed homes.

United States Geological Survey.

Figure 70.2 Severe slumping occurred in the Turnagain Heights region of Anchorage due to the March 27, 1964, earthquake. A combination of clay and water created unstable slopes that collapsed.

Translational Slide. In this type of slide, the movement occurs along a roughly planar surface with little rotation or backward tilting (**Figure 70.3**). A *block slide* is a translational slide in which the moving mass consists of a single unit or a few closely-related units that move downslope as a relatively coherent mass. Often called *rockslides*, the sliding surface is commonly a bedding plane, but rockslides can develop on other surfaces such as fractures that cut across layered rocks. One of the most common types of landslides, translational slides are found throughout the world. They generally occur close to the surface, and can range from small (residential-lot size) failures to very large, regional landslides that are kilometers wide. Movement may initially be slow (1.5 meters per month) but many are moderate in velocity (1.5 meters per day) and some extremely rapid. With increased velocity, the landslide mass of translational failures may disintegrate and develop into a debris flow. They are triggered primarily by intense rainfall, rise in ground water within the slide due to rainfall, snowmelt, flooding, or other inundation of water resulting from irrigation, or leakage from pipes or human-related disturbances, such as undercutting. These types of landslides can also be earthquake-induced. Occasionally the translating block will ride on a cushion of air, which greatly reduces the frictional forces, thereby causing material to move at high speeds. These situations are preceded downslope by an air blast that often flattens trees.

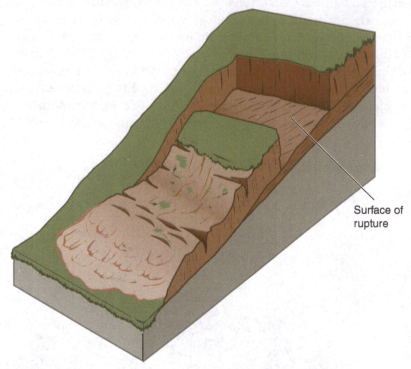

Source: United States Geological Survey.

Figure 70.3   A translational slide often has a solid block of rock sliding down a surface, which has been translated to a new, lower position on the hillside.

Falls

Falls are abrupt movements of masses of geologic materials, such as rocks and boulders, that become detached from steep slopes or cliffs. Rocks can separate along potential planes of weakness such as joints, fractures, or bedding planes. When material falls under the force of gravity, it comes to rest and accumulates at the base of the hillside or slope. The material at the base is termed **talus** (**Figure 70.4**). If the buildup is significant, the result is a talus slope.

talus: The natural pile of material that builds up at the base of a cliff or hillside by material falling from above.

Rockfalls. Rockfalls are abrupt, downward movements of rock, earth, or both, that detach from steep slopes or cliffs (**Figure 70.5**). The falling material usually strikes the lower slope at angles less than the angle of fall, causing bouncing. The falling mass may break on impact, or may begin rolling on steeper slopes, and may continue until the terrain flattens. They are common worldwide on steep or vertical slopes, in coastal areas, and along rocky banks of rivers and streams. The volume of material in a fall can vary substantially, from individual rocks or clumps of soil to massive blocks thousands of cubic meters in size. They are rapid to extremely rapid, free-fall, bouncing and rolling of detached soil, rock, and boulders. The rolling velocity depends on slope steepness. Rockfalls are commonly triggered by undercutting of slope by natural processes such as streams and rivers or differential weathering (such as

Source: David M. Best.

Figure 70.4 Material at the base of the cliff forms a talus slope. Weathered rock falls downhill and builds up a slope which usually unstable.

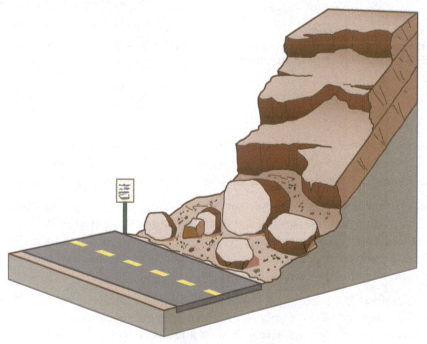

Source: United States Geological Survey.

Figure 70.5 A rockfall is often a free-fall of rock to the surface, where it can break into smaller pieces or roll along the surface.

the freeze/thaw cycle), human activities such as excavation during road building and (or) maintenance, and earthquake shaking or other intense vibration.

Topples. Toppling failures are distinguished by the forward rotation of a unit or units about some pivotal point below or low in the unit (**Figure 70.6**). Toppling is driven by gravity exerted on the weight of the displaced mass. Toppling can be caused by water or ice expanding in cracks or voids in the mass. Topples can consist of rock, debris (coarse material), or earth materials (fine-grained material). They are known to occur globally, often prevalent in columnar-jointed volcanic terrain, as well as along stream and river courses where the banks are steep. Velocities can range from extremely slow to extremely rapid, sometimes accelerating throughout the movement, depending on distance of travel.

Source: United States Geological Survey.

Figure 70.6 Topples result from rocks rotating about a pivot point at the base of a cliff and the rocks basically tilt over. The dashed blocks designate the position of the rocks prior to collapsing.

Flows

Flows are mass movements in which the material behaves like a viscous fluid. In many cases, mass movements that start as falls, slides, or slumps are transformed into flows farther downslope. Flows have broad characteristics and encompass wettest to driest, and fastest to slowest types of mass movement.

Debris Flow. A debris flow is a form of rapid mass movement in which a combination of loose soil, rock, organic matter, air, and water mobilize as a slurry that flows downslope (**Figure 70.7**). Debris flows typically consist of less than 50 percent fine grained material that has been carried along by extreme fluid flow generated by heavy rainfall or snowmelt. Debris flows are common in areas that have experienced wildfires, and are also associated with volcanic deposits.

These types of flows can exhibit a range of viscosities, being thin and watery or thick with sediment and debris, and are usually confined to the dimensions of the steep gullies that facilitate their downward movement. Movement across the surface is relatively shallow, and the runout is both long and narrow, sometimes extending for kilometers in steep terrain. The debris and mud usually terminate at the base of the slopes and create fanlike, triangular deposits called debris fans, which may also be unstable. They can be rapid to extremely rapid, depending on consistency and slope angle.

Figure 70.7 A debris flow is characterized by a low viscosity combination of soil, rock, and water that easily spreads out across the surface.

Debris and Rock Avalanche. Debris avalanches are large, extremely rapid flows formed when an unstable slope collapses and the resulting fragmented debris is rapidly transported away from the slope (**Figure 70.8**). In some cases, snow and ice will contribute to the movement if sufficient water is present, and the flow may become a debris flow and/or a lahar if volcanic conditions exist. They occur worldwide in steep terrain environments and on very steep volcanoes, where they may follow drainage courses. Rock avalanches can occur whenever shear rock walls collapse and can travel several kilometers from their source. They are rapid to extremely rapid and can move up to 100 meters per second (**Figure 70.9**).

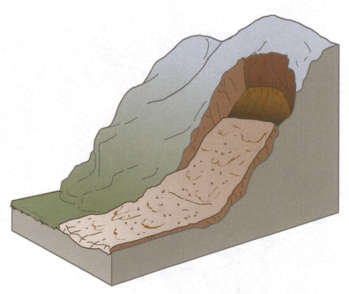

Figure 70.8 A debris avalanche consists of a viscous mixture of rock, soil debris, and a small amount of fluid that travel a short distance before becoming its own impedance.

Source: David M. Best.

Figure 70.9 Rock avalanches consist of large pieces of angular rock that have been broken off and moved downslope.

Earthflow. Spring and early summer are prime times for earthflows, when the upper layers of the regolith trap water. This reduces friction and grain cohesion, which allows the mass to make a slow downhill movement. Earthflows can occur on gentle to moderate slopes, generally in fine-grained soil, commonly clay or silt, but also in very weathered, clay-bearing bedrock (**Figure 70.10**). The mass in an earthflow moves as a plastic or viscous flow with strong internal deformation. Slides or lateral spreads may also evolve downslope into earthflows. Earthflows range from very slow (creep) to rapid and catastrophic. A *mudflow* is an earthflow consisting of material that is wet enough to flow rapidly and that contains at least 50 percent sand, silt,

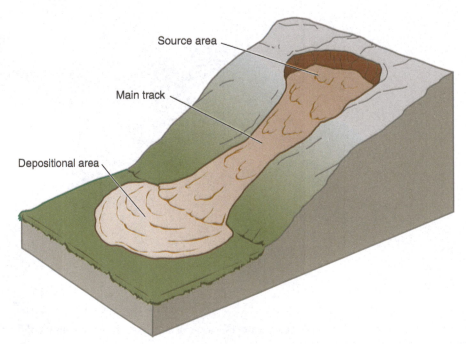

Source: United States Geological Survey.

Figure 70.10 Earthflows are relatively fluid masses that include a mixture of soils and weathered bedrock.

and clay-sized particles. These are often caused by earthquakes, volcanic eruptions and excessive rainfall. Newspaper reports often refer to mudflows and debris flows as "mudslides."

The most disastrous mudslides in recent history have been related to either volcanic eruptions or torrential rainfalls (Table 70.1). Four of the five deadliest events were caused by massive rainfall; only the 1985 Armero event was the result of the eruption of Nevado del Ruiz volcano in Colombia.

Table 70.1	**Five Deadliest Mudslide Disasters**		
Rank	**Mudslide Name**	**Location**	**Estimate Fatalities**
1	1999 Vargas Tragedy	Vargas, Venezuela	30,000
2	1985 Armero Tragedy	Tolima, Colombia	20,000
3	2013 India Monsoons	Uttarakhand, India	6,000
4	2010 Gansu Mudslide	Zhouqu County, China	1,471
5	2017 Sierra Leone Mudslide	Freetown, Sierra Leone	1,000

Source: World Atlas

Creep. Creep is the imperceptibly slow, steady, downward movement of slope-forming soil or rock. Evidence of creep is seen in curved tree trunks, tilted walls or fence posts, or small ridges or ripples on the surface of the soil. The driving forces acting on the slope are sufficient enough to permanently deform the material but not large enough to cause shear failure. There are three types of creep:

1. Seasonal, in which changes in soil moisture and temperature allow movement by expansion of the slope surface (**Figure 70.11**),
2. Continuous, where the driving force always exceeds the strength of the material (resisting force), and
3. Progressive, where slopes are reaching the point of failure as other types of mass movements affect the slope.

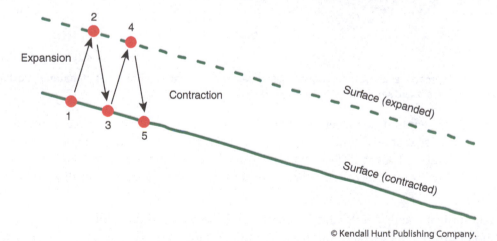

© Kendall Hunt Publishing Company.

Figure 70.11   **Creep occurs when a surface is expanded by the addition of moisture or frost action. Dessication or thawing causes the expanded surface to fall back under the force of gravity, with the result that a particle beginning at point 1 is repositioned at point 2 and so forth.**

Creep is widespread around the world and is probably the most common type of mass movement, often preceding more rapid and damaging types of landslides. Creep can occur over large regions (tens of square kilometers) or simply be confined to small areas. It is difficult to define the boundaries of creep since the event itself is so slow, and surface features representing perceptible deformation may be lacking. The rates of movement range from very slow to extremely slow, usually less than 1 meter per decade. Wherever seasonal creep is occurring, rainfall and snowmelt are the typical triggers, whereas other types of creep could have numerous causes, including chemical or physical weathering, leaking pipes, poor drainage, or destabilizing types of construction.

Lateral Spreads

Lateral spreads are easily identified because they usually occur on very gentle slopes or flat terrain (**Figure 70.12**). Movement is caused by horizontal tension that pulls the surface material and produces shear or tensile fractures. **Liquefaction** is the primary cause of lateral spreads, developing when loose sediments such as sand and silt become saturated. Water holds the particles in suspension so any movement will cause the liquid and suspended solids to move rapidly across the surface. Usually two layers are present; the overlying one is a more cohesive layer lying above material that can undergo liquefaction. If liquefaction occurs, the upper layer then responds by subsiding, rotating, or disintegrating, and begins to either flow or liquefy. The initial failure occurs in a small area that spreads rapidly as liquefaction intensifies.

liquefaction: A condition that exists when an overabundance of liquid, usually water, is present.

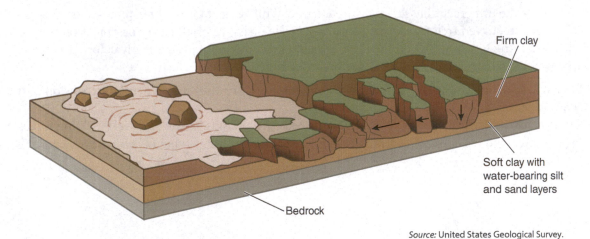

Source: United States Geological Survey.

Figure 70.12 **Lateral spreads result from the horizontal movement of a relatively firm layer of material moving over a deeper layer of water-rich silt and sand. Very little slope is required for this to occur.**

Lateral spreads are known to occur where there are liquefiable soils. The area affected may start small in size and have a few cracks that spread quickly, affecting areas hundreds of meters in width. Rate of movement may be slow to moderate; however, if the event is triggered by an earthquake or volcanic eruption, movement will be very rapid. The ground may then slowly spread from a few millimeters per day to tens of meters per day. Other causes could be:

- Natural or anthropogenic overloading of the ground above an unstable slope;
- Saturation of underlying weaker layer due to precipitation;
- Snowmelt, and (or) ground-water changes;
- Liquefaction of underlying sensitive marine clay;
- Disturbance at the base of a river bank or slope; and
- Plastic deformation of unstable material at depth such as thick deposits of buried salt.

Submarine and Subaqueous Landslides

Submarine and subaqueous landslides include rotational and translational landslides, debris flows and mud-flows, and sand and silt liquefaction flows that occur underwater in coastal and offshore marine areas or in lakes and reservoirs. The failure of underwater slopes can result from overloading by rapid sedimentation, release of methane gas in sediments, storm waves, current scour, or earthquake seismic shaking. Subaqueous landslides pose problems for offshore and river engineering, jetties, piers, levees, offshore platforms and facilities, and pipelines and telecommunications cables, as well as producing tsunami hazards.

On November 18, 1929 a magnitude 7.2 earthquake occurred about 250 km south of Newfoundland near the Great Bank. The entire region felt the event, with reports of ground motion as far away as New York and Montreal. The earthquake triggered a large submarine slump estimated at 200 cubic kilometers of material. At the time, major undersea transatlantic cables stretched from eastern Canada to Great Britain. An analysis of numerous cable breakages was used to determine that the slump moved across the ocean floor at 95 kilometers per hour as it rolled down the continental slope. The event also produced a tsunami that killed 28 people along the Canadian coast.

The much larger 1964 Alaska earthquake (M 9.2) resulted in almost instantaneous catastrophic failure of the steep submerged shore in the harbor of Kodiak, Alaska (**Figure 70.13**) The submarine slide retrogressed beyond the shoreline, submerging areas of coastal land and harbor facilities, and almost 75 million cubic meters of land of Valdez harbor disappeared into the sea.

United States Geological Survey.

Figure 70.13   The harbor at Kodiak, Alaska, was totally overrun by a tsunami that was generated by the Great Alaskan Earthquake of March 27, 1964. The city is located near the major fault that moved during the event.

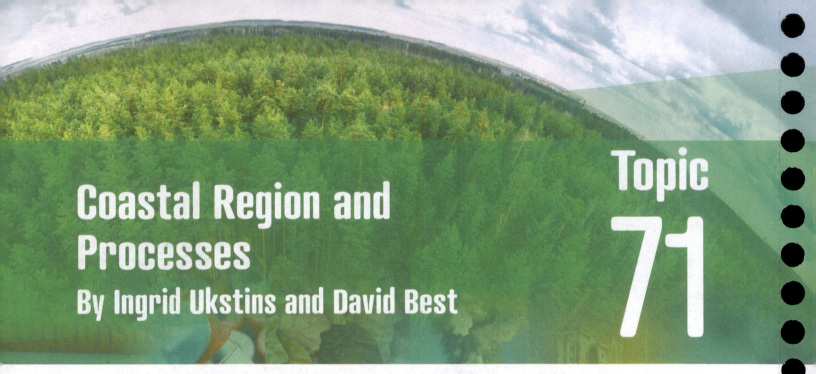

Coastal Region and Processes
By Ingrid Ukstins and David Best

© Bruce Ellis/Shutterstock.com

Figure 71.1 Morris Island Lighthouse, near Charleston, South Carolina, was built about 400 meters inland in 1876. By 1938 changes in nearshore currents had eroded the land so that the lighthouse was at water's edge. Today it is about 300 meters from the shoreline.

The Role of Oceans on Earth

Oceans cover 70 percent of Earth's surface. The three largest, the Pacific, Atlantic, and Indian, cover more than 100 million square miles (259 million sq km). All oceans are key to our weather and they, along with the atmosphere, serve as the main mechanism for heat transfer around the globe. Solar radiation provides the necessary energy to heat the water.

Ocean currents redistribute heat from the equatorial regions toward the poles. Thousands of kilometers of shoreline are the interface between the continents and the oceans. This contact zone is under constant stress, as it is being eroded by the energy of waves and storms. As sea levels continues o rise, we must be aware of the potential for increased flooding and erosion of coastal regions, areas that are home to several billion people on Earth.

The relentless attack by these waters will have devastating effects on the world's population in coming decades, as coastal areas demonstrate that the oceans are a dynamic part of our existence. Within our lifetimes there will undoubtedly be significant changes to coastal areas that were once thought to be stable.

Coastal regions, where the land meets the ocean or a large lake, are attractive places to live or vacation. Because large bodies of water moderate temperatures, coastal areas are cooler in summer and warmer in winter than land that is located farther inland. Humans have always been attracted to the coastal regions, to take advantage of the milder climate, abundant seafood, easy transportation, recreational opportunities, and commercial benefits. It is no wonder that the coastlines have become heavily urbanized and industrialized. Approximately 40 percent of the world's population lives within 100 kilometers of ocean coasts. Counties in the United States that lie directly on the ocean shoreline are about 10 percent of the total land area (excluding Alaska) but have 39 percent of the total population. The population density of these areas is more than six times greater than that of inland counties.

Increasing coastal development is of major concern to geologists (**Figure 71.1**) because they know that coastlines are among Earth's most geologically active and fragile environments. Water in oceans and lakes is constantly in motion due to winds, tides, currents, and, occasionally, tsunami. Therefore, coasts are dynamic and constantly changing from interactions between the energy in the water and the land. These coastal changes occur on two very different time scales. Short-term change (years to decades) is largely due to coastal erosion from waves, storms, and coastal flooding, while long-term changes (hundreds to thousands of years) are due to slower sea-level variation that causes a landward shift in the coastline. These natural processes posed no problem until people began to live along coasts.

Hurricanes and coastal storms are major hazards affecting most coasts in the United States, due to the high energy waves they bring to these areas. As these coasts become increasingly developed, they are highly vulnerable to these natural hazards, and storm damage continues to rise dramatically. Most of the coastline along the western United States is affected by multiple hazards including landslides caused by cliffs continually being undermined by large waves. Today, coastal erosion affects businesses, homes, public facilities, beaches, cliffs, and bluffs (cliffs along lakes) built close to the water's edge. It is estimated that within the next 60 years, coastal erosion may claim one out of four structures within 150 meters of the coastline of the United States.

Even though the coasts are dynamic environments, predicting future coastal change is often difficult, because of the many variables inherent in world climate, weather, nature of the coastline, and human activity. However, scientific studies show evidence that sea level will continue rising, and that storms will become more common and powerful in the coming years. If our present patterns and rates of development along coasts continue, then we are on a collision course with more disasters and catastrophes. It is therefore important to understand the nature of coastal processes and their inherent natural hazards as future development is planned along the coastal zone.

Coastal Processes

Coastal Basics

The **coastline** is a unique boundary where the geosphere, atmosphere, and hydrosphere meet and the systems interact (**Figure 71.2**). At this boundary, dynamic processes of erosion and deposition are constantly at work shaping and reshaping the landscape. Coastal processes active along the coasts are the result of interactions within the climate system and the solar system. Coastal surf and storms result from interactions between the atmosphere and the hydrosphere, with the Sun as the ultimate source of energy driving them. Wave activity that derives from blowing winds is the most important process acting along lake and marine coastlines. Gravity is also an important source of energy in producing rising and falling tides and currents that mostly affect oceanic coastlines. Tides are produced by gravitational interactions between Earth, the Sun, and Moon. Thus waves and tides are important processes in bringing energy to the coasts for erosion of the land and the transport and deposition of sediment along the shorelines.

> **coastline:** Unique boundary where the geosphere, atmosphere, and hydrosphere meet and the systems interact.

Source: Barbara H. Murphy.

Figure 71.2 **The shoreline serves as a boundary between the hydrosphere, geosphere, and atmosphere, where all three meet and interact.**

Waves

Wave Generation

A **wave** is energy in motion that is the result of some disturbance. The energy that causes waves to form in water is called a disturbing force. For example, a rock thrown into a still lake will create waves that radiate in all directions from the disturbance. Mass movement into the ocean, such as coastal landslides and calving glaciers (which creates icebergs) produce waves commonly known as splash waves. Sea floor movements change the shape of the ocean floor and release tremendous amounts of energy to the entire water column and create very large waves. Examples include underwater avalanches, volcanic eruptions, and fault movement, all of which can generate a tsunami. Human activities can also generate waves, such as when ships travel across a body of water and leave behind a wake, which is a wave. Wind blowing across a body of water disturbs surface waters and generates most waves that we commonly see on the surface of the oceans or large lakes. Wind-generated waves represent a direct transfer of kinetic energy from the atmosphere to the water surface. In all these cases, some type of energy release creates waves; however, wind-generated waves provide most of the energy that reaches land and shapes and modifies the coastlines.

> **wave:** Energy in motion that is the result of some disturbance that moves over or through a medium with speeds determined by the properties of the medium. Ocean waves are usually generated by wind blowing across the water surface.

Wave Characteristics

When wind blows unobstructed across the water, it deforms the surface into a series of wave oscillations. The highest part of the wave is the **crest**, and the lowest part between crests is the **trough** (**Figure 71.3**). The vertical distance between the crest and the adjacent trough is the **wave height**. The horizontal distance between any two similar points of the wave, such as two crests or two troughs, is the **wavelength**. The **wave period** is the time it takes for one full wavelength to pass a given point. The **wave speed** is the rate at which the wave travels and is equal to the wavelength divided by the period. **Wave steepness** is the ratio of wave height to wavelength. If the wave steepness exceeds 1/7, the wave becomes too steep to support itself and it **breaks**, or spills forward, releasing energy and forming whitecaps, often observed in choppy waters or in the surf zone along a beach. A wave can break anytime the 1:7 ratio is exceeded, either in the open ocean or along the shoreline.

> **crest:** The highest point of a wave.
>
> **trough:** The low spot between two successive waves.
>
> **wave height:** The vertical distance between the crest and adjacent trough of a wave.
>
> **wavelength:** The vertical distance between the crest and adjacent trough of a wave.
>
> **wave period:** The time it takes for one full wavelength to pass a given point.
>
> **wave speed:** The velocity of propagation of a wave through a liquid, relative to the rate of movement of the liquid through which the disturbance is propagated.
>
> **wave steepness:** The measured ratio of wave height to wavelength.
>
> **break:** When a wave steepness exceeds 1/7, the wave becomes too steep to support itself and it breaks, or spills forward, releasing energy and forming whitecaps often observed in choppy waters or the surf area along a beach.

Wave Motion

Waves are a mechanism by which energy is transferred along the surface of the water. Waves can travel many kilometers from their place of origin. Waves generated in Antarctica have been tracked as they traveled over 10,000 kilometers through the Pacific Ocean before finally expending their energy a week later on the shores of Alaska. But it is important to note that the water itself does not travel this great distance. The water is merely the medium for the waveform (the energy) to travel through, similar to earthquake seismic waves as they travel through solid rock.

In open water, water particles pass the energy along by moving in a circle (**Figure 71.4**). This is known as **circular orbital motion**. This motion can be observed easily by observing a floating object as it bobs up and down

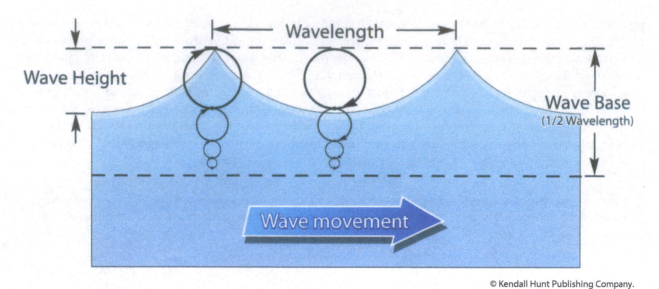

© Kendall Hunt Publishing Company.

Figure 71.3 Wave form with characteristics. Crests and troughs alternate across the wave form. Note that water motion dies off at a depth of about one-half the wavelength.

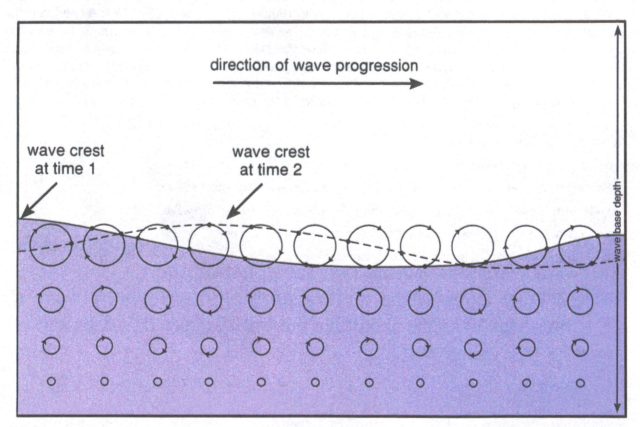

© Kendall Hunt Publishing Company.

Figure 71.4 Circular motion of a wave form in open water. Individual particles of water rotate in a circle, producing an overall lateral movement of a wave.

and sways back and forth as the wave passes. The object itself does not travel along with the wave. From the side, the object can be viewed moving in a circular orbit with a diameter at the surface equal to the wave's height. Beneath the surface, the orbital motion of the water particles diminishes downward with depth. At a depth equal to approximately one-half the wavelength, there is no movement associated with surface waves.

This bottom depth of orbital motion is known as the **wave base**. Submarines can avoid large ocean waves by submerging below the wave base. Even seasick scuba divers can find relief by submerging into the calm water below the wave base. If the water depth is greater than the wave base, the waves are called deep-water waves and do not contact the ocean floor. The motion of water in waves is therefore distinctly different from the motion of water in currents, in which water travels in a given direction and does not return to its original position.

> **circular orbital motion:** The movement of a particle by moving in a circle.
>
> **wave base:** Depth equal to one-half the wavelength where there is no movement associated with surface waves.

Wave Energy

As wind blows over the surface of a body of water, some of its energy is transferred to the water. The mechanism of energy transfer is related to frictional drag resulting from one fluid (the air) moving over another fluid (the water). Waves that crash along the coast in the absence of local winds are generated by offshore winds and storm events, sometimes thousands of kilometers away. The energy of a wave depends on its height and length. The higher the wave's height, the greater the size of the orbit in which the water moves. The height, length, and period of a wave depend on the combination of three factors: (1) wind speed (the stronger the wind speed, the larger the waves), (2) wind duration (the longer time the wind blows, the more time the wind can transfer energy to the water, and the larger the waves), and (3) **fetch**, the distance over which the wind blows (a longer fetch allows more energy to be transferred and form larger waves). If we compare waves on a lake to those on the ocean, when wind speed and duration are the same, the waves will be higher on the ocean because the fetch is far greater than on a lake. When the maximum fetch and duration have been reached for a given wind speed, waves will be fully developed and will grow no further. This is because they are losing as much energy breaking as whitecaps as they are receiving from the wind.

> **fetch:** The distance over which the wind blows across open water.

In areas where storm waves are generated, waves will have different lengths, heights, and periods. When the wind stops, or changes direction, waves will separate into waves of uniform length called **swells**. Swells moving away from the storm center can travel great distances before the energy of the wave is released by breaking and crashing onto the coast.

> **swell:** Wave of uniform wavelength moving away from a storm center. They can travel great distances before the energy of the wave is released by breaking and crashing onto the coast.

Interference Patterns and Hazardous Rogue Waves

When swells of deep water waves from different storms run together, the waves interfere with one another to produce different interference patterns (**Figure 71.5**). The interference pattern produced when two wave systems collide is the sum of the disturbance that each would have created individually. Constructive interference occurs when two waves having the same wavelength come together in phase (meaning crest to crest and trough to trough). The wave height will be the sum of the two, and if it becomes too steep, the wave may break forming whitecaps. Destructive interference occurs when waves having the same wavelength come together out of phase (meaning the crest of one will coincide with the trough of the other). If the wave heights are equal, the energies will cancel each other and the water surface will become flat. If the waves are traveling in opposite directions, the waves will return to their normal heights once they travel through the interference area.

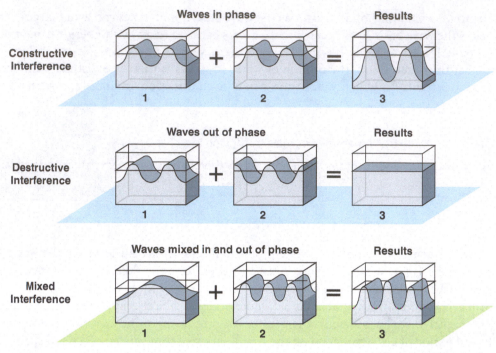

Figure 71.5 Interference patterns in water. Constructive interference creates larger waves, destructive interference reduces the waves to a flat surface, and mixed interference generates a mixture of small and large waves.

It is common for mid-ocean storm waves to reach 7 meters in height, and in extreme conditions such waves can reach heights of 15 meters or more. However, solitary waves called **rogue waves** can reach enormous heights and can occur when normal ocean waves are not unusually high. The word rogue means unusual, and in this case the waves are unusually large—monsters up to 30 meters in height (approximately the height of a 10-story building)—that can appear without warning. Rogue waves appear to be caused by an extraordinary case of constructive interference that can be very destructive and have been popularized in movies such as *The Poseiden Adventure* and *The Perfect Storm*.

> **rogue wave:** Large solitary wave caused by constructive wave interference that usually occurs unexpectidly amid waves of smaller size.

In 1942 during World War II, the RMS *Queen Mary* was carrying 15,000 American troops near Scotland during a gale and was broadsided by a 28 meter high rogue wave and nearly capsized. The ship listed briefly about 52 degrees before the ship slowly righted herself.

Waves Reach the Shore: Shallow-Water Waves and Breakers

When deep-water waves approach shore, the water depth decreases and the wave base starts to intersect the seafloor. At this point the wave comes in contact with the bottom and the character of the wave starts to change (**Figure 71.6**). In this zone of shoaling (shallowing) waves, they grow taller and less symmetrical. Because of friction at the bottom, the wave speed decreases, but its period remains the same, and thus, the wavelength will decrease. The circular loops of water motion also change to elliptical shapes, as loops are deformed by the bottom. As the wave moves farther shoreward, the wavelength shortens considerably and the wave height increases. The increase in wave height, combined with the decrease in wavelength, causes an increase in wave steepness. With continued forward motion at the top of the wave and friction at the bottom, the front portion of the wave cannot support the water as the rear part moves over, and the wave breaks as surf. Here in the **surf zone**, actual forward movement of water itself occurs within the wave as all the water releases its energy as a wall of moving, turbulent surf known as a **breaker**.

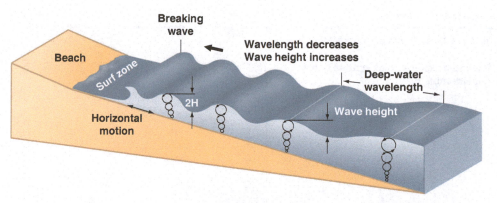

©Kendall Hunt Publishing Company.

Figure 71.6 Wave hitting the shore with the bottom of the wave intersecting the sea floor. Waves form breakers and move toward the beach as the surface of the water moves faster than the water underneath.

> **surf zone:** The nearshore zone of breaking waves.
>
> **breaker:** A wave in which the water at the top and leading edge falls forward producing foam.

There are three main types of breakers (**Figure 71.7**). **Spilling breakers** form on shorelines with gentle offshore slopes and are characterized by turbulent crests spilling down the front slope of the wave. **Plunging breakers** form on shorelines with steeper offshore slopes and have a curling crest that moves over an air pocket. Plunging breakers are prized waves for surfing. **Surging breakers** form when the offshore slopes abruptly and the wave energy is compressed into a shorter distance and the wave surges forward right at the shoreline.

> **spilling breaker:** Forms on shorelines with gentle offshore slopes and are characterized by turbulent crests spilling down the front slope of the wave.
>
> **plunging breaker:** Forms on shorelines with more steep offshore slopes and have a curling crest that moves over an air pocket.
>
> **surging breaker:** Forms when the offshore slopes abruptly and the wave energy is compressed into shorter distance and the wave surges forward right at the shoreline.

After the breaker collapses, a turbulent sheet of water, the **swash**, rushes up the slope of the beach (called the swash zone). The swash is a powerful surge that causes landward movement of sediment (sand and gravel) on the beach. When the energy of the swash is dissipated, the water flows by gravity back down the beach toward the surf zone as **backwash**. Therefore, as a wave approaches the shore, it breaks, and the stored energy in the wave is expended in the surf and swash zones, causing erosion, transport, and deposition of sediment along the coast (**Figure 71.8**).

> **swash:** A turbulent sheet of water that rushes up the slope of the beach following the breaking of a wave at shore.
>
> **backwash:** The flow of water down the beach face toward the ocean from a previously broken wave.

Wave Refraction

When waves approach an irregular shoreline, or at an angle to the shore, the wave base will initially encounter shallower water areas first and begin to slow down before the rest of the wave does resulting in **wave refraction**, or bending of the wave. Refraction of waves approaching coastlines concentrates wave energy on protruding **headland** areas and dissipates energy in the bays (**Figure 71.9**). The concentrated energy on

Fundamentals: Breakers

There are three types of breakers:

• Spilling breakers break gradually over considerable distance.

• Plunging breakers tend to curl over and break with a single crash. The front face is concave, the rear face is convex.

• Surging breakers peak up, but surge onto the back without spilling or plunging. Even though they don't "break," surging waves are still classified as breakers.

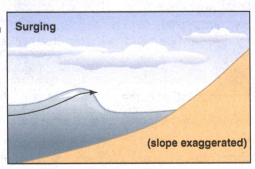

© Kendall Hunt Publishing Company.

Figure 71.7 Spilling, plunging, and surging breakers. Sea floor topography plays a key role in the type of breaker that forms.

headlands erodes them into cliffs and causes deposition of sediment in the bays; thus, headlands erode faster than bays due to stronger wave energy. The result of wave refraction is to erode headlands and smooth out the coastline. The eroded sediments are deposited offshore, on beaches, or in bays.

wave refraction: The process by which the part of a wave in shallow water is slowed down, causing it to bend and approach nearly parallel to shore.

headland: A steep-faced irregularity of the coast that extends out into the ocean.

Tides

Tides produce short-term fluctuations in sea level on a daily bases. **Tides** are the rising and falling of Earth's ocean surface caused by the gravitational attraction of the Moon and the Sun on the Earth. Because the moon is closer to the Earth than the Sun, it has a greater effect and causes the Earth's water to bulge toward it, while at the same time a bulge occurs on the opposite side of the Earth due to inertial forces. These different bulges remain stationary while Earth rotates and the tidal bulges result in a rhythmic rise and fall of the

© Dmitry Naumov/Shutterstock.com.

Figure 71.8 Swash forms as a breaking wave moves up the flat surface along the shore. Receding water produces the backwash.

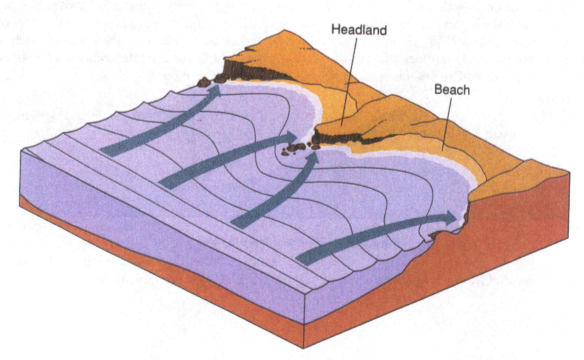

© Kendall Hunt Publishing Company.

Figure 71.9 Refraction of wave energy around a headland. Incoming wave energy is bent or refracted toward the headland due to changes in the sea floor topography that cause the wave to change speed. This produced a zone of high energy that erodes the protruding headland.

ocean surface, which is not noticeable in the open ocean but is magnified along the coasts. The changing tide produced at a given location is the result of the changing positions of the moon and Sun relative to the Earth, coupled with the effects of Earth rotation and the local shape of the sea floor.

tide: The periodic rising and falling of the water that results from the gravitational attraction of the moon and sun acting on the rotating earth.

The regular fluctuations in the ocean surface result in most coastline areas having two daily high tides and two low tides as sea level rises and falls along the shore. A complete tidal cycle includes a **flood tide** that progresses upward on the shore until high tide is reached, followed by an **ebb tide** falling off the shore until low tide is reached and exposing the land once again. Tidal ranges between high and low tides along most coasts range about 2 meters. However, in narrow inlets tidal currents can be strong and cause variations in sea level up to 16 meters. High and low tides do not occur at the same time each day but instead are delayed about 53 minutes every 24 hours. This is because the Earth makes a complete axis rotation in 24 hours, but at the same time, the moon is orbiting the Earth in the same direction, which means the Earth must spin an additional 53 minutes for the same point on Earth to be directly beneath the moon again (and thus in the bulge at its highest). This explains why the moon rises in the sky about 53 minutes later each day, and in the same manner, why the tides are also about 53 minutes later each day.

flood tide: The incoming or rising tide; the period between low water and the succeeding high water.

ebb tide: That period of tide between a high water and the succeeding low water; falling tide.

Because the Sun also exerts a gravitational attraction on the Earth, there are also monthly tidal cycles that are controlled by the relative position of the Sun and moon to the Earth. Although the Sun's gravitational pull on the oceans is smaller than the moon's, it does have an effect on tidal ranges (the difference in elevation between high and low tides). The largest variation between high and low tides, called **spring tides**, occurs when the Sun and the Moon are aligned on the same side of the Earth (new moon) or on opposite sides of the Earth (full moon) (**Figure 71.10**). Here the gravitational attractions of the Moon and Sun amplify each other and produce higher and lower tides. The lowest variation between high and low tides, called **neap tides**, occur when the moon is at right angles relative to the Earth and Sun (quarter moons).

spring tide: The highest high and the lowest low tide during the lunar month. The exceptionally high and low tides that occur at the time of the new moon or the full moon when the sun, moon, and earth are approximately aligned. Contrast with Neap Tide.

neap tide: A tide that occurs when the difference between high and low tide is least; the lowest level of high tide. Neap tide comes twice a month, in the first and third quarters of the moon. Contrast with spring tide.

The timing when hazards, such as storms or tsunami events, strike the shoreline, especially at spring tides phases, is very important. The intensity of a disaster is magnified when rising water from a storm surge or tsunami arrives at the same time as the highest high tides. The combination of these events often sends the destructive power of the water farther inland, producing much more catastrophic results.

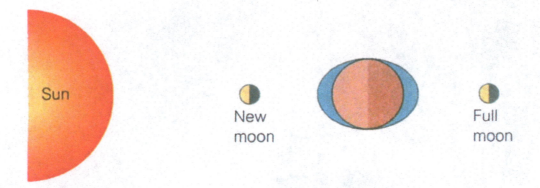

Sun

New
moon

Full
moon

Sun, Earth, Moon are aligned twice each month. Solar and lunar tides are additive to give the highest high tides and the lowest low tides—called spring tides.

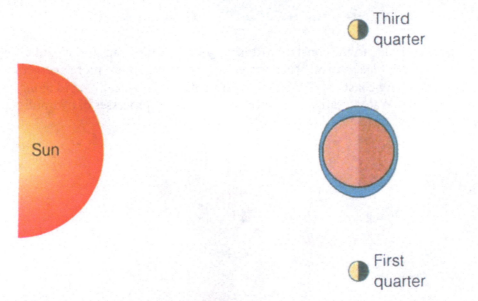

Third
quarter

Sun

First
quarter

At the first and third quarter of the month, the solar and lunar tides are nonadditive, producing the lowest high tides and the highest low tides—called neap tides.

Figure 71.10 The tidal ranges are affected by the position of the moon with respect to Earth. New and full moons produce spring tides, while first and third quarter moons produce neap tides.

Coastal Erosion
By Ingrid Ukstins and David Best

The erosion of the coastlines is due to the constant battering of waves which causes the land to retreat. Most coastal erosion occurs during intense storms when waves and storm surges are more energetic. The rate of wave erosion varies greatly along coasts of different compositions but is typically rapid along sandy coasts and slower along rocky coasts. Water weathers and erodes coastlines by processes of hydraulic action, abrasion, and corrosion.

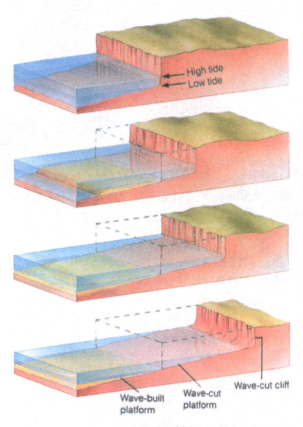

© Kendall Hunt Publishing Company.

Figure 72.1 Wave-cut platforms and wave-cut cliffs are formed by active erosion produced by incoming tides.

The force of water alone, called hydraulic action, is an effective erosional process. Breaking waves exert a tremendous force on the shores by direct impact of the water and are very effective on cliffs composed of sediment or fractured rocks. A large wave 10 meters high striking a 10-meter high cliff produces four times the thrust energy of a space shuttle's three main orbiter engines. A wave striking a cliff drives water into cracks or other openings in the rock and compresses air inside. As this happens, the water and air combine to create hydraulic forces on the surrounding rock that is large enough to dislodge rock fragments or large boulders. Repeated countless times, hydraulic action wedges out rock fragments from cliff faces which fall to the bottom of the cliff or sea bed. The debris can be picked up and used for another erosive wave action–abrasion. Loose sand is easily moved by wave action and by currents that run parallel to the shoreline.

Landforms of Erosional Coasts

Along coasts where erosion dominates, rocky coastlines are common. Since erosion occurs at sea level, abrasion and hydraulic action undercut exposed bedrock forming a wave-cut cliff. As the cliff continues to erode, it leaves behind a flat or gently sloping **wave-cut platform** (**Figure 72.1**). Farther offshore a wave-built platform can be formed by transported sediments that are deposited as the water moves seaward. Locally, refraction of waves onto a narrow headland can cut a cave into the rock which may eventually erode all the way through the headland forming a **sea arch**. When the sea arch collapses, a small prtion of the headland may be isolated from the retreating sea cliff and remain as a **sea stack** (**Figure 72.2**). As waves continue to batter the rocks, eventually the sea stacks crumble. The overall effect is to produce a rugged shoreline that is constantly being hit by incoming waves (**Figure 72.3**).

wave-cut platform: A gently sloping surface produced by wave erosion, extending far into the sea or lake from the base of the wave cut cliff.

sea arch: An opening through a headland caused by erosion.

sea stack: An isolated rock island that is detached from a headland by wave action.

Source: David M. Best.

Figure 72.2   Arches represent the remains of a once-protruding headland that reached out into the ocean. The sea stack in the distance seen through the arch is the remnants of a collapsed arch.

Source: David M. Best.

Figure 72.3 Waves strike a rugged shoreline in the Channel Islands National Park, California.

Coastal Sediment Transport

By Ingrid Ukstins and David Best

Sediment that is created by the abrasive and hydraulic action of waves, or sediment brought to the coast by streams, is picked up by the waves and transported. One of the most important processes of sediment transport within the shoreline area is longshore drift. Sediment is also transported by rip currents and tidal currents.

Longshore Drift and Currents

Longshore drift is the net movement of sediment parallel to the shore. The process starts when waves approach the shore at an oblique angle. Waves striking the shore at an angle, as opposed to straight on, will cause the wave swash to move sediment along the beach at an angle. The backwash brings the sediment directly down the beach slope, under the influence of gravity. This has the net effect of gradual movement of the sediment along the shore by the swash and backwash processes. The swash of the incoming wave moves sand up the beach in a direction perpendicular to the incoming wave crests and the backwash moves the sand down the beach perpendicular to the shoreline. Thus, with successive waves, the sand will move along a zigzag path along the beach parallel to shore. This process is known as **beach drift** (**Figure 73.1**).

> **longshore drift:** The net movement of sediment parallel to the shore.
>
> **beach drift:** The movement of sand along a zigzag path along the beach parallel to shore due to successive waves on the beach.

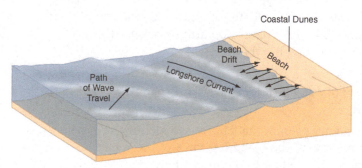

© Kendall Hunt Publishing Company.

Figure 73.1   Beach drift is created by incoming waves hitting the beach at an angle moving sand that returns to sea in a direction perpendicular to the beach. Longshore current is the overall movement of water parallel to the beach but in the same general direction as the incoming wave direction.

A related process, known as a **longshore current**, develops in the surf zone and a little farther out to sea (**Figure 73.2**). The movement of swash and backwash in and out from the shore at an angle creates turbulent water in the surf that transports sediments along the shallow bottom in the same direction as the longshore drift. Substantially more sediment is transported along many beaches as a result of longshore currents than beach drift. Thus longshore currents and beach drift work together as longshore drift to transport huge amounts of sediment along a coast (**Figure 73.2**). At Sandy Hook, New Jersey, approximately 2000 tons of sediment per day move past any given point on the beach. No wonder beaches are often referred to by coastal geologists as rivers of sand.

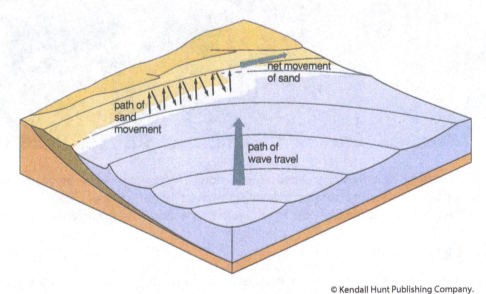

Figure 73.2   **Longshore drift creates a net movement of sand along the beach and shallow water zone.**

longshore current: A current that flows parallel to the shore just inside the surf zone. It is also called the littoral current.

Hazardous Rip Currents

There are times when waves can pile large volumes of water on the beach that are much greater than normal. The only way this water can return to the ocean is to ebb back in a channel through the surf zone. This creates a **rip current** that typically flows perpendicular to the shore and the strong surface flow can have sufficient force to be a hazard to swimmers (**Figure 73.3a**). It is often incorrectly called a "rip tide" or "riptide". However, the occurrence is not related to tides. Rates of return flow can range from 0.5 meter per second to as much as 2.5 meters per second. The position of rip currents can shift along the beach during the day as differing amounts of water are pushed up onto the shore.

rip current: Movement of water back into the ocean in narrow zones through the surf zone.

Often two characteristics are present that allow us to identify a rip current. As the water is receding toward the ocean, its force counteracts the incoming force of the waves, thereby canceling out the waves. Thus a relatively smooth surface will be flanked by incoming waves. Also the receding water can carry along large amounts of sand and silt, which will discolor the water. It is advisable to look for the existence of a rip current before heading into the water. Such currents can be extremely dangerous, dragging swimmers away from the beach and leading to death by drowning when they attempt to fight the current and become exhausted. The United States Lifesaving Association reports that rip currents cause approximately 30 deaths annually in the

United States, mainly on unguarded beaches. Over 80 percent of all rescues by beach lifeguards are due to rip currents, totaling almost 24,000 in 2017, a decrease of almost 40 percent in the past decade, as swimmers are paying more attention to posted warnings (**Figure 73.3b**).

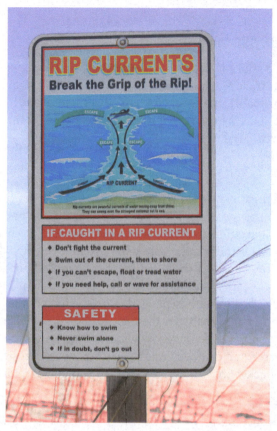

Figure 73.3a   Warning signs are posted in areas where rip currents are likely to form.

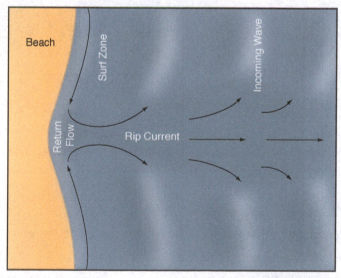

Figure 73.3b   Rip currents form when too much water is pushed up onto the beach. Its return to the sea causes a rapid movement of water seaward.

If a swimmer is caught in a rip current, one should not try to swim directly back to shore but rather swim parallel to the shoreline in order to get out of the current. Rip currents can be between 15 and 45 meters wide. If you see a person caught in a rip current, yell at them to swim parallel to the shore and you should move along the shoreline in a direction that leads them out of the current.

Coastal Sediment Deposition and Landforms

Sediment transported along the shore is deposited in areas of lower wave energy and produces a variety of landforms. Common landforms include beaches and barrier islands. Erosion of headlands and sea cliffs is the source of some sediment, but probably no more than 5 to 10 percent of the total. The primary source of sediment is that transported to the coast by rivers that drain the continents. This sediment is then redistributed along the shore by longshore drift.

Beaches

A **beach** is an accumulation of unconsolidated sediment along part of the coastline that is exposed to wave action. Beaches are formed from the wave-washed sediment along a coast, and represent interconnected zones of onshore and offshore sediment accumulation. Most beaches can be described in terms of three geomorphic zones (**Figure 73.4**). The **offshore zone** is the portion of beach that extends seaward from the normal low tide level. Strong backwash currents usually transport some sediment off exposed portions of the beach and deposit it offshore as submerged offshore bars. The **foreshore zone** represents the area between normal low and high tide levels. The **backshore zone**, which is commonly separated from the foreshore by a distinct ridge, called a **berm**, is the part of the beach extending landward from the high tide level to the area reached only during storms. Sediments in this zone are frequently redistributed by wind to form sand dunes (**Figure 73.5**). Dunes serve as a line of defense against storm water that can overrun the beach zone. Salt-resistant grasses are often planted to stabilize the dunes. Behind the backshore may be a zone of cliffs, marshes, or additional sand dunes.

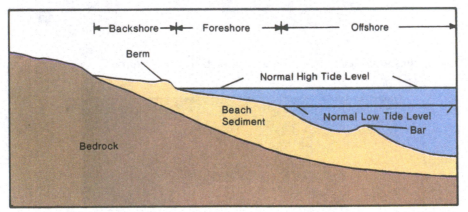

© Kendall Hunt Publishing Company.

Figure 73.4   The topographic features of a beach area show the different "shore" zones that extend from the land out to sea.

beach: An aggregation of unconsolidated sediment, usually sand, that covers the shore.

offshore zone: The portion of beach that extends seaward from the low tide level.

foreshore zone: The area located between the normal low and high tide levels.

backshore zone: The part of the beach extending landward from the high tide level to the area reached only during storms.

berm: A low, incipient, nearly horizontal or landward-sloping area, or the landward side of a beach, usually composed of sand deposited by wave action.

Source: David M. Best.

Figure 73.5 **Grasses stabilize sand dunes that serve as a break for storm waters.**

Even though beaches are areas of deposition, they are in a constant state of change, and dynamically responding to variations in the energy of waves and currents. The effect that waves have depends on their strength and duration. Destructive waves, occur on high energy beaches and are typical of winter storms (**Figure 73.6**). They reduce the quantity of sediment present on the beach by carrying it out to offshore bars under the sea. Constructive, weak waves are typical of low energy beaches, and occur usually during summer months. These are the opposite of destructive waves because they increase the size of the beach by removing sand from the offshore bars and piling it up onto the berm. This strong and weak wave activity alternates seasonally at most beaches. The weak wave activity produces a high and wide sandy beach at the expense of the offshore bars. The strong wave activity produces a narrow beach during the winter months and builds prominent offshore bars.

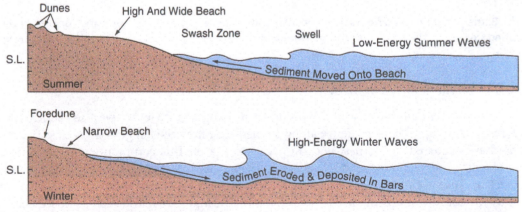

© Kendall Hunt Publishing Company.

Figure 73.6 **Beach profiles during the summer and winter differ in that the lower energy of the summer allows the beach to become enriched with sand. Winter storms erode most of the summer buildup and reduce the size of the beach.**

Barrier Islands

A **barrier island** is a long narrow offshore island of sediment running parallel to the coast and separated from it by a lagoon (**Figure 73.7**). These islands range between 15 and 30 kilometers long and from 1 to 5 kilometers wide. The tallest features are wind-blown sand dunes that reach heights up to 10 meters. Barrier islands are common features along the Atlantic and Gulf coasts of the United States which form the longest chain of barrier islands in the world. However, barrier islands are dynamic coastal features as they grow parallel to the coast by longshore drift and are often eroded by storm surges that often cut them into smaller islands. Despite their transient nature, many barrier islands are heavily populated with homes and resorts. Even several major cities occupy barrier islands, including Miami Beach, Atlantic City, and Galveston.

> **barrier island:** A long, usually narrow accumulation of sand, that is separated from the mainland by open water (lagoons, bays, and estuaries) or by salt marshes.

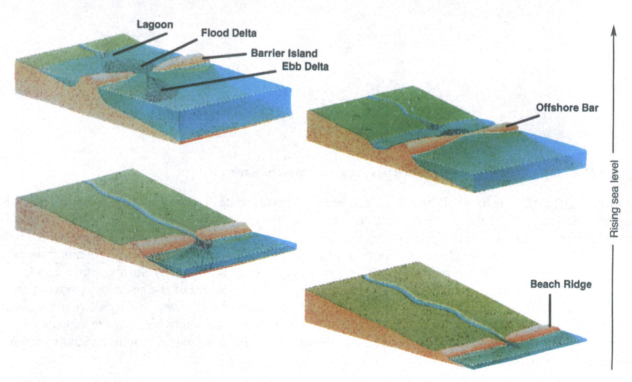

United States Geological Survey.

Figure 73.7   Barrier islands formed from sand deposited along low energy beachs. These islands provide protection of the shoreline during storms and period of high energy. During period of very high water these islands are overrun with water.

Emergent and Submergent Coastlines

While sea level fluctuates daily because of tides, long-term changes in sea level have also occurred. Such changes in sea level result from uplift or subsidence along a coastline. Many coastal geologists classify coasts based on changes that have occurred in the past with respect to sea level. This commonly used classification divides coasts into two categories: emergent and submergent. An **emergent coastline** is a coastline that has experienced a fall in sea level, because of global sea level change, local land uplift, or isostatic rebound. Emergent coastlines are identifiable by the coastal landforms which are now above the high tide mark, such as raised beaches or raised wave cut benches (marine terraces) (**Figure 73.8**). Alternatively, a **submergent coastline** is one that has experienced a rise in sea level, due to a global sea level change or local land subsidence. Submergent coastlines are identifiable by their submerged, or "drowned" landforms, such as drowned valleys and fjords.

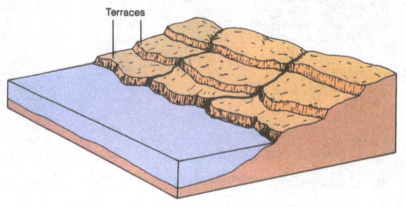

© Kendall Hunt Publishing Company.

Figure 73.8 **An emergent coastline often has wave-cut terraces that represent odler period of higher sea level.**

The type of coast produced is controlled mainly by tectonic forces and meteorological conditions (climate and weather). Tectonic processes can cause a coastline to rise or sink while lithospheric isostatic adjustment can depress or elevate sections of a continent. The tectonics at active plate margins can produce uplift or subsidence of a coast. In the northwestern United States, the coastline is slowly rising due to subduction of the Juan de Fuca plate beneath the North American plate. During the 1964 Alaska earthquake, large sections of the coast rose and other parts subsided beneath sea level from a single event.

During glacial periods, large continental ice sheets can displace the lithosphere into the plastic asthenosphere. About 18,000 years ago, a huge continental glacier covered most of Scandinavia, causing the land to sink isostatically. As the lithosphere sank, the displaced asthenosphere flowed southward, causing the region around modern-day Netherlands to rise. After the ice melted, the process was reversed and the asthenosphere flowed back north from below the Netherlands to Scandinavia today, Scandinavia is rebounding and the Netherlands is sinking. During the same glacial episode in North America, Canada was depressed by ice, and asthenosphere rock flowed southward. Today the asthenosphere is flowing back north and much of Canada is rebounding while much of the United States is now sinking.

Global changes in sea level can also occur. Such global sea level changes are called **eustatic** changes and can occur by three mechanisms: the growth or melting of glaciers, changes in water temperature, and changes in the volume of the mid-ocean ridges. During glacial periods large amounts of water that evaporated from the oceans became stored on the continents as glacial ice. This caused sea level to become lower, resulting in global land emergence. Similarly, when glaciers melt, water flows back into the oceans and sea level rises globally, causing land submergence.

Changes in the volume of mid-ocean ridges can also affect sea level. Growth of a mid-ocean ridge displaces seawater upward. If lithospheric plates spread slowly from the ridge they create a narrow mountain ridge system that displaces relatively small amounts of seawater, resulting in lower sea level. In contrast, rapidly spreading plates produce a high-volume ridge system that displaces more water upwards, resulting in higher sea level. At times during Earth's history, sea floor spreading has been relatively rapid, and as a result, global sea level has been higher.

Coastal Hazards
By Ingrid Ukstins and David Best

From the discussion of coastal areas it is apparent that diverse and complex processes are at work continually changing the coastal landscapes. Vast areas of coastal land have been lost since the mid 1800s as a result of natural processes and human activities (Table 74.1). The natural causes that have the greatest influence on coastal land loss are relative sea level rise, erosion from frequent storms, and reductions in sediment supply; whereas the most important human activities are sediment excavation, river modification, and coastal construction. Any one of these causes may be responsible for most of the land loss at a coast, or the land loss may be the result of several of these factors acting at the same time.

From a hazard point of view, coastal erosion is the most widespread and continuous process affecting the world's coastlines and contributing to land loss destruction. Global warming and sea level rise are slow-onset hazards that greatly contribute to the erosional process. However, catastrophic, rapid-onset events play a very significant role both for coastal erosion and human suffering. These include erosion and destruction from storms, landslides, and tsunami.

Table 74.1 Common Physical and Anthropogenic Causes of Coastal Land Loss

NATURAL PROCESSES	
Agent	**Examples**
Erosion	• waves and currents • storms • landslides
Sediment Reduction	• climate change • stream avulsion • source depletion
Submergence	• land subsidence • sea-level rise

Table 74.1 Common Physical and Anthropogenic Causes of Coastal Land Loss (*continued*)

NATURAL PROCESSES

Agent	Examples
Wetland Deterioration	• herbivory • freezes • fires • saltwater intrusion

HUMAN ACTIVITIES

Agent	Examples
Transportation	• boat wakes, altered water circulation
Coastal Construction	• sediment deprivation (bluff retention) • coastal structures (jetties, groins, seawalls)
River Modification	• control and diversion (dams, levees)
Fluid Extraction	• water, oil, gas, sulfur
Climate Alteration	• global warming and ocean expansion • increased frequency and intensity of storms
Excavation	• dredging (canal, pipelines, drainage) • mineral extraction (sand, shell, heavy mins.)
Wetland Destruction	• pollutant discharge • traffic • failed reclamation • burning

Source: United States Geological Survey.

Coastal Land Loss by Global Sea-Level Rise and Subsidence

A significant amount of coastal erosion presently plaguing today's coastlines is the result of gradual but sustained global sea level rise (**Figure 74.1**). Most of this rise is from the melting of polar continental ice sheets, coupled with expansion of the water itself as global temperatures increase. The sea level rise is currently estimated at about 0.3 meter per century. Although this amount does not sound very threatening, additional factors increase the risk. First, the slope of many coastal areas is very gentle so that a small rise in sea level results in a far larger inland advance of the coast than steeper sloping areas. The vulnerability of coastal regions along the eastern seaboard of the United States shows a wide range of potential risks (**Figure 74.2**). Estimates of the amount of coastline retreat in the United States due to sea level rise of 0.3 meters would be 15 to 30 meters in the northeast, 65 to 130 meters in California, and up to 300 meters in Florida. Second, the documented rise of atmospheric carbon-dioxide levels suggests that global warming from the increased greenhouse-effect will melt the glaciers more rapidly, as well as warming the oceans more, thereby accelerating the rise of sea level. Some estimates put the anticipated rise in sea level at 1 meter by the year 2100, which could affect as much as 50 percent of the population of the country. In addition to increased beach erosion, it would also flood 30 to 70 percent of coastal wetlands in the United States that protect shores from storm flooding events.

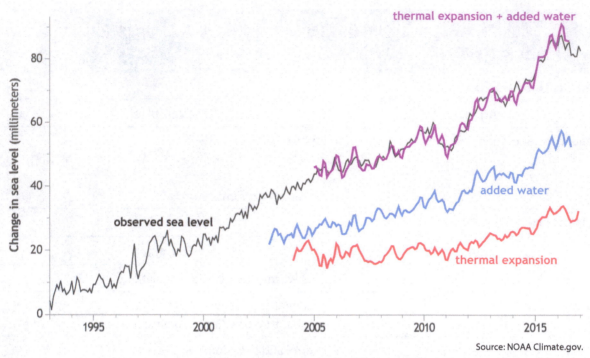

Source: NOAA Climate.gov.

Figure 74.1 Observed sea level as measured by satellite altimeter recordings since 1993 (black line). Added water (glacial melting; blue line) and thermal expansion (red line) match observed sea level very well.

The most widely assessed effects of future sea level rise are coastal inundation (submergence), erosion, and barrier island migration. The USGS estimates the primary impacts of a sea level rise on the United States to be: (1) the cost of protecting ocean communities by pumping sand onto beaches and gradually raising barrier islands in place; (2) the cost of protecting developed areas along sheltered waters through the use of levees (dikes) and bulkheads; and (3) the loss of coastal wetlands and undeveloped lowlands. The total cost for a one meter rise is estimated to be $270–475 billion, ignoring future development.

Coastal submergence refers to permanent flooding of the coast caused by either a rise in global sea level or subsidence of the land, or both. At many coastal sites, submergence is the most important factor responsible for land loss and as sea level rises, or the land subsides, it will inundate present unprotected low-lying coastal areas and cities such as: Boston, New York, Miami, New Orleans, and Los Angeles. Submergence also accelerates coastal beach erosion and landslides because it facilitates greater inland penetration of storm waves. In addition to accelerated land loss, coastal submergence causes intrusion of salt-water into coastal fresh water aquifers.

> **coastal submergence:** The permanent flooding of coastal areas by global sea level rise or land subsidence.

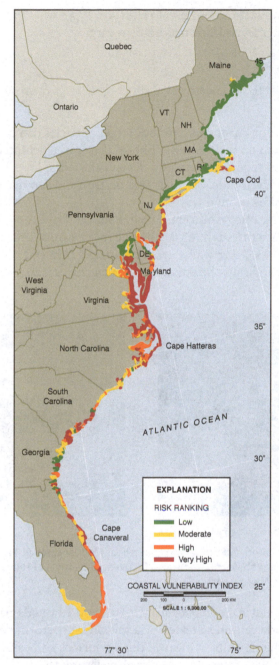

From Thieler and Hammar-Klose (2000). United States Geological Survey.

Figure 74.2 **Map of the Coastal Vulnerability Index (CVI) for the U.S. East Coast showing the relative vulnerability of the coast to changes due to future rises in sea level. Areas along the coast are assigned a ranking from low to very high risk, based on the analysis of physical variables that contribute to coastal change.**

Coastal Land Loss by Erosion: Impacts from Storms and Landslides

Superimposed on the slower sea level rise are shorter duration water fluctuations caused by storm events. The most damaging coastal storms for the United States are tropical cyclones (hurricanes) and extratropical cyclones (winter storms) that form around low-pressure cells.

Hurricanes form in the tropics during summer to early fall months and migrate northward and westward into temperate regions of the Atlantic and Gulf coasts. The extratropical storms occur mostly in winter (like nor'easters) and can cause erosion of the coastline at much higher rates than normal. Although each type of storm is unique, there are several factors common to all storm types which include strong winds, generation of large waves, and elevated water levels known as a storm surge.

Storm Impacts

Strong storms bring more energy to the coastline causing higher rates of erosion (**Figure 74.3**). Higher erosion rates are due to several factors:

- Wave velocities are higher during storms and thus larger particles can be carried in suspension causing sand on beaches to be picked up and moved offshore. This leaves behind coarser grained particles such as pebbles and cobbles, thus reducing the width of sandy beaches;
- Storm waves reach higher levels onto the coast and destroy and remove structures and sediment from areas not normally reached by normal waves;
- Wave heights increase during a storm and crash higher onto cliff faces and rocky coasts. Larger rock debris or debris from destroyed structures is flung against the rock causing rapid rates of erosion by abrasion;
- Hydraulic action increases as larger waves crash into rocks. Air and water occupying fractures in the rock becomes compressed and thus the pressure in the fractures is increased which causes further fracturing of the rock.

United States Geological Survey.

United States Geological Survey.

Figure 74.3 Upper figure shows erosion and property damage near Floridana Beach, Florida, caused by Hurricane Frances on September 4, 2004. The lower image shows the same area following the arrival of Hurricane Jeanne on September 25, 2004. Jeanne produced much greater beach erosion.

Storm surge is responsible for about 90 percent of all human fatalities and damage during storms (**Figure 74.4**). A **storm surge** is an onshore flood of water created by a low pressure storm system. The surge is caused primarily by the strong winds of the storm blowing over the sea surface and causing the surface water to pile up above sea level. Low pressure at the center of the storm also elevates the surface water upward and enhances the height of the mound of water. As the storm nears land, the shallower sea floor prevents the piled-up water from collapsing and it floods inland as a deadly storm surge. Storm surges are at their highest and most damaging when they coincide with high tide (especially at the high tides during the spring tide cycles), combining the effects of the surge and the tide.

storm surge: An onshore flood of water created by a low pressure storm system.

When Superstorm Sandy struck the coastal areas of the Mid-Atlantic states and southern New England in October 2012, its landfall coincided with a high tide. This caused extreme flooding and beach erosion throughout the region.

Figure 74.4 Damage along a pier in St. Mary's, GA, caused by Hurricane Irma in September 2017.

Dune and Beach Recession

High storm-generated waves erode large quantities of sediment from dune and beach areas. From March 5 through 8, 1962 a major coastal storm, known as the "Ash Wednesday" storm, moved northward and became stalled against the middle Atlantic coast through five high tides. The documented erosion that occurred at Virginia Beach, Virginia, showed that 30 percent of the beach and dune sand was removed. The crest of the dunes at Virginia Beach was reduced from an elevation of 4.9 to 3.4 meters thus enabling future storm surges to rise over the dunes and flood inland areas.

Dune and Beach Breaching and Overwash

Large amounts of sediment can be eroded from a beach and dunes during a major storm with some migrating along the shore by accelerated longshore currents and some moved offshore. In addition, storm waves may wash sediment through low areas between the dunes of islands and onto the back side it. This **overwash** is important because it maintains the barrier island's width as its front is eroded but can be devastating to homes and other structures. Overwash fans are lobe-shaped deposits eroded from the ocean side of a shore and deposited in the bays and lagoons behind barrier islands.

> **overwash:** A deposit of marine-derived sediments landward of a barrier system, often formed during large storms; transport of sediment landward of the active beach by coastal flooding during a tsunami, hurricane, or other event with extreme wave action.

El Niño Effects on Coastlines

Along the Pacific coast, winter storms and unusual oceanographic conditions such as El Niño cause the most erosion and land loss. Approximately every four to five years, El Niño conditions cause warm surface water of the Pacific Ocean to flow eastward piling up water along the west coast of North and South America. The elevated water levels and the unusually strong storms during El Niño events cause extensive flooding and erosion beaches and cliffs. In 1983, an unusually strong El Niño caused torrential rainfall, rapid beach erosion, and massive landslides along the Pacific coast of the United States. Land loss was concentrated along the southern California coast where numerous expensive homes built on cliffs were damaged or destroyed.

In the United States, barrier islands are found along the Gulf Coast and Atlantic Ocean coastline where they serve as a break between the sea and land. Examples of these islands include Padre Island, Texas, islands off the coast of Georgia, and a major set off the coast of North Carolina. Several barrier islands off the coast of Louisiana have been eroded as much as 30 meters due to a rise in sea level in the area.

These features are formed by the deposition of sand transported to the sea by rivers (**Box Figure 74.1.1**). Because the material is unconsolidated, the islands are dynamic and change their size and shape over time. The major factor that alters these features is an increase in the force of sea water driven ashore by storms. The storms cut through the unconsolidated sand of a barrier island to produce a tidal inlet or tidal delta. These features are often short-lived as they close naturally in a few weeks as the result of longshore drift deposits moving sediment along the shoreline.

The Outer Banks of North Carolina are a major series of barrier islands which are frequently affected by hurricanes and tropical storms (**Box Figure 74.1.2**). In September 2003, the islands were subjected to the forces of Hurricane Isabel which hit the coast with winds of 105 mph. Waves as high as 8 meters, coupled with a storm surge of 1.8 to 2.4 meters, breached the island, forming two inlets that were each more than 600 meters wide and 5 meters deep (**Box Figure 74.1.3**). One inlet was named Isabel Inlet to recognize the effect of the storm.

Several sections of State Highway 12, which runs the length of Hatteras Island, were destroyed. In addition, more than 30 beach houses and several motels were knocked off their pilings and foundations. One hundred thirty kilometers north in Dare County, North Carolina, surge flooding and strong winds damaged several thousand houses. In the end the storm caused more than $450 million in damage in North Carolina (2003 USD), with a total damage assessment in the United States of $5.5 billion. Fortunately, no deaths or injuries were reported in the Outer Banks.

© makasana photo/Shutterstock.com

Box Figure 74.1.1 Driving north on North Carolina Highway 12 large sand dunes and the Atlantic Ocean (not seen) are on the right.

Source: NASA.

Box Figure 74.1.2 **The Outer Banks of North Carolina are a major system of barrier islands between the land and Atlantic Ocean.**

United States Geological Survey

Box Figure 74.1.3 **Hatteras Island was breached by Hurricane Isabel, producing two new inlets along the barrier beach.**

Since 1900 at least 30 events have occurred. Two major El Niño storms hit California in October 1997 and six months later in April 1998. This series of storms was the largest of the 30 events and had significant effects on the weather in the central and eastern Pacific Ocean regions. Coastal areas were heavily eroded by these two events (**Figure 74.5**). Another set of storms struck the west coast of the United States in 2014 to 2016.

United States Geological Survey, Center for Coastal and Watershed Studies. United States Geological Survey, Center for Coastal and Watershed Studies.

Figure 74.5   **Coastal region near Ventura, California. Upper image shows the coast following the El Niño storm of October 1997; lower image shows a definite change in the coastal morphology following the April 1998 El Niño storm.**

Landslides and Cliff Retreat

Coastal landslides occur where unstable slopes fail and land is displaced down slope (**Figure 74.6**). Some of the fundamental causes of slope failures that lead to land loss are: (1) slope over-steepening (2) slope overloading, (3) shocks and vibrations, (4) water saturation, and (5) removal of natural vegetation. Sea level rise can elevate waves so they can erode and undercut cliffs at higher elevations, initiating mass movements. Cliffs may stay relatively stable and then retreat several meters in a single storm event which makes building structures near the edge of cliffs an especially risky during times of sea level rise.

Source: David M. Best.

Figure 74.6   **The small community of La Conchita, California, lies along the coast of the Pacific Ocean. Bluffs that overlie the town have collapsed twice since 1995 due to heavy rainfall that saturated the ground.**

Coastal Land Loss by Human Activities

There is increasing evidence that recent land losses in many coastal regions are largely anthropogenic, as the changes are attributable to human alteration of the coastal environment. Land losses indirectly related to human activities are difficult to quantify because they promote alterations and imbalances in the primary factors causing land loss such as sediment budget, coastal processes, and relative sea level changes. Human activities causing land loss are: transportation networks that tend to increase erosion, coastal construction projects that typically increase deficits in the sediment budget, subsurface fluid extraction and climate alterations that accelerate submergence and excavation projects that cause direct losses of land.

There are countless examples of human interference with coastal processes. The beach at Miami Beach must be restored periodically by sand pumped from offshore. In southern Louisiana the land is subsiding as sea level rises, thus causing loss of natural coastal wetlands. Many California beaches are eroding due to the damming of rivers for irrigation and flood control. The river-supplied sediment that normally replenishes the beaches is being trapped in reservoirs behind dams. Since this sediment is not being supplied to the ocean, longshore currents cannot resupply the beaches with sediment. Instead, longshore currents carry the existing sediment in the downdrift direction, resulting in significant erosion of the beaches.

Eliminating wetlands for development and agriculture removes the natural flood protection and storm-swollen estuaries now flood barrier islands from the bay side as storms move inland. Where beach dunes are removed, the most effective barrier to storm waters has been and lost overwashing becomes more common. Over the last 10,000 years, most of the state of Louisiana has formed from the deposition of sediments by the flooding of the Mississippi River. Humans, however, have prevented the river from flooding by building levee systems that extend to the mouth of the river. As previously deposited sediments become compacted they tend to subside. Since no new sediment is being supplied by Mississippi River flooding, the subsidence results in a relative rise in sea level. This, coupled with a current rise in eustatic sea level, is causing coastal Louisiana to erode at an incredible rate and experience more flooding.

Direction of Longshore Current

Barrier islands and beaches, since they consist of unconsolidated sediment, and sea cliffs, since they are susceptible to landslides due to undercutting, are difficult to protect from the erosive action of the waves. Human construction methods can attempt to prevent erosion, but cannot always protect against abnormal conditions. In addition, other problems are sometimes caused by these engineering structures.

BOX 74.2	One Case Study Made By the United States Geological Survey

OCEAN CITY, MARYLAND: AN URBANIZED BARRIER ISLAND

Relative recent changes in the shoreline have occurred in populated regions of the United States. USGS Circular 1075 provides a good summary of the events that have involved the shoreline near Ocean City, Maryland. For more than a century Ocean City, Maryland, has been a popular beach resort for vacationers from the Northeast and Mid-Atlantic States (**Box Figure 74.2.1**). During the Roaring 1920s, several large hotels and a boardwalk were built to accommodate visitors and development continued slowly until the early 1950s. Then a period of rapid construction began that lasted almost 30 years. Concerns about the coastal environment were raised in the late 1970s and led to Federal and State laws to limit dredging and filling of wetlands. The resort is built on the southern end of Fenwick Island, one of the chain of barrier islands stretching along the east coast (**Box Figure 74.2.2**). The Great Hurricane of 1933 (before names were assigned to hurricanes) opened the Ocean City Inlet by storm-surge overwash from the bay side. To maintain the inlet as a navigation channel, two large stone jetties were constructed by the U.S. Army Corps of Engineers. These jetties helped stabilized the inlet, but they have drastically altered the sand-transport processes near the inlet. The net longshore drift at Ocean City is southerly; it has produced a wide beach north of the jetty, but Assateague Island, south of the inlet, has been starved of sediment. The result is a westerly offset of more than 500 meters in the once-straight barrier island.

The most damaging storm to hit Ocean City within historic times was the Five-High or Ash Wednesday north-easter of early March 1962 which caused severe erosion and flooding along much of the middle Atlantic Coast. For two days, over five high tide cycles, all of Fenwick Island, except the highest dune areas was repeatedly washed over by storm waves superimposed on a storm surge measuring 2 meters high. Property damage in Ocean City was estimated at $7.5 million. Given the dense development of the island over the last 30 years, damage from a similar storm would today be hundreds of millions of dollars.

© Racheal Grazia/Shutterstock.com.

Box Figure 74.2.1 The beaches at Ocean City, Maryland, are a popular vacation spot for many people who live in the heavily populated Atlantic Seaboard of eastern United States. Large amounts of construction are at risk along the beaches of the United States from cyclonic storms.

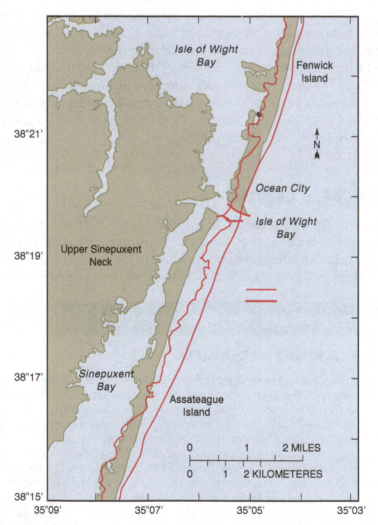

Box Figure 74.2.2 The natural sediment transport along the Fenwick Island-Assateague Island region has been altered by the construction of two large jetties at Ocean City Inlet. The landward shift of the southern barrier island was caused by a change in the longshore currents in the area.

Protection of the Shoreline

Shoreline protection can be divided into two categories: hard stabilization in which solid structures are built to reduce wave action, and soft stabilization which mainly refers to adding sediment back to a beach as it erodes.

Hard Stabilization

Two types of hard stabilization are often used. The first type interrupts the flow of sediment along the beach. These structures include **groins** (**Figures 74.7 and 74.8**) and **jetties** (**Figure 74.9**), built at right angles to the beach to trap sand and widen the beach. The second type interrupts the force of the waves. **Seawalls** are built parallel to the coastline to protect structures on the beach by allowing waves to crash against them and preventing water from running up the beach. **Breakwaters** serve a similar purpose, but are built offshore parallel to the beach (**Figure 74.10**), again preventing the force of the waves from reaching the beach and any structures.

groin: Solid structure built at an angle from a shore to reduce erosion from long shore currents, and tides.

jetty: A structure extending into the ocean to influence the current or tide in order to protect harbors, shores, and banks.

seawall: Massive structure built along the shore to prevent erosion and damage by wave action

breakwater: Structure built offshore and parallel to shore that protects a harbor or shore from the full impact of waves.

Source: David M. Best.

Figure 74.7 This groin along the English Channel is constructed along a beach that consists of pebbles and cobbles rather than sand.

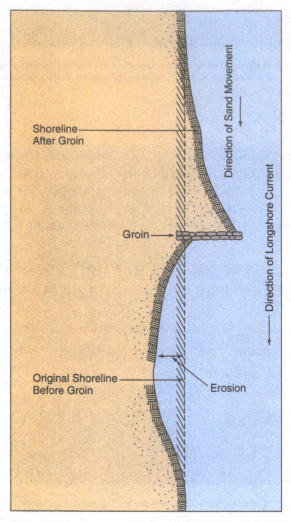

© Kendall Hunt Publishing Company.

Figure 74.8 The effects of constructing a groin on a beach are that portions of the beach undergo erosion while others down drift experience erosion.

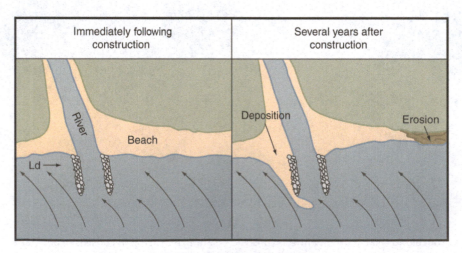

Jetties (ex., Santa Cruz, California)

© Kendall Hunt Publishing Company.

Figure 74.9 Jetties are extensions of pre-existing channels or river channels. Notice how depositional patterns change and sand begins to migrate around the jetty, eventually becoming a hazard to the entrance to the river.

While hard stabilization usually works for its intended purpose, it does cause sediment to be redistributed along the coast. A breakwater, for example, causes wave refraction, and alters the flow of the longshore current. Sediment is trapped behind the breakwater, and the waves become focused on another part of the beach where they can cause significant erosion (**Figure 74.10**). Similarly, because groins and jetties trap sediment, areas in the downdrift direction are not resupplied with sediment by the longshore current, and beaches are eroded and become narrower in the downdrift direction.

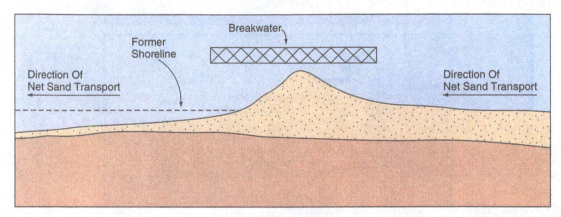

© Kendall Hunt Publishing Company.

Figure 74.10 The construction of a breakwater offshore and parallel to the shoreline creates a buildup of sand that extends from the shoreline out to the breakwater.

Arid Land Systems: Wind Shaping
By Dean Faribanks

The motion of air (wind) is like the flow of water, except that air is less dense. From earlier discussions in Module 3, the important forces include pressure and friction. Airflow across a land surface develops turbulent flow and eddies within the boundary layer (0–500 meters) due to the topography and vegetation causing friction. The ability of wind to pick up particles aloft is based on the wind velocity—distance air moves over a period of time.

Figure 75.1   Dust storm from the Sahara desert can carry across the Atlantic to the Caribbean islands. (USGS)

Wind transport increases 8x if wind velocity doubles, and by 27x if wind velocity triples. The threshold velocity increases with the square root of particle size. The larger the particle size the stronger the winds required to lift and move, even if by bouncing along. Fine dust and silt once lifted into the atmosphere can be carried from continent to continent (**Figure 75.1**).

- **Surface creep**—primarily for large particles/small pebbles. Accounts for 20% of eolian transport.

Eolian processes lead to erosion and landforms associated with wind. The two principal wind-erosion processes are **abrasion** and **deflation**. Abrasion is the mechanical erosion by passing sediment particles scrapping/sand blasting another surface. It usually occurs less than 2 meters above the ground (limited by sand saltation height). Its ability to erode a surface is affected by rock hardness, wind velocity, and wind constancy. Rocks that have evidence of abrasion are called **ventifacts** (Figure 75.2). These rocks represent a piece of rock etched and smoothed by abrasion. Winds also have the ability to sculpt features on the landscape. Rocks can be streamlined in parallel to the most effective wind direction, leaving behind distinctive features called **yardangs** (Figures 75.3–75.4).

Figure 75.2   **Ventifact in Antarctica. (USGS)**

Image © Shutterstock, Inc.

Figure 75.3   **Yardang in the Mojave desert, California.**

Figure 75.4   **Yardang in the Sahara Desert, Algeria. (USGS and Shutterstock)**

Deflation is a process of wind erosion that removes and lifts individual particles of unconsolidated sediment, leaving behind the larger coarse rocks and cobbles. When combined with the infrequent downpours of rain in a desert a swelling-shrinking process is initiated with the clays, which allows for the gravel and small rocks to displace upward and form **desert pavement**. The continuation of winds and further rains eventually leave no gaps between the rock surfaces that then remain clear of dust (**Figures 75.5–75.6**).

Figure 75.5 Desert pavement in the Gobi desert of China. (**USGS**)

Figure 75.6 Desert pavement in the Providence mountain region of the Mojave desert,

There are several uniquely recognized eolian depositional landforms. The scale of one of these forms ranges from small surface features to one much larger and more recognizable. The smaller feature is **ripples**. Ripples are the smallest sand dune–like features formed by saltating grains. When a wind current flows across loose *sand*, the *sand* is frictionally dragged along to pile up to form ripples. They form perpendicular to the wind's direction, and as the sand continues to collect the ripples grow larger to form sand dunes (**Figure 75.7**).

Figure 75.7 Ripples on a sand dune in northern Mexico. (**USGS**)

Figure 75.8 Effective wind direction and a supply of sand can develop sand dunes perpendicular to the wind direction. (**USGS**)

Sand dunes are much larger depositional features of sand grains deposited in transient mounds, ridges, and hills. Dunes are spectacular and dynamic landscapes that from not only in arid landscapes with sand supplies but also along coastal areas where ample sand supplies and wind prevail (**Figures 75.11–75.12**). Every sand dune has a **windward (stoss) slope**, crest (peak of the sand dune), **slipface**, and **leeward slope**. The windward side of the dune is perpendicular to the predominant wind direction. Saltating sand granules travel up the slope. The slipface forms right underneath the crest, where granules reach their maximum height and begin to fall steeply down the leeward side. The angle of repose (critical slope) for sand is 30–34 degrees, which is the steepest the slipface can form (**Figure 75.8**). Different types of sand dunes can form depending on the amount of sand supply, wind direction, and vegetation (**Figure 75.9**). Dunes can be classified on the basis of shape and include such forms as barchan, parabolic, transverse, dome, star, and longitudinal (**Figure 75.10**).

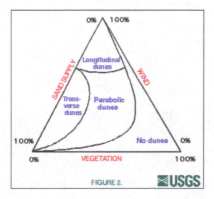

Figure 75.9 For actively moving sand dunes, there are three requirements: a source of sand, winds that are strong enough to move the sand, and a lack of stabilizing vegetation. Depending on the relative balance of these three variables, different dune types may result (**USGS**)

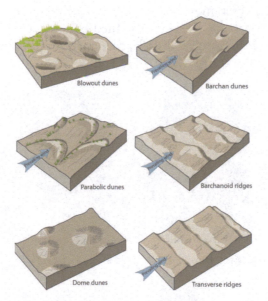

Figure 75.10 Types of sand dunes. Sand supply, wind direction, and interactions among groundwater salinity, topographic elevation, and vegetation growth affect dune morphology. (**NPS**)

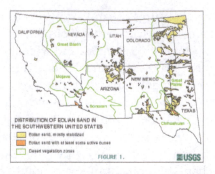

Figure 75.11 Sand dunes are widely distributed over the southwestern United States, particularly in the southern Great Plains and the southwestern deserts and high plateaus. **(USGS)**

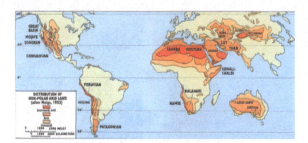

Figure 75.12 Non-polar arid regions of the world with large sand supplies. **(USGS)**

Loess represents large quantities of fine-grained clays and silts left as glacial outwash deposits; subsequently blown by the wind great distances and redeposited in great bodies in such places as the Great Plains of the United States and Northern China (**Figures 75.13–75.14**). In both these areas, they are especially thick but prone to erosion from disturbance by removal of vegetation. During drier periods, loess sediments are easily picked up by winds and transported as dust storms. The 1930s saw the Central United States go through severe dry period combined with poor agricultural practice to protect the soil. Subsequently large dust clouds—the Dust Bowl—plagued the entire region.

Figure 75.13 Loess deposits in the Badlands, North Dakota. **(USGS)**

Figure 75.14 Loess deposits of North America. (USGS)

Arid Land Systems: Water Shaping
By Dean Fairbanks

Desert or arid landscape systems have noneolian processes operating on them as well. The most important mechanism to shaping these landscapes is water. Wind is not the most important geomorphic agent of change. Even though fluvial erosion events are infrequent, when they happen they can be highly erosive, transporting, and depositing large amounts of sediment in a desert that can later be reshaped by wind. Typical events include **flash floods**, which occur in **washes** (dry creek beds, also called arroyos) (**Figures 76.1–76.2**). Other dynamic features include **alluvial fans** and the interlinking merger of them forming **bajadas** or piedmonts. Alluvial fans are sediments deposited by a stream issuing from a channelized canyon onto a valley floor. Once in the valley, the stream is unconfined and can migrate back and forth, depositing alluvial sediments across a broad area, which from above looks like an open fan. Typically the fans formed by multiple canyons along a mountain front join to form a continuous fan apron—bajada.

Figure 76.1   A flash flood in Capital Reef National Park, Utah. (**USGS**)

Figure 76.2   A desert wash in the Mojave Desert, California. (**USGS**

Finally two other water features can be found in arid environments, though one is more exotic or rare. There is the case where true perennial rivers run through arid landscapes acting as large dynamic erosional systems.

They are really creating fluvial landscapes in situations where the precipitation is much less than the evapo-transpiration of the environment. The Nile and Colorado Rivers, in Northern Africa and the Southwestern United States, respectively, play a large role in very arid environments as **exotic rivers**. For example, the Colorado River has carved the Grand Canyon in Arizona. An ephemeral water phenomena that can cover several square miles at a time on desert floors is a **playa** lake. These lakes cover vast areas of the Basin and Range province (Great Basin) in the Western United States. These lakes form from very infrequent heavy downpours in the desert, evaporate quickly and leave behind fine-grained sediments and salts. The infamous Black Rock Desert playa plays host to the Burning Man Festival each year in Northern Nevada.

Glaciation and Periglaciation

By Dean Fairbanks

Landforms that have been under past or present geomorphic influence of ice are very particular. Like water and wind, ice flows as well, albeit quite slowly. However, with an enormous amount of ice on top of a landscape the pressure and movement can scour a landscape in interesting ways. Why is ice an important factor to address in this era of global climate warming?

In the past, glacial ice covered large areas of the earth's surface. A little more than 10,000 years ago, 30% of the earth's surface was covered by **glaciers**. Even today most of the world's fresh water is stored in ice (~77%). Glacial landscapes have distinctive erosional and depositional landforms. In the era of climate change, glacial melting is accelerating, which is linked to sea level rise which affects human settlement activities (**Figure 77.1**).

There are two types of cold landscape environments: glacial and periglacial. A **glacier** is a large mass of perennial ice on land or floating shelf-like in the sea adjacent to the land. **Periglacial** areas are found on the edge of glaciers or ice masses and are characterized by permafrost and freeze–thaw action (summer temperatures can melt ice) (**Figure 77.2**).

Figure 77.1 *Glacier in Alaska. (USGS)*

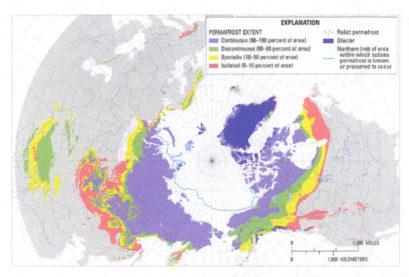

Figure 77.2 *Major periglacial (permafrost) regions of the Northern Hemisphere. (USGS)*

There are three types of periglacial regions: Arctic continental, Arctic maritime, and alpine. Each of these locations varies in terms of mean annual temperature and therefore the frequency and intensity with which processes operate. During glacial periods on the planet, periglacial environments extended, most profoundly in the Northern Hemisphere where they extended south and down in elevation on the advance front of glaciation. Periglacial areas are most associated with **permafrost**, which represents impermeable, permanently frozen ground (**Figure 77.3**). Three types of permafrost exist (**Figure 77.4**).

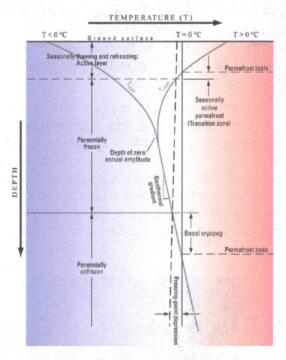

Figure 77.3 *Terms used to describe ground temperature relative to 0°C in a permafrost environment. (USGS)*

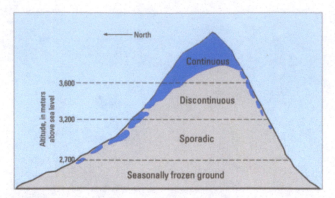

Figure 77.4 *Idealized diagram of altitudinal distribution of sporadic, discontinuous, and continuous permafrost. (USGS)*

- Continuous—mean annual temperatures of −5 to −50°C
- Discontinuous—mean annual temperatures of −1.5 to −5°C
- Sporadic—mean annual temperatures of 0 to −1.5°C

Above the permafrost is the active layer, which can seasonally thaw out and is associated with mass movements. The depth of the active layer varies with the amount of radiation it receives and therefore is highly latitudinally controlled.

Periglacial processes on landscapes are quite unique, and they all involve the phase changes in water from solid to liquid. As pointed out earlier, periglacial environments are highly controlled by **freeze–thaw** weathering actions. The freezing of water leads to expansion up to 10%, which causes pressure in jointed rocks with cracks and fissures. The splitting of rocks by freeze–thaw action is called **congelifraction**. **Cambering** is a process whereby segments of rock (usually along a cliff face) become dislodge from the main rock by congelifraction (splitting by freeze/thaw) and move downhill. **Frost heave** occurs when water freezes in the soil and pushes the surface upwards and churns it. Finally, **avalanches** are common where dry snow on north and east facing slopes steeper than 22° is unstable and fail causing a mass movement. Avalanches can also happen from a rapid melt of wet snow.

Periglacial landforms are also very unique owing to the processes outlined above. **Talus** (scree) slopes are a common feature, as well as large blockfields of large angular rocks. **Patterned ground** is a general term describing stone circles and polygons that are found in soils subjected to intense freeze–thaw action (**Figure 77.5**). Their exact process for forming is still unclear. Rivers and streams in periglacial areas are typically **braided**, with numerous small channels separated by small linear islands (**Figure 77.6**). Streams braid when they have little slope gradient and they have a much greater sediment load than they can carry. Finally, another unique feature only found in periglacial areas is a pingo, which is an isolated, conical hill up to 90 m high and 800 m wide. They form as a result of the movement and freezing of water under pressure (**Figures 77.7–77.8**).

Image © Shutterstock, Inc.

Figure 77.5 *Patterned ground in northern Alaska.*

Figure 77.6 *South-looking photograph showing diamondshaped bars and meandering braided stream channels, East Fork Toklat River, Alaska Range, Denali National Park, Alaska. (USGS)*

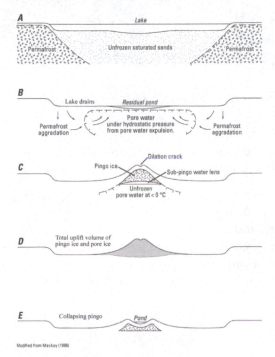

Figure 77.7 *Diagram illustrating the genesis and collapse of the closed-system pingos of the Tuktoyaktuk Peninsula area, Northwest Territories, Canada. (USGS)*

Figure 77.8 *Pingo on the Tuktoyaktuk Peninsula, Mackenzie River delta, N.W.T., Canada (USGS)*

Periglacial areas have hazards associated with them that may be intensified by human impact. These include mass movements, avalanches, rockfalls, ice heave, flooding, and land subsidence. The freeze–thaw cycle in the active layer can make it difficult for structures to remain in place without much stress and tension. Permafrost is becoming increasingly unstable as the climate warms. The increasing of the active layer and the subsequent settling of the soil causes subsidence.

Glaciation Processes and Landforms

By Dean Fairbanks

Glacial ice is the accumulation and recrystallization of snow. The snow falls as flakes and the accumulation is known as **alimentation**. With continued alimentation, the lower snow is compressed under more snow and gradually metamorphosed by the effect of pressure and melting into an aggregation of granular ice pellets called **neve**. Continued pressure (compaction) and melt/refreezing (recrystallization) changes neve into glacier ice. Glaciers are able to flow slowly under the pressure of their own weight and the pull of gravity.

Glacial systems are the balance between inputs, storage, and outputs (e.g., **mass balance**. Inputs include accumulation of snow, eroded rock debris, heat, and meltwater (**Figure 78.1**). The main store is ice, but glaciers also carry debris, called **moraine**, and meltwater. The outputs are losses due to **ablation**, the melting of snow and ice, as well as sediment. The accumulation zone tends to be at the top end of a glacier, and the ablation zone is usually apparent at the bottom end. A glaciers balance point or equilibrium line is where gain and loss are negated, usually about the middle of a glacier. Glaciers are typically described in terms of regime of advancing or retreating:

- Accumulation > ablation = advancing
- Accumulation < ablation = retreating (**Figure 78.2**)
- Accumulation = ablation = steady

Depending on the time scale, observed glaciers are constantly in some type of flux between advancing and retreating. Accumulation and ablation can be seen on a yearly time step between winter and summer, and global glaciations and warming in the past.

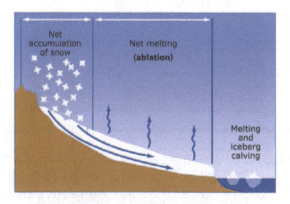

Figure 78.1 **Diagram of a glacier showing components of mass balance. (USGS)**

Figure 78.2 Retreat of South Cascade Glacier, Washington, during the 20th century and the beginning of the 21st century. (USGS)

Glaciers can be classified on the basis of scale and temperature. In size, glaciers can be considered continental covering hundreds to thousands of square kilometers or alpine/mountain covering hundreds to several thousand meters. Thermally, glaciers are either polar or temperate. In **polar** glaciers, the ice remains frozen at the base and therefore little movement occurs, resulting in little erosion. **Temperate** glaciers are warm- based glaciers where water is present in the ice mass and thus lubricates the ice to move freely and erode the rock they sit on. These glaciers also tend to be confined by topography in mountainous regions.

Figure 78.3 Glacial mass covering Greenland. (USGS)

Figure 78.4 **Crevasse on the Juneau ice field, Alaska. (USGS)**

The motion of glacial ice is slow, really glacial! The velocity of a glacier is controlled by gravity, the slope gradient of the topography, the ice thickness, and the temperature within the ice. The zone of brittle flow (upper portion) is caused by the velocity being different with changes in ice thickness across the glacier. These variations cause **crevasses**, fracturing, and faulting to occur on the surface (**Figure 78.4**). What is seen on the service of a glacier is compression and extension of the ice surface leading to the crevasses. Internally glaciers move by *plastic deformation* or creep (ice crystals rearrange in parallel layers and slide), while on the bottom there is **basal slipping** from the meltwater, which leads to a reduction of friction with the rock surface. The speed will increase with temperature and slope gradient.

The amount and rate of glacial erosion depends on the geological rock type, velocity of movement, the thickness (weight-pressure) of the ice, and the amount and character of the debris load carried. The two main glacial erosional processes are plucking (quarrying) and abrasion. **Plucking** occurs on the underside of the glacier as the freezing/thawing of water seeps into joints and cracks in the rock, causing the rock to break and be ripped out by the moving glacier. **Abrasion** is where the eroded debris carried by the glacier (embedded in the ice) drags and scraps the rock like rough sand paper, leaving behind **striations** (**Figure 78.5**).

Glacial erosion leads to landforms that have been carved and glacial deposits left behind that are also their own landform features. When glaciers melt and retreat, they leave behind the material they were carrying. These features are:

- **Moraines**—A general term for unstratified and unsorted deposits of sediment that form through the direct action of, or contact with, glacier ice. They are typically lines of loose rocks carried by the glacier from weathering. The most general is the **terminal moraine** which forms a crescent shape at the front end of a retreating glacier. Many different varieties are recognized on the basis of their position with respect to a glacier (**Figure 78.6**).
- **Erratics**—large boulders foreign to the local geology.

Figure 78.5 Striations left behind by abrasion from a glacier. (USGS)

Figure 78.6 Terminal moraine (USGS)

- **Eskers**—elongated ridges of coarse, stratified, sand and gravels created by subglacial meltwater tunnels under the glacier (**Figure 78.7**).
- **Drumlins**—small oval mounds up to 1.5 km long and 100 m high (**Figure 78.8**).

They are deposited from friction with the ice and geologic base causing the glacier to drop its load. They become shaped in the direction of the glacier as it continues to advance.

- **Kames**—irregular mounds of sorted sands and gravels, formed by supraglacial streams
- **Kettle holes**—shallow bodies of water formed by retreating glaciers (**Figure 78.9a**).

They are related to the form of kames as the blocks of ice become surrounded by the glacial outwash. The US/Canada Great Lakes could be considered very large kettle lakes as they were gouged out by glaciers <u>15,000 to 26,000 years ago.</u>

Figure 78.7 **Drumlin (USGS)**

Figure 78.8 **Meandering eskers (USGS)**

Figure 78.9a **Kettles (*USGS*)**

Figure 78.9b   **Cirques (USGS)**

Finally there are the landforms created by the erosive power of the glacier itself. These landscapes show the effects of large glaciers carving out valleys, creating sharp ridges or smoothing out landscapes. They include:

- **Cirques**—a bowl-shaped, amphitheater-like depression eroded into the head or the side of a glacier valley. Typically, a cirque has a lip at its lower end. The glacier in an alpine environment carves out by rotational manner, eroding the floor by plucking and abrasion (**Figure 78.9b**).
- **Arête**—a jagged, narrow ridge that separates two adjacent glacier valleys or cirques. The ridge frequently resembles the blade of a serrated knife (**Figure 78.10**).
- **Horns**—a pyramidal peaks caused by the headward recession of two or more cirques, the most famous being the Matterhorn in Switzerland (**Figure 78.11**).
- **U-shaped valleys**—Glacial troughs have steep sides and flat floors. The most famous one is the Yosemite Valley in Yosemite National Park in California. Flooded U-shaped valleys are called fjords in Norway (**Figure 78.12a–12b**).
- **Hanging valleys**—These are formed by tributary glaciers, which, unlike rivers, do not cut down to the level of the main valley, but are left suspended above. Waterfalls usually mark them. The waterfalls at the Yosemite National Park are some of the most famous (**Figure 78.13**).
- **Tarn lakes**—a lake that develops in the basin of a cirque, generally after the melting of the glacier (**Figure 78.14**).

Figure 78.10   **Arête (USGS)**

Figure 78.11   The Horn–Matterhorn, Switzerland. Image © Shutterstock, Inc.

Figure 78.12a   U-shaped valley—fjords, Norway. (USGS)

Figure 78.12b   U-shaped valley—fjords, Norway. Image © Shutterstock, Inc.

Figure 78.13 Hanging valleys, Yosemite National Park. Image © Shutterstock, Inc.

Figure 78.14 Tarn lake (USGS)

We are currently in an interglacial period when the glacial ice has retreated. Each glacial period is both preceded and followed by a periglacial period. As the massive sheets of ice melt and retreat, the pressure is released from the ground and isostatic adjustment commences. Some places in the northern United States and Canada are still rebounding 4 mm/year as the earth's crust flexes. Studying glaciation is complicated as there have been many past glaciations over the last 2 million years and each time the later glaciations remove the evidence of earlier ones. It is also not possible to study the processes happening under glaciers; we only have the evidence of what they have left behind to surmise how they shape landscapes.

In the era of global climate warming induced by human industrial processes, glaciers around the world are rapidly retreating. Much of the water from glaciers along Greenland, Iceland, Alaska, Canada, and Antarctic is dumped into the world's oceans, which is leading to rapid sea level rise. By the end of this century, the meltwater from land-based glaciers will have caused a 1-meter rise in the world's sea level. The role of ice on our planet is important, and understanding the cryosphere provides us with links to our climatic past and provides of our planet s glacial and interglacial future.

Introduction to Biogeochemical Cycles
By Ingrid Ukstins and David Best

The Biogeochemical Cycle

Besides water cycling through the biosphere in plants and animals, other materials (e.g., carbon and nitrogen) also have a high concentration in the biosphere. Cycles that involve the interactions between the biosphere and other reservoirs utilize biological processes such as respiration, photosynthesis, and decomposition (decay), which are referred to as biogeochemical cycles. A biogeochemical cycle is a pathway by which a chemical element or molecule moves through both biotic and abiotic components of an **ecosystem** (**Figure 79.1**). All chemical elements occurring in organisms are part of biogeochemical cycles, and, in addition to being a part of living organisms, these chemical elements also cycle through the other Earth spheres as well. All the chemicals, nutrients, and elements—such as carbon, nitrogen, oxygen, and phosphorus—used by living organisms are recycled. An important example is the carbon cycle, which is critical in regulating our global climate by affecting levels of greenhouse gases within the atmosphere.

ecosystem: A community of plants, animals and other organisms that interact together within their given setting.

© Guenter Albers/Shutterstock.com

Figure 79.1   A forest ecosystem contains trees, soil, moisture and the atmosphere, all of which interact with one another to provide a dynamic, living environment.

Carbon Cycle
By Ingrid Ukstins and David Best

Carbon Cycle

Carbon (C) is the basic building block of life. As the fourth most abundant element in the universe (after hydrogen, helium, and oxygen), carbon occurs in all organic substances, including DNA, bones, coal, and oil. Carbon moves through the Earth's spheres, each of which serves as a reservoir of carbon and carbon dioxide. The **carbon cycle** (**Figure 80.1**) involves the biogeochemical movement of carbon as it shifts between living organisms, the atmosphere, water environments, and even solid rock. The least amount of carbon is contained in the atmosphere while the largest amount is found in the lithosphere.

> **carbon cycle:** All carbon reservoirs and exchanges of carbon from reservoir to reservoir by various chemical, physical, geological, and biological processes. Usually thought of as a series of the four main reservoirs of carbon interconnected by pathways of exchange. The four reservoirs, regions of the Earth in which carbon behaves in a systematic manner, are the atmosphere, terrestrial biosphere (usually includes freshwater systems), oceans, and sediments (includes fossil fuels). Each of these global reservoirs may be subdivided into smaller pools, ranging in size from individual communities or ecosystems to the total of all living organisms (biota).

Carbon dioxide occurs in all spheres on Earth and as a gas is readily mobile. Our increased use of fossil fuels has increased the amount of carbon dioxide in the atmosphere, with some of gas being taken up in the hydrosphere and biosphere. Marine and freshwater organisms use carbon dioxide in their life cycles and through their incorporation into sedimentary rocks can increase the amount of carbon dioxide in the geosphere. An example is the coral reefs that are formed by corals extracting dissolved carbon dioxide from sea water.

Generally there is a dynamic equilibrium between and among the reservoirs of carbon. However, these are often perturbed by natural processes that contribute to the production and hence increased abundance of carbon dioxide. Volcanic eruptions and wildfires are two natural hazards that upset carbon equilibrium. These two examples are discussed elsewhere in the text.

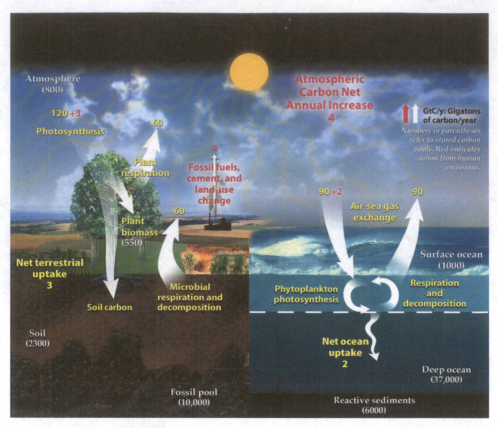

U.S. Department of Energy, Genome Management Information System, Oak Ridge National Laboratory.

Figure 80.1 The carbon cycle shows where carbon is contained in the geosphere, hydrosphere, biosphere, and atmosphere. Each year more than 3 gigatons of carbon are added to the cycle.

Biogeochemical Cycles
By Ingrid Ukstins and David Best

81

The Biogeochemical Cycles

Besides water cycling through the biosphere in plants and animals, other components, such as carbon and nitrogen, have high concentrations in the biosphere. Cycles that include the interactions between the biosphere and other reservoirs involve biological processes including respiration, photosynthesis, and decomposition (decay), which are referred to as biogeochemical cycles. A **biogeochemical cycle** is a pathway by which a chemical element or molecule moves through both biotic and abiotic components of an **ecosystem** (**Figure 81.1**). All chemical elements occurring in organisms are part of biogeochemical cycles and, in addition to being a part of living organisms, these chemical elements also move through the hydrosphere, atmosphere, and lithosphere. All the chemicals, nutrients, or elements, such as carbon, nitrogen, oxygen, and phosphorus, used by living organisms are recycled. An important example is the **carbon cycle**, which is important in regulating our global climate by acting as a greenhouse gas within the atmosphere.

> **biogeochemical cycle:** Natural processes that recycle nutrients in various chemical forms from the environment, to organisms, and then back to the environment. Examples are the carbon, oxygen, nitrogen, phosphorus, and hydrologic cycles.
>
> **ecosystem:** A community of plants, animals and other organisms that interact together within their given setting.
>
> **carbon cycle:** The movement of carbon between the biosphere and the nonliving environment. Carbon moves from the atmosphere into living organisms that then move it back into the atmosphere.

© Jacob Lund/Shutterstock.com

Figure 81.1 Mountain environments provide a complex and diverse array of habitats for a large range of plants and animals.

Carbon Cycle

Carbon (C) is the basic building block of life. As the fourth most abundant element in the universe (after hydrogen, helium, and oxygen), carbon occurs in all organic substances, including DNA, bones, coal, and oil. Carbon moves through the hydrosphere, geosphere, atmosphere, and biosphere, each of which serves as a reservoir of carbon and carbon dioxide (CO_2). The carbon cycle (**Figure 81.2**) involves the biogeochemical movement of carbon as it shifts between living organisms, the atmosphere, water environments, and even solid rock. The least amount of carbon is contained in the atmosphere, while the largest amount is found in the lithosphere.

Carbon dioxide occurs in all spheres on Earth and, as a gas, is readily mobile. Our increased use of fossil fuels has increased the amount of CO_2 in the atmosphere, with some of the gas being taken up in the hydrosphere and biosphere. Marine and fresh water organisms use CO_2 in their life cycles and can increase the amount of CO_2 in the geosphere. Coral reefs form by corals extracting dissolved CO_2 from sea water and creating their colonies in warm water environments.

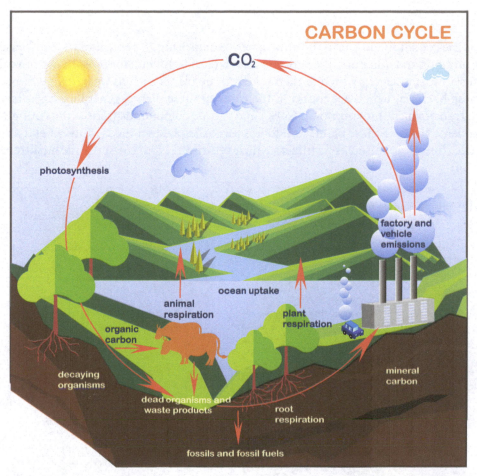

© danylyukk1/Shutterstock.com

Figure 81.2 The carbon cycle shows where carbon is contained in the geosphere, hydrosphere, biosphere, and atmosphere. Each year more than 3 gigatons of carbon are added to the cycle.

Generally there is a dynamic equilibrium between and among the four reservoirs of carbon. However, these are often changed by natural processes that contribute to the production and hence increased abundance of CO_2. Volcanic eruptions and wildfires are two natural hazards that can upset the carbon equilibrium. These two examples are discussed earlier in the text.

Oxygen Cycle

In the earliest stages of Earth's formation, no free O_2 existed in the atmosphere, and very little was present in the oceans. Approximately one billion years after Earth formed, bacteria began to produce O_2 and the amounts of this gas increased significantly. Oxygen is the second most abundant gas in the atmosphere and the most common element in Earth's crust by weight. As an anion, oxygen can actively combined with other cations and is a key constituent in numerous organic compounds. Oxygen is very prevalent in the lithosphere and is often freed up through chemical reactions to move into the atmosphere, hydrosphere, and biosphere (**Figure 81.3**). Additional amounts of oxygen and carbon are introduced to the atmosphere and hydrosphere by volcanic activity.

> **oxygen cycle:** This cycle is connected to the movement of carbon dioxide through the biosphere, lithosphere, and atmosphere as carbon dioxide is used by different organisms.

Nitrogen Cycle

Nitrogen (N_2) is the predominant gas in the atmosphere, comprising 78 percent of the total gases. Although it is a key part of proteins and other organic substances, nitrogen is difficult for most living organisms to assimilate. Soil bacteria must alter N_2, which is then taken up by plants such as peanuts, peas, beans, soybeans, and alfalfa. These bacteria, which are found in the roots of these plants, generate a chemical reaction that allows N_2 to be transformed into ammonia (NH_3) and nitrate compounds (those containing NO_3). Animals and other organisms that use these plants as a food source then take in the N_2 and expel it as organic waste. Waste products rich in N_2 are used as fertilizers which return the N_2 to the soil and the **nitrogen cycle** begins again (**Figure 81.4**)

nitrogen cycle: The movement of nitrogen through the biosphere and the atmosphere.

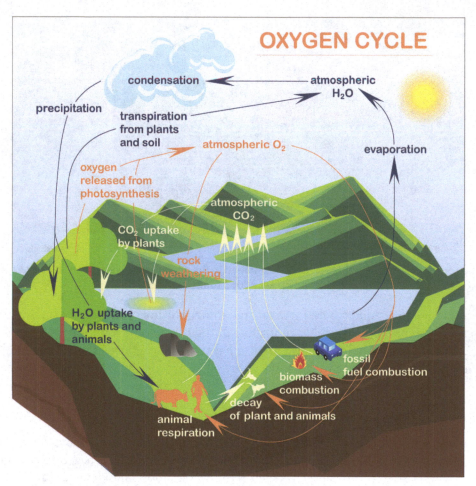

© danylyukk1/Shutterstock.com

Figure 81.3 Storage and movement of oxygen occur between the atmosphere, biosphere, hydrosphere, and lithosphere. Photolysis is the chemical decomposition of the atmosphere caused by sunlight.

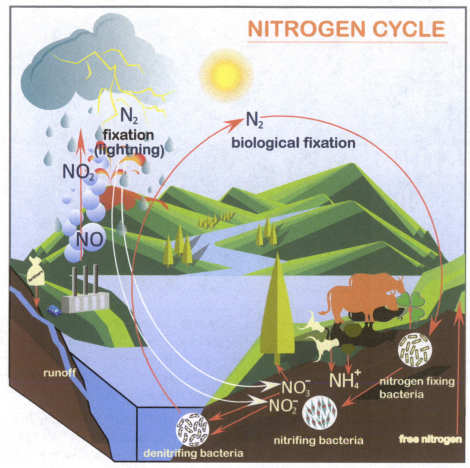

Figure 81.4 The Nitrogen Cycle, similarly to the carbon cycle and oxygen cycle, moves through the lithosphere, biosphere, atmosphere and hydrosphere.

Phosphorus Cycle

Phosphorus (P) is an important nutrient for plants and animals. It moves through the biosphere, hydrosphere, and geosphere but is lacking from the atmosphere, because it is in a liquid state at normal temperature and pressure. Phosphorus is part of DNA molecules and is a major component in animals and humans, as it makes up a significant portions of bones and teeth. Phosphorus is also a component of ATP, the energy molecule for all life forms. Due to its relatively low mobility, the **phosphorus cycle** is the slowest-acting of the ones described here.

phosphorus cycle: The movement of phosphorus through the biosphere and the geosphere.

Ecosystems
By Dean Fairbanks

The biosphere is the global sum of all ecosystems. An **ecosystem** is the association of biotic (organisms) components and their abiotic (physical) environment. It is linked to energy flow and nutrient cycling. The biosphere is an open system to insolation from the sun, but a closed largely self- regulating system with respect to materials used by biological life on the planet. The biosphere is highly dependent on the processes for the atmosphere, hydrosphere, and lithosphere, also called the four spheres.

The biological diversity of the earth is truly impressive. Tremendous amounts of time has allowed all the earth systems (the four spheres) and their variety of combinations to interact with the coevolving biological life themselves and the mechanisms of evolution to develop fascinating biological diversity across the planet. The biological diversity of the planet operates from the deepest ocean bottoms to at least the lower atmosphere. The variations in the diversity operates in ecosystems from micro to large scales all defined by unique geographic boundaries that support the interplay of living organisms with the abiotic processes. An **ecosystem** is an association of biotic components and their abiotic physical environment, linked to energy flow and nutrient cycling (**Figure 82.1**). Ecosystems are open systems to energy and matter, as their boundaries are not sharp blocking boundaries but only transitionary (fuzzy) ones. Distinct ecosystems are represented a forests, grasslands, deserts, lakes, ponds, oceans, etc.

Unpacking the complexity of ecosystems leads to the examination of **ecology**—the study of the relationships between organisms and their environment and each other among ecosystems. Biogeography is the section of physical geography that explores the spatial and temporal dynamics of the biosphere and the earth systems

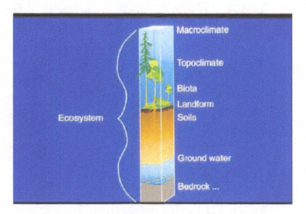

Figure 82.1   Ecosystem (USGS)

that support the biota. **Biogeography** is the past and present study of the spatial distribution of plants and animals, and the physical and biological processes, from local to global scales that produce earths species richness—the numbers of different kinds of life forms.

Ecosystems and ecology

Ecologists study ecosystems, while biogeographers use ecological principles to explain spatial distributions of organisms—where they are and why they are there. Unpacking ecosystems and the ecological relationships that in turn make them unique rely on the following components:

- **Individual**—living organism that can function on its own somewhat independently.
- **Population**—number of individuals of the same species. Relies on the method of **reproduction** to sustain the population, which is also related to the **carrying capacity** of the species in the ecosystem. There are always limitations to growth.
- **Community**—individuals of the same species and different species interacting together. Measured by presence, abundance, and competition/predation.
- **Biotic**—living plants (producers) and animals (consumers) along with waste materials
- **Abiotic**—air, rocks, soils and water that provide a home for the organisms.
- **Energy**—insolation is the main energy source that ultimately powers ecosystems.
- **Habitat**—the physical environment in which a species lives.
- **Niches**—the role of a species relative to other species in the ecosystem. What they do.

Ecosystems can be described based on their structures and functions. Structures most often include the species composition, the biomass within individuals/population/communties, available genetic resources, and then quantitative measurements of the abiotic components (i.e., pH of water, soil type and thickness, temperature, total insolation, precipitation, etc.).

The functions are dynamic processes in the environment that support the ecosystem. These include both land and atmospheric based processes such as cycling nutrients, plant growth (photosynthesis), decomposition, soil formation, water cycling, insolation seasonality, gas exchanges, wind, and evaporation/precipitation to name a number of the important ecosystem functions.

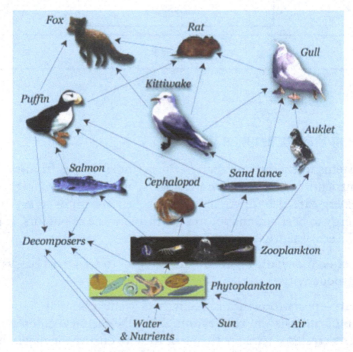

Figure 82.2 Food web (USGS)

The types of organisms that inhabit an ecosystem can be divided into **trophic levels** (position in a food chain):

- **Autotrophs** (producers)—organisms capable of converting sunlit energy into food energy by photosynthesis.
- **Heterotrophs** (consumers)—organisms that must feed on other organisms. These include herbivores (plant diet), carnivores (meat diet), omnivores (plant and meat), and detritivores (decomposers).

The complex chain of interdependencies is called a **food web** (**Figure 82.2**). Traditionally the dependence of on type of organism on another was arranged as a **food chain**, a top to bottom description of who consumes whom (**Figure 82.3**). For example:

Grass (*primary producer*) → Grasshopper (*primary consumer*) → Snake (*secondary consumer*) → Hawk (*tertiary consumer*) → Fungi (*decomposer*) → Nutrients to *primary producers*

However, the relationships among species in a community are much more complex, and so a network of links are described to for the food web. They are really a complex interaction of many food chains.

A final take on the trophic structure of ecosystems is to look the **trophic pyramid** or classification that is based on feeding patterns and describes the loss in energy efficiency with each step up the tropic pyramid. The bottom of the pyramid represents larger plant biomass, then the next step up is smaller consumer biomass, with the next steps representing higher trophic levels but smaller pyramid blocks. This occurs because energy at each level is lost as heat and available energy decreases. No energy transfer is 100% efficient (the transfer of insolation to plants-primary producers is only 1% efficient). The large losses in energy at each level are due to respiration, growth, reproduction, mobility, etc.

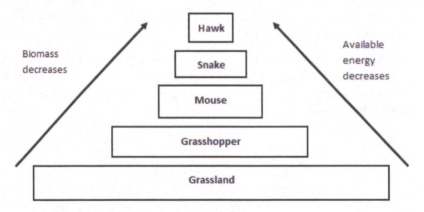

Figure 82.3 Food chain

Ecosystem Function and Biogeochemical Cycles

Ecosystems are open systems to energy and matter. Energy input comes from insolation and matter is in the form of water and nutrients. In the ecosystem biological organisms build structures thereby storing the matter. The matter and energy are also functional as the biological organisms are chemical processing entities transforming the energy to heat and matter to growth and waste as outputs. A portion of this output is feedback as a loop into the ecosystem as inputs.

The rate of energy production (normally on an annual basis) is referred to as ecosystem **productivity**. There are several measures of productivity:

- **Primary productivity**—refers to plant productivity. A plants essential biotic components to measure productivity are **photosynthesis** and **respiration**. Using the chemical chlorophyll in the leaves the process of photosynthesis creates sugars for the plant to store during the day light hours to then be consumed to build the plants structures at night as respiration (**Figure 82.4**).

Sunlight

Carbon Dioxide

Oxygen

Glucose

Water

Figure 82.4 Photosynthesis process (USGS)

<table>
<tr><td>FACT BOX</td><td></td></tr>
</table>

Photosynthesis –during daylight hours

$6CO_2 + 6H_2O + \text{Visible Light} \rightarrow C_6H_{12}O_6 + 6O_2$

Respiration –during night time hours

$C_6H_{12}O_6 + 6O_2 \rightarrow 6CO_2 + 6H_2O + \text{heat energy}$

- **Secondary productivity**—refers to that produced by animals.
- **Gross productivity** (photosynthesis)—the total amount of energy fixed.
- **Net productivity**—the amount of energy left after losses to respiration, growth, and so on, are taken into account. For an entire plant community, net productivity, is the amount of stored chemical energy.
- **Net primary productivity (NPP)**—the amount of energy made available by plants to animals at the herbivore level. It is normally expressed as grams/m2/year. NPP depends on the amount of heat, moisture, nutrient availability, number of sunlight hours, health of plants, age of plants, and competition among plants. NPP latitudinally increases towards the equator and declines towards the poles. The poles have limited insolation and cold air. Deserts also have low NPP, but they are limited by lack of moisture.

Biomes and Climates
By Elizabeth Jordan

Did You Know?

Certain pesticides, toxic chemicals, and heavy metals that are manufactured in warm climates can heat and evaporate into the atmosphere. With decreasing atmospheric temperatures, they cool and deposit in regions with colder climates—places such as the Arctic. This is called the **grasshopper effect** or, more technically, **global distillation**. This might explain why DDT (dichlorodiphenyltrichloroethane) and mercury are found in abundance in the otherwise pristine and untouched Arctic tundra. Here, they enter the ecosystem where they **bioaccumulate** until they are found in substantial quantities in caribou and polar bears. This unfortunate phenomenon is a harsh reminder that pollution and environmental issues are a *worldwide* problem.

Climate Influences Biomes

As a result of the curved surface of earth, prevailing winds in the northern hemisphere moving from high to low pressure seemingly deflect right (clockwise) while prevailing winds in the southern hemisphere moving from high to low pressure seemingly deflect left (counterclockwise). We call this the **Coriolis effect**. The Coriolis effect is believed to be responsible for large-scale weather patterns, but given its complexity and the many variables involved, scientists are not quite sure how this is so.

Because the earth rotates on an axis, the sun heats the earth's surface unequally, causing much variation in weather patterns. **Weather** is a result of this unequal heating and is defined as *short-term* fluctuations in such variables as temperature, humidity, precipitation, and cloud cover. We experience the variation of these *temporary* conditions on a daily basis.

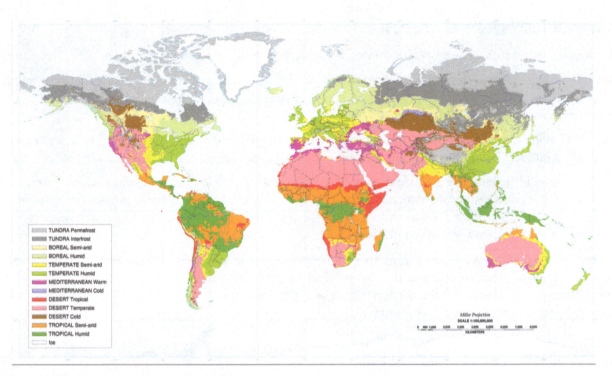

Figure 83.1 Major Climates

Source: Soil climate map, USDA-NRCS, Soil Science Division, World Soil Resources, Washington D.C.

The National Oceanic and Atmospheric Administration (**NOAA**) is the Federal agency that researches weather patterns, changes in our oceans, and in our climate

Climate is defined as *long-term* trends in weather patterns of a large given land area (Figure 83.1). Thus, long-term weather patterns directly determine climate. There is an observable correlation between latitude and climate. Because the earth spins on its axis, the equator gets the most net sunlight per year. The further distance on ecosystem is from the earth's equator, the colder its climate and average daily temperature. There is also an observable trend between altitude and climate—the higher the altitude, the colder the climate. That may explain why mountain peaks often resemble the Arctic tundra!

Biomes are observable trends of vegetation in terrestrial ecosystems that are directly correlated to climate. They are the highest level of the ecological system. Biomes are characterized by their dominant plant life which, in turn, is adapted to the climate of a given biome. For example, Southern California experiences a Mediterranean climate, marked by long, hot, dry summers and cool, wet winters. A dominant biome found in Southern California is chaparral—or Mediterranean woodland. Chaparral is comprised of vegetation that is adapted to the Mediterranean climate and therefore can endure long, hot summers and cool, wet winters (some adaptations include reduced waxy leaves—designed to reduce water loss). Other parts of the world that experience Mediterranean climates include central Chile, Italy, Greece, and Southern Australia. Each of these countries has the Mediterranean woodland biome. Although their vegetation does not share a recent evolutionary history with the chaparral found in California, they look very similar to the chaparral of California. The similarities are attributed to the fact that they have also individually adapted to the Mediterranean climate (this is an example of **convergent evolution**).

Ecotones are found where two distinct biomes overlap. They are a hotbed of biodiversity because a variety of plant and wildlife, adapted to a variety of environmental factors and two separate biomes, are found here.

Chaparral/Mediterranean Woodland

This biome, found in Mediterranean climates, is characterized by long, hot summers and cool, wet winters.

This biome has high species richness. Some of the plants have flammable essential oils that promote fires which, in turn, promote crown sprouting and fertility. Its soil is not considered nutrient-rich due to slow decomposition rates.

Chaparral vegetation is evergreen and typically has small, reduced waxy leaves to prevent and conserve water. Here (at least in California), you find sages, toyon, laurel, sumac, walnut, and prickly pear.

Much of the wildlife here are nocturnal; they burrow during the day to protect themselves from fire, escape the heat, and prevent water loss. An adaptation of some wildlife is to secrete concentrated urine in order to remain hydrated. Its wildlife includes hundreds of species of migratory birds, wood rats, lizards, deer, sheep, and in the Mediterranean woodland of Australia … kangaroos!

Chaparral is not without human disturbance. Many people prefer a mild Mediterranean climate to extreme long winters, and move or retire to this biome. Thus, high human density and development are common here. Humans clear the vegetation for agriculture and grazing. They also set fires (prescribed fires) to control pests and prevent more intense crown fires.

Climate Change Causes Biomes to Shift

Many terrestrial areas, or biomes, are at risk of shifting to a new type of natural vegetation as a result of changes in climactic patters. These changes in climate are associated with anthropogenic greenhouse gases.

Such climactic factors as wind, soil moisture, temperature, humidity, and precipitation influence or limit the geographic ranges of most animal and plant species. Any shift in the above variables may have a profound effect on plant and animal ranges and on biomes altogether. Why is this so?

Because Biomes are the highest level of an ecological system, a shift in these biomes can create a profound force, or ecological shift. When biomes change, wildlife or plants must adapt accordingly to the new variables. Wildlife or plants that cannot adapt to change may go extinct or disappear locally. For examples, Climate change, increasing temperatures and other factors, have shifted vegetation upslope, toward poles or toward the equator. It is predicted that the taiga, boreal forests (evergreen coniferous forests), which are adapted to colder temperatures, will move toward the poles. Research suggests that alpine plants have already shifted their range.[i] But the survival of some plant species may be at risk if climate changes faster than trees can seed, grow, and settle into new areas.[ii]

Many animals respond to changes in climate faster than plants. It is predicted that we may see many animals migrate north. Species with a narrow range of tolerance for temperature may seek out higher altitudes and latitudes as global temperatures get warmer Range shifts have already been reported in animal species such as birds,[iii] butterflies,[iv] and mosquitoes.[v]

Some animals are reportedly changing their diet with climate change and migration.[vi]

It is unpredictable how this will play out and how biomes will redistribute, but scientists are certain that there will be a major shift in most biomes as their plant species and animal species redistribute their ranges.[vii]

Earth's Biomes
By Elizabeth Jordan

Tundra

The tundra is a biome found in either high latitudes (the Arctic tundra) or high altitudes (the Alpine tundra on mountain tops). The Arctic summers are very short, lasting only a couple of months, but summers experience almost constant daylight. Precipitation varies from 200 to 600 mm yearly, and precipitation exceeds evaporation. This is the coldest biome and is distinguished from all others by the presence of **permafrost**.

Because the ground temperature is cold, decomposition rates are low creating this permafrost, a layer of frozen ground littered with layers of decaying or dead leaves and **detritus** (dead organic matter). The tundra is subject to **solifluction**—when frozen, water-saturated soil slowly moves down-slope. Dominated by grass, moss, sedges, lichen, and small shrubbery, the vegetation here is adapted to maximize heat absorption (they are mostly oriented toward the ground to absorb heat from the earth). Much of the vegetation has dark pigmentation to increase heat absorption. In addition, leaves orient themselves perpendicular to the sun to maximize sun exposure. Wildlife found in this biome includes polar bears, the Arctic fox, wolves, musk, ox, reindeer, ground squirrels, weasels, snowy owls, lemmings, and caribou.

Kekyalyaynen/Shutterstock.com

Permafrost

425

The tundra is being adversely affected by the increase in global temperatures. Due to these temperature changes, many animals inhabiting most southern biomes are migrating north to the Arctic tundra. The permafrost is melting, compromising the foundation of roads and infrastructure built on the frozen ground. Carbon dioxide trapped in the permafrost is released into the atmosphere as the permafrost melts, further impacting global temperatures. Furthermore, as the permafrost melts, the dark ground cover beneath it is exposed. Dark colors absorb more wavelengths of light, which also increases the overall temperature of the Arctic biome.

Sergey Krasnoshchokov/Shutterstock.com

Caribou in Russia

The Arctic has, until recently, been considered pristine, with little disturbance from human activity. However, recent years have seen more oil extraction, facilitating more human activity and disturbance. Furthermore, the tundra is vulnerable to the grasshopper effect, as pesticides and chemicals from distant populations accumulate in the Arctic. In addition to DDT and mercury (see the paragraph above), cesium-137, a radioactive element released in the devastating Chernobyl power plant disaster has been found in the tundra of Norway.

Taiga

The taiga (or the boreal forest or coniferous forest) is a biome found in the northern hemisphere in such places as Canada, Scandinavia, and Russia. A typical pattern of the taiga is picturesque freshwater lakes surrounded by dense rows of evergreens. The freshwater lake is what remains of a glacier that receded thousands of years ago.

Mark Herreid/Shutterstock.com

The taiga is found to the north and is marked by long severe winters (that last over 6 months) with little precipitation—about 20–60 cm a year. Temperature ranges from 70°F in the summer to 30°F in the winter.

The vegetation found here includes spruce, fir, and pine trees as dominant vegetation, but one also finds ferns, grass, and moss as the ecosystem producers. Since the ground is frozen for prolonged periods of time, the conifers (cone-producing trees like pine and spruce) are adapted to prevent water loss. Their leaves are thin, waxy, needle-like, and have minimal surface areas to prevent water loss through transpiration. The forest floors are nutrient poor and acidic. Pine needles drop to the ground, blanketing the forest floor, causing a decrease in soil pH. Acidity and cold temperature slows decomposition rate, causing a deficiency of available soil nutrients. Because soil quality is poor, and the long winters do not make for a suitable growing season, the Taiga is not suited for agriculture. However, they are a leading provider of lumber, paper, and other forest products. Drilling for oil, gas, and mining also occur in the Taiga.

Types of wildlife found here include moose, caribou, small rodents, birds, reindeer, bison, black and brown bears, wolves, lynx, and hares.

The taiga is being degraded by increased hunting and trapping of these animals and by pollution and disturbances associated with mining, logging, and oil drilling.

Deciduous Forest

Deciduous forest biomes are comprised of broad-leaved hickory, maple, beech, and oak trees, which lose their leaves annually. One also finds ferns and flowers as common producers. These cool and moist forests are home to some of the earth's largest organisms. Although winters are cold, temperatures are not as extreme as they are in the taiga. Its growing season (summer) is at least 4 months long and provides ample precipitation; annual rainfall reaches anywhere from 65 to 300 cm per year, with a temperate climate in which temperatures range from −30°C in winter to 30°C in summer. Deciduous trees lose their leaves in winter months to prevent water loss when the ground is frozen (and precipitation often comes down as snow). Some wildlife found here are deer, rodents, skunks, bear, fox, cougars, and wolves.

Petr Baumann/Shutterstock.com

Because these forest have an abundance of leaf litter, long growing seasons, and moderate temperatures, its soils experience rapid decomposition in warmer months, making soils rich and fertile with dead organic matter. Forest floors are often dominated by moss and mushrooms as well as microscopic bacteria that contribute to decomposition of the dead wood and other organic matter, thereby recycling nutrients.

Deciduous forests are being degraded as a result of human population growth and development. Many major cities such as Boston, DC, and London are built on what were once deciduous forests. They were also among the first biomes converted to agricultural land given the good soil quality and fertility. Urbanization, pollution, over-grazing, and deforestation have all disturbed this soil-rich biome.

Grasslands

Grassland biomes are found in the interior of continents where the annual rainfall is too little to sustain forest growth, but too abundant for a desert biome. They are characterized by fertile soils and herbaceous (nonwoody) vegetation.

Papa Bravo/Shutterstock.com

Zebra Running on the Savannah Grassland

Grasslands are dominated by either long or short grasses. Longer grasses thrive with more moisture and precipitation.

Tropical grasslands, such as the Savannah, found within 20° of the equator, experience both dry and wet seasons. Soil here is dense and does not retain water, making it difficult for trees to grow. Dry seasons occur in the summer and are often accompanied by fire which release nutrients from the soil that are imperative to plant growth, increased fertility, and resprouting of grasses. The African Savannah is home to many animals such as giraffes, zebras, lions, and elephants.

Savannah grasslands experience frequent fire which often kills its vegetation. The grass survives and resprouts. In this biome as well as many biomes, fire can have many ecological benefits.

Surface fires control pest and pathogen populations, release nutrients from the soil, enable vegetation to crown sprout, and nuts to crack open, releasing seeds (sequoias are an example). Periodic surface fires are also important for the Carbon cycle. Fire departments often utilize prescribed (or controlled) fires to mimic earth's natural fire ecology and gain its ecological benefits. **Crown fires** on the other hand are damaging. They burn hot, kill trees and wildlife, and cause soil erosion.

Temperate grasslands are the largest biome in North America and are even more expansive in Europe and Asia. The dominant vegetation is goldenrod, sunflowers, and clover. In temperate grasslands (such as the prairie grasslands of Montana), annual precipitation reaches 30–100 cm per year, with summer months receiving the most precipitation. Winters are cold, and summers are long and hot. Soils are very rich with nutrients and dead organic matter. Temperate grasslands are home to coyotes, bobcats, grey wolves, bison, and hundreds of bird species.

Jason Patrick Ross/Shutterstock.com

Prairie Grassland of Montana

Because their soil is rich and fertile, a large portion of prairie grasslands have been plowed over and farmed, causing them to lose up to 40% of their nutrients. Grasslands are also being degraded by overgrazing.

Desert

Desert biomes are often found ~30° north and south of the equator, just north and south of the subtropics. Here, tropical air descends, drying out and forming deserts. They are characterized by drought, flash-flooding, heat, bitter cold, and **lithosols**.

Sahara Desert

Deserts are expanses of land where *evaporation exceeds precipitation*. Their soil holds meager nutrients but often contain an abundance of stone, mineral, and salt. The vegetation that grows here are adapted to minimize exposure to sun and water loss. The leaves are usually small and waxy, and shed during dry periods to prevent water loss. Such examples include cacti, sagebrush, and yuccas. In the summer months, desert animals are usually active at dusk or night to avoid sun exposure.

Gobi Desert

Desert biomes are characterized by *hot desert, temperate desert,* or *cold desert.*

Hot deserts, such as the Sahara, are hot most of the year with temperatures reaching over 45°C. Temperate deserts, such as the Mojave in California, have almost no water vapor in the atmosphere, and the temperature can get hot during the day and very cold at night, yielding extreme variations daily. And finally the cold deserts, such as the Gobi, have long and cold winters.

The producers of desert biomes include cacti, ocotillo (desert coral), brush, and creosote bushes. For wildlife, the desert is home to lizards, snakes, scorpions, hawks, kangaroo rats, rodents, and foxes.

In all biomes, disturbance by humans, overexploitation, human settlements, mining, overgrazing, agriculture, etc., has caused desertification of the soil, increasing the number of desert biomes worldwide. Deserts are fragile ecosystems since their soils hold few nutrients. Disturbed desert ecosystems take many years to recover.

Tropical Rainforests

Tropical rainforests are found near the equator and experience almost daily rainfall. Comprising only about 6% of the earth's land mass, they contain an estimated 50% of the world's biodiversity on land!

Microstock Man/Shutterstock.com

Tropical rainforests experience little temperature fluctuation from month to month with the average daily temperature being 25°C. Annual rainfall can reach almost 400 cm per year. Rainforests are among the oldest forests found on earth. The tops of rainforest trees collectively create a dense impenetrable canopy, allowing little sunlight to reach the forest floor. As a result, trees and other vegetation must compete for sunlight. Many **epiphytes** live on rainforest trees, using the host tree as a facilitator in their vertical growth to access to sunlight. This relationship between the rainforest tree and the epiphyte is a type of **commensalism**.

The soil of the rainforests is nutrient poor with the constant rainfall leaching nutrients from the soil. Rapid decomposition from bacteria, ants, and fungi keeps soil nutrient content low as the nutrients are quickly absorbed by tree roots.

Trees (evergreen angiosperms) dominate this landscape and can grow up to 50 m in height with extensive canopies which themselves are homes to as many as 500 species—including plants, salamanders, birds (such as brilliantly colored parrots), snakes, frogs, and monkeys. Due to the extensive canopies, very little sunlight reaches the forest floor, so vegetation is not dense at ground level.

Deforestation, logging, mineral extraction, and agriculture destroy an estimated 100,000 square kilometers of rainforests per year.

In some areas, much of the Rainforest trees are cut down to make grazing land for cattle and to grow soy in order to feed them. Palm oil plantations are also a major driver of Rainforest destruction.

Temperate Rainforests

This cool, damp ecosystem stretches along the United States West Coast from southern Oregon to southeast Alaska. It also grows in Southern Australia, New Zealand, and Chile. Some of the common forest trees found in this ecosystem are very large evergreen angiosperms: Western Hemlocks, Douglas firs, Red Alder, and Bigleaf Maples that provide us with most of our lumber and wood pulp. They can grow over hundreds of years old, 60 feet in circumference, and 250 feet in height. They typically have epiphytes growing in the canopies; and moss, ferns, lichen and Huckleberry in the understory. These temperate rainforest receive plenty of annual rainfall—roughly 12–14 feet per year. Its mild temperatures rarely reach above 80°F or go below freezing. Temperate rainforests are typically found near the coast and the condensation from the marine fog add to the overall precipitation and dampness. Lots of dead and decaying wood, leaves and needles are found on the forest floor. This makes for new habitats for insects, moss, amphibians fungi and small mammals. Deer, elk, wood rats, and squirrels are common inhabitants.

tusharkoley/Shutterstock.com

A Temperate Rainforest

These old-growth forests have been over harvested for logging—to provide us with our pulpwood and lumber. At least in the United States, only a small percentage of this old growth remains. Most have been replaced with tree farms, which lack the biodiversity of the primary growth.

Human Impact on Ecosystems

By K.L. Chandler and Dean Fairbanks

Topic 85

Any successful effort to maintain ecosystem integrity and biological diversity depends on the support of its citizens. Ultimately, support for the conservation of biodiversity requires that the public understand the severity of the problem and support the needed planning and management actions. Accordingly, there are two issues that affect biological diversity: **degradation** and **habitat removal**.

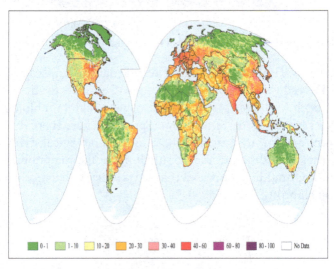

Figure 85.1   Percentage of human footprint on planet (**NASA**)

The adverse modification of habitat from its original condition causes the degradation of biological diversity (**Figure 85.1**). Degradation of biological resources may be subtler than the more apparent complete elimination of a vegetation community (and its animal associates) or ecosystem. Some examples of major current habitat degradation include: the reduction of stream flows from water diversion; the reduction in tree species diversity from timber production and intensive fire suppression efforts leading to catastrophic fires; the long-term reduction in tree cover as a result of air pollution resulting in acid rain; the disruption of natural ecosystem species linkages (food webs) by aggressive non-native alien plants and animals; and human disturbance through recreational developments. Of these, the impact of fire suppression has been the most degrading.

433

While fire prevention and containment has been a well-intentioned form of human management intervention, fire suppression has substantially altered the dynamics of vegetation communities (the need for ecosystem renewal through succession) (**Figure 85.2**). Natural fire once maintained vegetation communities in a mosaic of stands of different ages, and it created open, park-like stands and encouraged the growth of many plants and tree diversity in forests. Unfortunately, aggressive fire prevention and suppression programs have led many vegetation communities, especially in the US, to build up huge fuel loads and to become dangerous to other dependent plants and animals, and now human settlements.

Image © Shutterstock, Inc.

Figure 85.2   **Prescribed fires and suppression are two types of human management interventions.**

Habitat fragmentation and removal is the not-so subtle effect on natural landscapes. Human settlement expansion and agricultural activities reduce the habitat for plants and animals to survive, and the smaller remaining habitat patches are no match for larger pieces of landscape to maintain biological diversity (**Figure 85.3**). In the US, most of the species listed on the Endangered Species Act are there due to direct habitat removal. Ultimately, nearly all current and future threats to California's biological diversity from habitat loss and degradation are caused by the expanding human population, conflicting public attitudes toward biological resources and the poor decision-making processes that weakly try to balance economic development with the environment in land use planning (**Figure 85.4**).

Figure 85.3   California condor on the endangered species list in the US. (USGS)

Forestry clear cuts. Image © Shutterstock, Inc.

Figure 85.4 **All to often clear cutting is used in land use planning.**

At the end of the day, increased support for biodiversity and ecosystem conservation, as well as ecosystem restoration must come from expanded efforts to educate all segments of the world's population about the areas of the world that hold not only biological uniqueness but to the value of maximizing biodiversity inherent to any ecosystems in the world. The unique approach of geographical analysis and the synthesis of physical and human geography can be used to apply and demonstrate the benefits of resolving environmental and economic development conflicts for long term sustainability of the earth systems which in turn provide human's resiliency to future changes.

Ecosystems Effects and Impacts
By Ingrid Ukstins and David Best

Ecology is the study of the interactions between organisms and their environment (**Figure 86.1**). Organisms such as plants (producers), consumers (squirrels, foxes), and decomposers (fungi, bacteria) are the biotic components. Examples of the abiotic components are water, temperature, soil nutrients, pH of the substrate, elevation, latitude, aspect, slope, and amount of solar radiation received. If both the biotic and abiotic components are combined and studied, this discipline is referred to as ecosystem ecology.

> **ecology:** The study of the interaction between organisms and their environments.

A grassland ecosystem, for example, would include all of the grasses, insects, birds, rodents, earthworms as well as the soil nutrients, water availability, soil and ambient temperatures, and parent rock.

All of the ecosystems on the planet form the biosphere. With few exceptions, ecosystems are named after the dominant vegetation that occurs within the system. Some examples of ecosystems are: grasslands, savannahs, tundra, tropical rain forests, deciduous forests, deserts, coral reefs, oceans, and estuaries.

Each ecosystem has its own flora and fauna. Some of these are endemic, while others are found across several ecosystems. Ecosystems suffer from the effects caused by humans, directly and indirectly. An example of this would be the loss of a section of tropical rain forest. When the trees are removed, such as by burning or logging activities, animals and plants that used the trees as their habitat are immediately without a residence. The nutrients that were stored in the trees are removed entirely from that ecosystem, The soil, which is already very poor in nutrients, will not be replenished. When the seasonal rains come, there is no vegetation to intercept its flow, and soil erosion becomes rampant. If the surface has significant slope, landslides can result. Next the unchecked water with the eroded soil enters streams and rivers, causing sediment loads, which interfere and greatly reduce primary production of alga and aquatic vegetation. This loss of photosynthetic activity causes collapse of food chains within the associated aquatic habitats.

Figure 86.1   All organisms on Earth are related to one another and their physical surroundings.

Deserts are another ecosystem that has been greatly affected over the past century by humans. Deserts have their own very unique life forms that have evolved through adaptations to endure the hot, dry days and cold evenings (**Figure 86.2**). Deserts, especially in the southwestern United States, have suffered dramatic changes due to development of cities. Houses, roads, buildings, and infrastructure now rise up from the same place that cacti and other desert flora once flourished. Furthermore, the desert temperatures, coupled with atmospheric pollution from cars, trucks, and planes, plus the dust from unpaved desert roads, create a brown pall over many southwestern cities.

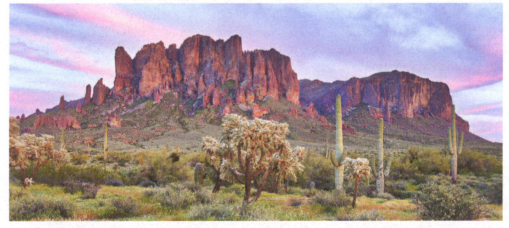

Figure 86.2   A desert ecosystem is made up of all the living and non-living components of a climate that gets less than 25 cm of rainfall a year. These harsh systems usually have poor soil as well, but organisms are adapted to survive and thrive with the limited supply of water.

Acidification, the process of lowering the pH of terrestrial and aquatic environments, begins with the combination of gases and water vapor in the atmosphere. Examples of three gases released into the atmosphere are CO_2, SO_2, and NO_2. Volcanoes, internal combustion engines, and various industries produce these oxides.

Geological activities such as volcanoes emit oxides of sulfur, and carbon dioxide is released from catastrophic forest fires. Once these gases are released and combined with water vapor the resulting acidic precipitation falls onto terrestrial and aquatic habitats. At first the buffering effects of various ions of the soil and water prevent the radical shift of pH. However, over time the acidification process gains momentum, resulting in losses of vegetation resulting in the loss of food and cover for the various organisms. In some lakes, acidification has rendered the water sterile of all life. If this sounds like the "Domino Theory"—it is. To paraphrase John Muir, everything is tied together in an ecosystem and therefore changing one aspect changes everything else within the system.

Since 75 percent of Earth's surface is covered by oceans and lakes, it is logical to some of our species (*Homo sapiens*) that the oceans are limitless with respect to being able to swallow the sewage, garbage, sludge, and other pollutants so frequently and directly shoved into the water. There is an indirect absorption by oceans of agricultural runoff from rivers that are fed by surface runoff, which are fed by farms. This agricultural "cocktail" contains fertilizers, herbicides, fungicides, insecticides, and in some cases fecal materials from feedlots. The oceans and their associated shores and intertidal zones are the largest ecosystem and as such touch every human in some way. The most obvious way is the utilization of the numerous foods that are harvested from the oceans. The pollutants, for example mercury, that have entered the oceans do come back to shore in the fish we eat. This is because of a process called **biomagnification**. Small amounts of a residual pollutant that are absorbed in the lower links of a food chain are increased in the tissues of the larger members of the food chain.

> **biomagnification:** The increase in amounts of toxins in animals higher in the food chain as a result of ingesting organisms that contain toxic materials.

Changes to Biodiversity

Biodiversity is a combination of two terms biology and diversity. The term means all of the diversity of life, from the unique gene combinations of individual species to the many ecosystems across the Earth where these species live. There is an intrinsic value bestowed on the megafauna and megaflora, as it should be. However, much of the biodiversity on Earth is hidden because it is either microscopic or in places that are difficult to explore. Consider the soil and its depths and the ooze of oceans and lakes and the aphotic zone of the abysses of the oceans. There is so much yet to be discovered, and much to be lost if we continue down the paths of ignorance and exploitation.

No one knows about the biodiversity that may have already been lost due to following the above two paths. The majority of humans have a soft spot for Bambi and other cute animals, but not that many have reasons to care about unseen down-in-the-ground microbes. These buried "biodiverse" organisms have unique adaptations that permit them to survive in otherwise uninhabitable places. "Buried" might also refer to species that live in or on another species, such as parasites. The presence of parasites often generates a negative response, yet these organisms are part of a larger ecosystem. Epiphytes, plants that attach and live on other plants, but do not derive anything except a place to attach, are also in a way buried since they are not as visible (**Figure 86.3**). Bromeliads are another type of epiphyte that live high on the trunks of tropical rain forest trees. Bromeliads collect water around their stems and leaves, creating little pools (**Figure 86.4**).

© Uwe Bergwitz/Shutterstock.com

Figure 86.3 Bromeliads are in a family of plants that contain over 3,000 different species, including the pineapple and Spanish Moss, which is neither Spanish nor a moss.

© Leonardo Mercon/Shutterstock.com

Figure 86.4 This bromeliad in the Guarapari rain forest of Brazil is the home of a white-banded tree frog and a bromeliadas tree frog.

There are tiny frogs that spend their entire life in the pool, from egg to adult. So who cares if they are lost; after all, out of sight out of mind, as the saying goes. We do assign labels to some species like endangered, threatened, species of concern—and once a label is appended to that species some protection is provided.

Good examples would be bald eagles and peregrine falcons; populations of both have recovered sufficiently to have their labels removed.

A common way for biodiversity to be reduced in an area is remove the habitat or alter it so significantly that there is no chance for certain species to hang on. The Dusky seaside sparrow fits into this category. Between flooding and then draining its habitat in Florida and with the addition of numerous pesticides in its environment, the "duskies" departed, going extinct in the late 1980s. The dodo, which was thought to have only existed on Mauritius in the Indian Ocean became extinct in the late seventeenth century (**Figure 86.5**). On rare occasions a species is "rediscovered in the wild." An example is the Banggai crow (Corvus unicolor), thought extinct from the early 1900s, was rediscovered in 2007 on Peleng Island in eastern Indonesia.

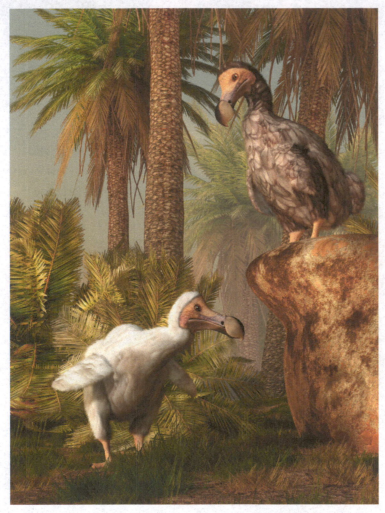

© Daniel Eskridge/Shutterstock.com

Figure 86.5   An illustration of two dodo birds, now extinct, on the island of Maurita.

Another reason for losses of endemic biodiversity, organisms that only live in a particular area, is the invasion of non-native biota, meaning organisms that did not evolve within the same ecosystem and therefore have no natural controls on their spread or population levels. There is a National Invasive Species Information Center in the United States. Some examples of invasive species are:

- The majestic American chestnut tree grew to great heights and diameters in the eastern United States until the chestnut blight (a fungus) was accidentally released in the United States on some imported wood from Asia. A few American chestnut trees remain, perhaps because of immunity to the blight.

- The European gypsy moth, brought into the United States in the 1860 for silk production is a notorious defoliator of hardwood trees across North America.
- The European starling was introduced into the United States in the 1890's for the sole purpose of having all the birds mentioned in any of William Shakespeare's writings to have a habitat in this country. These birds have displaced many native songbirds and have spread to all 48 contiguous states and Alaska. From the original 100 starlings released in New York's Central Park, hundreds of millions of these birds have resulted.
- Kudzu, a member of the bean family, was introduced into the United States around 1880 from Japan. Kudzu was originally touted as a good plant to control erosion. It found the southern United States to its liking and took off, covering everything in its path. Kudzu has grown around roads by traveling across electrical lines from one side of the road to another (**Figure 86.6**). As you can see, some organisms came in accidentally, other introduced on purpose, but either way each organism was not native and has disrupted native species as a consequence. This is one reason the United States Customs Service asked arriving travelers into this country if they have any plants or animals in their possession.

© Luke Ferguson/Shutterstock.com

Figure 86.6 **Kudzu has overgrown an abandoned building.**

You have seen the bumper sticker—"Extinctions Are Forever!" This is one slogan that speaks for itself and is hard to argue with.

Soil Characteristics
By Dean Fairbanks

The development, properties, and distributions of the world's soils are a complex endeavor for what for most people would seem so simple—dirt. How complex (and important) could the soil beneath our feet be? The role of soil as the finely developed interface between the activities of the geosphere and the living systems that use them directly, that is, vegetation, or indirectly, that is, fauna, implies why it is placed as a major component in the biosphere. Soils are the planet's skin, its eco tissue—the interface between the mineral and biological worlds. Nutrient rich, well developed, and biologically rich soils are the key to fully functional ecosystems (**Figure 87.1**).

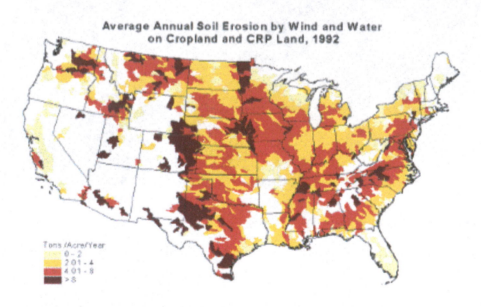

Figure 87.1　US soil erosion by wind and water on agricultural areas. (USDA)

Measurement of Soil Characteristics

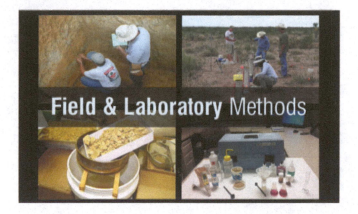

(USDA)

The following properties are commonly used to describe soils in the field and in the laboratory: **Color**— The color of a soil is the first impression one has when looking at the bare earth. It is used as an item to identify a soil type. The color is based on a property of the mineral makeup of the soil. A system called the **Munsell system of color notation** has been developed to describe soils based on their hue (a specific color), value (lightness and darkness), and chroma (color intensity). The soil color and Munsell notation can help connect the soil with the natural environment of the areas it was found. Among the many possible hues are:

- the reds and yellows (high in iron oxides–rusting);
- the dark browns to blacks (richly organic);
- white-to-pale hues (silicates and aluminium oxides);
- Gray and greenish-bluish (reduced iron from being inundated in water) and;
- White color (calcium carbonate or other water-soluble salts).

Texture—Soil texture refers to the size of the solid particles in a soil, ranging from gravel to clay. The proportions vary from soil to soil and between layers of the soils with depth. The texture is a three-component mixture based on the percent of sand, silt, and clay. Soil texture is important as it affects the moisture content and aeration, retention of nutrients, and ease of root penetration, thus agricultural cultivation.

- **Structure**— Soil structure refers to units composed of primary particles. Several basic shapes of structural units are recognized in soils: granular, blocky, platy, and prismatic or columnar. Soil structure is looked at to determine pore space, and how well water, air, and roots will move through the soil.

Granular from constant vegetated soils; blocky from large amounts of clay; platy from compaction; and columnar/prismatic from high sodium levels, usually in arid environments. (USDA)

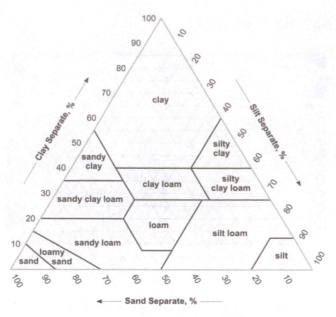

Figure 87.2 Soil texture triangle. (**USDA**)

Figure 87.3 Strong fine and medium granular peds.

Figure 87.4 Stong thin platy structure.

Figure 87.5 A cluster of strong medium columnar pads. The cluster is about 135 mm across.

- **Consistency**—The manner in which the soil material behaves when subject to compression with a measure of the soil water amount noted. It is also a measure of the strength of soil crust, through resistance to penetration. It provides information of the degree of cohesion of the soil.
- **Porosity**—The pore size of the soil provides the amount of void space for air and water to flow through. It provides a measure of the permeability of soil, that is, well drained or not well drained. For example, soils with a higher sand content will have greater porosity and thus will be well drained. Whereas soils with a higher clay content will have much lower porosity and tend to be not well drained.

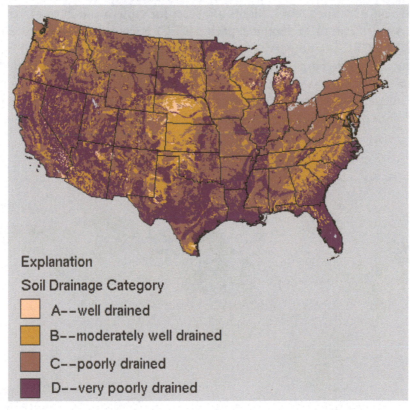

Figure 87.6 US soil drainage capacity. **(USGS)**

- **Moisture**—This provides a measure of the water-holding capacity of soils and is strongly linked to texture, structure, and porosity. Water- holding capacity is the amount of water retained in the pore spaces and attached to the soil particles. Coarse sand has the lowest ability to hold water, while silty clay loam has the highest (**Figures 87.6–87-7**).

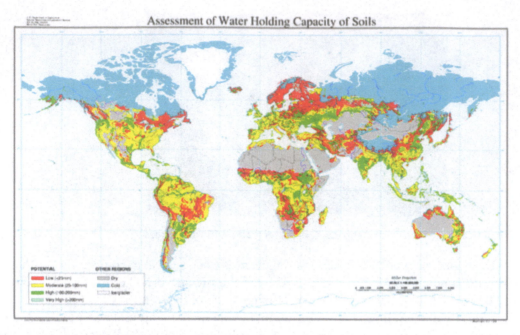

Figure 87.7 World water-holding capacity of soils. (USDA)

- **Organic matter and basic chemistry**—The basic chemical elements essential for plant growth include carbon (C), nitrogen (N), phosphorus (P), and potassium (K). These represent basic fertility of the soil. In soil chemistry, nutrients are called **bases**. Plants use bases for growth and in return provide the soil with hydrogen ions. A soil's **cation exchange capacity** (CEC) is considered the best single index of potential soil fertility. The CEC is a property of the soil **colloids** (clay content), which have a negative charge (anion) to hold positively charged cations (Ca, Mg, K, Na, H, trace minerals, etc.). A soil with a high CEC is more fertile than a soil with a low CEC. The colloids carrying the cations interface with the roots to provide nutrients. Soils become more acidic over time; however, bases can be returned via leaf debris, application of fertilizers or weathering of base-rich rocks such as limestone and chalk (**Figure 87.8**).

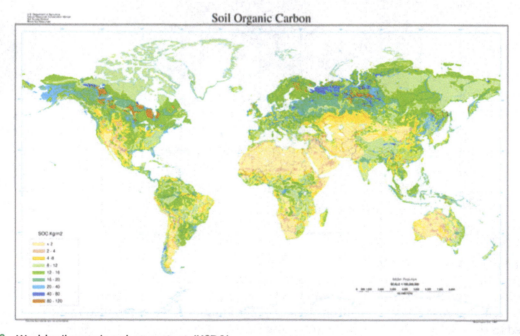

Figure 87.8 World soil organic carbon content. (USDA)

- **pH**—It is the measure of the concentration of exchangeable hydrogen ions, H+, present, or saturation of the soil colloids. It provides a measure of acidity or alkalinity of a soil. An ion is an atom or molecule that has lost or gained one or more electrons, making it positively or negatively charged. The more hydrogen ions, the more acidic a soil is (low pH), and the lesser the hydrogen ions, the higher the level of cation saturation is (more alkaline bases means higher pH). pH is measured on a scale from 0-14. Scale 0 means highly acidic and 14 means highly alkali with 7 being neutral (**Figure 87.9**).

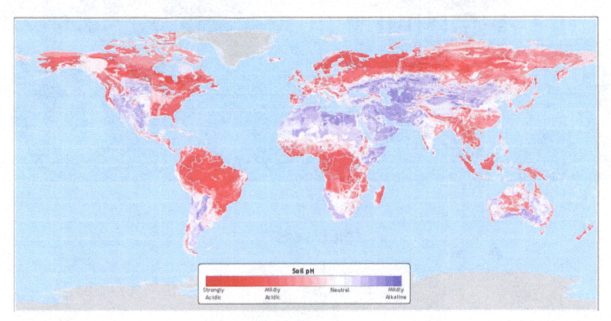

Figure 87.9   World soil pH. (University of Wisconsin, Madison)

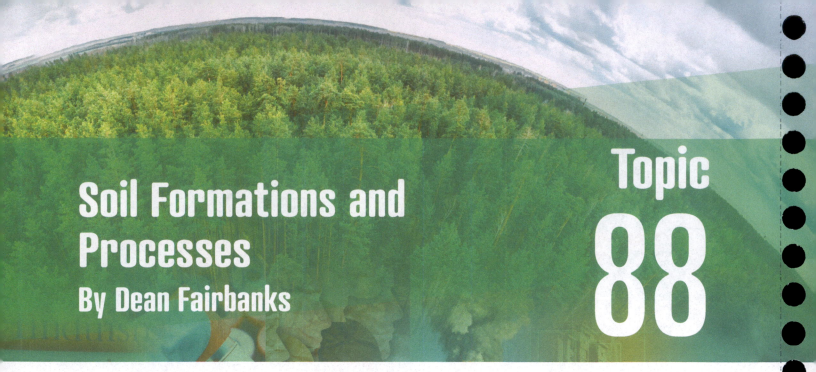

Soil Formations and Processes
By Dean Fairbanks

SOIL LAYERS

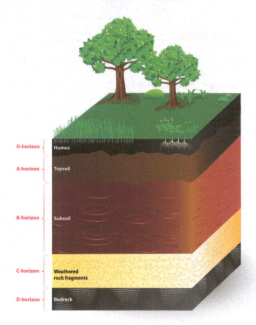

Image © Shutterstock, Inc.

Figure 88.1 Soil layers are made up of lining and non-lining material.

The composition of soils represents both inorganic materials (water, air, minerals—clays and colloids) and organic materials (animal and plant matter—humus). This leads the discussion to the processes of soil formation—**pedogenesis**. The main soil formation factors are notably climate, geology, biological organisms, and topography (linked to transport). These interact over time to produce distinctive soils and soil profiles (i.e., vertical soil layers).

The soil-forming process (pedogenesis) leads to **soil horizons**—distinct layers vertically organized in a **soil profile**. The soil horizons are linked to humidification, degradation, and mineralization that are the processes

Image © Shutterstock, Inc.

Figure 88.2   A soil profile showing various horizons.

whereby organic matter is broken down and the nutrients are returned to the soil. This occurs in the O horizon, which represents the surface to 5 to 10 cm down, composed of undecomposed litter and decomposed humus. The next horizon is A, which is a mixed mineral–organic layer, which tends to be darker from the humus content (also location for waterlogged or gleyed soil if found in wetter and cooler climates). Horizon E is the eluvial or leached horizon that is paler in color if heavily leached or light brown if weakly leached. Horizon B is the illuvial or deposited layer where iron oxides, clay, and humus are deposited. Finally, horizon C is the weather rock layer representing the regolith, just above the bedrock (**Figure 88.2**).

Climate in soil formation is concerned with two important mechanisms: precipitation and temperature. Precipitation is compared against potential evapotranspiration (PET) to determine its effectiveness. When there is more precipitation than PET, the soil materials tend to leach downwards (i.e., high infiltration and percolation). When precipitation is much less than PET, then materials move upward. Temperature provides a measure of the rate of biological and chemical action. Pedogenic regimes, as partly determined by moisture and temperature, are more illustrative when relating to processes of the movement of water through the soil. **Leaching** is the downward removal of materials in solution and in suspension. A further unpacking of this concept is informed by **eluviation**, which is the removal of inorganic and organic substances from a horizon driven by the downward movement of soil water, a distinctly erosional process. This is in contrast to **illuviation**, where downward-moving material, especially fine material, is accumulated/deposited.

Temperature with precipitation is a strong measure of chemical weathering of the parent material. The geological parent material has lasting effect on soils through texture, structure, and fertility. For example, soils based on sandstones produce freely draining soils, whereas clays provide much finer soils with less drainage. On a regional scale, soils often vary with geology quite strongly, but then topographic characteristics must be taken into account.

Topography in the form of slope relief plays strong role in informing transport processes within the soil: enrichment and erosion, translocation, and physical mixing. In general, steeper slopes have thinner soils. Soil erosion increases with slope angle as noted in Module 4 Section 3 on weathering, and aspect affects soil moisture and temperature via insolation and therefore the rate of chemical weathering. Relief can also help characterize the horizontal organization of soil in the context of the soil **catena**. A catena is the variation in soils developed from a similar parent material under similar climatic conditions but whose characteristics differ because of variations in slope angle, drainage, water table, and microclimate (**Figure 88.3**).

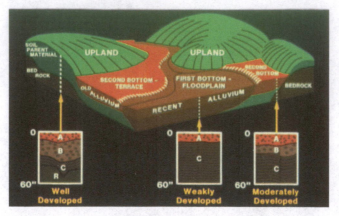

Figure 88.3 **Catena landscape. (USDA)**

Finally, all of the pedogenesis process happens over time, which is not a causative factor, but allows processes to operate. The time needed for soil formation varies by parent material and climate. Sandstones develop soils more quickly than granites and basalt. Places that had heavy glaciation in the recent past have soils that are evolving only since the removal of the ice; others are much older, but no soils are older than two million years (the Pleistocene).

Topic 89

Major World Soil Types
By Dean Fairbanks

The **twelve major soil types (soil orders)** that have developed around the world are classified based on the following factors (**Figures 89.1–89.2**):

- Azonal or recent youthful soils
  - Entisols—recent soils with undeveloped profiles, all climates
  - Andisols—volcanic material, largely ash
  - Inceptisols—weakly developed soils of humid regions
  - Gelisols—permafrost soils

- Podzolization—soils with intense leaching in low-temperature areas
  - Spodsols— cool, humid conifer forest soils
  - Alfisols— humid temperate forest soils
  - Histosols—organic rich, wet soils in bog areas

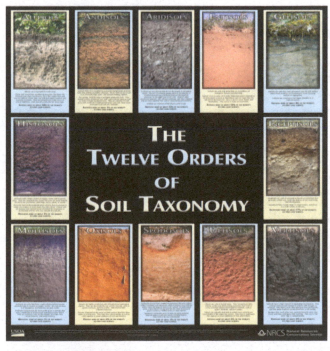

Figure 89.1   The diversity of the world's soils. (USDA)

- Calcification—results from ineffective leaching, in areas of low rainfall, causing accumulation of calcium in the soil
  - Aridisols—desert soils
  - Mollisols—grassland soils, semiarid
  - Alfisols—humid temperate forest soils

- Laterization—soils with intense leaching in high-temperature areas
  - Ultisols—highly weathered subtropical forest soils
  - Oxisols—tropical soils, hot and humid areas
  - Vertisols—expandable clay soils of the subtropics and tropics with a dry period.

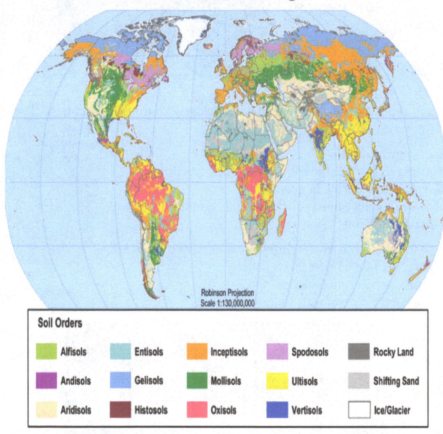

Figure 89.2 **Global soil regions. (USDA)**

Human Impact on Soils
By Dean Fairbanks

Humans, other animals, and plants all have effects on soil development. Microorganisms breakdown leaf litter, worms mix soils, burrowing animals turnover soils from deeper depths, vegetation returns nutrients to the surface, and human activities include adding fertilizer, irrigating, draining, compacting, and eroding soils. The three main problems with humans using soils for agriculture are **compaction, salinization, and erosion** (**Figures 90.1–90.3**).

Figure 90.1   Soil erosion (**USDA**)

Figure 90.2   Compaction (**USDA**)

453

Figure 90.3　Salinity (USDA)

Soil compaction is a change in structure by turning a freely draining soil or horizon into a compact impermeable soil. Compaction occurs with large heavy modern machinery and in the agricultural fields when ploughing damp soil. Salinization occurs when excessive irrigation causes the water table to rise to the surface. As the water evaporates, soluble salts are left forming a saline crust that can harm crops. This is a problem in California's San Joaquin Valley, where farmers must flush their field's soils at the end of the growing season in order to leach the salts to deeper layers in the horizon to not affect the next seasons crops. Finally erosion is a major problem. This occurs when the vegetation is removed like in deforestation, farming, overgrazing, and urban construction. The principals of good soil management should be adhered: revegetation, erosion control, crop management-cover crops, contour ploughing and compost application, and various levels of soil reclamation. Once we lose soils, it will take hundreds to sometimes thousands of years, depending on the location, to rebuild as few as only inches of new soil. This is why composting is an integral part of rebuilding soils in farming, especially organic farming practices. Soil is precious and it represents the true vitality of the plant kingdom and especially in our agricultural practices (**Figure 90.4**).

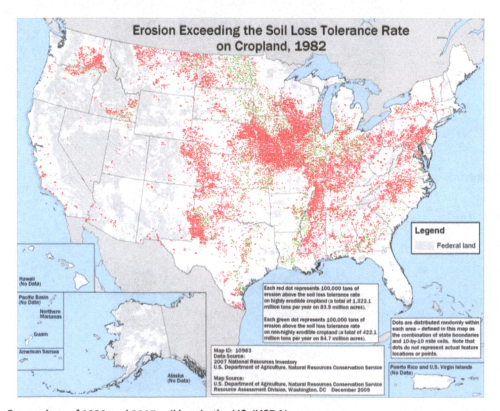

Figure 90.4　Comparison of 1982 and 2007 soil loss in the US. (USDA)

Soil Contamination
By Ingrid Ukstins and David Best

Soil Contamination

The introduction of hazardous materials either on the surface or into pore spaces in soil can produce harmful conditions for humans, animals, and plant life. Before the EPA was established in 1970, numerous instances of dumping of hazardous chemicals and other materials caused the soil and ground water to be filled with substances that had a dangerous effect of life. There was no concern or knowledge that materials placed in dump areas would produce negative effects on future generations.

Dangerous contaminants quickly work their way into the food chain as plants take up the material into their structure and then animals ingest the plants. Eating of the animals by humans and other animals then passes the substances along, where they can have a very deleterious effect on life.

Heavy minerals and liquids can contaminate the soil and lead to life-threatening conditions (**Figure 91.1**). The cleaning up of contaminated areas involves either: (a) treating the soil in place with chemical and other procedures that clean the soil, (b) leaving the soil in place and containing it to a small area, thereby preventing the contaminants from reaching animals, humans, or plants, or (c) removing the soil and treating it or disposing of it in a way that removes the harmful material.

Worldwide there are millions of sites that contain some amount of contamination. The proliferation of harmful chemicals and the thoughtless manner in which industrial wastes were handled in the early and middle twentieth century have created numerous harmful conditions. Two such examples in the United States are the incident at Love Canal in the state of New York and the mining of uranium in the western United States.

© Vladimir Melnik/Shutterstock.com

Figure 91.1   Soil contamination is caused by the presence of human-made chemicals that can be from industrial activity, agricultural chemicals, or improper disposal of waste material. The most common chemicals are hydrocarbons, solvents, pesticides, lead and other heavy metals.

Pollution and Toxins
By Elizabeth Jordan

Considering that there are over 7 billion people on the planet, it probably comes as no surprise that we generate exorbitant amounts of waste daily. This waste is causing harm to the existing human population as well as to wildlife and the whole natural world. Waste can take the form of pollution, contaminating the air we breathe, water we drink, and our agricultural soils, causing diseases and threatening the survival of many species and even an entire ecosystem. Toxic chemicals enter the food chain and accumulate in top level predators, decreasing their population size to the point of looming extinction. Plastic waste, both a land and water pollutant, has accumulated in the North Pacific taking form in a mass comparable to the size of Texas! This poses an obvious threat to aquatic ecosystems.

Plastic

Plastic is the by-product of hydrocarbons (organic molecules) derived from petroleum which has been chemically altered so they are durable, flexible, and do not degrade by natural means. They therefore may stick around for hundreds, if not thousands, of years. (Plastic was invented in the 1800s so we still do not have a lot of information on its long-term decomposition.)

Plastics (mostly some types of **PVC**—a popular form of plastic) contain phthalates that are toxic to humans. Phthalates have high human and environmental exposure and have been detected in the human body as well as in food we eat. Phthalates have been linked to developmental problems in the reproductive system (male feminization) and are carcinogenic. Plastics also contain BPA (bisphenol A) and compelling evidence indicates that it mimics estrogen, a sex hormone. This ability to mimic sex hormones can affect reproduction and development. Also of note, BPA has been linked to type II diabetes, obesity, neurological disorders, cancers, increased risk of miscarriages, and may compromise the immune system (all of this especially affects fetuses/infants during developmental stages). Plastics also contain HDPE (high-density polyethylene), the most common resin used in plastic bottles, and it is a suspected carcinogen.

We produce approximately 300 million tons of plastic annually, and only 5% of plastics are effectively recycled while most end up in landfills or the oceans. Keeping this in mind, it should not come as a surprise that in the North Pacific, due to the water currents and movement of gyres, there is an "island" made of plastic comprising almost 10 million square kilometers and is doubling in area each year. It is now estimated that 8 million metric tons of plastic enters the ocean each year.[i] As a result of this, millions of fish, sea turtles, and sea birds die each year from ingesting plastic. A report by the Ellen MacArthur Foundation and presented at the World Economic Forum predicts that by 2050, there will be more plastic than fish by weight.

Sun commonly melts plastic and causes plastic residue to dissolve in the oceans. Its effects on the food chain have not been determined. The good news is that all these health and environmental concerns have inspired research and projects that raise public awareness of plastic waste. This has lead to a reduction in the amount of plastic produced, and an increase in the reclamation of plastic and recycling. For instance, you may notice that many plastics are now labeled as recycled or BPA free.

JPL Designs/Shutterstock.com

Figure 92.1   BPA free plastics are common on plastics with resin identification codes 1-2 and 4-6

Sascha Corti/Shutterstock.com

Figure 92.2   Most plastics are not properly recycled and are littering shores and oceans

Air Pollution

In 1963, Congress passed the first Clean Air Act to address the escalating pollution problem and to authorize research in order to properly mitigate its adverse effects on the environment. In 1970, Congress passed another Clean Air Act which granted state and Federal governments unprecedented control over the amount of pollution created.

Rat007/Shutterstock.com

Figure 92.3 Reducing primary pollutants is one of the many goals of the Clean Air Act

The Clean Air Act of 1970 was seminal because it set standards for ambient air quality and limited pollution emissions of motor vehicles. As a result of the Clean Air Acts, the country has seen a dramatic increase in air quality since 1970.

Primary pollution is a pollution that is released directly into the atmosphere either by natural means such as volcanic activity and wildfire dust or by human activities such as automobile and coal-factory pollution. Examples of primary pollutants include sulfur dioxide, carbon monoxide, carbon dioxide, particulate matter, lead, and volatile organic compounds.

These primary pollutants often react with other pollutants in the atmosphere, causing both chemical and physical changes. The substances created by these processes are called **secondary pollutants**. Ozone, nitric acid, and sulfuric acid are examples of secondary pollutants.

Both primary and secondary pollution are harmful to human health and damage the environment. The detrimental effects of both types of air pollution are magnified by the occurrence of **temperature inversions**. A temperature inversion occurs when normal temperature patterns are reversed (warm, lighter air moves over cold, dense air) and cool air is trapped below a column of warm air. Pollution gets trapped below the inversion layer, sometimes for hours, and cannot dissipate easily into the atmosphere as a whole.

Common Outdoor Pollutants

The EPA (Environmental Protection Agency) was created by Richard Nixon's executive order in 1970 in response to first annual Earth Day demonstrations by Gaylord Nelson-protest is important![ii]

The EPA is a Federal organization through which all environmental legislatures move. Enforcement of these regulations is often done on a state level. The EPA has compiled a list of the most common and damaging outdoor pollutants whose concentrations must be controlled. These substances include ozone, VOCs (volatile organic compounds), carbon monoxide, particulates, sulfur dioxide, lead, and nitrogen dioxide. We will touch briefly upon the effects of these pollutants, their impacts on human health and the environment, and why their concentrations must be controlled.

Ozone (O_3) is a compound found in high concentrations in the stratosphere (approximately 10–30 miles from earth) and plays a critical role in protecting the earth from UV rays. These UV rays (UVA and UVB—literally cooks your skin, breaks down the structure, so wear your sunscreen!) damage biological molecules such as DNA and compromise our body's cells as well as the cells of the rest of the living world.

Ozone is also found at ground level close to earth's crust. Not naturally found in high concentrations close to earth, ozone is created at ground level when pollution called nitrogen oxides from fossil fuel combustion (via power plants, automobiles, and other forms of organic combustion) react with atmospheric oxygen gas and

Sai Yeung Chan/Shutterstock.com

Figure 92.4 Smog is a common outdoor pollutant in major cities.

VOCs in the presence of sunlight. Although helpful in the stratosphere, ground level ozone is damaging to both our health and the health of our environment. It weakens the immune system, aggravates heart disease, causes irritation to the nose and lungs, and harms plants. Ozone reacts with sunlight to form a **photochemical smog**, which acts as an irritant to eyes, nose (called a lacrimal response), and to lungs.

VOCs are organic molecules that transform into a gas at room temperature. (Examples are methane, paints, and solvents.) VOCs have been known to cause dizziness, memory lapse, respiratory impairment, and possibly cancer. This is a common phenomenon, especially in the workplace where it may be known as either new building syndrome or sick building syndrome.

Another outdoor pollutant (and also indoor) is **carbon monoxide**, a colorless, odorless gas that is released from car exhaust and forest fires (heaters, furnaces, ranges, etc.). It reacts with red blood cells, disrupting their ability to carry oxygen. It is extremely harmful to human health, takes a long time to exhale from your system, and is possibly deadly.

Particulates are air-borne solids or liquids such as dust, smoke, or even brake-pad debris that accumulate in the air we breathe. Particulates irritate our eyes and respiratory system and can cause bronchitis and asthma. Particulates cause an odd, dark coloration when you blow your nose.

Sulfur dioxide is a colorless gas with a noxious odor released as a result of fossil fuel combustion. It can irritate the cardiovascular and respiratory systems, sometimes causing bronchitis and asthma. (Remember CHNOPS and that fossil fuel used to be a living organism, thus sulfur is a byproduct of combustion.)

Lead is a heavy metal found both naturally and in manufactured goods, but most concentrations in the air are from automobile fuel. Exposure can lead to many health problems, including damage to the nervous system and possible developmental problems (such as female late puberty), and anemia.

Nitrogen dioxide is a highly reactive gas that comes from power plants and motor vehicles. It also contributes to the formation of ground-level ozone. Nitrogen can irritate the eyes, nose, and lungs, and overall respiratory function.

How Do We Reduce Air Pollution?

Cap and trade, sometimes called permit trading, is a practice in which companies receive a limit on the amount of carbon dioxide or other pollution they can release into the atmosphere. Over time, the limit is gradually reduced, and companies slowly decrease their emissions. The philosophy behind cap and trade is that the companies will slowly learn to cut down on their amount of pollution and improve overall environmental quality. If companies are under their limit or cap, and release less pollution than allowed, they receive **pollution credits**. These pollution credits may be used as a currency that companies can sell to other companies for money. The purchase of these credits allows the buyer to produce more pollution than their limit would otherwise allow.

Cap and trade has been considered successful in controlling sulfur dioxide, a pollutant which causes acid rain. But critics are dubious about whether it would be effective in controlling major pollutants such as carbon dioxide. Critics express concern that this system of cap and trade leaves too much room for dishonesty and corruption. Companies may fudge or embellish data about their emissions in order to receive unearned credits. Another common criticism of this practice is its underlying philosophy. To quote Albert Einstein, "We can't solve problems by using the same kind of thinking we used when we created them." In other words, the practice of cap and trade is a system that uses money as the ultimate bottom line—and is therefore the same kind of thinking that generated the pollution problem at the start.

Another possible solution to the air pollution problem is to tax companies based on how much pollution they release. This is the Polluter Pays idea, making companies accountable for their own waste so they evaluate existing practices to avoid paying high tax rates.

Another solution is to offer government subsidies and tax breaks to companies that are environmentally friendly, providing incentives to companies that are more environmentally responsible. Canada has recently imposed a carbon tax which has been pretty successful in curbing emissions. One example of this approach is shown by many companies that are buying and planting trees in order to offset their carbon footprint.

Indoor Pollution

Radon gas is a radioactive element that is found naturally in rock and soil. It is known to leak into homes through cracks or other openings in the foundation of homes and infrastructures. Radon gas is second to smoking for causes of lung cancer (a possible solution is to place plastic covers on house floors in areas that have high concentrations of uranium deposits). Other common sources of indoor pollution may be poorly ventilated stoves, tobacco smoke, and even the burning of wood and coal, all of which release particulate matter, nitrogen oxides, and carbon monoxide, to name a few. Tobacco smoke contains cancer-causing carcinogens. When homes are not ventilated properly, which is often the case especially in developing countries, the pollution is stagnant and is breathed unknowingly by the homes' occupants. For the above reasons, the Environmental Protection Agency considers indoor pollution the most dangerous type of pollution that exists today.

Soil Pollution

Millions of tons of trash are generated each year and placed into landfills. To prevent environmental contamination, the federally regulated landfills designate areas for disposal of hazardous waste. Such hazardous waste is produced often, believe it or not, from everyday household items.

By EPA standards, hazardous waste is classified as anything that is capable of causing harm, death, or illness to humans when released into the environment. Hazardous substances are typically corrosive, combustible, or toxic. **Corrosive items** are reactive chemicals that damage or compromise the surface of a material. Strong acids such as sulfuric acid, or battery acid, are examples of corrosive substances. **Combustible items** have a strong capability of igniting or burning at normal temperatures. Ethanol, solvents, and paint thinners are often combustible. **Toxic chemicals** may cause either long- or short-term harm to humans and animals when they are exposed to such chemicals. Examples of toxic chemicals are lead, mercury, and asbestos.

Landfills that are improperly managed or deteriorating may cause soil contamination when waste leaks out from the confines of the landfills. Soil can also be contaminated from chemicals and pollutants that get caught up in runoff. These chemicals often come from factories and agriculture farms. Some contaminants, such as BPA (commonly found in wastewater) may act as endocrine disruptors that compromise hormone production. Microorganisms found in wastewater are also known soil contaminants. Up to 25 million gallons of oil is spilled annually, killing plants and wildlife, among innumerable adverse impacts on human health and the environment.

DDT (dichlorodiphenyltrichloroethane) was a pesticide used in the 1960s and 1970s until it was banned in 1972. DDT was enormously effective in controlling the mosquito population. In the ecosystem, DDT may take up to 15 years to break down, so it has high **persistence** (amount of time it takes for a substance to break down). However, insects became resistant to DDT, and its efficacy decreased. Its toxicity to fish became increasingly problematic as well.

As it entered the food chain, it affected not just fish, but many birds such as the brown pelican and the American bald eagle. It compromised how birds use their body's calcium stores, resulting in thinner egg shells that could not sustain an incubating bird—the weight of the parent would crush the egg. Bald eagles became endangered, but aggressive preservation efforts proved successful and they were removed from the federal endangered species list in 2007. But the damage did not end there. DDT is a toxic chemical whose side effects in humans include cancer, liver problems, reproductive problems, and damage to the nervous system. It is still used in many tropical countries to control the spread of mosquitos.

The Montrose Chemical Corporation, a Los Angeles based company, started using DDT after World War II. Despite warnings and known health risks, Montrose did not stop using DDT until the early 1980s. It is estimated that they discharged over 1,500 tons of DDT waste into the Pacific where it settled in ocean sediment.

Many pesticides and other chemicals, such as DDT, do not degrade over time (high persistence) and are often fat soluble. These pesticides **bioaccumulate** in the food chain and are found in high concentrations in predators high up in the food chain. For this reason, toxic chemicals such as mercury and DDT are found in high concentrations in large predatory fish. This phenomenon is called **biomagnification**.

Figure 92.5   Many scientific studies have found high levels of lead poisoning in eagles

Water Pollution

Water that blankets the earth takes on various forms—from fresh mountain springs to underground aquifers, from saltwater to watersheds to wetlands. All these sources are major components of the hydrologic cycle. The Clean Water Act of 1948 was designed to control the pollution and contamination of U.S. water. Although strict measures are taken to procure clean water for our growing population, 40% of freshwater in the United States is still too contaminated for fishing or swimming. In 1974, the Safe Drinking Water Act was passed to ensure and protect the quality of drinking water—mostly from underground sources.

Some common biological water contaminants include pathogenic microorganisms, such as protozoa (dysentery), bacteria (such as cholera and fecal coliform bacteria), and viruses (polio and hepatitis). Natural materials such as sediments also act as contaminants. Sediments are nonbiological material derived from the weathering of rocks. Water pollutants also include mining wastes, such as acids, mercury, cyanide, and arsenic. Chemical disinfectants such as chlorine, sewage, runoff from factories and croplands, by-products from fire retardants, electronics, medicines, and corrosive household plumbing—all leach into our lakes and streams, making them unsafe for human use. They can also alter the pH of water (a measure of the buffer capacity of water), killing many organisms that have narrow habitable ranges. Methane from cow waste is another major cause for water pollution, and often ends up in bodies of water causing methane lagoons.

Pollution emitted directly into a surface through a confined, direct means such as a pipe, drain, or sewage is called **source pollution**, since it can be traced directly to its source. **Nonsource** pollution is water pollution that is not categorized as source pollution. Nonsource pollution tends to be more diffuse, less concentrated, and widespread. As runoff moves through a watershed, it picks up pollutants and deposits them into lakes and rivers, as well as underground sources of drinking water and surface water. There is no single point at which it is released, so it is considered nonsource pollution. Nonsource pollution can be sand, clay, and silt, but also human-made toxins.

Other forms of water pollution can be in the form of noise pollution—which can affect the sonar, communication, and life cycles of marine animals. In 1972, the Noise Control Act was passed. Enforced on a national level, it serves to protect the health and well-being of people, wildlife, and the environment by promoting a noise-free environment. Thermal pollution (heat) from global warming and power plants can also affect aquatic life, and so is also considered a form of water pollution.

WvdM/Shutterstock.com

Figure 92.6 Source Pollution

Chemicals and contaminants in groundwater (from http://water.usgs.gov/edu/groundwater-contaminants.html)		
Inorganic Contaminants Found in Groundwater		
Contaminant	**Sources to groundwater**	**Potential health and other effects**
Aluminum	Occurs naturally in some rocks and drainage from mines.	Can precipitate out of water after treatment, causing increased turbidity or discolored water.
Antimony	Enters environment from natural weathering, industrial production, municipal waste disposal, and manufacturing of flame retardants, ceramics, glass, batteries, fireworks, and explosives.	Decreases longevity, alters blood levels of glucose and cholesterol in laboratory animals exposed at high levels over their lifetime.

(continued)

Chemicals and contaminants in groundwater (from http://water.usgs.gov/edu/groundwater-contaminants.html) *(continued)*

Inorganic Contaminants Found in Groundwater

Contaminant	Sources to groundwater	Potential health and other effects
Arsenic	Enters environment from natural processes, industrial activities, pesticides, and industrial waste, smelting of copper, lead, and zinc ore.	Causes acute and chronic toxicity, liver and kidney damage; decreases blood hemoglobin. A carcinogen.
Barium	Occurs naturally in some limestones, sandstones, and soils in the eastern United States.	Can cause a variety of cardiac, gastrointestinal, and neuromuscular effects. Associated with hypertension and cardiotoxicity in animals.
Beryllium	Occurs naturally in soils, groundwater, and surface water. Often used in electrical industry equipment and components, nuclear power and space industry. Enters the environment from mining operations, processing plants, and improper waste disposal. Found in low concentrations in rocks, coal, and petroleum and enters the ground and	Causes acute and chronic toxicity; can cause damage to lungs and bones. Possible carcinogen.
Cadmium	Found in low concentrations in rocks, coal, and petroleum and enters the groundwater and surface water when dissolved by acidic waters. May enter the environment from industrial discharge, mining waste, metal plating, water pipes, batteries, paints and pigments, plastic stabilizers, and landfill leachate.	Replaces zinc biochemically in the body and causes high blood pressure, liver and kidney damage, and anemia. Destroys testicular tissue and red blood cells. Toxic to aquatic biota.
Chloride	May be associated with the presence of sodium in drinking water when present in high concentrations. Often from saltwater intrusion, mineral dissolution, industrial and domestic waste.	Deteriorates plumbing, water heaters, and municipal water-works equipment at high levels. Above secondary maximum contaminant level, taste becomes noticeable.
Chromium	Enters environment from old mining operations runoff and leaching into groundwater, fossil-fuel combustion, cement-plant emissions, mineral leaching, and waste incineration. Used in metal plating and as a cooling-tower water additive.	Chromium III is a nutritionally essential element. Chromium VI is much more toxic than Chromium III and causes liver and kidney damage, internal hemorrhaging, respiratory damage, dermatitis, and ulcers on the skin at high concentrations.
Copper	Enters environment from metal plating, industrial and domestic waste, mining, and mineral leaching.	Can cause stomach and intestinal distress, liver and kidney damage, anemia in high doses. Imparts an adverse taste and significant staining to clothes and fixtures. Essential trace element but toxic to plants and algae at moderate levels.

(continued)

Inorganic Contaminants Found in Groundwater

Contaminant	Sources to groundwater	Potential health and other effects
Cyanide	Often used in electroplating, steel processing, plastics, synthetic fabrics, and fertilizer production; also from improper waste disposal.	Poisoning is the result of damage to spleen, brain, and liver.
Dissolved solids	Occur naturally but also enters environment from man-made sources such as landfill leachate, feedlots, or sewage. A measure of the dissolved "salts" or minerals in the water. May also include some dissolved organic compounds.	May have an influence on the acceptability of water in general. May be indicative of the presence of excess concentrations of specific substances not included in the Safe Water Drinking Act, which would make water objectionable. High concentrations of dissolved solids shorten the life of hot water heaters.
Fluoride	Occurs naturally or as an additive to municipal water supplies; widely used in industry.	Decreases incidence of tooth decay but high levels can stain or mottle teeth. Causes crippling bone disorder (calcification of the bones and joints) at very high levels.
Hardness	Result of metallic ions dissolved in the water; reported as concentration of calcium carbonate. Calcium carbonate is derived from dissolved limestone or discharges from operating or abandoned mines.	Decreases the lather formation of soap and increases scale formation in hot-water heaters and low-pressure boilers at high levels.
Iron	Occurs naturally as a mineral from sediment and rocks or from mining, industrial waste, and corroding metal.	Imparts a bitter astringent taste to water and a brownish color to laundered clothing and plumbing fixtures.
Lead	Enters environment from industry, mining, plumbing, gasoline, coal, and as a water additive.	Affects red blood cell chemistry; delays normal physical and mental development in babies and young children. Causes slight deficits in attention span, hearing, and learning in children. Can cause slight increase in blood pressure in some adults. Probable carcinogen.
Manganese	Occurs naturally as a mineral from sediment and rocks or from mining and industrial waste.	Causes aesthetic and economic damage, and imparts brownish stains to laundry. Affects taste of water, and causes dark brown or black stains on plumbing fixtures. Relatively nontoxic to animals but toxic to plants at high levels.
Mercury	Occurs as an inorganic salt and as organic mercury compounds. Enters the environment from industrial waste, mining, pesticides, coal, electrical equipment (batteries, lamps, switches), smelting, and fossil-fuel combustion.	Causes acute and chronic toxicity. Targets the kidneys and can cause nervous system disorders.

(continued)

Chemicals and contaminants in groundwater (from http://water.usgs.gov/edu/groundwater-contaminants.html) (*continued*)

Inorganic Contaminants Found in Groundwater		
Contaminant	**Sources to groundwater**	**Potential health and other effects**
Nickel	Occurs naturally in soils, groundwater, and surface water. Often used in electroplating, stainless steel and alloy products, mining, and refining.	Damages the heart and liver of laboratory animals exposed to large amounts over their lifetime.
Nitrate (as nitrogen)	Occurs naturally in mineral deposits, soils, seawater, freshwater systems, the atmosphere, and biota. More stable form of combined nitrogen in oxygenated water. Found in the highest levels in groundwater under extensively developed areas. Enters the environment from fertilizer, feedlots, and sewage.	Toxicity results from the body's natural breakdown of nitrate to nitrite. Causes "bluebaby disease," or methemoglobinemia, which threatens oxygen-carrying capacity of the blood.
Nitrite (combined nitrate/nitrite)	Enters environment from fertilizer, sewage, and human or farm-animal waste.	Toxicity results from the body's natural breakdown of nitrate to nitrite. Causes "bluebaby disease," or methemoglobinemia, which threatens oxygen-carrying capacity of the blood.
Selenium	Enters environment from naturally occurring geologic sources, sulfur, and coal.	Causes acute and chronic toxic effects in animals—blind staggers" in cattle. Nutritionally essential element at low doses but toxic at high doses.
Sodium	Derived geologically from leaching of surface and underground deposits of salt and decomposition of various minerals. Human activities contribute through de-icing and washing products.	Can be a health risk factor for those individuals on a low-sodium diet.
Sulfate	Elevated concentrations may result from saltwater intrusion, mineral dissolution, and domestic or industrial waste.	Forms hard scales on boilers and heat exchangers, can change the taste of water, and has a laxative effect in high doses.
Thallium	Enters environment from soils; used in electronics, pharmaceuticals manufacturing, glass, and alloys.	Damages kidneys, liver, brain, and intestines in laboratory animals when given in high doses over their lifetime.
Zinc	Found naturally in water, most frequently in areas where it is mined. Enters environment from industrial waste, metal plating, and plumbing, and is a major component of sludge.	Aids in the healing of wounds. Causes no ill health effects except in very high doses. Imparts an undesirable taste to water. Toxic to plants at high levels.

Organic Contaminants Found in Groundwater

Contaminant	Sources to groundwater	Potential health and other effects
Volatile organic compounds	Enter environment when used to make plastics, dyes, rubbers, polishes, solvents, crude oil, insecticides, inks, varnishes, paints, disinfectants, gasoline products, pharmaceuticals, preservatives, spot removers, paint removers, degreasers, and many more.	Can cause cancer and liver damage, anemia, gastrointestinal disorder, skin irritation, blurred vision, exhaustion, weight loss, damage to the nervous system, and respiratory tract irritation.
Pesticides	Enter environment as herbicides, insecticides, fungicides, rodenticides, and algicides.	Cause poisoning, headaches, dizziness, gastrointestinal disturbance, numbness, weakness, and cancer. Destroys nervous system, thyroid, reproductive system, liver, and kidneys.
Plasticizers, chlorinated solvents, benzo[a]pyrene, and dioxin	Used as sealants, linings, solvents, pesticides, plasticizers, components of gasoline, disinfectant, and wood preservative. Enters the environment from improper waste disposal, leaching runoff, leaking storage tank, and industrial runoff.	Cause cancer. Damages nervous and reproductive systems, kidney, stomach, and liver.

Microbiological Contaminants Found in Groundwater

Contaminant	Sources to groundwater	Potential health and other effects
Coliform bacteria	Occur naturally in the environment from soils and plants and in the intestines of humans and other warm-blooded animals. Used as an indicator for the presence of pathogenic bacteria, viruses, and parasites from domestic sewage, animal waste, or plant or soil material.	Bacteria, viruses, and parasites can cause polio, cholera, typhoid fever, dysentery, and infectious hepatitis.

Some information on this page is from Waller, Roger M., Ground Water and the Rural Homeowner, Pamphlet, U.S. Geological Survey, 1982

How Do We Deal with Water and Soil Pollution?

Nanotechnology is one method used to prevent and to clean up water and soil pollution. Nanotechnology involves manipulating the environment on a molecular level. Current research suggests that nanotechnology is efficient in removing contaminants like chlorine from the soil; a fungus from the genus *Pestalotiopsis* has recently been discovered in the Amazon that is able to break polyurethane (a material found in rubber and plastic, and has very high persistence) down in anaerobic (no oxygen) conditions. Anaerobic conditions are similar to those at the bottom landfills. This is a breakthrough since many bacteria used in bioremediation require oxygen, and the study suggests that the fungus may be useful to break plastic down in landfills.[iii]

Green infrastructure is an overall approach to community living that integrates nature into developed communities. The infrastructure mimics natural processes and uses vegetation and soil to capture and enhance the hydrologic cycle to prevent excess runoff in urban areas.

Green technology is a means that uses natural organisms to clean up pollution.

Bioremediation is a form of both nanotechnology and green technology that uses microorganisms such as bacteria and protozoa to break down toxic chemicals in the soil and biotransform them into CO_2.

Dr. Morley Read/Shutterstock.com

Figure 92.7 **Bioremediation Pond for Soil Contaminated with Crude Oil**

Phytoremediation is a green technological practice used to remove toxic chemicals from the soil through plants, since plants have the ability to absorb heavy metals through their roots. A benefit of phytoremediation is localizing toxic chemicals in a controlled plant population. But care must be taken that they stay localized so plants do not hybridize with natural plant populations. Plants must later be removed in order to remove the heavy metals and other contaminants.

Waste Management

Each year, Americans generate millions of tons of trash from our everyday activities, whether we are aware of this or not. Almost half of all waste is considered either *agricultural* or *industrial* waste that comes from manufactures goods or the food we eat. Mining for ore and metals creates *mining waste*. And apartments, homes, small business, colleges, dormitories create *municipal solid waste*.

So how do we safely dispose of all our household and industrial items while maintaining ecosystem health—preventing air, soil, and water pollution? Long gone are the days when we used to drop our trash off at the open dump. These days, waste disposal is a lot more controlled and systematic. Thanks to stricter laws, our waste is now very carefully regulated so minimum damage is inflicted on the environment.

Most of our solid household waste goes into federally regulated landfills that contain areas designated for disposal of municipal solid waste. Such waste is monitored to ensure it is safe for disposal in our municipal landfills. Some of the items that get disposed are construction debris, industrial solid waste, municipal solid waste, and nonhazardous sludge. Hazardous household items, not safe for disposal, have their own drop-off site. Industrial wastewater that is contaminated with toxic chemicals cannot be disposed of by regular means, because it can contaminate groundwater and leak into the water table—so it is injected into deep wells. This sometimes leads to induced or man-made earthquakes.[iv]

Landfills are specifically selected for disposal in nonenvironmentally sensitive areas, which means that building on wetlands, floodplain zones, or fault lines, is prohibited (to prevent the aforementioned earthquakes).

Landfills are specifically designed and engineered to prevent pollution from leaching into the soil and contaminating the water table. This means that the site must have certain geological requirements such as a bed of clay soil lining the site (to protect groundwater from being contaminated). Landfills are constantly monitored for quality assurance, and groundwater is tested to make sure it is not contaminated. Landfills are periodically covered with soil to prevent odor and to prevent infestations from insects and rodents.

Huguette Roe/Shutterstock.com

Figure 92.8 **Bulldozer Working on a Landfill**

The EPA has outreach programs affiliated with some landfills to capture methane gas and carbon dioxide and eventually convert them into energy. This keeps carbon emissions down, improves air quality, and is a way of harnessing natural gas so that it is renewable.

According to the Environmental Protection Agency, of the material and trash that we throw out here in the United States, roughly 50% is discarded all together (meaning it ends up in landfills), 13% is combusted for energy (meaning it is incinerated), and 34% is composted or recycled. From this, we conclude that half of our waste is still being discarded! Landfills are filling up quickly as a result of our overwhelming generation of waste and refuse. Suitable sites for landfills are becoming increasingly scarce.

Logically, many people argue that we need to lessen our dependency on landfills and incinerators altogether. This happens when we cut down on the amount of trash and waste that we create and ramp up the amount of recycling. If we profoundly cut down on the amount of trash, we would also reduce our use of natural resources such as water, timber and minerals. We would also significantly reduce the amount of greenhouse gases and pollution associated with extraction of these natural resources.

We achieve less waste with the three 'Rs': Let us *Reduce, Re-use* and *Recycle* our waste.

Reduce: Reduce the amount of waste we create. Buy in bulk. Borrow or rent material that is not used often. This may include printers, ladders, bicycles, etc. This is also a great way to save money!

Re-use: Buy repurposed material or used goods like clothing, silverware, and building material. Buy products with less packaging. Repair or get maintenance done on appliances and electronics rather than buying them new. Donate your unwanted items and electronics to religious institutions, charities or thrift stores.

Recycle: Collection for recycling happens curbside or at deposit centers that offer refunds. Contamination can be high in recycle bins, so please take care that material is clean and sorted. When buying goods and products at the market, look for phrases such as: 'recycled-content product,' 'post-consumer product,' or 'recyclable product.' Many common household items such as paper towels, newspapers, nails, trash bags, cans, plastic bottles, aluminum and glass are increasingly being made with recycled material. The best way to support recycling is to vote with your dollar, purchase items that can be recycled or are made from recycled material.

Nonrenewable Energy
By Elizabeth Jordan

Did You Know?

Earth Day in the Gulf of Mexico, April 20, 2010, is when a deadly explosion occurred at the Deep Horizon Drilling Rig, releasing what would eventually become 5 million barrels and 206,000,000 gallons of crude oil into the Gulf of Mexico. You may be familiar with this event as the BP Oil Spill. Eleven people died from the explosion and the wildlife casualties—thousands of birds, fish, whales, and endangered sea turtles among many others—are staggering. The disastrous effects to the food web have been incalculable. Disturbing photographs of oil-soaked brown pelicans, formally a federally endangered species, were splashed against news sources as a macabre symbol of the devastation. Are deadly accidents such as the BP Oil Spill worth our dependency on oil as our major energy source?

We will be exploring different nonrenewable sources of energy such as oil and other fossil fuels. We will also be examining their pros and cons and the multiple issues associated with each. Keep in mind that fossil fuels used for transportation accounts for 13% of global greenhouse gas emissions while energy supply accounts for 26%.[i] Please try to examine how your everyday activities require energy and fossil fuel use and what you can do to reduce your carbon energy footprint.

Fossil Fuels

Fossil fuels have powered the United States for over a century and have been integral in the advancement of technology, health care, and modernization. They have also been profoundly instrumental in increasing the human carrying capacity. Fossil fuels have undoubtedly provided great value to economies, industries, culture, personal health, and individual interests.

However, there is a downside to using fossil fuel with which we are becoming increasing aware; especially as its use continues, expands, and evolves. As a growing population continues to place incessant demand on fossil fuels, it is becoming difficult to ignore the adverse effects of its widespread use. To put it bluntly, fossil fuels are dirty and cause profound negative impacts to the environment, wildlife, and on human health. They create air pollution, water pollution and are a major contributing factor to global warming.

We will be closely examining both the pros and cons of using fossil fuels and will also take an in-depth look at their impacts on human and environmental health.

Fossil fuels were formed millions of years ago as the by-product of either mostly dead plant matter (that would eventually become coal) or dead plant and animal matter (that would, for the most part, eventually become oil and natural gas). The remains of these living organisms have been exposed to excess heat and pressure over the years, forming what we recognize as coal and petroleum. Because these remains were once living organisms, coal and petroleum contain many hydrocarbons (carbon and hydrogen covalently linked into long molecules) as well as traces of sulfur, nitrogen, and other impurities. Hydrocarbons are found in liquid, solid, and gas form, as we will examine in subsequent paragraphs.

Fossil fuels are **nonrenewable** sources of energy because there is a finite amount of them on earth. Once their supply is used up, they cannot be regenerated in the amount of time necessary to facilitate our energy needs and consumption. Despite this, they account for 85% of all fuel use in the United States. The United States owns 3% of the world's oil reserves, but due to our high transportation demands (two-thirds of oil goes toward transportation), we consume a staggering 30% of it. Currently, between 50% and 70% of crude-oil supplies have been depleted. Half of the remaining reserves are in the Middle East.

Petroleum

Petroleum is a mass of hydrocarbons found in a *liquid* form. These hydrocarbons are the remains of ancient plants, animals, marine organisms, and other organic material buried and trapped over thousands of years. Their remains were exposed to very little oxygen, which resulted in slow decomposition. This process forms crude oil, or petroleum—oil with impurities. We use this energy source for heating, transportation, and creating plastic and petroleum jelly, paraffin, and other products.

Crude oil is found within pores inside rock. The drilling and pumping process draws out the oil. **Primary production** occurs during the initial drilling stages. Layers upon layer of rock put immense pressure on compressed oil, so during primary production, when the pressure is released through drilling, a large quantity of oil bubbles to the surface (**Figure 93.1**). Over time as the pressure starts to weaken, the oil flow ebbs and other more extensive measures must be implemented to extract oil. Drilling resumes, but now deeper into the ground using more energy and exertion, sometimes assisted by adding gas to expand and increase pressure, causing more oil to bubble to the surface. This stage is called **secondary production**.

iurii/Shutterstock.comOffshore Drilling

Figure 93.1   Oil rig

Crude oil is then shipped, via truck, rail, or pipeline, to refineries where it is distilled at a fractionating tower and purified to eventually become refined oil (**Figure 93.2**). Since our energy economy and our transportation are dependent mostly on oil, incentive is high to make this process efficient, so distillation technology is state-of-the-art. Oil is finally transported to its destination.

In the United States, prices of gas at the pump are artificially cheap due to large government subsidies. This means that gas prices do not contain environmental cleanup costs and all other adverse side effects from oil dependency. If you have traveled to Europe lately, you may have noticed that gas prices are almost three times as high as U.S. prices. Europe implements **full-cost pricing**, meaning gas prices include costs of cleaning up air pollution, oil spills, and all other side effects to the environment from oil. Some argue that if oil were more costly, or in other words, we were paying its "full" price up front; then we would use it more sparingly and look for alternatives.

Toman1111/Shutterstock.com

Figure 93.2 Oil Refinery

The Middle East contains the largest oil and natural gas reserves in the world and we rely on them for a large proportion of our oil. Being dependent on other nations for oil decentralizes our energy economy. Should international relations become tense with nations of **OPEC** for any reason, then the United States and any country dependent on them would be (and have been) in a precarious situation. Some Middle Eastern nations are not our allies so being reliant on them for our oil supply may cause unnecessary friction in international relations, putting our national security in jeopardy. Along with the many adverse environmental impacts, national security is another compelling argument to decrease our dependency on oil. It also leads to oil price instability, which is not favorable for the economy. Finally, at the rate we are consuming oil, it is expected that we have ample supply to last for the next 50–100 years. This number is an informed, but imprecise estimate, as records and information containing relevant data are unreliable.

Table 93.1

Pros of Using Oil	Cons of Using Oil
• Include: it is relatively cheap and readily available to meet our transportation and energy needs • we have an ample supply for 50 more years or so... • it is energy efficient • it is easy to transport from country to country • distillation technology is advanced.	• Include: national security is compromised • it is a nonrenewable resource that we cannot rely upon much longer • oil spills • oil prices are volatile and unpredictable, which can harm the economy • due to depleting crude-oil supplies, we are forced to resort to obscure sources such as tar sands containing **bitumen,** a semisolid and less pure form of petroleum • during secondary production, sediment accumulates in crude oil, making it more difficult to distill • drilling may cause subsistence and other subterranean abnormalities • large government subsidies (that we end up paying for) are costly • carbon emissions and other by-products produce air pollution • petroleum systems and natural gas are responsible for 29% of methane emissions in the United States[ii] • air pollution created through oil use has profoundly negative impacts to human health • developing nations that happen to be rich in oil supply can, and have been, exploited by wealthy nations looking to expand their drilling • energy economy is centralized.

Natural Gas

Found on coal beds, natural gases are hydrocarbons found in their gaseous form, mostly methane (CH_4), but sometimes natural gas is in the form of propane (C_3H_{12}) and butane (C_4H_{10}). Natural gas is the product of decomposed organic matter (such as marine microorganisms) that lived millions of years ago. This decaying matter eventually settled to the sea floor and became buried with sand, silt, and mud. In the presence of anaerobic conditions and an abundance of heat and pressure, the dead organic matter eventually turned into a hydrocarbon gas (methane, butane, or propane), or what we know as *natural gas.*

Because of the specific and controlled way natural gas was formed, it is often covered by porous sedimentary rock. Natural gas is low in density, and if rock is suitably porous, gas will naturally rise up out of the rock. *Conventional natural gas deposits* are associated with oil and make up 46% of natural gas resources. The rest—*unconventional natural gas deposits*—is associated with oil shale and coalbeds.[iii] Russia has the largest natural gas reserves, possessing five times that of the United States. Other countries with considerable natural gas reserves include: Iran, Qatar, Saudi Arabia, and Venezuela.[iv]

Natural gas is volatile and easily combustible, making it a valuable energy source. It is often harvested in a process called **fracking**. Fracking involves using water, sand, and chemicals to break up rock, build up pressure, and release natural gas from rock beds. It has allowed natural gas companies to access large quantities of natural gas in a relatively short amount of time. This is a controversial process since the chemicals used are potentially toxic, and therefore threaten to contaminate our land, air, and water supply. Furthermore, according to the USGS, fracking has on occasion contributed to earthquake activity (**Figure 93.3**).[v]

After it is pipelined to the power plant, natural gas is often heated to boil water and the steam is used by turbines to generate electricity. Natural gas contains impurities and must be refined to remove acid gases, such as hydrogen sulfide and carbon dioxide, and other toxic impurities such as mercury and cadmium.

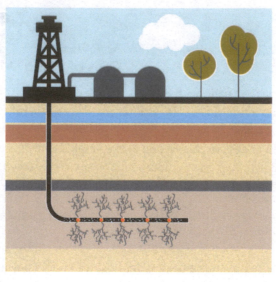

Figure 93.3 Hydraulic Fracturing is the Fracturing of Rock by A Pressurized Liquid

The U.S. Department of Energy is currently conducting research to examine how to utilize microbes for converting natural gas into a liquid form—without the input of more energy. This would ultimately lessen our reliance on petroleum.[vi]

Natural Gas in Landfills

Landfills are the third largest source of human-generated methane gas in the United States.[vii] The Environmental Protection Agency (EPA) has established an outreach program called Landfill Methane Outreach Program (LMOP) that recovers and reuses landfill gases—like methane—for energy. Methane is captured through a vacuum-like device and converted to electricity. This electricity can be sold to the power grid, used on-site, or used to fuel power plants, vehicles, homes, and neighboring towns. It is also used as a means for **cogeneration** (combined heat and water) in which heat is captured and used to produce steam.[viii]

The benefits of this outreach system are many. Currently 48 states use the program for heat, fuel, gas and electricity. It creates jobs through the marketing, design, implementation, and operation of the program. It also results in an overall decrease in greenhouse gas emissions because it uses up the methane instead of allowing it to descend into the atmosphere where it would ultimately act as a greenhouse gas. It would also theoretically improve the overall air quality by reducing the need for fracking or other fossil-fuel sources (coal and oil) which produce air pollutants such as sulfuric dioxide, nitrogen oxides, and particulates.

Critics of the Landfill Initiative are concerned about the amount of methane produced— which is a much more potent gas than carbon dioxide (20 times stronger than carbon dioxide at trapping heat over a 100-year period and 84 more times potent in a 20-year period). Since only a small fraction of the methane is utilized, much of it is released into the atmosphere or burned on-site. In fact, in the United States, landfills are responsible for 28% of methane emissions.[ix] This makes it hard to reconcile generating and using methane as an energy source. Critics say that perhaps we should be focusing on cutting down out amount of waste altogether and turning clean-burning renewable energy.

Table 93.2

Pros of Using Natural Gas	Cons of Using Natural Gas
• Include: it releases less CO_2 than oil or coal • it provides a high energy yield • a relatively ample supply is available • it is cheap and easily transported • it can be renewable if microbes in landfills are used to break down biomass • it can be used for cogeneration • because EPA standards are increasing, coal power plants have become more expensive to run, so there is a shift toward natural gas • fracking allows domestic fuel to be much cheaper than from other foreign sources • it uses less water and creates less water pollution than coal plants • it has greater flexibility with the power grid • it has given the economy a boost as we have shifted to a cheaper means of energy.	• Include: carbon emissions cause pollution • government subsidies are costly • it releases methane (a greenhouse gas) • fracking uses toxic chemicals and so causes air, water, and land pollution • fracking is also linked to earthquake activity • natural gas prices are volatile and may lead to a higher cost for electricity. This may harm the economy and, in extreme cases, cause industry to shift back to cheaper coal plants • Natural gas leaks methane gas, which is much more potent at trapping heat than carbon dioxide. Petroleum systems and natural gas are responsible for 29% of methane emissions in the United States[x].

Oil Shale

Unlike the other forms of fossil fuel, oil shale is still trapped in its original rock formation, formed from the accumulation of clay, silt, mud, and organic matter at sea beds. Oil shale is composed of compressed sedimentary rock that contains, not oil as the name suggests, but organic matter called kerogen. Unlike coal and oil that comes from dead plant and animal matter, oil shale is derived from decomposed cyanobacteria, marine algae, and other microbes. When it is heated, it releases gaseous kerogens that are combustible and may be used as an energy source. Oil shale garnered attention as we started looking for alternative fuels, but research was stalled due to the excessive energy used during extraction. In addition, little is known thus far about the environmental impacts of using oil shale as an energy source.

Table 93.3

Pros of Using Oil Shale	Cons of Using Oil Shale
• Include: it provides the benefits of using fossil fuels, but in a much shorter amount of time • it exists in ample supply • it is cleaner burning than other fossil fuels • advanced horizontal drilling and fracking allow access to large volumes of oil shale.	• Include: it is expensive • produces a large environmental impact such as air, water, and land pollution • experts are still unsure of how to deal with waste disposal.

Coal

Coal is a *solid* form of hydrocarbons that are the remains of plant matter. It was the major energy source of the Industrial Revolution of the 19th Century when pulverized coal was placed into coal power plants to produce energy. Today, coal is pulverized and burned in a furnace at coal power plants. Water runs through the furnace through a piping system, and then heated by means of the combustion process. Water is heated until it boils and creates steam, which is pushed through a turbine and produces electricity. The water cools and condenses and is placed back into the furnace.

Coal is currently one of the major sources of energy in the United States and worldwide. The Unites States' largest coal reserves are found in Wyoming and Appalachia. It contains many chemical impurities (noncarbon elements) depending on the type of coal (lignite, bituminous, anthracite—in order of increasing carbon purity); most coal also has hydrogen, nitrogen, oxygen, and sulfur (impurities) within its structure. Technology has progressed so as to reduce the amount of sulfur and nitrogen in coal as well as to limit its carbon emissions. Refining coal is not as intensive as refining petroleum. It is treated with water or chemical bath to remove impurities. This process can remove up to 30% of the sulfur content and reduces the amount of acid rain and other negative impacts that the combustion of fossil fuel has on the environment.

Coal seams are found either close to the earth or deep underground. Conventionally, coal is accessed by drilling into mountains and getting into greater depths by using explosives, which create major blasts, dispersing rock and debris. **Overburden** (rocky waste) is removed until a layer of coal is reached. This is called **surface mining** and is sometimes called *strip mining* (**Figure 93.4**). Often times, this entire process includes taking off the whole top of a mountain! The removal of overburden can be repeated several times or as needed to expose coal. The copious amounts of rubble and rocky waste are piled on to accompanying mountain valleys, causing lots of pollution and disturbance. The overburden contains toxic elements such as cadmium, lead, and mercury that can get absorbed into the water table, potentially and often contaminating it. Furthermore, sulfuric acid produced by this process can wash into the soil and water supply, resulting in **acid mine drainage**, a particularly toxic threat to the environment.

Another type of surface mining is **contour mining**, which is performed when coal seams are exposed on the mountain side. These mines are smaller but still generate considerable waste. **Pit mines** are created when when coal seams are deep in the mountain and deep pits must be dug to access coal.

Jim Parkin/Shutterstock.com

Figure 93.4 Surface Mining

Another type of mining is called **subsurface mining**. This form of mining is implemented when minerals are found deep in the ground, and mining usually carried out through the use of underground shafts (**Figure 93.5**). Sometimes tunnels are created to facilitate subsurface mining or even large rooms with pillars to support a shaky roof. Loose coal is collected and put on a conveyor. Sometimes dynamite is used to release coal. Subsurface mining comes with many risks: possible death or injury; and sink holes, or subsidence in the land.

An alternative to conventional mining is biomining, when microbes are used **in situ** to break down overburden and expose valuable minerals. This could potentially reduce the amount of waste associated with mining.

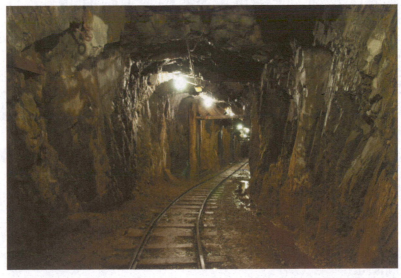

farbled/Shutterstock.com

Figure 93.5 Subsurface Mining

Table 93.4

Pros to Using Coal	Cons to Using Coal
• Include: it is energy efficient • coal emits less pollution and • CO_2 than oil • it is the cheapest to extract • it is safer and easier to ship • there is 10 times as much of it as natural gas and oil • we have an ample supply that could supply us with energy for a few hundred more years • the EPA has enforced stricter regulations on sulfur dioxide, nitrogen oxide, mercury, and particulates, producing less pollution and cleaner coal.	• Include: land disturbance and pollution is extensive, the extensive pollution and disturbance can destroy forest habitats and pollute streams • coal use places high demand on water resources • it releases sulfur dioxide that is linked to acid rain • it releases nitrogen oxide which is linked to acid rain • it releases carbon dioxide that is believed to be the leading cause of global warming • it releases carbon monoxide, a colorless, odorless toxic and often deadly gas • it releases toxic chemicals such as mercury and cadmium • it releases trace amount of uranium, which can produce radioactive emissions • it is responsible for 10% of methane emissions in the United States[xi] • it produces sink holes or subsidence and acid mine drainage • it causes black lung among other ailments, as mining practices take a human toll • carbon emissions can be catastrophic • because the EPA has enforces stricter regulations on sulfur dioxide, nitrogen oxide, mercury, and particulates, coal plants are more expensive to run.

The Impact of Fossil Fuel on Biodiversity and Human Health

We examined various pros and cons of fossil fuels above, but let us get a little more in-depth by examining how the use of fossil fuel specifically impacts ecosystem biodiversity and human health: As we mentioned above, the use of fossil fuels is linked to sulfur dioxide and nitrogen oxides that cause acid rain. This damages forests, buildings, and bodies of water. Fossil fuel use is linked to an increase in various types of air pollution which in turn can negatively impact agriculture and damage forests. This is especially damaging to primary growth forests which harbor a good amount of earth's biodiversity.

The use of fossil fuels has been linked to global climate change, which has many adverse effects on biodiversity, ecosystem function, and human health.

Using fossil fuels can lead to the risk of such catastrophic oil spills. Furthermore in 1989, the Exxon Valdez spilled 37,000 tons of crude oil in Prince William Sound in Alaska.[xii] This results in devastating loss of biodiversity. And the creation and expansion of oil fields encroaches on fragile ecosystems such as Tropical Rainforests and some that are typically used for conservation—such as the Arctic.

Fossil fuels produce particulate matter like soot smoke and dust. This air pollution may impact human health by increasing the risk of respiratory complications such as bronchitis and asthma, which can lead to long-term complications such as CPOD.

Nuclear Energy

In 1986, in Chernobyl, Ukraine (then, the Soviet Union), several explosions destroyed a nuclear reactor, releasing radioactive material, and contaminating the area for an estimated 100 years. As a result of such devastation, locals cannot drink the water or eat locally raised crops or meat from livestock. Dramatic increases in both cancer and birth defects have also plagued the area. Nuclear energy is a nexus of global controversy, not only because of the risk of accidents and subsequent health risks, but also because nobody can seem to agree how to effectively dispose of nuclear waste. Is the potential for accidents such as Fukushima (where the safety systems failed), Three Mile Island, and the Chernobyl disaster worth a dependency on nuclear energy? If so, what measures can we take to make the process safer and ensure public safety?

Nuclear energy provides 8% of the world's energy needs. France receives 80% of their electricity from nuclear power and the United States receives approximately 10% from nuclear power. Nuclear energy does not come from fossil fuel and is not produced from hydrocarbons. Rather, it is harvested from a process called **nuclear fission**. Nuclear fission produces 10 million times more energy than traditional energy sources such as coal or oil! This process involves the splitting of an atom, a radioactive element, such as uranium, with a heavy molecular weight that decays to become a radioactive **isotope**. This process is exothermic because it emits heat energy which is used to heat water, create steam, and generate electricity. The fission reaction must occur at a very specific rate, and must be stopped or adjusted when necessary. Large quantities of water are used in this process not only for steam generation but to act as a coolant as well. Water pressure and temperature must be heavily controlled to prevent a meltdown.

Uranium is collected by either surface mining or subsurface mining. After the mining process, it must be enriched to increase the concentration of the unstable isotope, U-235. Later, it is made into pellets and placed into fuel rods. Actual nuclear fission takes place in the reactor core—where the turbines generate electricity and condensers cool steam.

The average life span of a nuclear power plant is roughly 50 years. When they have reached the end of their run, nuclear power plants cannot be abandoned, they must be **decommissioned**. Radioactive waste must be carefully removed and placed in a storage facility for at least 100 years, then monitored for at least a thousand years to prevent radioactive leaks.

In 1982, the U.S. Nuclear Waste Policy Act was passed, positioning the Department of Energy in charge of nuclear waste disposal. The ideal site for disposal, according to The Act, must be geologically stable and located at a safe distance from groundwater. The site must be able to hold waste for tens of thousands of years, or at least until the waste is no longer radioactive. Currently the United States is stalemate concerning how to dispose of its radioactive waste (which is piling up at power plants). The consensus was that we would store our waste underground at Yucca Mountain in Nevada. As scientists started learning more about this landscape and the multiple risks involved, not to mention the resistance from concerned citizens, the project stalled and has yet to move forward.

Table 93.5

Pros of Nuclear Energy	Cons of Nuclear Energy
• Include: carbon emissions are low • a vast energy supply is conveniently produced in such a small amount of material, 1 kilogram of uranium produces similar amounts of energy as 100,000 kilograms of coal • land remains relatively undisturbed • without accidents, it is very safe • security and safety measures have increased significantly since September 11, 2001 • the technology is already established • some consider it green energy because it does not release greenhouse gases and so is not directly linked to global warming.	• Include: lending institutions (for construction purposes) need loan guarantees because they are so expensive to build and risks are high • power plants are more expensive to build than coal power plants • there is great disturbance and water contamination associated with mining for Uranium • Nuclear power plants are vulnerable to natural disasters • knowledge use by terrorists would be devastating • accidents and spills prove to be cataclysmic • no clear solution on how to dispose of waste is evident • the net energy yield is low compared to coal • the risk of reprocessing or removing plutonium from nuclear waste could cause a security threat since it is used to build nuclear weapons • concerned scientists and critics believe that the NRC (Nuclear Regulatory Commission—the agency that oversees safety of the power plants) is too lenient and therefore ineffective • some argue that the standards for relicensing are too loose • tritium (radioactive hydrogen) has been known to leak from nuclear reactors and into the water table (tritium is linked to cancer) • some argue that the risk/reward system is distorted since an accident or nuclear meltdown can eradicate all benefits of using nuclear power. It is estimated that the costs to clean up Chernobyl, for example, far outweighed any economic benefits.

The Impact of Nuclear Energy on Biodiversity and Human Health

The construction of nuclear power plants releases small amounts of greenhouse gases during its initial construction, so impacts would include most of the ecological impacts of using fossil fuels; Water used to cool the reactors is dramatically above ambient temperatures. This may accelerate or compound the ecological impacts of climate extremes, and people and wildlife exposed to radiation can suffer disease and genetic changes, and in extreme cases death. Tritium leaks and exposure to tritium can lead to cancer.

On a lighter note, because of the significant risks associated with nuclear power plants, the surrounding area usually serves as protected areas for wildlife conservation and research. Provided there are no radiation leaks or meltdowns, this would be beneficial to the ecosystem and advancements in science.

How Do We Make Nuclear Energy Safer?

Let us examine some final thoughts on how to make nuclear energy safer for the general public. Some possible course of action may include enforcing strict earthquake and fire regulations for nuclear power plants, especially for older and aging facilities. We also need to design a solid plan of action in the case of disasters like floods—especially in coastal areas. We should determine a safety storage plan once and for all and remove the accumulating waste from the facilities, and devise rigorous security plans to make these power plants less vulnerable to terrorism. And finally, the NRC must enforce stricter regulations and their oversight should be consistent and thorough.

Are You a NIMBY?

Have you ever heard or uttered the phrase, "Yes, but not in my backyard?" (Not In My BackYard?) If you have, or have supported this notion in any way, then you might be what people may refer to as a 'NIMBY.'

What exactly does this mean? Often times, people will support a particular cause or action, even though there may be possible environmental or health risks associated with it, because the benefits seemingly outweigh the costs.

For instance, a person may be pro-Nuclear Energy, or supportive of a pipeline that transfers oil, or in favor of the concept of hydraulic fracking. They may back all of these as possible purveyors of energy—UNLESS the nuclear reactor is in *their* town, unless the pipeline goes through *their* backyard, unless the hydraulic fracking happens in *their* community. It is acceptable to carry on as long as it does not affect their families directly, or the health of their family directly. If these are in fact happening 'in your backyard,' or when one's family is directly affected, the costs then outweigh the benefits. In other words—"not in my backyard."

If it is health or environmental risks that people object to affecting them in the "not in my backyard" scenario, then perhaps it is time to take an honest assessment of the situation. Concede to the notion that it will, in fact, affect another person's family. Is this fair and just? Why is it acceptable to subject another family to the same risks that you or your family is adamantly opposed to? If you take the, "Yes but not in my backyard" approach, then perhaps in fairness you should reconsider your stance on the matter (as long as it is a real risk and not just a cosmetic concern).

Ecuador is also rich in oil supply and happens to be an OPEC country. Present day, oil is Ecuador's number one export, followed by bananas, shrimp, cacao, coffee, and tourism. Oil exploration in the Jungle of Ecuador began in 1950—commencing with seismic activity (explosives). In the year 1960, the first oil wells were drilled. From this point forward, oil companies expanded their quest throughout the Amazon Jungle in search of more oil.

Oil exploration has resulted in the construction of cities in the otherwise pristine Jungle, as there is great monetary incentive for people to relocate and work in what was once uninhabited land but now rich with oil. Currently, the target areas for oil drilling are home to Indigenous tribes that live in isolation (such as the Tagaeri Taromenane and the Waoranis).

This is an obvious reason to pause and develop a deeper understanding on the gravity of what is happening below the surface in the jungles of Ecuador. This is a multi-tiered issue that is complex and carries with it many environmental, social, and socioeconomic considerations.

Let us explore some pros and cons:

Some possible pros and arguments *in favor* of oil exploration in Ecuador include the generation of money and revenue for public interests such as roads, hospitals, schools in rural areas, and academic scholarships for higher education; the creation of jobs and subsidies for the elderly who need monetary assistance. Because oil is so lucrative and brings money in to the economy.

There are many cons to the issue that need to be addressed as well. Such cons fall into both the environmental and social considerations. Let us first explore the environmental cons; oil exploration causes a lot of air pollution because the natural gas is burned all year long in many locations. The vehicles and machinery used cause major land disturbance; ground pollution is a result of oil extraction and oil spills that occur during different stages of the extraction process (extraction, transport, and waste) and the disposal by oil companies often does not always use these safety measures; perforation water, which is reinjected into the ground, contains a lot of toxins that pollute water and contaminate drinking water that plants and animals depend on. The creation of roads is one of the main impacts because access roads invite illegal loggers, illegal hunters and settlers who are attracted to the resources, causing big impacts to the environment and the local communities; Unsustainable agriculture (such as the African oil palm) is one of the main causes for the jungle being cut down for crops (enabled through the construction of access roads) causing an overall loss of biodiversity.

Negative social ramifications include: a loss of local cultures because their life is changing by exposure to material and alcohol—the closer natives are to roads and cities, the worse it is. In addition, health problems to local people have been increasing (possibly because the air and water is polluted causing respiratory problems, cancer, etc.) And finally, there is an observable change in traditional activities of the indigenous people. They have stopped hunting animals in favor of grazing animals such as chickens and pigs.

Renewable Energy
By Elizabeth Jordan

Did You Know?

In the northern hemisphere, the south-facing side of homes and buildings receives the most net sunlight throughout the day. Many homes today are being built to orient windows to the south-facing side, maximizing the amount of sunlight throughout the winter to capture the sun's radiant energy for warmth. These homes also deliberately have few east- and west-facing windows in order to shield the home from the sun in hot summer months. This clever architectural design is quickly gaining popularity and being used in order to save on escalating heating costs as well as participate in the green movement and reduce carbon emissions and energy bills.

Why Renewable Energy?

Renewable energy is defined as energy originating from a source whose supply is indefinite or renewed (almost) as quickly as it is used. Most renewable energy supply comes either directly or indirectly from the sun. Solar energy, as illustrated in the above example to heat homes, is one example of renewable energy, but in this section, we will be exploring several other sources of renewable energy as well.

Benefits of renewable energy can be categorized with political, economic, social, or environmental considerations. From a political standpoint, using renewable energy improves our national security by relying less on **OPEC nations**. Reducing our dependency on these countries might potentially improve national safety by avoiding unnecessary conflict. Using more renewable energy sources also decentralizes our energy economy and helps reduce our financial debt to oil-exporting countries. It creates new jobs and a new economy around renewable energy and its science, innovation, sales, technology, and marketing strategies. Renewable energy is reliable and is becoming more affordable with the availability of tax cuts, rebates, and reasonable financing options. It provides increasingly affordable energy with stable and predictable energy prices, which is beneficial for the economy. The shift is becoming socially popular and gaining momentum within the green movement.

Environmentally, it burns cleaner and reduces carbon emissions and pollution. Fossil fuel use has been linked to cancer, breathing problems, heart attacks, and neurological damage. Shifting away from fossil fuels and toward renewable energy therefore improves public health by improving overall environmental quality and potentially reducing overall healthcare costs.[i] And by using renewables, we do not have to worry about running out of energy supply. About 25% of the world's energy supply is renewable and 18% of the United

States' energy supply is from renewable sources. Renewable energy is the fastest growing energy sector, with growth projected at 2.3 % a year. Most of this growth is in the form of solar and wind energy (Source eia. gov). Many communities that would like greener energy or more energy independence are moving toward **MUNICIPAL AGGREGATION or CCA's** - Community Choice Aggregation. Communities with CCAs choose their power supplier which gives them more control over energy sources and increases competition among energy suppliers.

Solar Energy

Sunlight does not take up space or have mass, so it is not considered **matter**. Rather, sunlight is classified as **electromagnetic energy** that can be captured, transmuted, and used for our energy needs. Some of its many uses include heating our homes and heating our water supply, as well as generating electricity. (**Photons** are light acting as matter-like particles; but for simplicity, we will consider light as having wave-like properties.)

Active and Passive Solar Energy

According to the legend, the Greek scientist Archimedes used a collection of solar reflectors to concentrate solar energy and use it to burn Roman ships during the 212 BC Battle of Syracuse. This feat has been reconstructed recreationally by volunteers and proven to be a possibly viable (albeit diabolical) endeavor. Historians are dubious of the historical accuracy regarding this event; however, it reminds us that there are many creative and innovative uses of solar energy.

Solar energy is growing, and it is growing rapidly as it becomes more affordable and accessible and efficient. In the United States, solar jobs have increased 123% since 2010, with over 200,000 workers (Source: National Solar Jobs Census).[iii,iv] Solar is not only providing energy for residential homes; many companies and industries are shifting toward rooftop solar cells to improve their reputation, company profile, marketing, and for reducing their operating costs. In 2017, PV accounted for 55% of newly installed energy capacity, more than nuclear energy and fossil fuels combined (Source: REN21).

Active solar energy is a process that uses conductors such as metal to trap sunlight. These conductors are usually a dark color, since dark colors absorb more wavelengths of light than lighter colors. This ensures that more electromagnetic energy is available for energy use. Sunlight is then transferred to a water supply, surrounding air, or pipes to heat homes. The heat may also be stored in a liquid form to be used later. We will be examining the different types of active solar energy in the following paragraphs.

Concentrated Solar Power (CSP) is a form of active solar energy that uses hundreds and sometimes thousands of mirrors to concentrate sunlight and amplify its reflective properties. A typical CSP farm is roughly the size of 200 football fields. The largest CSP plant is located outside Boulder City, NV, and was constructed in 2007 (**Figure 94.1**).

In using CSP, a solar collector is used to convert sun's energy to steam through heating a mineral oil fluid, which is eventually transferred into electricity with use of steam turbines. Solar receivers track the sun's movement to maximize energy production. It is a relatively simple process, and because of its simplicity, it generates renewable energy in quantities that are increasingly becoming comparable to coal or other fossil fuels energy sources.

Figure 94.1   **Concentrated Solar Power**

Another form of active solar energy is **photovoltaics**, often referred to as solar cells. Smallscale PV systems comprise the majority of solar installations and typically found on residential rooftops. Along with CSP, large-scale PV systems are a huge component of solar energy's generation of electricity. Photovoltaics use elements like silicon to capture sunlight and convert it into electricity right in the cell. During this process, solar energy energizes electrons on the silicon collector and the electrons flow to produce energy. The solar panels are connected to a circuit that provides a path for the flow for electrons, providing an electricity supply (**Figure 94.2**).

A concern for many people is how the amount of sunshine, or lack thereof, would impact the availability of solar energy. Is solar energy effective in places that do not experience an abundance of yearly sunshine? This is a reasonable concern, and it turns out that many solar systems are energy effective all over the globe, even where there is not an abundance of sunshine. A solar panel in Portland, Maine (which experiences long harsh winters) for example, would generate 95% as much power as the same system in Miami, and 85% as much power as the same system in Los Angeles.[v] This is a very generous amount of energy in comparison.

When businesses or homes generate more electricity than they use, often the electricity is placed back to the grid.

Figure 94.2   **PV Solar Cells**

If you have ever stepped into a greenhouse on a sunny day, you are aware of the powerful energy of the sun. Greenhouses are an example of **passive solar energy**—or a means of harnessing sun light without the use of a collector or solar cell. Passive solar energy typically uses natural materials such as water, stone, and brick to store the sun's heat (and typically passive solar energy uses no moveable parts). The contextual example at the beginning of this section—houses architecturally designed to maximize sunlight—is another example of passive solar energy.

Table 94.1

Pros of Using Solar Energy	Cons of Using Solar Energy
• Include: it lessens our dependency on oil and OPEC nations (many of whom support and fund terrorism) • it decentralizes our energy economy • it is renewable • it is not very costly • it creates jobs • it is convenient for those who live in rural areas, not in close proximity to the energy grid • solar panels are relatively easy to install • it has no carbon emission and little pollution • tax credits and rebates are available • reasonable financing is now available, PV rooftop installations are becoming more affordable and more cost efficient • unlike nuclear energy and coal, PV systems do not require water to generate electricity • CSP can store sun's energy as heat and may use it to make electricity even when the sun is not shining • price of solar energy is stable and predictable.	• Include: the energy transfer is not 100% efficient and so energy yield is not high • sunlight may not be available at peak times of need • fuel cells can be expensive to install in homes • solar farms that are constructed in the desert or other fragile ecosystems cause large disturbances and compromise wildlife • water is used to manufacture PV solar components • PV manufacturing involves some toxic chemicals such as sulfuric acid, nitric acid, and hydrochloric acid. If not disposed of or handled carefully, these chemicals could be harmful to human and environmental health • although there are no fossil fuels associated with the generation of electricity, there are fossil fuels associated with other stages of the solar energy life cycle, such as manufacturing, installation, maintenance, materials transportation, decommissioning, and dismantling • solar farms often produce more electricity than is needed at off-peak energy times, and the U.S. power grid is limited in its storage capacity • solar collectors use mirrors that focus sunlight onto a receiver. This focused concentration of heat in the desert can accidently cause burns to birds and other wildlife.

Wind Energy

The sun's unequal heating of earth's surface causes a difference in pressure gradients which we experience as wind. Wind energy has been used by humans for 2000 years for various purposes such as grinding grain and pumping water. In the United States, between 1870 and 1930, farmers used wind energy to pump water. With the arrival of the "New Deal" in the early part of the 20th century, windmills phased out, and grid-connected electricity was introduced to the countryside and rural areas. Wind energy in Europe has been more long term and consistent than that of the United States.

majeczka/Shutterstock.com

Figure 94.3 **Wind turbines**

Wind energy is one of the cleanest, cheapest forms of energy, and is one of the fastest growing sources of electricity in the entire world. In the year 2000, there was 2.53 GW of wind energy installed across the US. In 2013, it was 60.72, and 404.25 is projected for the year 2050. The state of California is currently working toward its goal to make 100% renewable by the year 2045. of its energy renewable by the year. To reach this goal, Southern California Edison is overseeing and making contracts with private construction companies erecting wind farms in the desert to connect to its current electrical system, providing neighboring metro areas with renewable wind energy.

Wind farms are comprised of wind turbines (**Figure 94.3**). Wind turbines have rotating blades, or sails, that when rotate, produce an electrical current that is transformed into electricity. Turbines are either horizontal-axis turbines or vertical-axis turbines. They can vary on size depending on whether they meet residential needs or are part of a wind farm. Off shore turbines tend to be large and generate the most power. Turbines typically have three parts: the *sails*, the *tower*, and the box behind the blade called the *nacelle*. The nacelle is where motion is transformed into electricity. Transmission lines carry the electricity to the power grids close to the urban areas where energy needs are high. Wind farms are typically built where there are high wind speeds—such as the desert or in natural wind tunnels. Unfortunately, wind tunnels are also often wildlife corridors, passages that connect one parcel of a species' land with a contiguous parcel of land. This can disrupt wildlife corridors and species' range patterns. Wind farms are more productive in oceans where wind speeds are far greater than that on land.

But off-shore wind farms are not without controversy. High-value property owners do not want their ocean view compromised by off-shore wind farms.

Table 94.2

Pros of Using Wind Energy	Cons of Using Wind Energy
• Include: it lessens our dependency on oil and OPEC nations • it decentralizes our energy economy • it is renewable • it is the cheapest of all renewable energy • moderately high energy yield • stable energy prices • can be built offshore • possibilities of multiple use—wind farms can also be used for agriculture • it has no carbon emission or other pollution.	• Include: energy transfer is not efficient • wind farms that are constructed in the desert may cause a disturbance to the ecosystem • they disrupt birds' migratory patterns • transmission lines attract crows that are pests to sensitive species • bats spontaneously die when they come in to proximity to wind farms • wind farms often produce more electricity than is needed at off-peak energy times, and the U.S. power grid is limited in its storage capacity • they cannot be built near military base • they can be lethal to eagles and birds of prey • cannot have wind farms near human settlement for risk of blades falling off at high speeds and causing major damage or injury • turbine syndrome—not yet recognized by the CDC (The Centers for Disease Control and Prevention)—people living near wind farms claim to experience nausea, headaches, and dizziness. It is perhaps caused by the constant hum or the physics of the turbines. More research needs to be conducted regarding this[ix] • loss of economic of value of land located near wind farms • NIMBY—most people do not want a wind farm in their back yard.

Geothermic Energy

Geothermic energy is one type of renewable energy that cannot be traced back to the sun. It makes up 0.8% of the world's renewable energy source. Deep underground in the earth's core, radioactive decay of isotopes (typically Uranium or Potassium) emits tremendous amounts of energy and heat with temperatures up to 9,000°F! At such extreme temperatures, earth's rock (**mantle**) becomes liquid **magma**. Eventually, magma flows to the earth's surface heating subterranean water and creating geysers and hot springs. This intense heat energy can be harvested as an energy and electricity source.

Places of tectonic plate boundaries are ideal for geothermic energy, or where earth's crust is thin and allows heat to penetrate. The country of Iceland lies on a fault line and experiences a relatively high amount of volcanic activity. This geography makes Iceland ideal for creating geothermic energy power plants and using geothermic energy as a primary energy source. Geothermal plants account for 25% of the electricity production in Iceland (and El Salvador for that matter),[x] and Iceland receives 50% of its energy from geothermal (for heating homes and water.)[xi] China and Turkey are the 2 countries that rely most on geothermal (up to 80% according to iea.org) In the United States, some ideal areas for harvesting heat from earth's crust include Oregon, Alaska, parts of Nevada, and California. As a result of geothermic technology becoming advanced and more suitable sites being discovered and utilized, this type of renewable energy is growing. Although the US has large geothermal resources, it is not growing as quickly as solar or wind energy.

But if the geography is not ideal for geothermic energy, then capturing heat from deep in the earth's crust and transferring it, even with new technology may be a tricky endeavor and the idea abandoned and replaced with other renewable resources.

Geothermic power plants are used to convert the energy in earth's crust into electricity. So how does this work exactly? Cool water is pumped into the earth's crust, picks up earth's heat through a convection process, and brings heat to the surface. Or wells are drilled into an underground reservoir that can be up to 700 degrees Fahrenheit (Source: www.eia.gov). Heat and steam cause turbines to spin, generating an electrical current that connects to the power grid.

There are even still more ways to gain access to earth's natural subterranean heat stores. More than half a million homes around the world have a **Geothermal heat pump** that enables people to use less electricity. The earth's temperature underground remains consistently at 55°F. Engineers place air or antifreeze-filled pipes underground that naturally takes on the same temperature as the ground. The air or antifreeze is eventually pumped and circulated throughout homes with the increased temperatures. Granted, 55°F is still a relatively chilly temperature and people tend to feel most comfortable at 70°F. Indoor air temperatures need to be heated up from 55°F to a comfortable temperature of about 70°F, so additional energy would be used to cover the difference (ultimately lessening reliance on electricity).

Table 94.3

Pros of Using Geothermic Energy	Cons of Using Geothermic Energy
• Include: it lessens our dependency on oil and OPEC nations • it decentralizes our energy economy • it is renewable and not very costly • it is energy efficient • technology is developing and becoming more available • it causes little carbon emission and little pollution and pumps are available all day, every day.	• Include: energy from geothermic power plants must come from a relatively local source, otherwise it becomes expensive to transport • relatively few suitable environments exist • toxic gases like hydrogen sulfide may be released into the air, accompanied by a strong unpleasant sulfur odor, and some other air pollution • arsenic and minerals may be released into the steam • salt can build up in pipes.

Impacts of Using Geothermal, Wind, and Solar Energy on Biodiversity and Human Health

We discussed some general pros and cons of using geothermal, wind and solar energy above. Now let us get a little more in-depth in examining the specific impacts that these energy sources have on biodiversity and human health.

First of all, the expansive photovoltaic and wind farms require a large land mass, and so compete with agricultural land, protected wildlife areas, and forests. Solar farms mean less available land for wildlife, forests, and agriculture and can lead to land degradation. The disposal of wastewater from geothermal plants may contaminate groundwater and freshwater supplies affecting human and wildlife populations; and toxic chemicals that are used for the creation of solar cells can be harmful if they are handled improperly, or if they leak into the ecosystem during either use of solar cells or disposal.[xiii] These toxins released into the environment may cause various human health problems.

Water Power/Hydroelectricity

Created by the Hoover Dam, Lake Meade in Nevada supplies electricity and water supply to California. With high water demand, increasing global temperatures, and increased evaporation, Lake Meade's water level has decreased dramatically since the 1970s (**Figure 94.5**). As a result, the turbines have become increasingly inefficient at generating electricity. Therefore, fossil fuels have been used increasingly as an alternative energy source to hydroelectricity. Burning of fossil fuels effectuates the increase in global temperatures. This results in a positive feedback loop of less water and more fossil fuels.

Hydroelectricity is generally generated through dams which, like the Hoover Dam, contain turbines. When water pressure becomes high, the pressure spins the turbines, generating a flow of electrons that creates electricity. Hydropower stations can come from flowing water (rivers), stored water (reservoirs), or recycled water from reservoirs. Water turbines were a 19th-century invention. This technology was seminal in advancing the efficiency of water power. By the 1920s, 40% of the world's energy came from water. Not too long after, the emergence and efficiency of fossil fuels caused hydroelectricity to lose its edge in the energy economy. Present day, 4.5% of the world's energy supply is from water, and countries such as Norway and Costa Rica use it as a major energy source. Globally, it the leading source of renewable energy. Countries that generate the most hydropower are the US, China, Brazil, Canada, Russia and India according to worldenergy.org.

Heide Hellebrand/Shutterstock.com

Figure 94.5 **Lake Meade in Nevada**

Dams are not the only source of hydroelectricity. The gravitational ebb, flow of the ocean's tides, and the strong resulting energy pull are also used as a means to generate electricity. And finally, steam engines were a big power source of the industrial revolution, powering factories and trains. In steam engines, water is heated to its gaseous form, steam, which enters the engine and is expanded through a turbine or piston. The "choo choo" you hear coming from a train is steam being released as exhaust.

Table 94.4	
Pros of Using Water Power	**Cons of Using Water Power**
<ul><li>Include: it decreases our dependency on oil and OPEC nations</li><li>it decentralizes our energy economy</li><li>it is renewable, so not very costly</li><li>it has less carbon emission or other pollution than fossil fuels</li><li>high energy efficiency</li><li>dams can have multiple uses—flood control, electricity, and recreation—and have a long life span. Hydropower is the most versatile renewable energy. It is able to meet peak demand and it is consistent and predictable.</li></ul>	<ul><li>Include: it is costly to build</li><li>it has high environmental impact</li><li>loss of land can result if reservoirs are created</li><li>the construction of dams can potentially cause masses of people to relocate from their homes</li><li>high evaporation rates (as observed in still water) causes water loss and seepage into rock beds</li><li>salinity increases with evaporation</li><li>sediment build up, causing dams to become less functional over time, and resulting in a less volume of water in reservoirs</li><li>fish spawning and other ecological processes become impacted.</li></ul>

Impacts of Using Water Power on Biodiversity and Human Health

There is an abundance of rotting vegetation in dam reservoirs, and this leads to the emission of greenhouse gases; the construction of dams leads to the destruction of other natural habitats such as forests and grasslands; and disruptions to river flow—dam construction can alter the availability of freshwater to humans and wildlife (depending on the dam, it can either increase or decrease).

However, the construction of dams can also inadvertently lead to the creation of wetlands which act as conservation areas for various species of waterfowl and fish.

Biomass

Biomass, at 79%, is the most abundant source of worldwide renewable energy and 11% of all energy sources worldwide. Biomass constitutes 1.5% of electrical sales in the United States, and in the United States, it made up more than 35% of renewable energy generation in 2009. Biomass energy was surpassed by wind energy as the major renewable energy source shortly thereafter.[xiv]

Biomass is considered an important component of the carbon cycle and is the remains of what was once living organisms. In the form of anything ranging from wood to corn stalk to algae to animal waste, these remains can be burned for fuel.

Why is Biomass an energy source? Through photosynthesis, the pigment chlorophyll captures sunlight to make carbohydrates—a complex organic compound containing Carbon, Hydrogen and Oxygen. These carbohydrates store a lot of potential energy in their bonds. When they are burned, they release the energy that was initially captured from the sun—and release water and carbon dioxide back into the atmosphere. If done correctly, this form of energy can grow quickly and produce little waste and use little water. It can reduce air pollution and carbon emissions that would otherwise be created through fossil fuel use.

Some examples of biomass energy include: *energy crops* that can be grown in large quantities and also are able to grow on marginal soil. Growing them on marginal soil would keep the energy crop from directly competing with agriculture for land. Many perennial grasses can be used, and sometimes can be harvested for up to ten years before being replanted. Many grasses, like switchgrass that grows in the Midwest and the South and on the Great Plains, is drought resistant, flood resistant, and can grow on poor soil (**Figure 94.6**). This means that it does not compete with agriculture and is ideal to grow on marginal soil. It addition, it does not require too much fertilizer.[xv] Today switchgrass is mostly used for soil cover and livestock feed, but can also be used as biopower. *Crop residues* include biomass left over from agriculture is often collected and used as biomass.

Manure—or waste from livestock—has many uses. It contains nutrients and minerals (some cultures cook with it!), and so helps maintain soil fertility by putting it back into the soil. In some cases, it can be used as an energy source through gasification, anaerobic digesters, or combustion. *Woody biomass* is byproduct of milling timber such as sawdust and bark and is often used to power mills. *Forest Debris*—such as branches, limbs, tree tops, dead trees, left over from timber milling can also be used; trees that take a short time to grow and easily rotated can be used as biofuel. And *urban waste*: waste from construction wood, tree trimmings, methane from landfills, and biodegradeable garbage are all possible candidates for bioenergy.

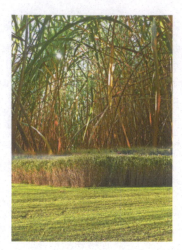

hjschneider/Shutterstock.com

Figure 94.6 Switchgrass

Bioethanol is a form of energy derived from biomass. Bioethanol is made from sugarcane or other plants with high sugar/starch content. Microorganisms **ferment** by feeding on the sugar and produce bioethanol that is used as fuel and even to power vehicles in some parts of the world. Catalysts for this process include yeasts and bacteria. Biodiesel, another fuel, is biomass made from plants with high oil content such as soybeans. Biodiesel is the only biomass fuel to meet the standards of the **Clean Air Act** and to be officially approved by the EPA (Environmental Protection Agency).

Ways that Biomass is Converted to Energy

In the above paragraph, we discussed the many potential and various sources of biomass. Traditionally, biomass was used to produce steam power, but that has changed with advances in technology and now has many different uses. Now we will briefly examine an overview of new various ways that biomass is utilized for energy:

Direct combustion: This is the most common and the oldest way to use biomass to create electricity. This involves burning biomass to produce steam. Steam is used to spin the turbines which generate electricity.

Repowering: is the ways in which coal plants (and natural gas plants) are converted to run entirely on biomass.

Anaerobic digestion: In a controlled setting, microorganisms can be used to break down biomass to produce methane and carbon dioxide. Methane can be used for heat and power.

Combined Heat and Power (CHP): CHP uses the heat from the combustion of biomass to power industrial processes or heat buildings located in close proximity.

Co-Firing: This involves using biomass energy and mixing it with coal energy. This can reduce harmful emissions such as mercury and sulfur, and reduce operating costs because it only requires small adjustments to already-standing coal operating facilities.

Biomass gasification: This is when biomass is heated under controlled conditions and in the presence of Oxygen. Regulated and under pressure, it can be converted to syngas, a mixture of carbon monoxide and hydrogen. Syngas is typically refined to remove impurities and is put through a turbine to generate electricity.

Energy density: Most biomass has a large water mass content, and so is less concentrated in potential energy than fossil fuels. This means that per pound, biomass carries less energy. To make it more efficient, biomass is often dried up and made into pellets to increase energy density and efficiency.

Impacts of Using Biomass on Biodiversity and Human Health

Burning wood indoors can cause many respiratory and health issues. The agricultural demands for using biomass create the need for heavy machinery, pesticides, and herbicides. This requires the use of fossil fuels on the onset, and so all of the negative ecological impacts of using fossil fuels that we discussed would be factored in to using biomass power. Direct combustion can waste a lot of energy and create pollution. The expanding need for land to grow monocultures (such as sugarcane and fast-growing trees) takes habitats away from possible cropland and other agricultural land. Clearing land for energy cops is harmful in areas such as clearing forests, savannas, or grasslands. This is a damaging practice because it often leads to displacing food production for biofuel and compromises and depletes a carbon-rich ecosystem where carbon remains safely in the ground. This may also indirectly increase the amount of carbon in the atmosphere by taking it out of the ground; Loss of land mass and the agricultural demand of growing biomass can lead to an overall shortage of food. The agricultural demands of raising crops for biomass can contribute to chemical pollutants in the air and soil, which have adverse effects on wildlife and human health; and the burning off crop residue (slash and burn) leads to lack of water retention and nutrients in the soil (**Figure 94.7**).

Algaculture

Algae (collectively called plankton and a type of biomass) are photosynthetic organisms found in all aquatic habitats that form the base of marine food webs. They are also among the most diverse organisms on earth. Algae are formed as a result of a symbiotic relationship with cyanobacteria (unicellular photosynthetic microbes) that release oxygen gas and absorb carbon dioxide. Please note that when earth was in its infancy, there was very little free oxygen gas in the atmosphere. Many scientific historians credit cyanobacteria as being instrumental in creating free oxygen gas as a major atmospheric gas that it is today.

Additionally, algae can be used as a fuel source when cultivated, separated from water, and dried. In fact, there is a movement called algaculture supporting a core belief that algae could be used for jet fuel to meet our air travel needs. Consider that algae are able to produce up to 300 times more oil than soybeans.[xvi]

Nagy-Bagoly Arpad/Shutterstock.com

Figure 94.7   Wood Biomass as Fuel

Table 94.5

Pros of Using Biomass	Cons of Using Biomass
• Include: it decreases our dependency on oil and OPEC nations • it decentralizes our energy economy • it is renewable, and not very costly • energy sources can be grown locally and with little overhead cost • energy is renewable with less pollution and carbon emissions than fossil fuels • sources can be grown in poor soil acting as groundcover to prevent runoff • there is no net increase in carbon dioxide in the atmosphere since biomass is already part of the carbon cycle • it is possible to grow some of the agriculture on marginal soil, preventing runoff.	• Include: deforestation to make room for suitable agricultural sites • air pollution from burning agriculture • low energy yield (so a lot of biomass must be used) • large water and carbon footprint created for watering biofuel crops • soil erosion from raising crops • biofuel requires a lot of land use and cannot be harvested in large quantities without displacing land used for other crops.

Hydrogen Fuel Cells

Hydrogen fuel cells are much more efficient than internal combustion engines and, possibly, a better alternative to fossil fuels, especially since they have lower and cleaner emissions. They are also conveniently small in size. Hydrogen fuel cells are based on electrochemical reactions (the flow of ions) and work in a similar fashion to batteries, since the cells have both negative and positive electrodes. They can be used to run cars, boats, mass transit, and now, possibly, aircraft.

This process uses the Hydrogen cation (H^+) for energy. Hydrogen is a ubiquitous atom found in nearly all organic molecules, so this is considered a renewable energy source. A catalyst is used to remove a hydrogen ion that is stored in water from the hydrogen atom and eventually use it to generate an electrical current which is then converted to electrical or thermal power.

The energy for the catalyst can be produced by renewable or nonrenewable energy sources. One of the obstacles of the hydrogen fuel cell is that it requires a lot of energy to remove Hydrogen, which could, theoretically, use a great deal of fossil fuel in its preparation stages, nullifying the benefits of using renewable energy.

The hydrogen must be stored either as a gas (this is potentially dangerous because it is explosive and large in volume), as a liquid (which must be stored at very low temperatures), or as crystals. Pros: energy efficient; waste is water; clean emissions; can fuel most types of travel.

Cons: It is expensive, requiring a great deal of energy in preparation stages to isolate the hydrogen cation; the technology is not widespread; more research is required to make it viable, less costly, and convenient for every-body; and storage can be complicated and dangerous.

The Future of Renewable Energy

According to NREL, the United States could get 80% of its energy from renewable sources by 2050. In addition to adopting CCAs, many cities are moving toward more sustainability and carbon neutrality. Cities are moving toward renewable energy and adopting net zero energy codes for buildings allocated for residential dwelling. and incentivizing public transportation and alternative transportation with bike paths. Their buses have transitioned to natural gas sourced from landfills. And some cities like Santa Monica have adopted ordinances that solar energy is required for all new construction. What we know about renewable energy is that it is growing. Wind power tripled from 2007 to 2012.[xvii] In 2012, wind power provided almost 25% of the energy for South Dakota and Iowa, and more than 10% in seven other states.[xviii] In the United States, 29 states and the District of Columbia have all adopted renewable electricity standards, and 17 states and the District of Columbia are required to get 20% of their energy from renewable resources by the year 2025. Hawaii and Maine have the highest renewable energy standards.[xix]

With the above statistics, it appears are though we are moving toward a decentralized energy economy in which we rely less on fossil fuel and more on cleaner, renewable energy. And the National Renewable Energy Laboratory (NREL) agrees! The NREL is working closely with the U.S. Department of Energy to make renewable energy competitive with fossil fuels and other nonrenewable sources of energy.

All over the United States, photovoltaic research is conducted at universities, institutions, and laboratories in conjunction with the NREL and the U.S. Department of Energy. Their efforts are directed at decreasing capital costs of installing wind farms (which can be costly to construct) in addition to improving their energy output. They are also looking to expand hydroelectricity. Research includes innovative means to harness energy from the ocean tides, currents, and waves. Their initiatives also include working to make geothermal energy more widespread, cleaner, and affordable (nrel.gov).

In November 2015, members of the United Nations Framework Convention on Climate Change (**UNFCC**) met in Paris to discuss how to fight climate change and how to take aggressive measures to ensure an international, low carbon future. Part of this agreement includes improving and sharing technology for renewable forms of energy.

There are many other things we can do to become energy efficient and make renewable energy more widespread, like upgrade the energy grid so they are more efficient and can provide energy at peak times—and adapt to changing supply and demand. We could improve the technology to enable storage of energy for use when energy is in high demand. We could invest in the technology of batteries, thermal storage, etc. And we can look into building new transmission lines to link regional energy grids and to connect areas with lots of potential solar energy and wind energy to areas with high energy demand. The EPA is promoting a RE-powering program in which contaminated landfilled and mines are used as a site for solar and wind farms.

Fun Fact

Human ingenuity seems to know no limits when it comes to new and innovative ideas to meet our growing energy demands. Chemist Daniel Nocera, who received funding from the U.S. Department of Energy's Advanced Research Projects Agency (ARPA-E), has created an artificial silicon leaf that is able to mimic photosynthesis—using sunlight to split water into oxygen and hydrogen. This mechanism does not require access to the centralized power grid, enabling people who do not live near an energy grid to have access to electricity.

What measures can we take to move our energy economy toward a place where renewable energy sources are as widespread, cheap, and energy-efficient as nonrenewable? Instead of government subsidies for fossil fuels, the government could make renewable energy artificially cheap so that consumers have incentive to explore and use these options. Furthermore, tax breaks and concessions for renewable energy would increase incentive to shift toward a renewable energy economy. And finally, education is always a key component to becoming a responsible and informed consumer.

Renewable and Nonrenewable Resources
By Elizabeth Jordan

Did You Know?

The tropical rainforests are one of the earth's oldest ecosystems. In fact, many fossil records estimate them to be a few hundred million years old! As far as ecosystems go, this is considered extremely old. Take into account that North America was covered in ice until approximately 10,000 years ago, and so many of the ecosystems that exist today on the continent were created after the glaciers receded. The tropical rainforests never experienced glaciation, which helps explain their longevity.

Tropical rainforests contain over 50% of the world's biodiversity—including plants, animals, fungi, and microbes. Within this biodiversity are valuable goods that we use every day. Such goods include plants that provide us with medicinal value—antiviral, antitumor, as well as disinfectant and antiinflammatory properties; furthermore, they provide a vast variety of food from coffee to bananas. It is estimated that 80% of plants used to treat cancer are found here.

It would seem that we have great incentive to preserve these forests, but despite their immeasurable value, we cannot prevent them from being cut down faster than they regenerate. It is estimated that a rainforest area, the size of a football field, is cut down each day, and, along with it, 50 of its species become extinct. Is there anything we can do to slow down deforestation of the tropical rainforests?

Fifian Iromi/Shutterstock.com

A Tropical Rainforest

Natural resources are defined as anything that comes from the earth and is useful to humans in some capacity. Whether or not natural resources are considered renewable depends on how often we use them and how quickly they are regenerated. Air, water, some types of energy (solar, wind, etc.), plants, and animals are, for the most part, considered renewable resources because they regenerate almost as quickly as we consume them. Therefore, they are able to evade long-term depletion. Freshwater, land, and minerals are considered nonrenewable because they do not regenerate as quickly as their supply is used by humans, and are at risk of diminishing returns, if not, complete depletion.

Forests

Forests have both economic and ecological value that is difficult to quantify. Economically, they provide crops, timber, medicine, and eco-tourism, among many other things. Ecologically, they moderate climate, are instrumental in chemical cycling, provide habitats for both plant and wildlife, and act as a major carbon sink.

Timothy Epp/Shutterstock.com

An Old-Growth Forest

Rainforests, as well as other forests (deciduous, old-growth, coniferous) provide timber that is used to build homes, ships, furniture, paper, etc. Most forests— especially **old-growth forests** that have not been touched or disturbed by humans in at least 100 years—are nonrenewable resources. Old-growth forests (sometimes called primary growth forests) are characterized by having little human activity, older trees, indicator species, rich and fertile soil with a healthy fungal ecosystem, and lots of dead, decaying biomass. They also often have epiphytes such as moss growing on them and contain an abundance of biodiversity.

Dr. Morley Read/Shutterstock.com

The Effects of Clear Cutting

In the last 8,000 years, 46% of the world's forests have been cut down, most of this occurring since the 1950s in tropical climates. The tropical rainforests with their tall stature and large canopies provide us with timber and crops like coffee, tea, nuts, and fruit as well as the plants with pharmaceutical properties mentioned above. Besides the unfortunate direct loss of wildlife and biodiversity, another side effect of deforestation is the **edge effect**. This occurs when, as a result of deforestation, a larger proportion of land receives excessive sunlight. Edge effect causes **desertification** of the land, when once-fertile semi-arid lands turn infertile and unable to grow crops or vegetation. Desertification makes it difficult for plants and wildlife to grow and flourish. It has also been linked to reduced rainfall in the tropics, as transpiration rates dwindle. Could deforestation lead to the tropical rainforests drying up?

One method in which trees are harvested is **clear cutting**, when a large proportion of trees are cut down in one fell swoop and the ground is left exposed. Clear cutting is the most common type of deforestation and perhaps the most controversial. The pros of this practice include generating a high timber yield (and therefore profit) in a short amount of time. It also enables a fast-growing replacement tree farm to be grown in its place. Unfortunately, tree farms do not compensate for the biodiversity lost during deforestation, but they can be useful as a means to control runoff and provide future timber. As a con, clear cutting greatly reduces biodiversity and also causes major runoff that pollutes lakes and streams. Furthermore, after the trees are cut down, they are often burned, releasing the greenhouse gases, carbon dioxide and carbon monoxide, into the atmosphere. Another method is **strip cutting** in which a corridor is cut through a patch of forest land, usually upslope from a body of water (so intact land absorbs runoff and prevent water pollution). This prevents massive runoff from occurring but compromises biodiversity and disrupts species' ranges. And finally, **selective cutting** is the process in which only the most desirable trees are targeted and cut down. Selective cutting is considered the most sustainable tree harvesting method, as it causes the least amount of disturbance to an intact ecosystem. But it is not without its own controversy. The selected trees are often the oldest, and home to many species.

There are many adverse effects to logging. After deforestation, forests are unable to offer their ecological services—since there is no biomass to retain water flooding may occur, contributing to runoff and desertification of the soil. Forestry and logging account for 17% of Global Greenhouse Gas emissions.[i] Furthermore, the access roads that are built to facilitate logging act as conduits for other activities and factors that are harmful to the ecosystem and indigenous people who live there. The ready-built-roads often bring illegal loggers, illegal hunters, and bring in disease and pathogens to people and animals that they have never been exposed to before. This means that they have not developed and do not have the proper immunity.

Christopher Kolaczan/Shutterstock.com

Strip Cutting

Tree farms are an alternative to deforestation. This is the deliberate planting of trees with uniform size and age for commercial use. Usually, the tree of choice is fast growing, such as pine or eucalyptus. Although they do not have as much biodiversity as old-growth forests, tree farms can be beneficial to the environment because they cover the soil, preventing runoff and desertification, and they provide commercial use of wood as well as wildlife habitats.

JOEYSTUDIO/Shutterstock.com

A Eucalyptus Tree Farm

Conservation
By Elizabeth Jordan

Did You Know?

The original national parks, Yellowstone, Yosemite, and Sequoia, were originally open lands in need of protection from vandals, poachers, and trespassers (**Figure 96.1**). In 1872, a group of civilians had been appointed to protect Yellowstone. Despite this noble gesture, they had little legal jurisdiction or firepower to back them up. Furthermore, Congress saw no need to fund protection for such national treasures.

In 1886, the U.S. Cavalry, still without funding from Congress, stepped in and started patrolling these vast lands—first Yellowstone, and later the other abovementioned areas, until 1918 when the U.S. Park Service was implemented. If you ever stop to take note of National Park Service Ranger uniform, he is wearing a modern-day version of a U.S. cavalry hat!

Today the national parks, an agency of the U.S. Department of the Interior, provide multiple recreational purposes for people as well as overseeing plant life and wildlife management. Encompassing 85 million acres, the national parks contain and protect the majority of the U.S. forests, permit logging, mining, and oil drilling, and provide various recreational opportunities, such as hiking and camping.

The National Parks also have a Conservation Association that is committed to protecting wildlife such as wolves and bears and also to reintroduce species into many new habitats in order to protect their population numbers (gray wolves were successfully introduced into the Pacific Northwest for example).

The role of the National Park today is also to take conservation measures in order to protect many different species' from habitat loss—the number one cause for extinction. They are committed to maintaining as pristine an environment as possible to help preserve all species and their respective ecological roles.

The national parks are not without threats. Such threats include pollution and graffiti, crime, eroded trails, invasive species, and disturbance from off-road vehicles.

EastVillage Images/Shutterstock.com

Figure 96.1 Yosemite National Park

Conservation efforts are dedicated to preserving species and biodiversity. Currently the world is experiencing an alarming loss of species, loss of biodiversity, and accelerated rates of extinction. Now we will examine various ways to preserve and protect individual species and ecosystems, and possible reasons for species decline.

Extinctions and the Decline of Species

It has been estimated that 99% of all species that roamed this planet are now extinct.

It is normal for species to go extinct eventually with time. In fact, most species exist for approximately 10 million years, then succumb to extinction resulting from moderate changes in the environment—this is called **background extinction**.

However, due to rapid climate change, overexploitation, deforestation, and general loss of habitat for many species, some experts predict that we are currently experiencing 45 times the normal extinction rate, and that could increase to almost 100–1,000 times[i], resulting in **mass extinction**.

According to the International Union for the Conservation of Nature (IUCN) nearly 22% species of mammals are threatened or going extinct.[ii] In 2014, the IUCN assessed roughly 75,000 species and found over 20,000 to be threatened and almost 7,000 to be endangered.[iii] And our oceans are not faring any better. It is estimated that 20–25% of our common species living in the oceans are threatened with extinction.[iv] This is the same percentage of land plants and animals in the same predicament.[v]

A **biological extinction** occurs when a species is no longer found anywhere on earth. The dodo bird, eastern cougar, and passenger pigeon have all succumbed to biological extinction and no longer walk this earth. **Ecological extinction** occurs when a species' population size becomes so reduced in size, that members of the species become ineffective at performing their ecological role or having significant community relations with members of other species. The guanaco population of the Patagonia region of South America, for example, has reduced drastically as a result of introduced species. Present day, guanacos no longer serve their ecological role as prey for larger species, disrupting the food web (**Figures 96.2-96.3**).[vi]

If population does indeed get drastically reduced, the species may succumb to the **Allee effect**, which is in essence a self-thinning rule. The Allee effect occurs when species numbers become so low that it has a negative effect on the population size which, in turn, keeps reducing. Population reduction occurs because either members of a species do not have suitable access to mates, or there is ineffective cooperation in food gathering, etc., and the population ultimately experiences a decline in population growth and thus becomes vulnerable to extinction.

Pichugin Dmitry/Shutterstock.com

Figure 96.2   Guanacos in Patagonia, Chile

gallimaufry/Shutterstock.com

Figure 96.3   An Extinct Dodo

A **local extinction** occurs when a species is **extirpated**; goes extinct in a specific area but is found elsewhere in the world. Rhinoceros, for example, have gone extinct in Mozambique, but are found elsewhere in Africa—although threatened and almost driven to the brink of extinction.

A **deterministic extinction** implies that a species will go biologically extinct unless things change, such as preserving habitats and conducting aggressive conservation efforts. Species facing imminent extinction are sometimes referred to as a **vulnerable species**. Typically, **endemic species** are vulnerable to extinction. Causes for deterministic extinction include the following:

- **Habitat destruction:** Habitat destruction is the number one cause for extinction. Cutting down trees for livestock grazing and agriculture, as well as building homes and roads are all unsustainable practices that take habitats away from plants and animals.
- These activities also cause **habitat fragmentation**, a dramatic reduction in the size of an area or habitat of a species. Large mammals such as male mountain lions have a habitat distribution of one male mountain lion per *at least* 50 square miles (it is important to note that this is a conservative estimate).
- **Overkill.** This occurs when a species is overhunted or overharvested and few species remain in the wild. Animals that are large, tasty, and slow in movement and have parts perceived as useful to some humans, are naturally vulnerable to overkill. An example is the African Elephant which only procreates every

3–9 years and is killed for its tusks. Fearing overkill and thus the imminent extinction of elephants if drastic measures were not taken, conservationists lobbied for a worldwide ban on ivory. In 1989, their efforts were actualized.

- Bluefish tuna, a desirable type of fish, has been depleted at alarming rates due to illegal fishing. It is predicted that they may become endangered if the unsustainable fishing continues.
- **Poaching** is the illegal human killing and commerce of animals for their valuable parts. Rhinoceros and elephants, for example, are hunted for their tusks. Lions and leopards are hunted for their beautiful fur coats. Poaching is also responsible for the decline in chimpanzees and gorillas in Western Africa. They are illegally sold as **bushmeat**. Again, these vulnerable animals usually do not leave a lot of offspring, so they are particularly vulnerable to extinction.
- **Introduced species**: Introduced species, also known as exotic species, are introduced into a given area from other parts of the world. They are usually generalists, thriving in a variety of environments, and easily adapted to their new niches. Take into account they are no longer in their natural habitats, which means they no longer have their natural population controls such as predators or competitors. Without such population controls, they become a possible pest by growing in population and driving native species out of their habitats. Introduced species are responsible for 40% of all extinctions. The Nile perch, for example, was introduced to Lake Victoria in Africa, and has caused extinction of 200 species of cichlid fish.
- **Chains of extinction**: All species are connected through a food web. Therefore, it stands to reason that large predators disappear when their prey go extinct (since they lose their food source). For example, the black-footed ferret became endangered when the population of prairie dogs, their prey, began to decline.

MVP, or *minimum viable population*, is the minimum number of individuals in a population of species in the wild that will provide enough genetic variability to sustain its population in the wild. The MVP for mammals is generally about 500 individuals, but this number varies considerably depending on the species. Recent research has estimated that at least 20,000 members of a specific species must be present in their respective population in order for biological evolution to occur.

When a species is about to go extinct, it is listed as **endangered** by the U.S. Fish and Wildlife Service. Many factors could cause a species to become endangered. Often their habitat is compromised or they are too few in number to play an ecological role or have a healthy amount of genetic variability. Northern spotted owls, humpback whales, black-footed ferrets, and African and Asian elephants are all examples of endangered species, but there are many, many more.

According to the **Endangered Species Act (ESA) of 1973**, it is illegal to harm or kill an endangered species or to remove it from its natural habitat. The ESA gave the U.S. Wildlife Fish and Game authority to protect endangered and threatened species both at home and abroad. For example, no endangered or threatened species, nor their body parts, can be imported into the United States without ample proof that in doing so, the well-being of the species improves. A **threatened species** is a species that is about to be endangered in the foreseeable future, and conservation efforts need to be implemented to preserve their habitats for their survival. Twenty five percent of known mammals are considered threatened.

Currently, 25% of all mammals are threatened, mostly due to habitat loss. If extinction rates continue as they are, it is estimated, according to a 2011 study in the journal *Nature*, that 75% species that exist today could go extinct.

Under the Lacey Act of 1900 (amended in 2008) endangered species may not be imported by any means:

It is unlawful to import, export, sell, acquire, or purchase fish, wildlife or plants that are taken, possessed, transported, or sold: (1) in violation of U.S. or Indian law, or (2) in interstate or foreign commerce involving any fish, wildlife, or plants taken, possessed, or sold in violation of State or foreign law.[vii]

CITES

Implemented in 1973, **CITES** (Convention on the International Trade in Endangered Species of Fauna and Flora) is an international agreement between 179 countries (referred to as parties) participating on a volunteer basis. The agreement monitors the international trade of wildlife and plants by issuing permits and licensing for those legally imported and exported. If and when they are transported, CITES oversees the process to ensure no harm or injury occurs. Roughly 30,000 species of plants and 5,600 species of animals are protected by CITES. Although parties participate on volunteer basis, agreements are legally binding.

In terms of *illegally* traded animals or poaching, CITES cooperates with governments to fight against wildlife crime and assists in designing strategies to combat poaching. Animals— sometimes those that are endangered—are readily exploited for their valuable parts, making the wildlife trade a billion dollar industry. Despite rigorous efforts and cooperation among the international community, elephant and rhinoceros poaching is at an all-time high. (According to CITES.org, in the year 2000, seven rhinoceroses were poached and in the year 2013, there were over 830.) In 2013, The UN Security Council declared The Central Republic of South Africa in crisis due to its rampant poaching. Given the amount of money at stake (kilogram per kilogram, as of the year 2013, rhinoceros tusk was worth more than gold), poachers have amassed an arsenal of sophisticated weaponry and have been known to open fire on park rangers who try their best to thwart the slaughter of elephants and rhinoceros. In turn, park rangers in South Africa have been receiving training by former British and other International Special Forces, so they now have infantry-style training to fight against the poachers.

Criticism of CITES is that they are not doing enough to fight poaching and protect vulnerable, threatened, and endangered species. Consider that recent research estimates that a staggering 10,000 elephants were killed from the years 2010–2012.[viii] The African Lion population has decreased 80% from its historical range, and conservationists estimate that there are only 24,000 left living on the African Continent.[ix] Rhinoceros are in the worst shape of all. There are only about 5,000 Black Rhinos left off in the wild on the African Continent and 20,000 White Rhinos.[x] Despite these staggering numbers, poaching continues to climb.

Legal government-issued permits are allotted for trophy hunting, another controversial practice, for all of the aforementioned species. Although trophy hunting is legal, many question the ethical considerations of hunting sensitive and rare species for sport. And many are concerned that trophy hunting is a smokescreen for poaching, as such activity occurs in countries with little governance and with widespread poverty. Such places are unable to create or enforce conservation laws, since many of its inhabitants live in impoverished conditions and all funding and energy goes toward helping its people. The temptation to surreptitiously "up" the lucrative hunting quotas—especially given little oversight from government agencies—may be great.

BOX 96.1	Case Study

THE AKASHINGA

A legion of female rangers are leading the charge against poaching in the lower Zambezi Valley of Zimbabwe. They are called the Akashinga, or 'Brave Ones' and put their lives on the line every day to protect African animals from the illegal wildlife trade. Many come from disadvantaged backgrounds such as poverty, human trafficking, or abusive relationships. But they have found their calling as well as their independence in protecting these majestic animals.

The inspiration for the Akashinga came from Damien Mander, founder of IAPF (International Anti Poaching Foundation). He is a former Australian special forces commando turned anti-poaching ranger – turned animal welfare and environmental advocate. He put the women through a grueling selection process and to his surprise, very few dropped out. He is inspired by their commitment.

The Akashinga make excellent wildlife rangers because, according to Mander, they are not as corruptible as their male counterparts. Females tend to be better skilled at gathering intelligence, saving months of surveillance. Females are also good at deescalating tense situations, minimizing the risks of armed conflict that may place local communities in crosshairs. Avoiding conflict helps instill confidence in local communities, ensuring their cooperation with intelligence gathering. Females also tend to put more money into the local communities and economy. This is essential, as the success of local communities in strongly aligned with conservation success.

Having been met with incredible success and support, Mander plans to expand this conservation model throughout the African Continent. The future of conservation could very well be female rangers.

© Kat Webb/Shutterstock.com

Box Figure 96.1.1

***(Featured Picture is Sergeant of the Akashinga, Vimbai Kumire)

Preserving Species

Conservation of species is either **in situ** or **ex situ**. In situ conservation is done in nature, in species' natural habitats: in parks, wildlife reserves, and protected areas. Ex situ conservation is not performed in natural habitats, but is typically performed in human controlled areas: zoos, botanical gardens, animal parks, seeds banks, captive breeding, etc (**Figure 96.4**).

Often times, wildlife conservationists create **wildlife corridors** that connect two patches of wildlife habitat that have been interrupted by human encroachment, such as roads and developments (**Figure 96.5**). This makes the separated wildlife areas contiguous, increasing the overall intact land area. Wildlife corridors have succeeded with panthers, bobcats, cougars, and deer. In fact, a large-scale conservation project underway in Florida creates and conserves a wildlife corridor from the Everglades to Atlanta. This ensures that various species have an intact and available means for migration.

Baloncici/Shutterstock.com

Figure 96.4 **Botanical Gardens**

Robert Crum/Shutterstock.com

Figure 96.5 **Highway Crossing Bridge for Animals, Banff National Park, Canada**

Efforts to create and conserve wildlife corridors are carried out by multiple agencies, such as the national parks, the state park system, U.S. Fish and Wildlife, and the U.S. Bureau of Land Management.

Parks, zoos, aquariums, and botanical gardens often employ conservation efforts designed to sustain a species that is threatened or close to extinction.

Conservation organizations, such as zoos and aquariums, breed species in captivity. Captive breeding for conservation purposes is intended to promote healthy genetic diversity in a population, especially for populations that are near extinction. Over time, many of these species are released into the wild and their progress is monitored closely. This has been successful for some species such as the Giant Panda, Whooping Cranes, and Woodrats.

Unfortunately, the overall success rate is low—only about 10%. Considering the many unknown factors that are involved in reintroduction, it is almost impossible for wildlife and conservation biologists to determine which species will be viable and will thrive after reintroduction. Captive bred animals appear to be missing essential survival skills evident in their wild-raised counterparts. For some reason, this knowledge is neither learned nor passed down ex situ. For example, captive-bred lemurs, native to Madagascar, were raised in California and reintroduced into the wild. Shockingly, several of them turned up dead, having poisoned themselves by sampling all the plants and eating berries—not realizing that they were poisonous to their systems.[xi]

Thus, it seems logical that animals learn what is medicinal and what is harmful by observing members of their own species. For reasons that are still unclear, this knowledge must be learned in the wild, of their native habitat.

Captive bred animals also have numerous reproductive issues that are still not explained: elephants do not display prolonged interest in sex, the testes of male gorillas have been known to shrivel and ultimately become useless, and it is difficult for many pandas to breed. These reproductive obstacles suggest that the stress of living a life of confinement takes a physical toll on captive-bred animals.

Many wildlife biologists believe that we should expand the field of **ethology**—the scientific study of animal behavior. Research done in the wild through observation of animals and their behavior in their natural environment may fill in necessary gaps and help determine what needs to be adjusted or refined in captive-bred environments.

Another possible solution is *Assisted Migration and Colonization*—that is, facilitating the transplantation of a species. This idea was first proposed by Camille Parmesan, an Assistant Professor of Integrative Biology of the University of Texas at Austin. Dr. Parmesan is considered a leader in Conservation Biology as a result of her extensive research on Edith's checker spot butterfly as well as many other species (**Figure 96.6**).[xii]

Many see this as a controversial proposition, since risks of assisted migration may include unintended introduction of new pests and diseases. Another risk is that the species may become invasive. Furthermore, species reintroduced to a new area will have lack of knowledge of the surrounding vegetation relative to habitats in which they co-evolved. Thus, they may inadvertently ingest a poisonous plant as seen with the Madagascar lemurs transplanted from California mentioned above. Several other unintended consequences are possible as well.

A way to help preserve endangered plant species is to create **seed banks** (also known as gene banks). Seed banks are storage facilities where seeds are stored in low temperatures and low humidity (to protect them from drying out). Although seed banks offer seed protection from habitat loss (and loss of genetic variability from plant hybridization in the wild), many seeds do not withstand being dried out for long. Furthermore, since seeds are ex situ, they are not in the wild, and therefore do not undergo the mechanisms that would

Figure 96.6 **The Checkerspot Butterfly**

enable them to adapt to a changing environment. For example, there is no natural selection or biological evolution, which would ultimately increase their fitness in their natural environment. Therefore, seed banks are not a viable long-term option.

The Problem with Zoos and Living in Captivity

Zoos have come under fire in recent years as critics say that although they are effective in educating on the value of biodiversity, they have little conservation value. Many of the animals in zoos do not need conservation to begin with, or they are sterile hybrids that render conservation ineffective. Then there is the ethical question of whether it is humane to confine animals to captivity for the duration of their lives.

Animals in zoos live longer, but this does not necessarily mean they enjoy better health in their longevity. Many captive animals display typified behavior such as pacing, weaving, and rocking, and have on occasion been known to inflict self-abuse. Many zoo gorillas are infertile, often lose interest in mating, suffer eating disorders, and die from cardiovascular disease. Captive elephants often lose interest in mating as well. A considerable number of captive rhinos suffer and die from hemolytic anemia. These trends are not observed in the wild.

Animals also suffer innumerable health issues because they are not in their natural environment and cannot self-medicate. Their new environment is devoid of the soil, water, microorganisms, food, and climate they are adapted to in the wild, while wound-tending and grooming, observed in the wild and important for survival, become lost behaviors. In captivity, they are exposed to new pathogens and diseases to which they have not yet built up immunity. It is difficult for captive animals to get adequate amounts of exercise or even the right vitamins and minerals (calcium, phosphorous, and vitamin D). Even getting the right amount of sunlight and shade is imperative to wild animal health. Many animals sunbathe in the wild to synthesize vitamin D and to self-medicate (even some bacteria have been known to use sunlight for correcting damaged DNA). Zoos do not necessarily provide ample sunlight and without it, many animals suffer from deficiencies and can have bone abnormalities. For all the above concerns, the trend is to make zoo environments as similar to the species' natural environment as possible, but this has obvious limitations.

Finally, the consumer market seems to be shifting as people are less inclined to be entertained by viewing animals in captivity and prefer to observe them in their natural habitat. This is reflected in circuses deciding to phase-out using animals as entertainment and in the growing trend in eco-tourism.

What Factors Influence the Success of Conservation?

Why are some particular conservation efforts successful while others are not? According to a 2014 study in *Nature*, most marine protected areas do not achieve their conservation goals, while many conservation efforts, such as the Florida Wildlife Corridor initiatives, have been successful. What accounts for this difference?[xiii]

Economic growth may have an impact, albeit complicated, on conservation. Economic growth often means a relative abundance of disposable income, translating to more philanthropic endeavors. This in turn translates to ample donations to nonprofit organizations dedicated to conservation agencies, such as the Nature Conservancy, The Sierra Club, and the NRDC. This is significant since, in the United States alone, nonprofit agencies donate billions of dollars to conservation and preservation of species. In this regard, economic growth may *positively* impact conservation.

On the other hand, economic growth may also lead to unsustainable development of houses, roads, and infrastructure, as well as deforestation as consumer demand increases. This ultimately encroaches on species' habitats, undermining conservation efforts. In this regard, economic growth may *negatively* impact conservation.

Science is a major component of conservation and leads to invaluable information about species and the environment. But there are other factors that contribute to conservation success. Some such factors include planning, education, public relations, and political action. Blogs, Letters to the editor, opinion pieces in newspapers, education, meetings with government officials, activism, petitions, and engaging the general public in all areas of the conservation movement are influential, and sometimes imperative, to the success of a conservation movement. In other words, grassroots efforts are a major contributing factor to the success of environmental conservation.

It is also often more impactful when multiple agencies make jointed efforts toward a particular conservation effort. Agencies such as state fish and wildlife agencies, coastal conservancies, departments of forestry, and state and federal park services are more effective when they work together for the same overarching goal.

Countries with corrupt or little governance may find it difficult to successfully conserve land and preserve its species. Developing countries (as we saw in the case study of Ecuador) may be willing to take money from oil companies and allow drilling for an economic boost to the economy, instead of committing to the long-term efforts to build an economy on ecotourism or sustainable agriculture.

Corrupt government officials may look the other way when it comes to unsustainable hunting for sport. Or well-intentioned governments may lack the resources needed to successfully combat poaching. This is a contentious issue in Africa where much of its wildlife (elephants, rhinos, lions) is poached, hunted for sport, or raised on farms for recreational hunting. Governments may be unable to curb poaching, or may concede to wealthy hunters, who are willing to pay several hundred thousand dollars to hunt such rare and endangered species.

Namibia for instance has auctioned off the lives of endangered Black Rhinoceros (of only a few thousand left in the wild) to recreational hunters who pay to shoot and kill them in the wild. A better alternative is to keep the animals alive and build an economy through walking safaris and eco-tourism, which has been done with success in Botswana. Eco-tourism is estimated to bring in 15 times more revenue from livestock than does trophy hunting.[xiv,xv] In other words, animals provide more economic value when alive than when hunted (not to mention their innate, intrinsic, and ecological value).

Some of the most effective conservation efforts are community-based. The underpinnings of this philosophy are that effective wildlife conservation and the fight against poaching will not be complete without the consent and the cooperation of the local community. In some conservation areas, many locals themselves are suffering from food insecurity and are living in abject poverty. Wildlife and environmental concerns therefore take a backseat to the welfare of the people (if you recall, poverty is a cause of environmental problems). On the other hand, if conservation efforts assist in breaking the cycle of poverty of the local people though microbusiness, micro-lending and education; then possible extended efforts can be directed toward the wildlife and other conservation concerns. This also allows the opportunity for locals to establish a long-term economy built on eco-tourism. A good example of this Conservation model is LEWA Conservancy, a UNESCO organization that promotes wildlife conservation in Kenya through education, microlending, and community outreach. LEWA is considered by many as one of the most successful conservation models.

BOX 96.2	Case Study

COSTA RICA: A PARAGON OF CONSERVATION AND SUSTAINABILITY.

Costa Rica, a small country in Central America—home to lush tropical rainforests and vibrant coral reefs—contains 3% of the world's biodiversity. A common practice in many developing countries fortunate to be rich in such natural resources is to engage government subsidies encouraging the sale of private property in order to create open space for plantation crops (like palm oil) and livestock. Costa Rica's government, on the other hand, firmly believes that Costa Rica's rich biodiversity is a currency and would ultimately be a long-term economic investment. They designed a long-term strategy that would ultimately stabilize their economy and improve overall standard of living. This strategy included preserving biodiversity, not destroying it. The government disposed of their military and started paying private citizens to conserve their land. About 25% of their land became a designated conservation area, thus enabling their economy to be driven by ecotourism.

Pavalena/Shutterstock.com

Box Figure 96.2.1

To preserve the environment and biodiversity—now their biggest currency— Costa Rica proactively promotes sustainability and stewardship. It uses hydroelectricity as major energy source and their other energy sources are renewable as well. Their hotels are even rated on sustainability!! It is not unheard of for some hotels to use recycled gray water for their toilets. This practice has proven itself a remarkable success. Costa Rica's economy and standard of living is much higher than other developing nations.

Eco-tourism, however, is a bit conundrum for Costa Rica. On one hand, the country relies heavily on eco-tourism for its well-being. On the other hand, tourism creates the biggest ecological footprint (with air travel, pollution, etc.). This may explain why the country emphasizes sustainable practice for everybody, including its tourists.

Types of Forests

By Cynthia McKenney, Ursula Schuch, and Amanda Chau

Boreal Forest

The boreal forest or **taiga** is located between 50°N and 70°N latitude, south of the arctic tundra and north of the temperate coniferous forest or grasslands. The boreal forest extends across Alaska, most of Canada, Northern Europe, and Russia, comprising the largest terrestrial biome and one third of the world's forest area. Summers are short with a growing season up to 4 months. Temperature fluctuations are extreme with summers warmer than in the arctic tundra. Annual rainfall varies from 9.8" (25 cm) to over 39.4" (100 cm) in boreal forests of Western North America, creating a moist, cool to cold climate where little water is lost to evaporation. The boreal forest is dominated by a few species of conifer trees, which can tolerate the environment (**Figure 97.1a**). Soils are very high in organic matter and the forest floor is covered in a thick layer of litter. Decomposition rates are slow and soils are often acidic and have mineral deficiencies. Some areas have waterlogged soils and ponds are often found in summer when soils thaw in colder regions (**Figure 97.1b**). Fires caused by lightning or humans are common and vital to the regeneration of the boreal forest. Fires release nutrients from vegetative litter and increase growth of subsequent vegetation. Fires change the vegetation composition and contribute to biodiversity of the flora and fauna in the ecosystem (Tyrell, 2018).

taiga: Wide belt of conifer forest located south of the polar ice. See *tundra*.

The main genera of trees growing in the boreal forest are spruce (*Picea*), pine (*Pinus*), fir (*Abies*), larch (*Larix*), birch (*Betula*) and aspen (*Populus*). The evergreens are well adapted and have the ability to resume photosynthesis whenever temperatures are favorable. Understory vegetation is sparse in the densely shaded forest with low growing shrubs of wild berries and some herbaceous species. Productivity is low as trees grow slow and live for a long time. Common conifers in the North American boreal forest include black spruce (*Picea mariana*), white spruce (*Picea glauca*), jack pine (*Pinus banksiana*), lodgepole pine (*Pinus contorta*), tamarack (*Larix laricina*), white cedar (*Thuja occidentalis*), and balsam fir (*Abies balsamea*). Some species emerge especially after a disturbance such as timber harvesting, insect outbreaks, storm throw, or fire. These early colonizers include deciduous birch and aspen trees. Cones of jack pine and lodgepole pine are covered with a waxy coating and rely on fire to discharge their seeds.

(a)

A and B: © Pi-Lens/Shutterstock.com

(b)

Figure 97.1a–b (a) **Boreal forest in Yukon Territory, Canada. (b) Pond from thawing permafrost in the boreal forest.**

Boreal forests differ in species composition and fire ecology based on their geographic location. Boreal forests influenced by the moderating climate of the Pacific or Atlantic oceans are generally more productive due to greater availability of moisture and less extreme temperatures. Interior boreal forests withstand extreme cold and dry conditions and are often afflicted by widespread intense fires. Trees in boreal forests in Eurasia have a longer lifespan, 400–600 years on average, than boreal forest species in North America. They live on average about 150–200 years, which is primarily due to the more frequent and widespread wildfires (Tyrell, 2018).

Temperate Coniferous Forest

The temperate climate with annual temperatures from 41°F to 68°F (5°C to 20°C and higher rainfall supports the temperate coniferous forest (**Figure 97.2a**). In North America, two distinct areas are in the Pacific Northwest and in the Southeast (Pidwirny, Draggan, McGinley, & Frankis, 2007, Revised on January 2012). This biome is also found in Europe, Asia, and South America with trees of similar phenotypes to those found in North America, but different species. Temperate coniferous forests produce the highest amount of biomass of any terrestrial biome. They can consist entirely of conifers, or they can be a mix of broadleaf evergreen and conifers. In general, these forests are composed of trees in the overstory and of small shrubs and herbaceous plants in the understory.

Coniferous forest in the Northwest close to the ocean enjoy ample rainfall with more than 250 cm and boast highly productive evergreens such as Douglas fir (*Pseudotsuga menziesii*) (**Figure 97.2a**), red cedar (*Thuja* *plicata*), western hemlock (*Tsuga heterophylla*), sitka spruce (*Picea sitchensis*), and redwood (*Sequoia sempervirens*). This area is sometimes referred to as temperate rainforest.

Further inland on the east side of the mountains, the climate becomes more continental and drier supporting forests dominated by ponderosa pine (*Pinus ponderosa*), Engelmann spruce (*Picea engelmannii*), and lodgepole pine (*Pinus contorta*). Summer drought and fires occur regularly. Giant sequoias (*Sequoiadendron giganteum*), the largest organisms in the world, grow in a small area on the western Sierra Nevada in California. In the Southeastern United States, low productivity conifer forests grow on nutrient poor, sandy soils. Typical plants adapted to this environment are pitch pine (*Pinus rigida*), longleaf pine (*Pinus palustris*), and slash pine (*Pinus elliottii*), all adapted to the frequent fires in this habitat.

© Robert Crum/Shutterstock.com

(a)

© Alexander Studentschnig/Shutterstock.com

(b)

© Marilyn barbone/Shutterstock.com

(c)

Figure 97.2a–c (a) temperate conifer forest of Douglas fir (*Pseudotsuga menziesii*); (b) temperate mixed forest in fall; (c) bluebells (*Hyacinthoides*) in the broadleaf forest in England flower in early spring before the canopy of emerging leaves casts too much shade on the forest floor.

In the southern hemisphere, the temperate coniferous forest covers the lower elevations of the south-central Andes in Chile into Argentina, and the monkey puzzle tree (*Araucaria araucana*) is the dominant conifer. The Valdivian temperate rain forest extends from the Pacific coast to the base of the Southern Andes and the

Patagonian cypress (*Fitzroya cupressoides*) is one example of the characteristic large conifers growing in this cool, moist climate.

Temperate Broadleaf and Mixed Forests

The temperate broadleaf and mixed forest is also known as the temperate deciduous forest. This biome is located north and south of 30° latitude. The climate is characterized by warm summers, cold winters, and relatively high amounts of rainfall throughout the year. In many areas, rainfall is distributed throughout the year, although in some regions a dry season can occur in summer or winter. This temperate climate has less annual temperature fluctuation than the biomes located further north and supports many species of deciduous broadleaf trees. The agreeable climate has led to much deforestation for urban development and agriculture.

Temperate broadleaf and mixed forests dominate in the northern hemisphere and cover large areas of the Northeastern United States and Southeastern Canada, Western Europe into Western Asia, and Eastern Asia including Japan. In the southern hemisphere, this biome is found in the southern coastal region of South America, along the eastern coast of Australia and in Tasmania and New Zealand. In Europe and Eastern North America, deciduous trees dominate this biome, although conifers are mixed among the broadleaf trees (**Figure 97.2a**). Broadleaf tree species include maple (*Acer*), beech (*Fagus*), oak (*Quercus*), elm (*Ulmus*), poplar (*Populus*), and willow (*Salix*), and conifer species include pine (*Pinus*), fir (*Abies*), and spruce (*Picea*). In the southern hemisphere, broadleaf deciduous trees dominate the overstory of this biome almost exclusively. Dominant large trees include many Eucalyptus species in the warmer areas and beech (*Nothofagus*) in the cooler areas. Commonly up to 25 different tree species are found, generally more in warmer locations and fewer in cooler regions of this biome. Distinct seasons are marked by the drop of leaves in fall and flowering of the rich herbaceous understory plants in spring before the leaves of deciduous trees and shrubs cast a dense shade (**Figure 97.2c**). Smaller trees form a subcanopy under the taller trees. The large quantities of litter in fall are quickly decomposed into humus. The soil in this forest is nutrient rich and supports diverse plant communities.

Tropical and Subtropical Moist Broadleaf Forests

This biome is also known as the rainforest and is rich in plant and animal species. Tropical and subtropical moist broadleaf forests occur near the equator within 23.5° northern and southern latitude. Due to this location, there is almost no deviation from the 12-hour daylength throughout the year. Temperatures in the tropical rainforest are almost constant year-round between 68°F and 86°F (20°C and 30°C), and the difference between day and night temperature can be larger than the average change in annual temperature. Annual rainfall is very high, ranging from 79" to 236" (200 to 600 cm) precipitation. Distribution throughout the year can vary with no dry months in the tropical rainforest where trees maintain their evergreen canopy throughout the year. The largest areas covered by tropical moist broadleaf forests are found in the Congo basin in central Africa, the Amazon basin in South America, and Malaysia and Indonesia. In seasonal rainforests, dry periods in winter cause leaf drop of some species although temperatures remain warm. These forests occur in coastal West Africa, parts of the Indian subcontinent and Southeast Asia, and South and Central America.

© Dr. Morley Read/Shutterstock.com

(a)

Daniel Sambraus/Science Source.

(b)

© Robyn Mackenzie/Shutterstock.com

(c)

Figure 97.3a–c (a) Aerial view of tropical rainforest in the Amazon. (b) Kapok tree (*Ceiba pentandra*) grows to 197 to 230 feet (60 to 70 m) above the general canopy in the Amazon rainforest and is estimated to be more than 800 years old. (c) Australian temperate rainforest.

Tropical rainforests house the largest number plant and animal species of all the biomes (**Figure 97.3a and b**). Up to 300 tree species may be found in one hectare (10,000 m²) while in a temperate rainforest (**Figure 97.3c**) between 20 and 30 tree species are growing in a comparable area. Some very tall trees grow in the tropical rain forest with just a few towering up to 197' (60 m) above the canopy of trees below (**Figure 97.3b**). Most trees are evergreen and have broad leaves. Multiple layers of larger and smaller trees, shrubs, vines including lianas, and plants such as orchids, bromeliads, ferns, and mosses contribute to the extraordinary diversity of this environment. With so many layers of canopy, the forest floor is often dark with few plants. Low growing plants on the forest floor are often shade tolerant and a greater number of plants is found where the canopy has been broken up, admitting more light. Litter on the ground does not accumulate because it is quickly decomposed by microorganisms. Soils in the tropics are exposed to a fast weathering process with abundant rains leaching nutrients and soils turning acidic.

Rainforests house more than half of the plant and animal species living on earth. These forest ecosystems are fragile and are often destroyed for agriculture, logging, and mining. Plants from rainforests show great promise for medicinal purposes and more than 2000 species show anticancer properties. This number is less than 1% of all the plant species found in the tropical forest.

Impacts of Deforestation at the Brazilian Amazon Frontier

Amy Parsons, April 2019

The Amazonian frontier is characterized by the area that has seen a dramatic transition in land cover over the past half-century. Specifically, in the area known as the "arc of deforestation" which has experienced over 80% of the cumulative and current effects of forest clearing to date. The peak rate of Brazilian Amazonian deforestation occurred in 2004 and did see a steady decline over the next few years. However, 2018 saw the largest rate of deforestation in a decade, solidifying the fact that the problem has not gone away and still poses a major threat to forests in the region. In the Brazilian Amazon, the large-scale shift to mechanized agriculture, extensive logging, and intensive cattle ranching has led to a more vulnerable transitional ecosystem between moist, tropical forests and seasonally dry, savannah vegetation.

Impacts of disturbance

Increased rates of forest loss along the arc of deforestation have created a regional positive feedback cycle that is predicted to worsen in the coming decades. After forest is cleared, the surface albedo (or reflectivity) drastically increases allowing for a shift in the absorption of incoming solar radiation. In turn, the air above that land heats up faster than it would if the energy was being absorbed by dark tropical vegetation and draws away moisture, therefore lessening the amount of potential rainfall over forests. Along newly-formed forest edges, the microclimate at the forest floor changes with increased amounts of incoming sunlight and wind, making normally moist conditions much drier.

Forest edges also see the mortality of large trees increase and an invasion of native and non-native pioneer grasses resulting in a homogenous landscape and a decrease in overall biodiversity. As grasses take over a landscape, they create a blanket of available fuels, which typically burn at high intensities especially under extremely warm and dry conditions. These fire-favourable conditions are becoming more common thanks to anthropogenic climate change which is increasing the length and severity of droughts. These conditions become exacerbated by a positive feedback cycle in which fires burn through these areas opening the canopy further, thus increasing the size of forest edges which cause that forest to become more susceptible to burning. As a result, the arc of deforestation in Brazil is at the highest risk of fire due to the coalescence of compounded anthropogenic and environmental variables.

Solutions

With demands for a shift in land management practices at the Amazonian agricultural frontier, the general consensus seems to be a push for transforming the already cleared areas into ones that are higher-yielding but also maintained in a long term, sustainable way. Other experts have recommended a change in economic markets and political drivers that influence forest degradation at the frontier boundary. Seemingly simple solutions have emerged that directly call for harsher enforcement of already established environmental policies through mechanisms such as the Brazilian Forest Code. However, the Brazilian Forest Code has seen the removal of important regulations in 2018, making it easier for landowners to legally clear forests. Improvements to the general monitoring of annual forest degradation and newly established protected forests and indigenous preserves have begun to take effect on reducing the fire-related devastation to forests. However, emphasis on local education could be extremely beneficial in order to see longer-term changes to the way people are using their land. Overall, it will take a combined and extensive approach to ease the positive feedback cycle involving deforestation, agricultural intensification, pasture expansion, increased fire frequency, and climate change that has the potential to drive further degradation in other parts of the continent and possibly the globe.

Tropical Deforestation

By Jon Turk, Terrence Bensel, and Rebecca Lindsey

Tropical forests make up less than 10 percent of the Earth's land surface, yet these ecosystems are home to roughly half of the planet's biodiversity. The rapid loss of tropical forests in Latin America, Asia, and Africa is therefore of great concern to ecologists and environmental scientists. In the following article, Rebecca Lindsey, a technical writer for the National Aeronautics and Space Administration (NASA), examines the causes and impacts of deforestation in tropical regions, as well as possible approaches to slow or reverse this process.

Tropical, moist forests have the highest rates of biodiversity and primary productivity of any of the world's terrestrial ecosystems. This high level of diversity provides direct benefits to humans in the form of various useful products (nuts, fruit, latex, medicines). This specialized diversity also represents a store of genetic information that could be useful to medical advances for current and future generations. Tropical forests have been likened to the Library of Congress in that they contain a vast store of information. And like with this massive library, we may never know even a fraction of the information that is contained within them.

The causes of deforestation are complex and vary from region to region, but it is generally clearing forests, or cutting down trees, to open land for crops and livestock that has the biggest impact. Efforts to conserve tropical forests have focused on finding markets for products that can be harvested from standing forests. Such a sustainable use approach is based on the idea that if managed properly, forests may yield greater economic and social value over time instead of simply cutting them down. Indeed, tropical forests provide many critical ecosystem services— services provided by ecosystems that are necessary to sustain life. Perhaps the most important of these is the role that forests play in the carbon cycle and in maintaining climate.

Tropical deforestation adds carbon dioxide to the atmosphere, diminishes freshwater supplies, and leads to biodiversity and species loss—all planetary boundaries identified in the previous reading. To the extent that growing human populations in tropical countries can be accommodated sustainably in urban areas or megacities, it might be possible to reduce some of the pressure on forests in those regions. That issue will be addressed in section 4.4 when we discuss megacities.

Stretching out from the equator on all Earth's land surfaces is a wide belt of forests of amazing diversity and productivity. Tropical forests include dense rainforests, where rainfall is abundant year-round; seasonally moist forests, where rainfall is abundant, but seasonal; and drier, more open woodlands. Tropical forests of all varieties are disappearing rapidly as humans clear the natural landscape to make room for farms and pastures, to harvest timber for construction and fuel, and to build roads and urban areas. Although deforestation meets some human needs, it also has profound, sometimes devastating, consequences, including social conflict, extinction of plants and animals, and climate change—challenges that aren't just local, but global. Ecologists have identified certain areas of the Earth as biodiversity "hotspots." These areas are important centers of biodiversity that contain a large number of native species that are not found anywhere else. Identifying and saving these areas is critical to protecting our planet's biodiversity (**Figure 99.1**).

Based on Conservation International Foundation. Retrieved from http://www.conservation.org/where/priority_areas/hotspots/ Documents/cl_Biodiversity-Hotspots_2013_Map.pdf

Figure 99.1   **Biodiversity hotspots**

Impacts of Deforestation: Biodiversity Impacts

Although tropical forests cover only about 7 percent of the Earth's dry land, they probably harbor about half of all species on Earth. Many species are so specialized to **microhabitats** within the forest that they can only be found in small areas. Their specialization makes them vulnerable to extinction. In addition to the species lost when an area is totally deforested, the plants and animals in the fragments of forest that remain also become increasingly vulnerable, sometimes even committed, to extinction. The edges of the fragments dry out and are buffeted by hot winds; mature rainforest trees often die standing at the margins. cascading changes in the types of trees, plants, and insects that can survive in the fragments rapidly reduces [sic] biodiversity in the forest that remains. People may disagree about whether the extinction of other species through human action is an ethical issue, but there is little doubt about the practical problems that extinction poses.

First, global markets consume rainforest products that depend on sustainable harvesting: latex, cork, fruit, nuts, timber, fibers, spices, natural oils and resins, and medicines. In addition, the genetic diversity of tropical forests is basically the deepest end of the planetary gene pool. Hidden in the genes of plants, animals, fungi, and bacteria that have not even been discovered yet may be cures for cancer and other diseases or the key to improving the yield and nutritional quality of foods—which the U.n. [United nations] Food and Agriculture organization says will be crucial for feeding the nearly ten billion people the Earth will likely need to support in coming decades. Finally, genetic diversity in the planetary gene pool is crucial for the resilience of all life on Earth to rare but catastrophic environmental events, such as meteor impacts or massive, sustained **volcanism**.

Soil Impacts

With all the lushness and productivity that exist in tropical forests, it can be surprising to learn that tropical soils are actually very thin and poor in nutrients. The underlying "parent" rock weathers rapidly in the tropics' high temperatures and heavy rains, and over time, most of the minerals have washed from the soil. nearly all the nutrient content of a tropical forest is in the living plants and the decomposing litter on the forest floor.

When an area is completely deforested for farming, the farmer typically burns the trees and vegetation to create a fertilizing layer of ash. After this **slash-and-burn** deforestation, the nutrient reservoir is lost, flooding and erosion rates are high, and soils often become unable to support crops in just a few years. If the area is then turned into cattle pasture, the ground may become compacted as well, slowing down or preventing forest recovery.

Social Impacts

Tropical forests are home to millions of native (indigenous) people who make their livings through subsistence agriculture, hunting and gathering, or through low-impact harvesting of forest products like rubber or nuts. Deforestation in indigenous territories by loggers, colonizers, and refugees has sometimes triggered violent conflict. Forest preservation can be socially divisive, as well. national and international governments and aid agencies struggle with questions about what level of human presence, if any, is compatible with conservation goals in tropical forests, how to balance the needs of indigenous peoples with expanding rural populations and national economic development, and whether establishing large, pristine, uninhabited protected areas—even if that means removing current residents—should be the highest priority of conservation efforts in tropical forests.

Climate Impacts: Rainfall and Temperature

Up to thirty percent of the rain that falls in tropical forests is water that the rainforest has recycled into the atmosphere. Water evaporates from the soil and vegetation, condenses into clouds, and falls again as rain in a perpetual self-watering cycle. In addition to maintaining tropical rainfall, the evaporation cools the Earth's surface. In many computer models of future climate, replacing tropical forests with a landscape of pasture and crops creates a drier, hotter climate in the tropics. Some models also predict that tropical deforestation will disrupt rainfall pattern far outside the tropics, including china, northern Mexico, and the south-central United States.

The Carbon Cycle and Global Warming

In the Amazon alone, scientists estimate that the trees contain more carbon than 10 years worth of human-produced greenhouse gases. When people clear the forests, usually with fire, carbon stored in the wood returns to the atmosphere, enhancing the greenhouse effect and global warming. Once the forest is cleared for crop or grazing land, the soils can become a large source of carbon emissions, depending on how farmers and ranchers manage the land. In places such as Indonesia, the soils of swampy lowland forests are rich in partially decayed organic matter, known as peat. During extended droughts, such as during **El Niño** events, the forests and the peat become flammable, especially if they have been degraded by logging or accidental fire. When they burn, they release huge volumes of carbon dioxide and other greenhouse gases.

© Elena Sherengovskaya/Shutterstock.com

Figure 99.2 When forests are cleared with fire, carbon stored in the wood returns to the atmosphere and the soil's nutrient reservoir is lost.

Causes of Deforestation: Direct Causes

People have been deforesting the Earth for thousands of years, primarily to clear land for crops or livestock (**Figure 99.2**). Although tropical forests are largely confined to developing countries, they aren't just meeting local or national needs; economic globalization means that the needs and wants of the global population are bearing down on them as well. Direct causes of deforestation are agricultural expansion, wood extraction (e.g., logging or wood harvest for domestic fuel or charcoal), and infrastructure expansion such as road building and urbanization. Rarely is there a single direct cause for deforestation. Most often, multiple processes work simultaneously or sequentially to cause deforestation.

The single biggest direct cause of tropical deforestation is conversion to cropland and pasture, mostly for subsistence, which is growing crops or raising livestock to meet daily needs. The conversion to agricultural land usually results from multiple direct factors. For example, countries build roads into remote areas to improve overland transportation of goods. The road development itself causes a limited amount of deforestation. But roads also provide entry to previously inaccessible—and often unclaimed—land. Logging, both legal and illegal, often follows road expansion (and in some cases is the reason for the road expansion). When loggers have harvested an area's valuable timber, they move on. the roads and the logged areas become a magnet for settlers—farmers and ranchers who slash and burn the remaining forest for cropland or cattle pasture, completing the deforestation chain that began with road building. In other cases, forests that have been degraded by logging become fire-prone and are eventually deforested by repeated accidental fires from adjacent farms or pastures.

> ### Consider This
>
> If people have been deforesting the Earth for thousands of years, why are we so worried about deforestation occurring today? What might be different about the pace and scale of current deforestation trends than that of earlier time periods?

Although subsistence activities have dominated agriculture-driven deforestation in the tropics to date, large-scale commercial activities are playing an increasingly significant role. In the Amazon, industrial-scale cattle ranching and soybean production for world markets are increasingly important causes of deforestation, and in Indonesia, the conversion of tropical forest to commercial palm tree plantations to produce bio-fuels for export is a major cause of deforestation on Borneo and Sumatra.

Underlying Causes

Although poverty is often cited as *the* underlying cause of tropical deforestation, analyses of multiple scientific studies indicate that that explanation is an oversimplification. Poverty does drive people to migrate to forest frontiers, where they engage in slash and burn forest clearing for subsistence. But rarely does one factor alone bear the sole responsibility for tropical deforestation.

State policies to encourage economic development, such as road and railway expansion projects, have caused significant, unintentional deforestation in the Amazon and central America. Agricultural subsidies and tax breaks, as well as timber concessions, have encouraged forest clearing as well. Global economic factors such as a country's foreign debt, expanding global markets for rainforest timber and pulpwood, or low domestic costs of land, labor, and fuel can encourage deforestation over more sustainable land use.

Rates of Tropical Deforestation

The scope and impact of deforestation can be viewed in different ways. one is in absolute numbers: total area of forest cleared over a certain period. By that metric, all three major tropical forest areas, including South America, Africa, and Southeast Asia, are represented near the top of the list. Brazil led the world in terms of total deforested area between 1990 and 2005. The country lost 42,330,000 hectares (163,436 square miles) of forest, roughly the size of california. Rounding out the top five tropical countries with the greatest total area of deforestation were Indonesia, Sudan, Myanmar, and the Democratic Republic of congo.

Another way to look at deforestation is in terms of the percent of a country's forest that was cleared over time. By this metric, the island nation of comoros (north of Madagascar) fared the worst, clearing nearly 60 percent of its forests between 1990 and 2005. Landlocked Burundi in central Africa was second, clearing 47 percent of its forests. The other top five countries that cleared large percentages of their forests were Togo, in West Africa (44 percent); Honduras (37 percent); and Mauritania (36 percent). Thirteen other tropical countries or island territories cleared 20 percent or more of their forests between 1990–2005.

Sustaining Tropical Forests

Strategies for preserving tropical forests can operate on local to international scales. on a local scale, governments and non-governmental organizations are working with forest communities to encourage low-impact agricultural activities, such as shade farming, as well as the sustainable harvesting of non-wood forest products such as rubber, cork, produce, or medicinal plants. Parks and protected areas that draw tourists—ecotourism—can provide employment and educational opportunities for local people as well as creating or stimulating related service-sector economies.

On the national scale, tropical countries must integrate existing research on human impacts on tropical ecosystems into national land use and economic development plans. For tropical forests to survive, governments must develop realistic scenarios for future deforestation that take into account what scientists already know about the causes and consequences of deforestation, including the unintended deforestation that results from road-building, accidental fire, selective logging, and economic development incentives such as timber concessions and agricultural subsidies.

Several scientists are encouraging the conservation community to re-consider the belief that vast, pristine parks and protected areas are the holy grail of forest conservation. In 2005, for example, scientists using satellite and ground-based data in the Amazon demonstrated that far less "unfettered" deforestation occurred in recent decades within territories occupied and managed by indigenous people than occurred in parks and other protected areas. The year before, scientists studying Indonesia's tropical forests documented a 56 percent decline in tropical lowland forests in protected areas of Borneo between 1985 and 2001. They concluded that the deforestation in the protected areas resulted from a combination of illegal logging and devastating fires that raged through logging-damaged forests during the 1997–1998 El niño-triggered drought. While some might argue that these losses could be prevented in the future through better enforcement of environmental laws, it may also be true that inhabited forest reserves are a more realistic strategy for preserving the majority of biodiversity in larger areas than parks alone can accomplish.

<table>
<tr><td>Consider This</td></tr>
<tr><td>Why might tropical forest areas inhabited by indigenous, or native, people have less deforestation?</td></tr>
</table>

Finally, on the national and international scale, an increasing value in the global marketplace for products that are certified as sustainably produced or harvested—timber, beef, coffee, soy—may provide incentives for landowners to adopt more forest-friendly practices, and for regional and national governments to create and enforce forest-preservation policies. Direct payments to tropical countries for the ecosystem services that intact tropical forest provide, particularly for carbon storage to offset greenhouse gas emissions, are likely to become an important international mechanism for sustaining tropical forests as more countries begin to seriously tackle the problem of global warming.

Adapted from Lindsey, R. (2007). Tropical Deforestation. NASA Earth Observatory Feature Article. Retrieved from http://earthobservatory.nasa.gov/Features/Deforestation/

Wildfires
By Ingrid Ukstins and David Best

Approximately 85 percent of all wildfires are started by humans either through carelessness or premeditated arson. The majority of these occurs in the southern United States. The remaining 15 percent are the result of the more than 4 million lightning strikes that occur daily on Earth. The mountainous parts of the western United States have the most lightning-generated fires due to warmer climates that dry out the forests, not that the regions have more lightning (Box 100.1)

BOX 100.1	Causes of Wildfires–Humans versus Mother Nature

Data provided by the National Interagency Fire Center (NIFC) in Boise, Idaho, for the period 2001 to 2017 show that humans caused an annual average of 61,952 wildfires. The greatest number of these (67 percent) occurred in the southern and eastern regions of the United States.

Source: NIFC.

Box Figure 100.1   The United States is divided into 10 fire management regions under the U.S. Forest Service.

The five leading regions with the most acreage burned due to humans are

Southern	989,355 acres
Southwest	331,300 acres
Southern California	296,281 acres
Rocky Mountain	215,787 acres
Northwest	196,335 acres

During the period 2001 to 2017 lightning caused an annual average of 10,143 wildfires. The five leading regions that were affected were as follows:

Alaska	1,497,520 acres
Great Basin	474,237 acres
Northwest	413,851 acres
Northern Rockies	342,226 acres
Southwest	304,596 acres

An interesting observation is that Alaska only experienced an annual average of 180 wildfires as the result of lightning strikes per year, but the wildfires they created destroyed more than three times the acreage of the Great Basin region.

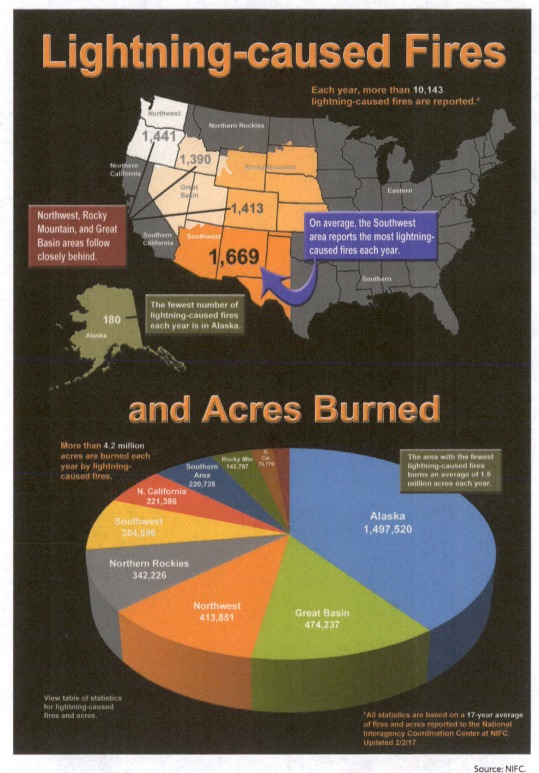

Lightning-caused Fires

Each year, more than 10,143 lightning-caused fires are reported.*

Northwest
1,441

Northern Rockies

Northern California

1,390

Great Basin

Rocky Mountain

Eastern

1,413

Northwest, Rocky Mountain, and Great Basin areas follow closely behind.

Southern California

Southwest

On average, the Southwest area reports the most lightning-caused fires each year.

1,669

Southern

180
Alaska

The fewest number of lightning-caused fires each year is in Alaska.

and Acres Burned

More than 4.2 million acres are burned each year by lightning-caused fires.

The area with the fewest lightning-caused fires burns an average of 1.5 million acres each year.

Rocky Mtn
142,787

S. Cal.
76,776

Southern Area
220,728

N. California
221,386

Southwest
304,596

Northern Rockies
342,226

Alaska
1,497,520

Northwest
413,851

Great Basin
474,237

View table of statistics for lightning-caused fires and acres.

*All statistics are based on a 17-year average of fires and acres reported to the National Interagency Coordination Center at NIFC. Updated 2/2/17

Source: NIFC.

Box Figure 100.1 Almost 60 percent of all lightning-caused fires occur in the western United States.

With changes in global climate conditions and weather patterns, wildfires have become larger and more destructive. When they occur near urban areas or in developed sections in the forests, homes can be adversely impacted.

Types of Wildfires
By Ingrid Ukstins and David Best

There are three main types of wildfires, depending on the location of the burning material.

1. **Ground fires**. **Ground fires** consist of slow-burning material on or just below the surface and include plant roots and buried vegetative matter such as leaves or needles (**Figure 101.1a**). These fires can smolder below the surface so the damage they do is not readily seen. Ash from these and other fires can fill spaces in the soil and create a **hydrophobic** layer that impedes the downward movement of moisture. This condition contributes to erosion and other problems.

2. **Surface fires**. **Surface fires** consume fuel lying on the surface, such as ground litter, limbs, fallen trees, and grasses (**Figure 101.1b**). The intensity of surface fires is quite variable depending on how dry the material is and how much fuel is present. Some of these surface fires can become very intense. Many of the wildfires in California are located in areas where fuels have dried out and fires are fanned by **Santa Ana winds**, which result from high pressure over the Great Basin of Nevada. Common from October

United States Forest Service.

Figure 101.1a   Ground fire.

Figure 101.1b   **Surface fire.**

of each year to the following March, they are produced by cool air in the eastern deserts of California being pushed westward over mountains. At the top, the cold, dry air descends, heating up and generating strong winds that are often associated with wildfires in southern California.

3. **Crown fires**. In forested areas, fires that begin on the surface often leap into the upper portions of trees to produce **crown fires** (**Figure 101.1c**). The ladder concept causes low-burning fires to climb up into the crowns of trees, where winds easily fan the flames from tree to tree (**Figure 101.2**). Unlike surface fires, which can be fought with fire personnel and equipment, crown fires are nearly impossible to control and the results are disastrous (**Figure 101.3**).

Figure 101.1c   **Crown fire.**

© Kendall Hunt Publishing Company.

Figure 101.2   A progression of fuel heights produces a ladder effect to get the tops of trees burning.

© BGSmith/Shutterstock.com.

Figure 101.3   A wildfire destroyed a once healthy forest in Glacier National Park, Montana.

ground fire: Fire that moves along the ground and rises several feet above the surface.

hydrophobic: Incapable of absorbing water.

surface fire: Fire that moves along the ground and consumes ground litter.

Santa Ana winds: Seasonal winds that originate from high-pressure over the Great Basin of Nevada, pushing dry air from east to west across southern California.

crown fire: A fire that moves through the upper portions of trees; these are very difficult to extinguish.

Fire Behavior Variables

Fire behavior is controlled by several variables. Topography, fuel types, and weather each play a major role in how a fire will spread.

Topography

Topography, or the lay of the land, can influence how rapidly a fire moves along the surface. The direction that a slope faces in terms of its exposure to sunlight is critical as the amount of vegetation can be much greater on those slopes receiving a high amount of sunlight. In the northern hemisphere, south-facing slopes produce thicker vegetation, which can become dried out in periods of low humidity by constant exposure to the sun. This is especially true from May to September in the western United States.

Healthy forests can be quickly destroyed by wildfires (**Figure 101.3**). As heated air from a fire rises, it will dry out vegetation at the higher elevations, thereby producing preheated, dry material that can subsequently be burned. A general rule is that the steeper the slope, the faster the fire advances and it spreads upslope. Convection of hot gases causes updrafts that fan the flames from below, causing the fire to spread rapidly.

topography: The general configuration of the land surface.

Fuel Types

Fire behavior is dependent on the type of fuel that is burning. Dried grasses and low shrubs can produce hot fires but these are often short-lived events due to the lack of material. These fires often creep along the ground as low-intensity fires.

Taller trees generate more intense fire activity due to their larger size. Branches long with leaves and needles will ignite and spread the fire upward. When the tops of trees ignite, a crown fire can result. One reason we have seen more large fires in recent years is that for many years the prevailing philosophy was that all fires should be put out immediately. This forest management style resulted in an overall buildup of forest floor fuels, which served to generate large fires once the material was ignited. In recent years, the US Forest Service, National Park Service, Bureau of Indian Affairs, and Bureau of Land Management, stewards of most of the forests and grasslands in the United States, have allowed large fires that were ignited naturally to burn while being monitored. These fires are considered to be in the wildland fire use (WFU) category. Such a designation was used in the summer of 2006, when a large fire on the north rim of Grand Canyon National Park in Arizona was started by a lightning strike. Of the more than 58,000 acres consumed by the fire, almost 19,000 acres were considered as the wildland fire use phase of the incident, while the remaining 39,000 acres burned as a wildland fire.

Governmental fire agencies consider wildland fire use as the management of naturally ignited wildland fires in order to accomplish resource management goals for a given area. These goals include the reduction of likelihood of unwanted fires, the maintenance of natural ecosystems in a given area, thus enhancing the health and safety of firefighters and the public. The purpose of wildland use fires is the same as that of prescribed burns.

Weather

Weather plays an obvious role in the spread of fires. Dry, windy conditions will greatly enhance the spread of wildfires, while moisture-laden storms will help quench a fire. Firefighters are aware of the amount of relative humidity in the air. A higher relative humidity means more moisture is present that can help impede the advance of a fire. As a general rule, relative humidity increases at night, which helps slow the growth of fires. Air temperature is also important as higher daytime temperatures will dry out the fuel.

Rapidly spreading fires are often accompanied by strong winds that tend to flatten out the flames and also transport glowing embers ahead of the fire, making it difficult to fight the fire along a well-defined line. Wind increases the rate of spreading of a fire.

Teams that are sent to fight wildfires are in contact with nearby weather service personnel to keep informed about upcoming conditions. Oftentimes one member of the team assigned to direct firefighting operations serves as the liaison with weather officials.

Besides drying out fuel, drought can also stress trees and reduce their ability to ward off insects because there is less moisture in the trees themselves. This condition makes the trees susceptible to insect damage and diseases. In recent years, the pine bark beetle has killed thousands of acres of trees in North America, particularly conifers such as ponderosa pine and Douglas fir (**Figure 101.4**).

These standing dead trees are then fuel for raging wildfires that can rapidly move through an area, consuming healthy trees as well. Sometimes even without the presence of dead trees, healthy trees will burn, particularly if they are stressed from experiencing dry conditions for several years (**Figure 101.5**).

US Fish and Wildlife.

Figure 101.4   Conifers destroyed by bark beetles. These trees become fuel for fast moving wildfires.

Source: David M. Best.

Figure 101.5 **Results of a wildfire in a young forest that has experienced a decade of drought conditions. The 125-acre fire was started by a thrown radial tire along Interstate 40 outside Flagstaff, Arizona.**

Many areas in the United States have experienced significant drought conditions over the past twenty or more years. This has greatly reduced the moisture content of trees, grasses, and other vegetation, thus making them primary targets for wildfires. Major wildfires have devastated the landscape in the several states in the central and western parts of the country since 2010. These are areas that do not receive much normal precipitation, so recovery will be very slow.

Firestorms

Very large wildfires are capable of producing their own weather by generating massive updrafts that can rise to altitudes of 20 km or more. The upward movement of heated air causes surface winds to be drawn inward, generating a **firestorm**. These self-produced weather systems will have winds in front of the progressing fire and can also generate lightning that can produce more fires. Wind also increases the flow of oxygen to the fire, and the variable nature of wind can make it extremely difficult to forecast where a fire will move.

firestorm: A widespread, intense fire that is sustained by strong winds and updrafts of hot air.

Fire whirls are a phenomenon sometimes observed at locations of extreme heat. Fire whirls are small tornado-shaped currents that consist of fast, rotating flames. They are capable of carrying embers and other burning debris well above the ground surface and can deposit them ahead of the fire line. Some embers have been carried more than 5 km by wind and convection associated with a fire.

Fires have become larger in the past twenty years although the number of fires has decreased. Changes in climate have contributed to more dry fuel, which burns faster than wet trees.

Another factor possibly causing larger fires is that the initial response is sometimes slow as the fire's potential growth is assessed. Sudden changes in weather conditions, especially in conjunction with thunderstorms, can cause the fires to expand rapidly.

Minimizing Wildfires and the Result

By Ingrid Ukstins and David Best

Land Clearing

Areas of urban expansion often have an interface with forested land. Many municipalities have put plans in place that require the removal of vegetation around homes and other structures. This causes its own problem, however, as it increases soil erosion, and runoff can be up to several hundred times greater than the normal amount after surface fuels are cleared.

BOX 102.1	The Deadly Hazards of Fighting Wildfires

SOUTH CANYON FIRE, GLENWOOD SPRINGS, COLORADO, 1994

On July 2, 1994, lightning started a fire several miles west of Glenwood Springs, Colorado, in an area that was extremely dry because of a prolonged drought. High temperatures and low humidity contributed to the tinder-box conditions. Because of other fires in the vicinity, resources were not sent to this fire until it had burned for two days. By July 6 several small crews had been assigned to the fire, including two groups of smoke jumpers who had been dropped into the fire zone. During the midafternoon of July 6, winds up to forty-five miles per hour resulted in sixty- to ninety-meter flames causing the fire to jump a drainage area, where it rapidly overtook and killed twelve members of a hotshot crew along with two members of a helitack crew. Fortunately, thirty-five firefighters survived the onslaught. The combination of extremely steep terrain and catastrophic weather conditions with an explosive fuel source, along with poorly established escape routes, led to the disaster. The entrapment of these firefighters in this 2,115-acre fire marked the greatest loss of life in a wildfire in more than forty years. Two primary causes of the personnel disasters in the Mann Gulch fire and the South Canyon tragedies were that the firefighters were working downhill toward the fire and were inattentive to the weather and fire behavior.

YARNELL HILL FIRE, YARNELL, ARIZONA, JUNE 2013

June 2013 was a typically dry period in central Arizona, but monsoon activity had begun to build toward the latter part of the month. The region around Yarnell, located 100 kilometers northwest of Phoenix, is situated in the Weaver Mountains, which consist of rugged granitic outcrops and interspersed chaparral (Box Figure 102.1). The chaparral, ranging from one to three meters in height was extremely dry due to drought, which had lasted more than fifteen years.

Thunderstorm activity in the late afternoon of June 28 ignited a fire that drew a small crew the next day to contain it. Within twenty-four hours it had expanded to about 100 acres, and by the early morning of June 30, it covered almost 500 acres. Its rapid growth prompted the assignment of a Type 1 Incident Management Team to oversee operations. Several crews were dispatched, including the Granite Mountain Hotshots, based in nearby Prescott, Arizona. They were on the fire less than twenty-four hours when they were overrun by extreme fire conditions. The fire was moving at an estimated ten to twelve miles per hour, being driven by winds of thirty-five to

© Tim Gray/Shutterstock.com.

Box Figure 102.1 Terrain and chaparral vegetation typical of that found at the South Canyon and Yarnell Hill Fires.

forty miles per hour. Their planned escape route to a designated safety zone was cut off. They deployed their fire shields but nineteen firefighters died. Only their lookout survived.

The South Canyon and Yarnell Hill fires bring home the fact that more than 950 personnel have died fighting wildfires in the United States. There have been instances when more than ten firefighters died in single burnovers due to rapidly changing conditions, inadequate communication, and a lack of awareness of the terrain and the fuel conditions in which they were operating. With each unfortunate episode the governmental agencies conduct thorough reviews and strive to improve field operations.

Unfortunately, most homeowners living in residential neighborhoods and forested areas like having trees and vegetation near these buildings. The establishment of defense zones surrounding the homes and other structures lessens the chance that wildfires will destroy buildings.

Prescribed Burns

The continual buildup of surface fuels such as leaves, needles, branches, and low undergrowth and grasses provides the fuel for wildfires. Fire management agencies use prescribed burns to remove these surface fuels so as to lessen the likelihood of future uncontrolled fires. A **prescribed burn** is an intentionally-set fire that takes place under a set of defined parameters and predictable weather conditions. Prior to 2000 the term **controlled burn** was used to describe pre-set fires. Such fires are monitored and actively managed by fire personnel to prevent their getting out of control. Established control lines and some suppression techniques aid in keeping the fire within desired boundaries. In addition to burning material as it lies on the ground, fire managers will also often ignite slash piles consisting of small, rotten, or otherwise undesirable wood discarded during logging (**Box Figure 102.1**). There have been times, however, when slash has provided the fuel for devastating fires such as the fires in Michigan in the nineteenth century and the 1910 fires in northern Idaho.

prescribed burn: A controlled fire purposefully set to remove fuel from an area.

controlled burn: A fire set to remove fuel. Term is no longer used, as not all such fires were kept under control.

National Wildfire Coordinating Group.

Figure 102.1 Slash piles are burned to reduce fuel in forests.

Prescribed burns reduce the fuel side of the fire triangle by removing ground litter and undergrowth. The resulting ash returns nutrients to the soil, helping to sustain the growth of new plants.

There are times, though, when a prescribed burn can turn disastrous, as in the case of the Cerro Grande fire in the summer of 2000. This fire, begun as a controlled burn, quickly got out of hand and inflicted major damage on the city of Los Alamos, New Mexico. Following this catastrophe, the term controlled burn is no longer used by fire management agencies.

Firebreaks and Burnouts

The spread of wildfires is slowed whenever firebreaks are established. **Firebreaks** are areas that have been cleared of surface and low-lying vegetation and ground litter. Clearing small areas involves actual removal of the fuel sources. However, a fire can generate airborne embers that can spread the fire past a narrow firebreak. Larger areas are cleared by setting prescribed burns on dozens or often hundreds of acres.

There are times when fire crews will conduct a **burnout**, a tactic that removes the nearby fuel before a fire gets to it. When a wildfire is threatening a community or if valuables are at risk, firefighters will sometimes deliberately set a back burn, a rather risky decision that is often a last resort. **Backburns** are carried out along a natural break, such as a roadway or near threatened buildings. The burn is drawn into the larger fire because of the convective heat rising from the larger fire.

firebreak: A clearing that provides a gap across which fire should not burn.

burnout: An intentional fire that is lit to burn fuel that lies in the path of a spreading fire.

backburn: A fire that burns back toward the main part of a wildfire.

Results of Wildfires

Rejuvenation

Fires do have a useful purpose. They can help regenerate plants and other flora by opening seeds and by adding nutrients to the soil. This is evident in areas that have sustained severe burns when plants and trees rejuvenate themselves, often within a few months after the fire if moisture is present. In cold climates where there is little chemical breakdown, fire can speed up the process of enriching the soil with nutrients. Nitrogen is released from the burned plants and helps accelerate future plant growth. Large wildfires in Alaska are very beneficial to that state's high-latitude ecosystems because soils are slow to form naturally in cold conditions.

One excellent example of natural rejuvenation is seen in the regeneration of aspen trees following a wildfire. The root systems of aspen are relatively shallow and put out numerous sprouts that penetrate the surface in a fire area. This often occurs very soon after a fire (**Figure 102.2**).

Source: USFS.

Figure 102.2   Aspen quickly return within two years of a fire that burned the predominately ponderosa pine forest.

Erosion

In the time following a large wildfire, the potential for major surface erosion is a primary concern. Steep slopes that were once covered by lush vegetation that could hold water and impede its rapid flow across the surface no longer slow the flow. Rapid runoff removes top soil and cuts deep channels and rivulets into the surface (**Figure 102.3**). In regions near highways, road cuts, and other steep slopes, mass wasting is accelerated. California experiences significant slides every year in burned areas. This was also the case with the Thomas Fire that adversely affected the city of Montecito and surrounding communities.

New Mexico Environment Department.

Figure 102.3   **Removal of surface vegetation by wildfires enhances surface erosion.**

Extreme temperatures exceeding 1500°C transform surface soils into welded particles that produce an impermeable zone which is termed a hydrophobic layer. The **hydrophobic layer** causes more sheet erosion and flooding. Seasonal rains and snowfall will produce debris flows and landslides, along with flooding that can be catastrophic to low-lying areas. Replanting hillsides with grasses and seedlings helps alleviate the problem, but the landscape will not be the same for many years, especially in arid areas where plant life is slow to grow.

hydrophobic layer: A layer of material that cannot absorb water; usually forms in dry climates or in regions that have experienced intense fires.

Fiscal Costs

There are two main monetary costs related to wildfires: the direct cost of fighting the fires and the secondary cost of their economic impact. Over the past 30 years the cost of fighting fires has risen 500 percent. This is due to fires being much larger than in the past and often occurring in developed area requiring more resources to put them out.

In forested areas where timber is harvested for sale, the revenue source is lost. The profits from the sale is used to help cleanup and rehabilitated burned areas and to plant and reseed new areas. In previous years, when wildfires were less damaging, the revenue from these timber sales could be used by the affected communities to rehabilitate and strengthen their forests. These steps aid in the recovery process to minimize erosion problem.

Fracking
By Ingrid Ukstins and David Best

BOX 103.1	Case Study: Fracking

Fracking is a process for extracting natural gas from shale layers thousands of feet below the surface of the Earth. These shale layers contain hydrocarbons, but they are not very permeable, so the gas is trapped in the rock. The process of extracting that gas is called hydraulic fracturing, or fracking, and allows industry to extract natural gas from rock reservoirs that otherwise wouldn't be economically viable resources. The process involves drilling a well to the target rock and pumping down a fracking fluid made up of water and thickening agents with a solid material like sand suspended in it, which pressurizes the rock and cracks it. The fluid flows in and the sand grains or other solid material, called a proppant, moves into the fractures, holding them open (or propping them apart). Once the rock is fractured and the fractures are kept open with the proppant, the trapped gas can more easily flow into the fractures and be extracted for commercial use (**Box Figure 103.1.1**).

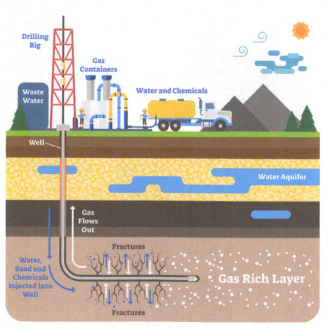

© VectorMine/Shutterstock.com.

Box Figure 103.1.1 Hydraulic fracturing is a process that allows petroleum companies to extract hydrocarbons from shale reservoirs that would not otherwise be economically viable resources. However, it is highly controversial, with proponents arguing for the economic benefits of the petroleum resources and opponents arguing that there are significant and long-term environmental impacts.

Fracking began commercially in 1950, and current estimates of how many wells have been fracked are difficult to make because reporting requirements are variable in different places, and data are difficult to access. A 2017 analysis of petroleum production in the United States suggests there are as many as 1.3 million fracked wells in thirty-four states that are actively producing oil or gas (**Box Figure 103.1.2**). The largest number of these are in Texas, which has almost 400,000 facilities alone. Other states with significant numbers of fracked wells are Pennsylvania (more than 100,000), California, Kansas and Ohio (more than 90,000 each), New Mexico (almost 60,000), and Colorado and Illinois (about 50,000 each).

In late October of 2018, the energy firm Cuadrilla began drilling for gas reservoirs in shale rocks in Lancashire, United Kingdom. Two days later, three earthquakes occurred in the area of the drilling, and after nine days there had been seventeen earthquakes. These earthquakes measured up to 0.8 on the Richter Scale and were too small to be felt on the surface but can be detected by seismic monitoring equipment. Any earthquake over 0.5 magnitude requires an immediate stop to drilling for eighteen hours. Environmentalists tried to block the fracking in court but were not successful. After the earthquakes, Manchester's metro mayor Andy Burnham said, "The Earth is telling us something. Stop this process now." These induced earthquakes are one of the reasons fracking is so controversial; they can be generated by any over-pressurization of layers under the surface of the Earth because pumping fracking fluid into shale reservoirs pressurizes these layers. Fracking waste-water can have significant public health and environmental impacts, such as ground and surface water contamination.

© Anton Foltin/Shutterstock.com.

Box Figure 103.1.2   After the well is drilled, it is connected to pumps that push the fracking fluid into the well to pressurize it.

In 2008, people living in Pavillion, Wyoming, began to notice a bad taste and smell in their drinking water. They live in the middle of a natural gas basin that was being drilled and fracked for petroleum resources. Wastewater from the fracking wells was being stored in unlined pits dug in the ground, where it leaked into surface waters and contaminated the drinking water. Fracking fluids can also contaminate subsurface water reservoirs if the wells are drilled through those layers and the the fracking fluids aren't sealed off from the water reservoirs. Documenting groundwater contamination from fracking is difficult because a lot of the compounds used for hydraulic fracturing are not commonly analyzed in commercial labs. Chemicals like methanol can be part of the fracking fluid and can trigger permanent nerve damage and blindness when consumed at high enough levels. However, it degrades rapidly and virtually disappears from samples over a few days. New testing methods confirmed that groundwater wells from Pavillion contained methanol, high levels of diesel compounds, salts, and other compounds used in fracking fluids. New tests have detected nineteen concerning chemicals in the water supply, half of which are unstudied and do not have known safe exposure levels. Residents of Bradford County in Pennsylvania have methane contamination in their drinking water, which is attributed to fracking that began in the area in 2008. Bradford has more wells than any other county in Pennsylvania, and some residents near the wells have noticed chemical smells and tastes in their drinking water, including methane levels high enough to light their water on fire Chesapeake Energy, who drilled in the area, reached a $1.6 million settlement with three families in 2012 after they sued the company for contaminated drinking water from their wells. According to an EPA study: "The most important findings are that drilling, fracking, and the use of hazardous chemicals necessary to frack have caused groundwater contamination."

The Geography of World Population
By Tim Anderson

104

The tripartite structure of the world-economy (core, semi-periphery, and periphery) is a useful model for understanding the world's human geography. Human population varies across the planet, especially regarding distributions, structures, and core-periphery relationships. The news media and various international organizations often remind us that our world confronts population "problems" today. Invariably, these problems are presented as relating to either overpopulation or how the growing world population affects the supply and use of various natural resources. If there are so-called problems related to population, what are they? Do such issues vary between the core and periphery? How do the structures of populations differ in the zones of the world-economy? How is population distributed around the world?

World Population Distribution

If we examine a map of global population distribution, one of the first things that is readily apparent is that the world's population is not evenly distributed. While some regions are very densely populated (Europe and much of Asia, for example), large parts of the earth (such as the arctic regions, Australia, and Siberia) are very lightly populated. In general, if indeed problems relating to overpopulation exist, those problems are not found everywhere around the world; it is not that there are too many people on the planet (that is a value judgment), but that there are too many people in certain places. The densest population clusters tend to be located in two main types of natural environments around the world. The first is the fertile river valleys of the tropics and subtropics, and the second is the coastal plains of the mid-latitudes, which generally are temperate regions. More precisely, we can identify four major concentrations of population:

- East Asia (China, the Koreas, Vietnam, and Japan)
- South Asia (India, Pakistan, and Bangladesh)
- Europe (The British Isles to western Russia)
- North America (Boston to Washington, D.C.; West Coast)

Roughly 75 percent of the world's population lives in these four areas; four out of ten people in the world live in just two: East and South Asia. Other smaller concentrations of dense populations occur on the island of Java (part of Indonesia), the Nile Valley of Egypt, central Mexico, and southeastern South America (southern Brazil and eastern Argentina).

Factors In World Population

Density

We have described the way populations "look," the way they are structured and their rates of growth, and how that varies significantly between the core and the periphery. It is possible to compare and contrast different populations by comparing various statistics (examples of these statistics for various countries are listed in Table 1.1). One of the most elementary of these factors is population density. Density can be measured as crude population density, the total number of people per unit area of land in a place or region, or as physiologic population density, the total number of people per unit area of arable (agriculturally productive) land. The latter is actually a more telling figure because it measures the density of populations with respect to how much of the land on which they are living is productive enough to produce enough food for that population. When the difference between a country's crude and physiologic densities is very large, it is a sure sign that that country has considerable marginally productive agricultural areas. This can be observed in the following list:

Country	Crude Density/km²	Physiologic Density/km²
Japan	862	6,637
Bangladesh	2,124	3,398
Egypt	142	7,101
Netherlands	1,041	4,476
USA	67	335

Growth Rates

A second method for comparing populations is by examining population growth rates.. Since the Industrial Revolution, the world's population has been growing at an exponential rate (2, 4, 8, 16, 32, 64 —). Currently, the world rate of natural increase is about 1.8 percent per annum. This means that 1.8 percent of the current population is being added each year. This figure translates into a current doubling rate of 40 years, but this doubling rate will decrease with added population each year. Given this growth rate, the world's population will exceed 7 billion by 2012. It is clear that peripheral populations are growing at a much faster rate than those in the core and semi-periphery. That is, on a broad global scale, direct correlation exists between economic "development" and population growth rates:

Country	Rate of Natural Increase	Doubling Time
Poland	0.5%	141 years
Australia	0.75%	94 years
China	1.5%	46 years
Kenya	4.0%	17 years

Structure

A third way of comparing populations around the world is by observing differences in the structure of populations. By structure, we mean the relative number of men and women in different age cohorts in a population. A population's structure is most clearly seen by constructing a population pyramid that charts both male and female populations in five-year age cohorts on a y-axis and the percentage of total population on an x-axis. The term population pyramid is used to refer to such age-sex diagrams because the shape of these diagrams is pyramidal in developing countries, that is, in the peripheral regions of the world-economy. This pyramidal shape indicates a population that is "young" and growing. Birth rates and fertility rates (discussed below) are relatively high, and life expectancies are relatively low. Thus, a substantial proportion of the population in the peripheral countries is very young, under 15 years of age, while the number of people in higher age cohorts, above 60, is very low. By contrast, in the core and parts of the semi-periphery, age-sex diagrams tend to have a rectangular shape. These countries have low birth and fertility rates, and higher life expectancies, and thus, the population is more evenly distributed among age cohorts. These populations are "old" and stable.

Demographic Cycles

If we examine past patterns of population growth rates in different parts of the world, we can identify demographic stages through which populations tend to pass. Where a country is concerning this cycle (that is, what stage the country is in) tends to mirror economic development. We can discern these cycles and stages by examining the relationship between three major indictors: the crude birth rate, defined as the number of live births per 1,000 persons per year; the crude death rate, defied as the number of deaths per 1,000 persons per year; and the total fertility rate, defined as the average number of children born to women of childbearing age (roughly 15–45) during their lifetimes. We can compute the rate of natural increase for a given country by subtracting the crude death rate from the crude birth rate. For example, if the crude death birth rate of a country is 20/1,000 and the crude birth rate is 5/1,000, then the natural increase is 15/1,000, or a rate of 1.5% per annum. Death rates do not vary substantially from core to periphery. Indeed, some core countries have higher death rates than some of the poorest countries in the world. Only in areas of famine or economic and political unrest (and such occurrences are usually short- lived) are death rates inordinately high. This is largely due to advancements in medical technology, especially immunizations for diseases that used to kill millions of people every year. Such technology has become available in the last 50 years even in some of the poorest countries in the world. On the other hand, birth rates and fertility rates, and thus rates of natural increase, vary substantially between the core and periphery.

Core-Periphery Population Patterns

As mentioned above, populations tend to pass through stages, and these are revealed by comparing long-term historical patterns for the relationship between birth rates, death rates, and the rate of natural increase. This historical model of population change is usually called the Demographic Transition Model, of which there are four stages (**Figure 104.1**). In Stage 1, birth rates and death rates are both very high, resulting in relatively low or fluctuating rates of natural increase. Until the Industrial Revolution, when societies around the world were still agricultural in nature, all world populations were in Stage 1, but today, there are virtually no populations in this stage. In Stage 2, death rates fall o substantially but birth rates remain high, resulting in very high rates of natural increase. In Europe, this stage began around the middle of the eighteenth century, as economies and societies began to industrialize. New medical technology greatly reduced death rates, but birth rates and fertility rates remained very high due to advances in medicine and improved agricultural yields because of more efficient agricultural techniques and tools. With enough food to go around, most people saw little need to alter traditional conceptualizations and norms concerning reproduction. This rapid and exponential growth in population in Stage 2 is the "transition" in populations that the Demographic Transition Model refers to. Today, most countries in the periphery, and some in the semi-periphery, are in Stage 2 of this demographic transition.

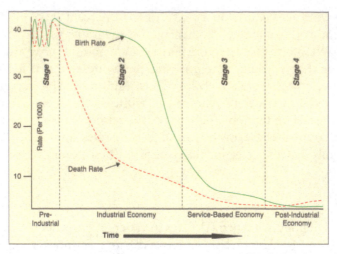

© Source: Timothy G. Anderson

Figure 104.1 The Demographic Transition Model.

In Stage 3 of the Demographic Transition Model, death rates remain quite low and birth and fertility rates begin to drop dramatically, resulting in decreasing rates of natural increase. Most of Europe and North America went through this stage during the late nineteenth and early twentieth centuries as these regions developed mature industrial economies. Today, most of the semi-periphery of the world-economy is in Stage 3. By Stage 4, which began sometime in the late twentieth century in most of the core, birth rates had fallen so much that some countries had approached zero population growth. Indeed, in a handful of core countries today (mainly in northern and eastern Europe), populations are actually declining. The populations in the core regions of the world-economy have passed through each of these stages, and today, they are the only regions in Stage 4 of the model.

Why is there such a strong correlation between economic development and fertility, and what factors account for these global patterns? These are extremely complex questions, for which there are few easy answers. We can, however, identify some of the most probable explanations. To be sure, traditional values and customs concerning reproduction and conceptualizations of femininity and masculinity, as well as traditional religious customs, in the folk cultures of the periphery are part of the explanation. Access to modern forms of birth control (expanded in the core, more limited in the periphery) may also help to explain these patterns. But, we could argue that both of these explanations fail to consider the power that women have in most societies around the world about reproductive choice. They also fail to address differences in economies and lifestyles between the core and periphery. The most likely and most plausible explanation for the correlation between fertility and economic development is that the role of women in the societies of the core and periphery are quite different. In the subsistence agricultural societies of the periphery, the role of a woman is often what we might call traditional—they are not only mothers, but also farmers. In such societies, traditional conceptualizations of women and children predominate, and in these traditional economies, children are an economic asset—the more hands for the fields, the better. On the other hand, in the core regions of the world-economy, post-modern ideas have led to radical critiques of such traditional roles for women. In most of the core societies, the role of a woman is not seen as just a mother, but also as a breadwinner. Also in these post-industrial economies where very few people farm for a living and where the costs of living are substantial, children are in fact an economic liability. In the core, then, women have embraced other roles and put o having children until later in life. This change has resulted in drastically lower fertility rates, since waiting to have children until later in life statistically reduces how many children a woman can have.

In summary, population problems in the periphery and parts of the semi- periphery involve those of an ecological nature. These populations are in Stage 2 of the demographic transition, with very high rates of natural increase. But, at the same time, these are precisely the places that are least able to cope with young and growing populations, mostly due to weakly developed political and economic infrastructures. In short, there are increasingly too many people and not enough resources to go around—not only food resources, but other resources such as fuel and clean water. In the core and parts of the semi-periphery, the issues are quite different. These societies are in either Stage 3 or Stage 4 of the demographic transition, with low birth and fertility rates and increasingly older populations. While the advanced post-industrial economies of the core would be able to cope with larger populations, they are precisely the places where rates of natural increase are the lowest. Here, the most pressing issues related to population involve questions of how to cope with an aging population in which more and more older people who are not working must be supported by fewer and fewer people of working age. This is an especially significant problem in core societies having substantial social welfare systems, where governments are in charge of funding retirement and pension plans.

Different societies are approaching such problems in different ways, with varying results. In India, for example, the population is now over 1 billion and is growing quite rapidly, at just under 2% per annum. In the 1970s, India's federal government attempted to take an active role in population reduction by opening family planning clinics, dispersing contraceptives, and appealing to the patriotism of the population through public relations campaigns and advertisements. These e orts, however, have been met with much public resistance because they do not dovetail well with traditional Indian ideas about reproduction and the family. The results of the government's e orts have been mixed at best, and India's population continues to grow rapidly. Another example of governmental intervention in population growth is China. In the early 1980s, the Chinese government, a very powerful one-party system, took an active and rather forceful role in population reduction. Laws were enacted that gave tax breaks and other incentives to couples who chose to have no children. A one-child-only policy was also enacted and rigidly enforced; it limited each couple in the country to only having one child. The government also used public relations campaigns and advertisements to appeal to the patriotism of its citizenry. The results of such policies were quite different from those in India. In 1970, the rate of natural increase was 2.4%, but it had dropped to 1.2% by 1983 and 1.0% in 1997. This success, however, came with some significant social costs. For example, a heavy male gender imbalance now exists in China because of an increase in abortions of female fetuses due to traditional Chinese ideals about inheritance.

Population Theory

The issue of population has attracted the attention of large numbers of writers and social scientists over the past two centuries. Probably the most famous of these was the English writer Thomas Malthus, whose Essay on the Principle of Population (1798) set in motion a long-running debate regarding population growth that continues to this day. Malthus was writing at a time when England was in Stage 2 of the demographic transition and was experiencing exponential population growth. Malthus argued that while population was growing exponentially, food supplies were only growing arithmetically, and therefore, they would not be able to keep up with the demand for food. This, he wrote, would at some point result in a crisis punctuated by famine and social collapse.

Obviously, the crisis that Malthus predicted did not come to fruition, for he failed to predict new agricultural technologies and techniques that revolutionized agriculture in the nineteenth and twentieth centuries. These new technologies (such as crop rotation schemes, irrigation technologies, and scientific genetic hybrids) greatly increased the amount of food that could be produced, even in some of the poorest countries in the world. Malthus' failings attracted many critics. Marxist thinkers, for example, argue that the real problem facing the world is not overpopulation, but the fact that the world's resources are not equally shared or distributed and are coopted by the capitalist class. Another critic, Esther Boserup, has argued that population growth does not necessarily produce significant problems; it could in fact stimulate economic growth and better food production technology as it did in Europe in previous centuries. But, populations in peripheral countries are increasing at unprecedented rates, and many of these countries have more poor people than ever even though food production has in general increased substantially over the past few decades. These alarming trends have caused some experts to reevaluate Malthus' theory, taking into account not just food, but a variety of other natural resources. These so-called Neo-Malthusians argue that Malthus erred in the sense that he wrote only of food and not of other natural resources, but that his overall idea was correct. They contend that population growth in the developing world is a very real and very serious problem because the billions of very poor people that will be added to the world's population in the coming centuries will result in an ever-increasing desperate search for food and natural resources punctuated by more wars, civil strife, pollution, and environmental degradation.

Demographics and the Demographic Transition Model

K.L. Chandler

Demography

Demography uses statistics about age, gender, race, occupation, income, education, and family status to study the population of people, people groups, and geographic regions. As of May 2022, there is an estimated 7.94 billion people unevenly distributed around the world, many of them settled in the Northern Hemisphere's mid-latitude coastal areas. (**Figure 105.1**) Demographers study these nearly eight billion people and make predictions to improve planning for the future.

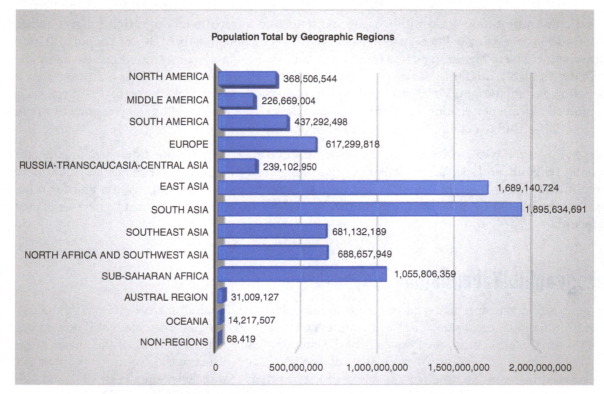

Population Total by Geographic Regions

NORTH AMERICA	368,506,544
MIDDLE AMERICA	226,669,004
SOUTH AMERICA	437,292,498
EUROPE	617,299,818
RUSSIA-TRANSCAUCASIA-CENTRAL ASIA	239,102,950
EAST ASIA	1,689,140,724
SOUTH ASIA	1,895,634,691
SOUTHEAST ASIA	681,132,189
NORTH AFRICA AND SOUTHWEST ASIA	688,657,949
SUB-SAHARAN AFRICA	1,055,806,359
AUSTRAL REGION	31,009,127
OCEANIA	14,217,507
NON-REGIONS	68,419

Source: K.L. Chandler

Figure 105.1 Grouping populations by geographic region is a common practice by demographers.

Population change

Population change is the difference between the number of people in a particular place from the beginning to the end of a given time period, usually one year. Three variables change the population: *birth, death, and migration*. Births and deaths are causes of **natural change**. When the number of births is higher than the number of deaths in a given year, the population experiences a natural increase. Conversely, it goes through a natural decrease when the year's death rate is higher than the birth rate. Many factors influence these natural changes, including global events, such as wartime and economic shifts. The COVID-19 pandemic impacted many areas, increasing death rates among many people groups around the world since 2020. Some of the deaths were a direct result of the virus, while others were indirect results of isolation and reduced medical care.

Migration

Migration, the movement of people from one permanent home to another, creates unnatural change that affects two populations: the one that loses people and the one that gains. An **immigrant** enters a new region or country with the intent of living there (immigration), while an **emigrant** participates in the act of leaving one's country to live in another place, called emigration. People migrate for many reasons that can be categorized into push or pull factors. **Pull factors** are positive motivations that draw people to a new location – food, safety, employment opportunities, quality education, better climate, lower taxes, and even square footage. **Push factors** are the negative reasons that people leave a location, including poverty, famine, drought, poor climate, depleted or poisoned lands, natural disasters, floods, job loss, political change, violence, war, and terrorism. Many people participate in **voluntary migration**, choosing to leave a geographic area without the use of force, while **forced migration** includes the historic transatlantic slave trade and modern human trafficking.

Domestic migration (migration within a political state) and **international migration** (migration between political states) help many political states offset negative population growth. MDCs are popular destinations for immigrants due to higher standards of living, as well as accessible healthcare and greater financial security. Understandably, migrants, as well as **refugees** (individuals who legally flee their political states, due to persecution or human rights violations), and **asylum seekers** (individuals who leave their political states and seek protection from persecution and human rights violations but are not yet legally recognized) would look for protection within the confines of more stable political states.

However, migration is expensive and difficult especially due to varying cultural differences in language and, especially, religious, and societal dissimilarities that often create insurmountable challenges, making assimilation difficult. Many migrants' stories often include experiences of ethnic segregation and for some may even include **return migration** (migrants' returning to their country of origin) due to a new set of push and pull factors.

Demographic Terminology

Demographers define specific terms to create absolute data sets. Much of the collection of these data sets comes from national census data, as well as surveys and statistical models. The data is still only as accurate as the information disclosed, the frequency of the enumeration, and the reliability of the record keepers. A decennial census (enumerating the population of a political state each decade) is common for most political states around the world, with a few notable drastic exceptions that include Somalia in 1985 and Lebanon in 1932.

Demographic Transition Model

The **demographic transition model (DTM)** uses birth and death rates to demonstrate shifts in a population's demographics during economic and social development. Developed in 1929 by American demographer Warren Thompson (1887-1973), the DTM is based on historical data and trends and surmises that birth and death rates correspond with stages of industrial development. (**Figure 105.2**)

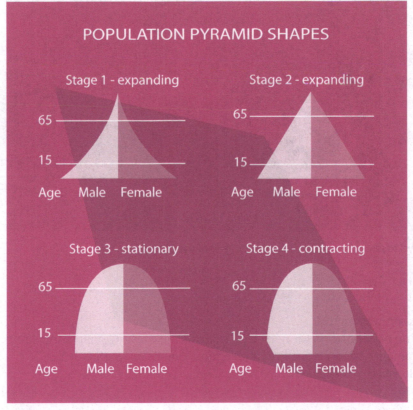

© Iamnee/Shutterstock.com

Figure 105.2   The first four stages of the Demographic Transition Model.

Stages of the Demographic Transition Model

Stage 1: high population growth potential

> economy: pre-industrial
> society: lacks food, adequate medical care, and effective sanitation and hygiene
> birth rate: high
> death rate: high

Stage 2: population explosion

> economy: industrial revolution
> society: more food supply; better medical care; more effective sanitation and hygiene
> status of women: increase in female literacy; available public health education programs
> birth rate: high
> death rate: falling and lifespans are longer

Stage 3: population growth starts to level

> economy: post-industrial revolution (**Figure 105.3**)
> society: reduction of child labor; further improvement in health and sanitation
> status of women: increase in social status and education; available contraception
> birth rate: falling
> death rate: low

Figure 105.3   A crowded street in India, a country in Stage 3 of the DTM.

Stage 4: stationary population

> economy: stabilization
> society: aging of population born in stage 2; shrinking working population; lifestyle diseases increase
> status of women: widely available contraception; desire for small families
> birth rate: lower than replacement level
> death rate: low

**Stage 5: transition

      economy: declining population
      society: increases in lifestyle disease
      birth rate: low, possibly well below replacement level
      death rate: low with possible slight increase

**Theorists debate the legitimacy of a true Stage Five. Some people believe that a new model is emerging, rather than adding a stage to the existing model. Demographers agree, however, that the direction of fertility rates in the next phase is uncertain and that more study must occur before reliable projections can be made.

Demographic transition models do not include migration, only the natural change data sets are considered. Some studies predict an impending population implosion, a drastic population decline. Falling birth rates can be the result of deliberate choice. People might decide that they cannot afford or do not want to have children; other people choose not to continue with a pregnancy that occurs. Countries that face population implosion are starting to use cash incentives and creative marketing strategies, hoping to encourage people to have more children. In 2005, the governor of Ulyanovsk, Russia, declared September 12 an unofficial Day of Conception holiday, granting couples the day off work to help produce the next generation. Couples who gave birth on June 12 of the next year were awarded prizes. (In 2007, the grand prize was a sports utility vehicle.) Population implosion can be devastating for a country that faces the care of an aging, non-employed population when government programs depend on a thriving workforce to support them. This **demographic winter** (negative natural change) creates a societal challenge of significant proportion, and nearly twenty political states currently navigate the challenges of negative population growth in the present day or projected future.

CASE STUDY

CHINA'S POPULATION GROWTH

The country of China has the world's largest population, which has increased tremendously since the twentieth century. A growth rate of 3% in the 1950s, when women bore an average of 5.9 children, prompted advisors to warn Chairman Mao that numbers were accelerating too quickly. Citing the Rule of 70, they projected the political state's population would double in less than 24 years – a number the economy could not sustain. As a result, the government strongly encouraged a *fewer-later-longer childbearing practice* in the early 1970s, hoping to inspire three things: fewer births, the decision to have children later in life, and a longer period between children. Numbers began to drop in the late '70s, but, after Mao's death, the government manipulated a new population growth rate by enforcing the one-child policy in 1979. The effect of the two agendas reduced population growth by more than 50% by the 1980s and curbed growth rates to .05% by early 2000.

Combined with traditional Chinese society's preference for male children, these policies also produced the unintended consequence of an imbalanced gender ratio. Globally, the average population gender ratio is 101 males to 100 females at birth. In China, that ratio disparity has been as great as 107.5 males to 100 females, suggesting that sex-selective abortions occur, despite their illegality. Differentials of male to female children in the 0 to 4 age-bracket are even more stark, showing 113.6 males to 100 females – indicating the possibility of female infanticide. Many female babies were historically placed in orphanages or left unregistered, so couples could legally attempt to have a male child. An estimated 13 million babies were born outside the policy and have no identifying documents, and technically do not even exist. Decades of imbalanced gender ratios have affected a dearth of women of marriageable age in this century. Some estimates project that up to thirty million men will be unable to find women to marry, and this supply and demand issue fuels the horrific human trafficking throughout East Asia and its impoverished neighbors.

The Chinese government changed its childbearing policy in 2016 to allow for two children per family, and again in 2021 allowing for three children. These new policies are an attempt to adjust the gender imbalance, increase the population, and preemptively anticipate the need for caregivers for the state's aged population as children are considered to be a safety net to help aging parents. (Figure 105.4) However, the generations that lived under the one-child policy are not culturally adjusting to the idea of having larger families, so the population growth rate remains below the necessary 2.1 TFR for population replacement. China has closed abortion and sterilization centers, in response to abortion rates that fell from 14 million in 1991 to 9 million in 2020. Fertility clinics are expanding, as up to 18% of the population report conception issues (compared to a global average of 15%), due to older maternal age and the consequences of multiple abortion procedures, some of which were government-mandated. In stark contrast, other citizens receive compensation and subsidies from the Chinese government after the deaths of their only children left them alone in old age.

Figure 105.4 In 2016, China changed the one-child policy to a two-child policy

Human Population Growth
By Ingrid Ukstins and David Best

Biological Annihilation and the 6th Mass Extinction: Us

Planet Earth is in the middle of the 6th wave of mass extinction of plants and animals in the last half-billion years—this one caused almost entirely by humans. Right now, 99 percent of threatened species are at risk from human activities, mainly those causing habitat loss, global warming, and from invasive species. We are presently living through the highest extinction rates since the dinosaurs 65 million years ago. Extinctions are part of the evolution of our planet, and 'background' rates are from one to five species each year—now, however, we're losing species at 1,000 to 10,000 times this background rate—a minimum of 16,928 species threatened with extinction, and that is from an analysis of only 2.7 percent of the 1.8 million species on Earth, according to the International Union for Conservation of Nature. The dodo, the great auk, the thylacine, Chinese river dolphin, passenger pigeon, imperial woodpecker—all have been driven extinct, some deliberately. Using cloning to 'de-extinct' species, or bring them back, sounds like science fiction but could become a reality. However, putting species back after they've gone extinct is loaded with its own problems. The Earth is a delicately balanced system and our actions have a profound impact on all aspects of our planet.

Global population was estimated at five million inhabitants 10,000 years ago. With the domestication of animals and the development of agriculture that allowed people to move away from hunter-gatherer societies, the number of people on Earth began to rise. Growth rates are estimated to have been fairly low for the next 11,000 years, but by the Middle Ages (from the years 1100 to 1500 AD) almost 500 million people were living on the planet. A setback occurred between 1348 and 1350 AD when bubonic plague (the Black Plague) struck Europe, killing tens of millions of people. The world population rebounded in the 1400s and by 1800 it was slightly less than one billion.

The United Nations Population report *The World at Six Billion* provides the following facts about changes in the world population over the past two hundred years:

- It reached one billion in 1804;
- It took 123 years to reach 2 billion in 1927;
- 33 years to reach 3 billion in 1960;
- 14 years to reach 4 billion in 1974;
- 13 years to reach 5 billion in 1987; and
- 12 years to reach 6 billion in 1999.

This report estimated that global population would reach 7 billion in 2013, but according to National Geographic, the planet reached 7 billion people on October 31, 2011. (**Figure 106.1** and U.S. Census population clock). There has been a decline in the annual growth increment from about 2 percent in 1970 to 1.1 percent in the early twenty-first century. However, it is still a positive growth rate. Current global human population growth is about 1.1% per year, or almost 83 million people annually.

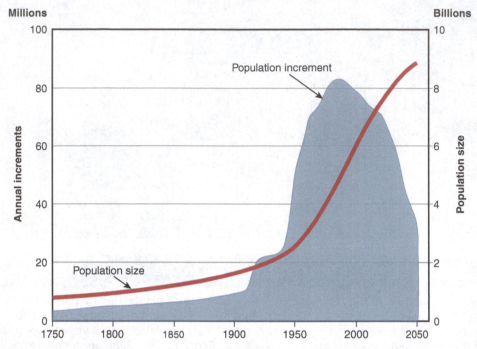

Source: "An Inuit Reality", Duane Smith in UN Chronicle © United Nations, November 2007 (Volume XLIV). Reproduced with permission.

Figure 106.1   The dark line shows global population as measured on the right hand vertical axis. Annual incremental changes are shown by the rectangles which are scaled on the left.

Contributors to the exponential growth in population worldwide include increased industrialization in Europe in the eighteenth century; increased food sources and improved nutrition; the eradication of certain diseases; and development of medicines to improve longevity. Unfortunately diseases such as cholera, influenza, malaria, and tuberculosis continue to kill hundreds of millions of people each year, mainly in underdeveloped countries that lack sanitation and good health services. Childhood diseases also take a huge toll in poor nations, where infant mortality exceeds 10 percent of the live births.

Average life expectancy varies widely throughout the world. Many countries in Africa and Asia have life expectancies of less than 50 years. The more-developed countries have the highest expectancies. The position of the United States is number 43 in the ranking, well below other countries that have considerably higher longevities.

Effects of Overpopulation and Land Use Change

By Ingrid Ukstins and David Best

Land Use

The continued exponential population growth worldwide has placed additional demands on Earth's resources, both those that occur naturally such as water and oil, and on resources that are produced, such as agricultural goods. Destruction of the landscape has had a profound effect on the environment. Our natural resources have not increased, only our ability to find them. Essentially no new land is created (tens of thousands of years are often necessary for volcanic material to form productive soils).

Table 107.1 **Average Life Expectancy***

Rank	Country	Life Expectancy (in years)
1	Monaco	89.4
2	Japan	85.3
3	Singapore	85.2
4	Macau	84.6
5	San Marino	83.3
6	Iceland	83.1
7	Hong Kong	83
8	Andorra	82.9
9	Guernsey	82.6
10	Switzerland	82.6
35	United Kingdom	80.8
43	United States	80

Source: Central Intelligence Agency, *World Fact Book*.

*Measured in Years. Entries contain the average number of years to be lived by a group of people born in the same year, if mortality at each age remains constant in the future. The entry includes *total population* as well as the *male* and *female* components. Life expectancy at birth is also a measure of overall quality of life in a country and summarizes the mortality at all ages. Data represent estimates made in 2017.

Urbanization has placed a higher percentage of people into compact cities (**Figure 107.1**). In the United States only 3 percent of the area of the country is categorized as urban (60 million acres). Often these municipalities are built at the expense of farmland, areas once used to produce crops and raise animals.

© igorstevanovic/Shutterstock.com.

Figure 107.1 Urbanization means that people are become concentrated in small areas, more than half of the world's population is currently living and working in cities. This results in huge social, economic and environmental transformations.

The Department of Agriculture (USDA) reports that 97 percent (2.2 billion acres) of the land in the United States is classified as agricultural, forest, or other use land. In the United States the population explosion in the Southwest has transformed once **desert** areas into bedroom communities. In recent years the overbuilding in cities such as Phoenix and Las Vegas has resulted in thousands of unsold homes and a significant decrease in home values due to a glut of housing. Parts of southern California and southern Arizona that were once rich agricultural areas have been reshaped into suburbs populated by large numbers of people.

> **desert:** A region that receives less than 25 cm (10 in) of annual precipitation.

Agriculture

The USDA reports the following major uses of agriculture land in the United States in 2012:

- Cropland: 392 million acres (17 percent of the land area)
- Grassland pasture and range: 655 million acres (29 percent)
- Forest-use land (total forest land exclusive of forested areas in parks and other special uses): 632 million acres (28 percent)
- Special uses (parks, wilderness, wildlife, and related uses): 352 million acres (14 percent)
- Urban land: 70 million acres (3 percent)
- Miscellaneous other land (deserts, wetlands, and barren land): 196 million acres (9 percent).

Proportions vary across regions of the United States due to differences in climate, geographic setting, and population densities. For example, the Northeast has 12 percent of its area in cropland, compared with 58 percent in the Corn Belt. However, nearly 60 percent of the Northeast is in forest, compared with only 2 percent in the Northern Plains. The land in Alaska, the largest state in the Union, skews the data due to large amounts of forest and very little cropland, so it is not included in these values.

Small changes have occurred in land use, in percentage terms, for the 48 contiguous States. As reported by USDA, the largest acreage change from 1997 to 2002 was a 13-million-acre decrease in cropland (a drop of 3 percent), which continued to 2012 with an additional 50 million acres and 3 additional percent drop. This continues a long downward trend from 1978, when cropland totaled 470 million acres. Total cropland area in 2002 was at 442 million acres, its lowest point since in 1945. Cropland has been relatively constant from 1945 to 1997, ranging between 442 and 471 million acres and averaging about 463 million acres (**Figure 107.2**). The USDA reports an estimated 70 million acres in 2012 as urban use, compared to an estimated 66 million acres in 1997.

© Fotokostic/Shutterstock.com.

Figure 107.2 The amount of crop land has remained relatively constant over the past sixty years. At the same time, global crop production has expanded by 300% through increasing yields and more intensive farming practices.

Factors Contributing to Land Use Change

Land has shifted into crop production from other nonurban uses in response to rising commodity prices. However, land-use changes are gradual, due to conversion costs.

Between 1945 and 2018, the population of the United States more than doubled from 133 million to 326.5 million people. More land was converted to urban uses, especially for homes (**Figure 107.3**). New residential uses also require land for schools, office buildings, shopping sites, and other commercial and industrial uses. The amount of land converted to urban use rose steadily from 15 million acres in 1945 to an estimated 70 million acres in 2012. These increases came mostly from pasture, range, and forest land, a shift that decreased the amount of land available for farming.

© Jeff Whyte/Shutterstock.com.

Figure 107.3 An increase in the urban population has led to the construction of many new homes in areas that were once used for agricultural purposes. This housing development outside Calgary, Canada, is built on land previously used for farming.

Effects of Human Population Growth on the Environment

Rapid increases in global population during the past two centuries have affected cycles related to the environment and ecosystems. These alterations and disruptions have changed the amounts of important elements and components of the environment, such as carbon, oxygen, nitrogen, phosphorus, and water. The disruption and pollution of the atmosphere and hydrosphere have long-ranging effects on ecosystems and **biodiversity**.

> **biodiversity:** All the diversity of life on Earth, ranging from single cell organisms to the most complex ecosystems.

The Concept of Human "Culture"

By Tim Anderson

Traditional Definitions

Culture is one of those words that many of us often use without thinking very deeply about what it exactly means. In the post-industrial era, punctuated by the so-called "culture wars" and post-modern dialogue, it is often used as a catch-all term to describe attributes relating to such things as race, ethnicity, and gender. Today, culture is a highly elusive term that means a lot to some social scientists but nothing at all to others. That is, its meaning and importance are highly debated, and in academia today, it is one of those ideas that is being rigorously critiqued and deconstructed. But, however one might approach or de ne human culture, there is no denying that the study of how it varies geographically and how it shapes and influences cultural landscapes is central to the field of human geography. It is altogether necessary, then, to have a basic understanding of what culture is, how different academic traditions de ne it, and how it is expressed in the landscape in different parts of the world-economy.

Although the concept of human culture in its various forms is a central focus of most of the humanities and social sciences, it tends to be defined in different ways by various academic disciplines. Sociology, for example, stresses the codes and values of a group of people. Sociologists would argue that to really know the culture of a group, whether it be an ethnic group or a class of people or an entire society, one must understand the rules of conduct members of that group have agreed upon. These rules of conduct might include laws, social mores and traditions, and codes relating to family and societal structure. Sociologists tend to look at how order is achieved and how society is organized to understand the values and traditions of that society. Anthropology, a discipline defined as the study of human cultures, tends to focus on the everyday ways of life of a group or society. Such ways of life might include linguistic norms, religious ideals, food, dress, music, and political structures. To uncover and understand the cultural values and traditions of a group or society, anthropologists study these ways of life.

While there is a long tradition of debate about the meaning and nature of culture in the academic literature of many of the social sciences, cultural geographers, until recently, have traditionally spent little time defining what culture is. Instead, traditional human and cultural geography in the United States focused on how culture was expressed in the landscapes of places— the landscape, especially its physical manifestations such as houses, fields, settlement patterns, neighborhoods, etc., were "read" and analyzed to uncover clues about the cultural values and traditions of the people that built them. Until post-modern thought began to influence the field in the 1970s, most cultural geographers were content to rely on definitions of culture that sociologists and cultural anthropologists had developed in the twentieth century. For these cultural geographers, then, human culture consists of many aspects of a group or society that, when combined together, results in a distinctive way of life that distinguishes that group or society from others:

- Beliefs (religious beliefs and political ideals, for example)
- Speech (language and linguistic norms and ideals)
- Institutions (such as governmental and legal institutions)
- Technology (skills, tools, use of natural resources)
- Values and Traditions (art, architecture, food, dress, music, etc.)

As such, cultural values and traditions are not biological in nature. Such traditions are learned, not genetically inherited (that is, we are not born with these values), and are passed on from generation to generation through a mutually intelligible language and a common symbol pool or iconography. For cultural geographers, who seek to identify the spatial expression of culture, a **culture region** is an area in which a distinctive way of life (as defined above) is dominant.

The New Cultural Geography

Post-modern thought has had a dramatic effect on the field of cultural geography over the past 20 years, especially in Great Britain and the United States. In what has come to be known as **The New Cultural Geography**, a new generation of scholars is turning upside down traditional notions of culture and its expression on the landscape. Heavily influenced by post- modern literary and philosophical traditions, and by neo-Marxist thought, one of the leading voices in this new movement has gone so far as to argue that culture does not even exist and that we learn little about the nature of the world and its societies by approaching culture in the traditional ways outlined above. Rather, it is argued by the new cultural geographers that what we might call "culture," for lack of a better term is not a thing, but rather a process that shapes values, traditions, and ideals. These processes and their accompanying values and traditions differ significantly, not from society to society or country to country, but from person to person; they are influenced by such things as an individual's class, gender, race, and sexuality. In this line of thought, it follows that our perception of the world is influenced by the same factors, and it is argued that cultural landscapes hold clues to such factors working in society. They can be read, deconstructed, and analyzed in the same way that a literary text can. The cultural landscape, then, is not seen as simply the built environment or the human imprint on the physical landscape. Instead, the New Cultural Geography conceptualizes it as a place or a stage upon which, and within which, societal problems and processes are worked out, especially with respect to struggles relating to class, race, ethnicity, gender, and politics.

Core-Periphery Relationships

Folk Cultures

Both traditional and post-modern conceptualizations are valuable for a broad understanding of how culture varies around the world. Employing these ideas alongside the core, semi-periphery, periphery model of the world-economy from world-systems analysis, we can understand basic, general differences in ways of life around the world. These basic differences do not translate well down to the local or individual level. To understand the cultural processes at work at such scales, we must analyze cultural processes and patterns at those scales. Here, we are concerned with broad global patterns.

At the global scale, and in a broad sense, we can distinguish between two primary types of culture operating today. At one end of the spectrum are so- called folk cultures. This term describes human societies and cultures that existed in most parts of the world until the Industrial Revolution. At this point, as the core, semi-peripheral, and peripheral areas of the modern world-economy became better defined, a major divergence occurred. Folk cultures remained the norm in the periphery and parts of the semi- periphery, but in the core, and today in some parts of the semi-periphery, cultural values and traditions came to be increasingly modified by "popular" tropes, fads, and ideas (this is discussed in more detail below).

A **folk culture** refers to a way of life practiced by a group that is usually rural, cohesive, and relatively homogeneous in nature regarding traditions, lifestyles, and customs. Such groups and societies are characterized by relatively weak social stratification; goods and tools are handmade according to tradition that is passed on by word of mouth through tales, stories and songs,; and nonmaterial cultural traits (e.g., stories, lore, religious ideals) are more important than material traits (e.g., structures and technologies). The economies of folk societies are most characteristically subsistence in nature —farming or artisan activities are undertaken not to necessarily make a pro t, but rather to simply survive—and the markets for such products are usually local or regional in nature. Finally, order in folk cultures is based around the structure of the nuclear family, ancient traditions, and religious ideals. If we de ne a folk culture by these characteristics (and this is a very conservative definition), then such cultures and societies are practically nonexistent in the core of the world-economy. Instead, they describe most societies that are tribal or "traditional" in nature in the periphery and in some remote, rural parts of the semi-periphery (the Amazon Basin of Brazil, for example, which is part of the semi-periphery). Even so, some folk culture traits almost always persevere even in the post-industrial societies of the core; they are holdovers of our folk cultural roots from hundreds or even thousands of years ago. Some examples would include the popularity of astrology or tarot card reading, the fairy tales that each of us learns as children, and folk songs from long ago that are still passed on today ("Auld Lange Synge" or "Yankee Doodle").

Popular Cultures

While folk cultural traditions dominate most societies in the periphery, the societies of the core and parts of the semi-periphery today are best described by the term *popular culture*. Although some ethnic groups in core regions of the world-economy attempt to live in a traditional or folk manner, or practice a traditional lifestyle, it is nearly impossible for such groups do so in the core because popular culture is so pervasive and far-reaching. For example, even though many Americans see the Amish as a distinctively folk society, closer inspection reveals that, compared to true folk cultures in the periphery, Amish society today is not truly folk in nature. Although most Amish do not use electricity and they do employ rather simple machinery in their agricultural systems, that machinery is mass-produced, material for barns and houses is purchased from retail stores, and their agricultural endeavors are capitalistic, pro t-making undertakings.

Compared with folk cultures, **popular cultures** are based in large, heterogeneous societies that are most often ethnically plural, with a concomitant plurality of values, traditions, and ideals. While folk cultures are by definition conservative (that is, resistant to change), popular cultures are constantly changing. This is due to the power and influence of fads and trends that change rapidly and often in core societies, as well as to the dominance and influence of mass communication in the core. While ideas and trends are slow to move from place to place in folk cultures (usually through hierarchical diffusion), they can move around the planet instantaneously by means of mass communication technology (satellites, the internet, television, radio, etc.) in a popular culture (by means of contagious diffusion). This, in fact, is the central defining characteristic of popular cultures—such fast change and quick diffusion is what makes a culture subject to "popular" (read trends and fads) ideas. In the post-modern era, such trends and fads have significantly shaped how people in core societies receive news, what music they listen to, what books they read, what movies they see, what food they eat, and what clothes they wear. In a folk culture, such things are dictated by tradition that has been passed on by word of mouth over many generations.

Other characteristics of popular cultures include the use of material goods that are invariably mass-produced, and societies in which secular institutions (government, the lm industry, MTV, multi-national corporations employing advertisements to entice people to buy their products) are of increasing importance in shaping the "look," the landscapes, and ways of life in core societies. The power of such popular fads, trends, and ideas is expressed in the standardized landscapes that are a hallmark of the core and parts of the semi-periphery.

That is, popular culture tends to produce standardization that is reproduced *everywhere* in such societies. This can be seen in styles of architecture, music, clothing, dialects, etc., that are the same throughout large, populous societies and over large distances. Currently, the strongest popular culture in the world stems from the United States. Things "American" (music, food, films, styles and the like) affect nearly every place on the planet, including even traditional folk societies in the periphery. Because the diffusion of popular ideas and fads occurs via mass communication, even traditional societies in far corners of the developing world are not immune to the influences of popular culture from the core.

Delimiting and defining the geography of popular culture presents a challenge to human geographers because such cultures tend to produce "placelessness" that challenges unique regional expression. For example, ranch-style homes became popular in the United States in the 1950s. Although the style probably originated in the eastern part of the country, such houses became so popular so fast that they soon could be found everywhere around the country, including Alaska and Hawaii. Another example would be popular music. When a song goes out over the radio or on television, it is heard by millions of people at once, all over the country, or even the world. That song, then, becomes known by millions of people of varying ethnicities, cultures, and nationalities—it has become a song known to millions, not just a few members of a specific tribe or ethnic group, as is the case with a folk song. In this way, popular culture fads and trends are extremely powerful. Popular culture supersedes ethnic and national boundaries, and spreads rapidly across large distances, often at the expense of local or regional folk cultures. Even so, regional expression often still exists in the form of such things as regional dialects, accents, and food preferences, even in societies such as the United States where strong popular cultures predominate. For example, many people in the South today continue to speak English with a strong regional accent even though most people there are exposed to standardized, accent-neutral English in schools and on television news programs and the like.

The post-industrial world-economy of today is punctuated by stark divisions within it. We have seen how this plays with respect to vast differences between the core, semi-periphery, and periphery in such things as standards of living, population structure, modes of production, and social relations of production. These differences are also seen in the types of cultures operating in the world-economy: popular cultures in the core, folk cultures in the periphery, and a mix of each in the semi-periphery. Other aspects of culture vary with respect to location in the world-economy.

Cultural Geography
K.L. Chandler

Cultural Geography

Culture includes a society's entire way of life. It includes **cultural universals**, patterns or traits that appear in every people group – such as economy, architecture, gender roles, education, music, diet, government, medicine, settlement, family structure, agriculture, dress – and the two most prevalent - language, and religion. Often, culture develops as a group of people begins to observe certain practices that deliberately continue through subsequent generations. Those traditions may isolate, change, diffuse, or evolve into something richer or more complex – or they may cease to be practiced altogether and become extinct.

People and societies leave an imprint on the natural landscape. How humans interact with their physical environment, organizing their settlements, and using the area's natural resources, reveals their cultural values and establishes their **cultural landscape** (**Figure 109.1**). Many historic cultures left little evidence or remnants behind of their existence, while today's more developed countries have left cultural legacies behind that will take centuries to biodegrade or erode.

© Jimmy Tran/Shutterstock.com

Figure 109.1   Terraced rice fields in Vietnam are an example of how the cultural landscape has impacted the natural landscape for the necessity of agricultural production.

Settlement

Although, Earth's topography varies across continents and even across square miles choices for settlement are still limited by climate, mountains, rivers, seas, and concrete. People's lifestyles are shaped by the land that surrounds them and the borders that imprison them. However, physical boundaries can restrict or aid political and societal relationships.

Geographers identify and analyze types of land, so they can understand why some areas are more occupied than others. Consider plains, plateaus, hills, and mountains. More people live on plains than any other landform on Earth because the flat expanse offers an easier surface for building and cultivation. Two people groups who live on the same stretch of land come into contact with each other more often than others who live near more forbidding terrain. Their cultures and languages blend more. Mountains, on the other hand, are the least populated areas. Although the land is easier to defend, its lack of accessibility and less arable features tend to welcome fewer inhabitants. As a result, people groups may choose to live on opposite sides of the mountain, separated by the natural barrier; their cultures and language will drift, and over time, new people groups will be created due to isolation. The **uncontacted tribes,** Indigenous people groups who have no sustained connections with the outside for trade nor any other purpose, are an example of isolated cultures today. It is estimated that there are up to 100 uncontacted tribes still in existence globally.

Languages
K.L. Chandler

Language

Understanding a people group's language is one of the best ways to study its culture. Language is considered a cultural universal (patterns or traits that appear in all people groups). Linguists estimate the world speaks about 7,000 languages today, grouping them into fifteen main **language families**, a broad assortment of languages related by common ancestry. The largest of these families is the Indo-European group, spoken by nearly three billion of the world's population (**Figure 110.1**). The largest of the major language families include:

1. Indo-European
2. Sino-Tibetan
3. Niger-Congo
4. Austronesian
5. Afro-Asiatic
6. Dravidian
7. Turkic
8. Japonic
9. Austroasiatic
10. Kra-Dai

The grouping of language families is divided into language subfamilies. Two of the ten known language subfamilies of the Indo-European language family are the Germanic and Romance languages, but, although they began in the same cultural hearth, their similarities largely end there. Further subdivisions of language subfamilies create language groups. It is within language groups that root-word commonalities may begin to appear in vocabulary.

Dialects

Dialects, regional differences in vocabulary, spelling, pronunciation, grammar, and syntax, regularly appear within a single language, whether it happens between countries or within a country's borders. For example, asking for biscuits and gravy in Birmingham, England, will not get the same response as it does in Alabama. The United States uses at least thirteen regional dialects. On a given road trip, the same drive-through order for Sprite could be for a soda, pop, coke, tonic, cocola, or soft drink.

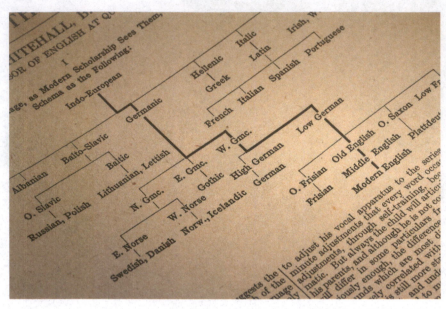

© Benoit Daoust/Shutterstock.com

Figure 110.1   This vintage dictionary shows the English language's evolution from Indo-European to Modern English.

Pidgins and Creoles

A **pidgin** develops when people who do not share a common language come into contact with each other and need to communicate. It is a simplified form of a language that often blends two or more languages and usually has a small vocabulary with simple grammatical structures. Its limited function results in a short lifespan – usually only a few decades – but a **creole** evolves when a pidgin becomes the mother tongue of a particular place.

English and French are common base languages for many of the creoles that are used today, especially in the Caribbean islands (West Indies). Many of the creoles that developed started as pidgins for simplistic communication during times of cultural convergence. The transatlantic slave trade, shipbuilding, and long-distance trade were all historic starting points for language divergence. Over time, these pidgins became the mother tongue of the next generation and, in places like Haiti, are now the primary language. Haitian Creole, the most common creole in the world, has an estimated 12 million users. Some of the islands also speak a **patois**, which is similar to a creole but employs more incorrect and nonstandard language.

Lingua franca

Also described as a bridge language, a **lingua franca** is designated when people who do not share a native language need to communicate with each other. English, the first global lingua franca, gradually replaced French as the lingua franca of international diplomacy after the end of World War I. Vatican City's lingua franca was Latin, considered a **dead language** (a language that is no longer a spoken first language) for generations, until the Pope changed it to Italian in 2014.

Official language

An **official language**, also called a state language, has special legal status in a jurisdiction and is assigned by the political state to be used in government and court. Some political states declare multiple official languages: Canada uses both French and English, and the political state of South Africa has eleven. The United States has no official language at the federal level, despite ongoing, lively debate on the matter since the 1750s.

As the vast majority of the people speak English there, it is considered to be the country's de facto (in practice, instead of law) language. Thirty-one states legally declare English to be the official language, meaning it must be used for government communications, and three states (Alaska, Hawaii, and South Dakota) have other official languages, in addition to English.

Mega languages

As the world becomes more interconnected, a few languages have become known as **mega languages**, the most widely spoken languages in the world: Arabic, Chinese, English, Hindi, and Spanish. One consequence of this phenomenon is that some people are beginning to turn away from their first languages. Linguists believe that approximately 1,000 languages could disappear in this century, so linguists are working hard to research and record endangered tongues, so they can be preserved and studied, in the event of extinction.

The Geography of Language
By Timothy Anderson

Students in an introductory human geography course might wonder why the analysis of the distribution of languages and religions is such a prominent component of most such courses, given that language is the central theme of linguistics and religion is a major theme of philosophy and history. The answer really is very simple: language and religion are the defining cornerstones of human culture (at least as it is traditionally defined) and identity (ethnic and national, or even individual, identity). They are two of the most important characteristics that distinguish human beings from all other animals, given that no other species possesses the capacity for speech or the ability to perceive of one's own mortality in a spiritual sense (at least as far as most biologists know). Many other kinds of animals can communicate with each other, sometimes in very complex ways, but none possesses the capacity for language. Likewise, it is clear that many other species exhibit emotions and feelings, but so far as we know, no others can think and philosophize about what is going to happen to them when they die. So, if a main goal of human geography is to delineate and understand how human cultures vary over space, then it behooves us to know how the two major facets of culture vary over space. That is, a map of religious and linguistic regions is, in many ways, a map of culture regions.

The Classification and Distribution of Languages

Classification of Languages

Language is defined as an organized system of spoken and/or written words, words themselves consisting of symbols or a group of symbols put together to represent either a thing or an idea, depending on the kind of writing system in use. In **syllabic languages**, the symbols that are used (e.g., letters) represent sounds. German, English, Arabic, and Hindi are examples of syllabic languages. **Ideographic languages** employ ideographs as symbols to represent an idea or thing. Examples of ideographic languages include Chinese, Japanese, and Korean. All human beings are biologically "hard- wired" for language. This means not that we are born knowing a language, but rather that we are born with the *ability* to learn a language or languages because it is imbedded in our genetic makeup. The linguist Noam Chomsky calls this "deep structure."

With respect to other forms of animal communication, human language is unique in two primary ways. First, human languages are *recombinant*. This means that words and symbols can be taken out of order in a sentence and recombined to form a different sentence and thus communicate a completely different, or even subtly different, idea. Dogs, for example, are not capable of arranging barks in different orders to communicate extremely intricate ideas. Second, word formation in all human languages is almost completely *arbitrary*.

This means that there really is no rhyme or reason about why a certain word, symbol, or ideograph is used to stand for something. There are, of course, exceptions to this rule. One is *sound symbolism,* in which the pronunciation of a word or the shape of an ideograph suggests an image or meaning. For example, many words in English that begin with gl- have something to do with sight (glimmer, glow, and glisten). Another exception to this rule is *onomatopoeia,* an instance in which a word sounds like something in nature that it represents (e.g., cuckoo, swish, cock-a-doodle-doo). But, such exceptions are very rare. The vast majority of words in all languages have simply been made up and then passed down over generations, although, to be sure, words and languages change over time.

Roughly 6,700 different languages are in use around the world today. But, the vast majority of these languages are spoken by a relatively small number of people. This means that a relatively small number of languages have thousands or even millions of speakers. Consider the following list of the top ten languages by number of native (mother-tongue) speakers in 2019, according to SIL International, one of the leading organizations that collects and publishes linguistic data Ethnologue:

Top Ten Languages, 2019 (Native Speakers)

Language	Number of Native Speakers	Primary Locations	Language Family
Mandarin	917 million	China; Chinese Diasporic Communities Worldwide	Sino- Tibetan
Spanish	480 million	Spain; Latin America; United States	Indo- European
English	379 million	U.K.; North America; Australia; New Zealand	Indo- European
Hindi	341 million	India; Indian Diasporic Communities Worldwide	Indo- European
Arabic	315 million	Southwest Asia; North Africa	Afro-Asiatic
Bengali	228 million	Bangladesh; Northeast India	Indo- European
Portuguese	221 million	Brazil; Portugal	Indo- European
Russian	154 million	Russia; Central Asia	Indo- European
Japanese	128 million	Japan	Japonic
Western Punjab	93 million	Pakistan	Indo-European

Of the nearly 6,700 languages in use today, 389 (about 6%) account for 94% of the world's population. The remaining 94% of languages in the world are spoken by only 6% of the world's population.

Linguists have devised classification schemes that describe and account for similarities and differences between and within different languages. At the broadest level, a **proto-language** describes an ancestral language from which several language families (described below) or languages are descended. No proto-languages are spoken today, but they are theorized to have been in use thousands of years ago. For example, Proto Indo-European was the language theorized to have been spoken in eastern Anatolia (present-day Turkey) and the Caucasus Mountain region 5,000 years ago. These people were the original Indo-Europeans (probably some of the first Caucasians), and the language they developed became the basis for all of the languages linguists classify as Indo-European. If we use the analogy of a tree to represent a group of languages that are linguistically related, then a proto- language is the roots and trunk of the tree.

In this analogy, each branch of the tree represents a **language family.** A language family is a group of languages descended from a single earlier language whose similarity and "relatedness" cannot be the result of circumstance. How do we know that certain languages are related to each other? Linguists employ two main methods to determine linguistic relatedness. One is genetic classification, in which it is assumed that languages have diverged from common ancestor languages (proto- languages), and therefore, languages that diverged from the same proto- language will have inherent similarities. Compare, for example, the following words for "mother" in selected Indo-European languages:

Engish	mother
Dutch	Moeder
German	Mütter
Irish Gaelic	mathair
Hindi	mathair
Russian	mat
Czech	matka
Czech	mater
Spanish	madre
French	mére

It is obvious that all of these words sound very similar to each another. Given the rule that word formation is arbitrary, it is impossible that such strong similarities are the result of mere coincidence, especially given the fact that some of these populations are separated by thousands and thousands of miles. When we add to this list hundreds or thousands of other words that display such similarities, it is clear that these languages have a common ancestor, common linguistic roots. When languages are shown to have a common ancestor, such as those above, they are said to be **cognate languages.**

Reviewing the list above once again, some of the languages clearly have even more commonalities to others in the list. Compare, for example, the even more clearly defined similarity between Latin and Spanish, English and Dutch, and Russian and Czech. These groupings of languages whose commonality is very definite are called **language subfamilies.** Think of these groupings as twigs of larger branches on the language tree, while individual languages are the leaves of the tree. Proceeding even further with respect to similarity, linguists recognize **dialects.** A dialect is de ned as a recognizable speech variation *within the same language* that distinguishes one group from another, both of which speak the same language. Sometimes these differences are based on pronunciation alone (the different varieties of English spoken around the world or a Southern or New England accent) and sometimes they are based on slightly different words for the same thing (British English "lorry," American English "truck"). Similarly, **pidgins** are languages that develop from one or more "mother" languages that have highly simplified sentence and grammatical structures compared to the mother languages. When such a language becomes the native language of succeeding generations, it is known as a **creole language.** Most creoles are either English-, French-, or Spanish-based and are spoken in the periphery of the world-economy, in former colonial areas, where two or more groups of people speaking mutually unintelligible languages were forced to communicate with each other during the colonial era.

Political and ethnic fragmentation is characteristic of many former colonial regions in the periphery of the world-economy. In Nigeria, for example, at least 500 different tribal languages are spoken, many of them not even in the same language family. In such places, governments and businesses often make use of a *lingua franca* to carry out official business. A lingua franca refers to a language that is used habitually among people living in close contact with each other whose native tongues are mutually unintelligible. English and French are common lingua francas in much of sub-Saharan Africa, Arabic is the lingua franca of much of North

Africa and Southwest Asia, and English can be thought of as the lingua franca of the internet and of air traffic control and airline pilots around the world.

Linguistic Diffusion and Change

Why does the map of world language families look the way it does? What spatial processes have led to the present-day distribution of languages? How do languages change over time and space? In general, cultural diffusion (both hierarchical and relocation diffusion) and geographical isolation over time and space have resulted in the linguistic patterns we observe today (**Figure 111.1**). Relocation diffusion on a massive scale since the advent of the capitalist world-economy in the fifteenth century, together with the displacement and subjugation of native populations, has resulted in very large linguistic regions, especially in the Americas.

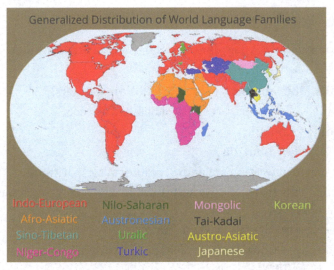

Source: Timothy G. Anderson

Figure 111.1 Generalized Distribution of World Language Families

These two processes (relocation diffusion and displacement) explain the fact that the vast majority of populations in North, South, and Middle America speak one of three languages (all of them Indo-European languages): English, Spanish, and Portuguese. Before Columbian contact, probably as many as 30 million people were living in the Americas, speaking literally hundreds of distinctive languages in at least a dozen different language families. In other words, the linguistic map was highly complicated and extremely diverse. Massive relocation from Europe and the decimation of native populations over a 300-year period resulted in a vastly simplified and less complex linguistic map in the Americas. This is not to say that no Native American languages survive. Many do, but with few exceptions, the number of people who speak these languages is very small compared to the number of English, Spanish, and Portuguese speakers. The colonial era also ushered in a period in which many European languages, such as English, Spanish, Portuguese, and French, acquired many more new speakers in their overseas colonies than they ever had at home. In part, this was a result of the outright extermination of African and Native American languages in the Americas through either severe population decline or cultural subjugation through slavery. In the process, European languages were of course deemed as "superior" to others, and Europeans forced Africans and Native Americans to learn the language of the colonizers. This was the case not only in the Americas, but also in colonial sub-Saharan Africa as well.

The present-day world linguistic map also is the product of centuries of linguistic change over time and space. In the pre-industrial era, for example, migration and spatial isolation and segregation gave rise to separate, mutually unintelligible languages. As populations diverged over time and space, populations became isolated from each other. And, as these migrating populations encountered new natural environments and human societies, they were forced to invent new words to describe new circumstances, places, and things. It has also been shown that languages change naturally in place over time, even in the absence of outside cultural forces such as immigration or hostile invasion. Take, for example, the case of English and how significantly the language changed between the ninth and seventeenth centuries. The Old English of ninth-century Britain (Beowulf is the most famous literary example) would hardly be recognizable to most English speakers today. But, 500 years later, due to in infuences from Latin, French, and Danish, as well as to natural linguistic evolution over time, the language had evolved into the Middle English of Chaucer (The Canterbury Tales), a language that most English speakers today can understand. By the seventeenth century, the language had evolved into the Modern English of Shakespeare. As one of the world's most widely spoken languages and as a world-wide lingua franca, English is changing more rapidly today than ever before as it incorporates words from a variety of different cultural sources around the world.

The Distribution of the World's Major Language Families

This section lists the world's major language families and important subfamilies, and maps the distribution of their speakers

1. **Indo-European Family (386 languages; about 2.5 billion speakers)**
   - Albanian Subfamily—Albania, parts of Yugoslavia and Greece
   - Armenian Subfamily—Armenia
   - Baltic Subfamily—Latvia, Lithuania
   - Celtic Subfamily—parts of western Ireland, Scotland, Wales, Brittany
   - Germanic Subfamily—northern and western Europe, Canada, USA, Australia, New Zealand, parts of the Caribbean and Africa
   - Greek Subfamily—Greece, Cyprus, parts of Turkey
   - Indo-Iranian Subfamily—India, Pakistan, Bangladesh, Afghanistan, Iran, Nepal, parts of Sri Lanka, Kurdistan (Iran, Iraq, Turkey)
   - Italic (Romance) Subfamily—France, Spain, Portugal, Italy, Romania, Brazil, parts of western and central Africa, parts of the Caribbean, parts of Switzerland
   - Slavic Subfamily—eastern Europe, southeastern Europe, parts of south- central Asia
2. **Sino-Tibetan Family (272 languages; about 1.1 billion speakers)**
   - Chinese Subfamily—China, Taiwan, Chinese communities around the world
   - Tibeto-Burman Subfamily—Tibet, Myanmar (Burmese), parts of Nepal and India
3. **Austronesian Family (1,212 languages; 269 million speakers)**
   - Formosan Subfamily—parts of Taiwan
   - Malayo-Polynesian Subfamily—Madagascar, Malaysia, Philippines, Indonesia, New Zealand (Maori), Pacific Islands (e.g., Hawaii, Fiji, Samoa, Tonga, Tahiti)
4. **Afro-Asiatic Family (338 languages; 250 million speakers)**
   - Semitic Subfamily—North Africa (Arabic), Israel (Hebrew), Ethiopia (Amharic), Middle East
   - Cushitic Subfamily—Ethiopia, Kenya, Eritrea, Somalia, Sudan, Tanzania
   - Chadic Subfamily—Chad, parts of Nigeria, Cameroon
   - Omotic Subfamily—Ethiopia
   - Berber Subfamily—parts of Morocco, Algeria, Tunisia
5. **Niger-Congo Family (1,354 languages; 206 million speakers)**
   - Benue-Congo Subfamily—central and southern Africa
   - Kwa Subfamily—bulge of west Africa
   - Adamaw-Ubangi Subfamily—northern part of central Africa
   - Gur Subfamily—between Mali and Nigeria

- Atlantic Subfamily—extreme western part of the bulge of west Africa
- Mande Subfamily—western part of the bulge of west Africa

6. **Dravidian Family (70 languages; 165 million speakers)**
   - Four Subfamilies—southern India, parts of Sri Lanka, parts of Pakistan

7. **Japanese Family (12 languages; 126 million speakers)**
   - Turkic Family (60 languages; 115 million speakers)

8. **Turkic Subfamily—Turkey, Uzbekistan, Turkmenistan, Kazakhstan, Azerbaijan, eastern Russia (Siberia)**
   - Mongolian Subfamily—Mongolia, parts of adjoining areas of Russia and China
   - Tungusic Subfamily—Siberia, parts of adjoining areas of China

9. **Austro-Asiatic Family (173 languages; 75 million speakers)**
   - Mon-Khmer Subfamily—Vietnam, Cambodia, parts of Thailand and Laos Munda Subfamily—parts of northeast India

10. **Tai-Kadai Family (61 languages; ca. 75 million speakers)**
    - Tai Subfamily—Thailand, Laos, parts of China and Vietnam

11. **Korean Family (1 language; 60 million speakers)**

12. **Nilo-Saharan Family (186 languages; 28 million speakers)**
    - Nine Subfamilies—southern Chad, parts of Sudan, Uganda, Kenya

13. **Uralic Family (33 languages; 24 million speakers)**
    - Finno-Ugric Subfamily—Estonia, Finland, Hungary, parts of Russia
    - Samoyedic Subfamily—parts of northern Russia (Siberia)

14. **Amerindian Languages (985 languages; ca. 20 million speakers)**
    - As many as 50 different language families, hundreds of subfamilies
    - North America = ca. 500,000 speakers; 150 languages (top languages = Navajo and Aleut)
    - Central America = ca. 7 million speakers; (top language = Nahuatl)
    - South America = ca. 11 million speakers; (top language = Quechua)

15. **Caucasian Family (38 languages; 7.8 million speakers)**
    - Four Subfamilies—Georgia, surrounding region on western shore of the Caspian Sea

16. **Miao-Yao Family (15 languages; 5.6 million speakers)**
    - Southern China, northern Laos (Hmong), northeast Myanmar

17. **Indo-Pacific Family (734 languages; 3.5 million speakers)**
    - The most linguistically complex place on earth—Papua New Guinea and surrounding islands

18. **Khoisan Family (37 languages; 300,000 speakers)**
    - Three Subfamilies—parts of Namibia, Botswana, Republic of South Africa

19. **Australian Aborigine (262 languages; ca. 30,000 speakers)**
    - Only five languages have over 1,000 speakers
    - At time of European contact, 28 language families, 500 languages spoken by over 300,000 people

20. **Language Isolates (296 languages; ca. 2 million speakers)**
    - Languages that have not been conclusively shown to be related to any other language; some examples are:
      - Basque (Euskara)—southern France, northern Spain
      - Nahali—5,000 speakers in southwest Madhya Pradesh in India
      - Ainu—island of Hokkaido, Japan; now probably extinct
      - Kutenai—less than 200 speakers in British Columbia and Alberta

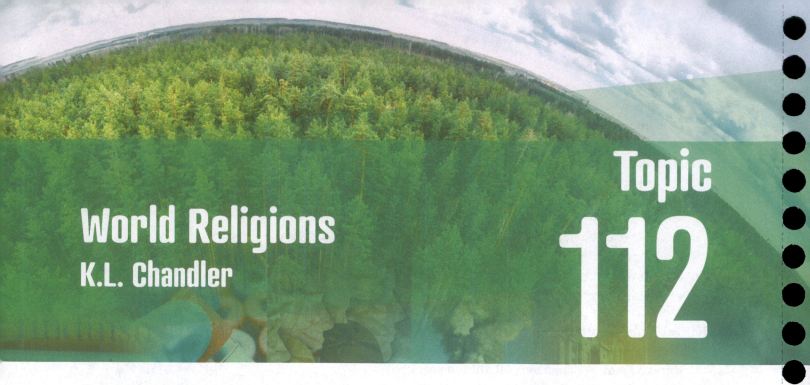

World Religions
K.L. Chandler

World Religions

Five main religions are practiced around the world: Christianity, Islam, Hinduism, Buddhism, and Judaism (**Figure 112.1**). More people claim to be practicing participants of Christianity than of any other faith in the world. Also the most globally prevalent, the Christian church spreads across the Americas, Europe, Russia, sub-Saharan Africa, and the Austral region.

In second place for both adherents and geographic presence, Islam is currently the fastest growing world religion, likely surpassing Christianity in several decades, if it continues to grow at its present rate. Southeast

© Miloje/Shutterstock.com

Figure 112.1 **Symbols of the five main world religions.**

Asia's Indonesia boasts the largest population of Muslims, followed by the geographic regions of North Africa and Southwest Asia, South Asia, and the geographic subregion of Central Asia.

By population, Hinduism ranks third of the world's five major religions. It is the main belief system in India, the second-largest populated country in the world (becoming the world's most populated country in the next decade). It is also common in India's neighbor, Nepal. Common in East Asia and throughout the mainland countries of Southeast Asia – Myanmar, Thailand, Cambodia – Buddhism is the fourth largest world religion, which is also represented globally after spreading to the western world over the last century.

Judaism may not be the fifth-largest faith in terms of population numbers, but its role as the hearth of both Christianity and Islam earns its place on the list. Practiced worldwide, it is most geographically concentrated in Israel (**Figure 112.2**).

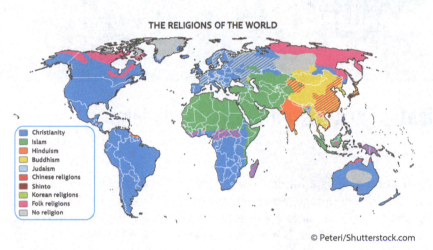

© Peteri/Shutterstock.com

Figure 112.2   **The religious landscape of the world.**

Classification of Religions

Nearly 10,000 religions are practiced in the world today, a number further multiplied by sects and denominations. Theologians often group religions into broad classifications. One of these classifications is by eastern or western origin. This idea organizes a religion by its place of origin or concentrated practice. Buddhism, Confucianism, Hinduism, Shinto, and Taoism are largely recognized as Eastern religions; Christianity, Islam, and Judaism are considered to be the largest Western religions.

Religions can also be organized by belief in a singular god or multiple gods. **Monotheistic religions** like Christianity, Islam, and Judaism teach the belief in one god. Religions that practice **polytheism** believe in many gods or spirits. Buddhism, sometimes considered to be agnostic or philosophical, is neither monotheistic nor polytheistic. Buddhists seek to achieve enlightenment, or nirvana, which is an independent search, realized on an individual basis.

The last classification of world religions is based on spread or diffusion. **Universalizing religions**, common in Western belief systems, are spread by missionaries, who travel away from their homes to add converts to their respective faiths. **Ethnic religions** are observed and practiced locally often as cultural and philosophical worldviews.

Classification and Distribution of Religions

By Timothy Anderson

The Classification and Distribution of Religions

Classification of Religions

Together with language, religion is a human characteristic that distinguishes us from every other animal species on the planet. As with language, we are also most likely biologically "hard-wired" with the capacity for abstract thought about spiritual matters, our own mortality, and the nature of the universe and our place in it. We are born with these capacities, but not with a certain set of beliefs concerning these things—these beliefs are learned. By definition, a **religion** is a system of either formal (written down and codified in practice) or informal (oral traditions passed from generation to generation) beliefs and practices relating to the sacred and the divine.

Religion helps us answer questions such as: Who am I? Why am I here? What is my purpose in life? What is my place in the universe? What will happen to me when I die?

Every human being ponders these questions because we have the biological capacity to think about and philosophize about such things. Human religions attempt to answer such questions through systems of beliefs, practices, and worship. As the answers to these questions vary, so do religious belief systems; as the answers to these questions vary from place to place, to a large extent, so does human culture. As is the case with human languages, literally thousands of religions are practiced around the world today. In most of the peripheral regions of the world-economy, there are as many religions as there are ethnic groups, and they can vary substantially over relatively short distances. For human geographers, religion is an extremely important aspect of the cultural landscape because religious beliefs and customs have a significant physical manifestation in the form of religious structures. Although religious practices, beliefs, and traditions vary substantially around the world, most religions share the following characteristics:

- Belief in one or many supernatural authorities
- A shared set of religious symbols (iconography)
- Recognition of a transcendental order—offers a divine reason for existence and an explanation of the inexplicable
- Sacraments (prayer, fasting, baptism, initiation, etc.)
- Enlightened or charismatic leaders (priests, shaman, prophets)
- Religious taboos

- Sacred structures (temples, shrines, cathedrals, mosques, etc.)
- Sacred places (pilgrimage sites, holy cities, etc.)
- Sacred texts

Human religions can be divided into two broad categories and two narrower sub-categories (**Figure 113.1**). At the broadest level, we can distinguish between monotheistic and polytheistic religions. **Polytheistic religions** involve a belief in many supernatural (that is, not of this world) beings that control or influence some aspect of the natural or human world. The vast majority of human religions are polytheistic in nature, numbering in the thousands. Such religions usually, but not always, have a very small geographic distribution, often coinciding with tribal or ethnic boundaries. For most of human history, polytheistic belief systems have been by far the most common.

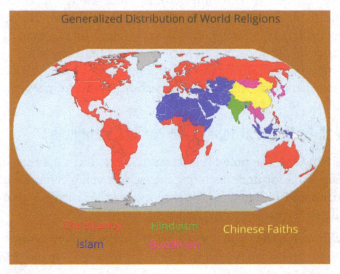

Source: Timothy G. Anderson

Figure 113.1   **Generalized Distribution of World Religions.**

Monotheistic religions appeared on the human stage quite late, probably not until around 1,500 B.C.E. Monotheism involves the belief in one omnipotent, omniscient, supernatural being who created the universe and everything in it, and thus controls and influences all aspects of the natural and human world. There are in effect only three monotheistic religions today: Judaism, Christianity, and Islam. In terms of number of adherents and believers, however, two of these (Christianity and Islam) have over one billion adherents each. That is, out of a total world population of around 7 billion, fully one-third are either Christians or Muslims. The geographic boundaries of Christian and Islamic beliefs, then, do not coincide with political boundaries but rather supersede and overlap them. We can account for such distributions in much the same way that we can account for the very large distribution of Indo-European speakers around the world: relocation diffusion on a massive scale during the colonial era, and the acquisition of new members either by force or through missionary activities.

We can also identify two sub-categories of religions. First, **ethnic religions** are those in which membership is either by birth (one is "born into" the religion) or by adopting a certain complex ethnic lifestyle, which includes a certain religious belief system. That is, an ethnic religion is the religious belief system of a specific ethnic or tribal group and is unique to that group. Most (but not all) of these kinds of religions are polytheistic in nature and have very small geographical distributions, sometimes no larger than a village or group of villages. These religions, therefore, have very strong territorial or ethnic group identity. In most such religions, there is no distinction made between one's ethnic identity (i.e., one's culture) and one's religion: one's religion is one's culture. Examples of ethnic religions include Judaism, Hinduism, and various tribal belief systems that are ubiquitous throughout the periphery of the world-economy. Second, **universalizing**

religions are those in which membership is open to anyone who chooses to make a solemn commitment to that religion, regardless of class or ethnicity. Membership in these religions is usually relatively easily obtained, and usually involves some sort of public declaration of one's allegiance to the belief system (baptism, for example). Universalizing religions are also distinguished by the fact that they are often characterized by strong evangelic overtones in which members are admonished to spread the faith to nonbelievers. For these reasons, universalizing religions have very large geographic distributions that cover vast regions of the world, the boundaries of which overlap the political boundaries of individual states. There are only three universalizing religions: Christianity, Islam, and Buddhism, although Buddhism rarely carries with it evangelic activities and therefore has a much smaller distribution than Christianity and Islam.

Finally, it should be noted that the influence of popular culture and post-modern thought and philosophies, especially in core regions of the world-economy, have significantly influenced the growth of secularism. **Secularism** refers to an indifference to, or outright rejection of, a certain belief system or religious belief in general. In its extreme, such "beliefs" may become like a religion. It is increasingly characteristic of many post-industrial societies, and thus influences core societies more than any other societies around the world. At least one-fifth of the world's population, by this definition, is secular, and this figure is even higher in parts of northern and western Europe, where the figure approaches 70 percent in some instances.

A Comparative Approach to Understanding the World's Religions

In his recent book God is Not One, the noted religious scholar Stephen Prothero employs a four-part comparative model as a starting point for understanding the world's major religious belief systems and for contextualizing their similarities and differences. Prothero argues that each religion's theology and system of beliefs identify and articulate four primary concerns:

1. A *central problem* with which the theology and belief system is chiefly concerned
2. A *primary solution* to this problem, which in most cases functions as the primary goal of the religion
3. A *technique or set of techniques* for achieving this solution/goal
4. An *exemplar or exemplars* who demonstrate a path from problem to solution

As we approach a comparative analysis of the world's major religions, we will employ this useful framework.

Attributes and Distributions of the World's Major Religions

This section lists the world's major religions, identifies their major characteristics, and maps the distribution of their adherents.

1. Hinduism (ca. 740 million adherents concentrated in India, Nepal, and Sri Lanka)

- One of the world's oldest extant religions
- The ethnic religion of the Hindustanis
- Hearth in the Indus Valley ca. 1500 BC, then spread to India, Nepal, Sri Lanka, and parts of SE Asia
- Beliefs and practices:
  - A common doctrine of *karma*, one's spiritual ranking, and samsara, the transfer of souls between humans and/or animals
  - A common doctrine of *dharma*, the ultimate "reality" and power that governs and orders the universe
  - The soul repeatedly dies and is reborn, embodied in a new being
  - One's position in this life is determined by one's past deeds and conduct
  - The goal of existence is to move up in spiritual rank through correct thoughts, deeds and behavior, in order to break the endless cycle and achieve moksha, eternal peace

- Life in all forms is an aspect of the divine—hundreds of gods, each controlling an aspect of the natural world or human behavior
- One need not "worship" a god or gods
- **The Caste System**—a social consequence of the Hindu belief system
  - The social and economic class into which one is born is an indication of one's personal status.
  - In order to move up in caste, one must conform to the rules of behavior for one's caste in this life.
  - This thus highly limits social mobility.
- Sacred texts
  - The *Rig Vedas*, hymns composed by the Indo-Aryans after the invasion of the Punjab; the oldest surviving religious literature in the world, written in Sanskrit
  - *Brahmanas*, theological commentary, defined different castes
  - *Upanishads*, defines karma and nirvana, etc.
- Cultural landscapes
  - Shrines, village temples, holy places, and rivers (the Ganges), pilgrimage sites and routes

© Stakes/shutterstock.com

The Hindu World-View

The Problem:

- Samsara
- The endless cycle of life, death, and rebirth
- "Wandering on" or "flowing"

The Solution(s):

- *Moksha*
  - "Release"
  - Spiritual liberation
  - The freeing of the soul from the bondage to reincarnation

- *Kama*
  - "Sensual pleasure"
- *Artha*
  - Wealth and power
- *Dharma*
  - "Duty"

The Technique(s):
- *Yogas* ("disciplines")
  - Karma Yoga (the discipline of ritual action)
  - Jnana Yoga (the discipline of wisdom)
  - Bhakti Yoga (the discipline of devotion)

The Exemplar(s):
- Gurus ("teachers")
- Holy men and women
- Ancient Sanskrit holy texts (the Vedas; the Sutras; the Bhagavad Gita, the Upanishads)

2. Buddhism (ca. 300 million concentrated in East and Southeast Asia)

- Founded by Gautama Siddhartha in the sixth century B.C.E. in northeast India
- Diffusion was mainly to China and Southeast Asia by monks and missionaries
- The primary religion in Tibet, Mongolia, Myanmar, Vietnam, Korea, Thailand, Cambodia, Laos; mixed with native faiths in China and Japan
- A universalizing religion
- Beliefs and practices:
  - Retains the Hindu concept of *karma*, but rejects the caste system
  - More of a moral philosophy than a formal religion
  - The ultimate objective is to reach nirvana by achieving perfect enlightenment
  - The road to enlightenment, Buddha taught, lies in the understanding of the four "noble truths":
    1. to exist is to suffer,
    2. we desire because we suffer,
    3. suffering ceases when desire is destroyed,
    4. the destruction of desire comes through knowledge or correct behavior and correct thoughts (the "eight-fold path")

- Sects:
  - Theravada (Sri Lanka, Myanmar, Thailand, Laos, Cambodia)
  - Mahayana (Vietnam, Korea, Japan, China, Mongolia)
  - Zen (Japan)
  - Lamaism (Tibet)

- Cultural landscapes:
  - Shrines and temples
  - Holy locations where the Buddha taught

© Anne Mathiasz/Shutterstock.com

The Buddhist World-View

The Problem:

- *Dukkha*
  - "Suffering"
  - The undesirable "wandering" from rebirth to rebirth
  - Desiring something other than "what it is"

The Solution(s):

- *Nirvana*
  - "Blowing out" [of suffering]
  - "Awakening"

The Technique(s):

- Understanding of the Four Noble Truths
- The Eightfold Path (ethical conduct; mental discipline; wisdom)
- *Dharma* (understanding things as they really are)
- Meditation
- Chanting

The Exemplar(s):

- Arhats (Theravada monastic tradition)
- Bodhisattvas (Mahayana "greater vehicle" tradition)
- Lamas (Vajrayana "guru/teacher" tradition)

3. Chinese Faiths (ca. 300 million adherents in China)

- Two main forms: Confucianism and Taoism, both date from the sixth century B.C.E.
- The goal of both is moral harmony within each individual, which leads to political and social harmony.
- Chinese religion combines elements of Buddhism, Animism, Confucianism, and folk beliefs into one "great religion"; each element services a different component of the self.

- The Taoist approach to life is embodied in the Yin/Yang symbol; stresses the oneness of humanity and nature; people are but one part of a larger universal order.
- Confucianism is really a political and social philosophy that became a blueprint for early Chinese civilization; it teaches the moral obligation of people to help each other, that the real meaning of life lies in the here and now, not in a future abstract existence; Kong Fu Chang taught that the secret to social harmony is empathy between people.

© casejustin/Shutterstock.com

The Confucian World-View

The Problem(s):

- Chaos
- Disharmony
- Disorder

The Solution(s):

- Order
- Self-cultivation
- Social Harmony through "Correct" Relations with other Human Beings

The Technique(s):

- Education
- The Study of Ancient Classic Texts
- Learning and Practicing Proper Etiquette and Rituals
- Practicing the "Five Virtues" (human-heartedness; justice; propriety; wisdom; faithfulness)

The Exemplar(s):

- Confucius
- *Junzi* ("exemplary persons")

The Taoist World-View

The Problem(s):

- "Lifelessness"
  - A life that is not fully enjoyed
  - Blind adherence to social conventions
  - Not living life "to its fullest"
  - To be led around by the noose of social conventions and ritual propriety is to be alienated from oneself, from other people, and from the natural environment around us
  - The dictates of social convention, moral rules, formal education, ritual prescriptions
  - All of this destroys social and natural harmony and keeps us from "flourishing" in the way that all living things are meant to

The Solution(s):

- Human "Flourishing"
  - "Nurturing" one's life
  - A vital and genuine life
  - Living life to the fullest
  - In the end, "physical immortality" in this world
  - Living in harmony with the natural rhythms of the Tao (the life force which governs the universe)

The Technique(s):

- Returning to the creativity of the Tao
- Preserving and circulating one's *qi* ("vital energy")
- Balancing one's yin and yang "Sitting and forgetting"
- "Free and easy wandering"
- Dietary regimes; breath control; visualization exercises; purification rites; meditation techniques; physical exercises

The Exemplar(s):

- Early Taoist sages (Lao Tzu, for example)
- Taoist texts (Tao Te Ching)

4. Judaism (ca. 18 million adherents mainly in North America and Israel)

- The ethnic religion of the Hebrews
- The oldest religion west of the Indus (ca. 1,500 B.C.E.)
- Founder regarded as Abraham (the patriarch)
- Sacred text = the Torah (the five books of Moses)
- Beliefs and practices:
  - God is the creator of the universe, is omnipotent, but yet merciful to those who "believe" in Him.
  - God established a special relationship with the Jews, and by following his law, they would be special witnesses to His mercy.
  - Emphasis is on ethical behavior and careful, ritual obedience.
  - Among the traditional, almost all aspects of life are governed by strict religious discipline.
  - The Sabbath and other holidays are marked by special observances and public worship.
  - The basic institution is the Synagogue, led by a rabbi chosen by the congregation.
- Cultural landscapes:

  - Synagogues
  - Sacred sites (e.g., the Wailing Wall, Jerusalem, sites of miracles, etc.)

© Creative icon styles/Shutterstock.com

The Judaic World-View

The Problem:

- Exile (from God)
  - "Distance" from God and where we "ought" to be
  - In both a literal and metaphorical sense
  - A "chronic" problem in Judaic thought

The Solution:

- Return
    - A return back to God and to our "true" home (both literally and metaphorically)
    - The completion of a long journey, from Paradise to desert wilderness, to the New Jerusalem
    - To make things ready and to make things "right"

The Technique(s):

- To Remember
    - To remember the story of one's people and to tell that story;
    - To Obey
    - To obey God's law

The Exemplar(s):

- The Patriarchs
- Abraham, Noah, Moses, Elijah, Esther, among others

5. Christianity (ca. 1.6 billion adherents worldwide, but especially in Europe, North America, Middle and South America)

- A universalizing religion
- A revision of Judaic belief systems
- Founder regarded as Jesus, a Jewish preacher believed to be the savior of a sinful humanity promised by God; his main message was that salvation was attainable by all who believed in God (died ca. 30 C.E.)
- Sacred text = the Bible; Old Testament is based on the Hebrew Torah and is the story of the Jews; New Testament is based on the life of Jesus and his teachings
- Mission: conversion by evangelism through the offering of the message of eternal life and hope
- Reform movements:
    - Split in the fifth century between the western church at Rome (Catholicism) and the eastern church at Constantinople (Orthodoxy)
    - Protestant Reformation in the fifteenth and sixteenth centuries, led mainly by northern Europeans over moral and political issues
    - Protestantism took hold in northern Europe and spread to North America, Australia and New Zealand
- Cultural landscapes:
    - Churches, cathedrals, graveyards, iconography
    - Sacred sites (e.g. Marian apparition pilgrimage sites)

The Christian World-View

The Problem:

- Sin

The Solution:

- Salvation
- Redemption

The Technique(s):

- Some combination of faith and good works
- Practicing the sacraments of the faith

The Exemplar(s):

- The Saints (Roman Catholicism and Eastern Orthodoxy)
- "Ordinary" People of Faith (Protestantism)

6. Islam (ca. 1 billion adherents worldwide, but especially in North Africa, Southwest Asia, South-Central Asia, Indonesia, Malaysia)

- Founder: Muhammad ("Prophet"), born 571 C.E.; believed to have received the last word of God (Allah) in Mecca in 613 C.E.
- Diffusion: rapidly throughout Arabia, SW Asia, North Africa, then to South and Southeast Asia
- Organization: theoretically, the state and the religious community are one in the same, administered by a caliph; in practice, it is a loose confederation of congregations united by tradition and belief
- A universalizing religion
- Sacred Text: the Koran—the sayings of Muhammad, believed to be the word of God
- Divisions: two major sects—Sunni (Orthodox) and Shi'a (Fundamentalist); Shiites mainly in Iran and parts of Iraq and Afghanistan; Sunni are the majority worldwide

- Beliefs: mainly a revision of both Judaic and Christian beliefs; those who repent and submit ("Islam") to God's rules can return to sinlessness and have everlasting life; religious law as revealed in the Koran is civil law; smoking, gambling and alcohol are forbidden
- The faithful are admonished to practice the five "**pillars of Islam**"
  - Public profession of faith
  - Daily ritualistic prayer five times per day
  - Almsgiving
  - Fasting during daylight hours during Ramadan
  - A pilgrimage to Mecca at least once in one's lifetime if physically and economically possible
- Cultural landscapes:
  - Mosques, minarets, religious schools, iconography

© Georgios Kollidas/Shutterstock.com

The Islamic World-View

The Problem(s):

- Pride
- Disobedience (to God)
- The Fallacy of "Self-Sufficiency" ("the idol of the self")

The Solution(s):

- Submission (to God and God's laws)
- The Achievement of Having a "Soul at Peace" in This Life and the Next

The Technique(s):

- A Combination of Faith and Good Works
- Practicing the "Five Pillars" of the Faith
- Public profession of the faith
- Daily prayer
- Charity

- Fasting
- Pilgrimage (to Mecca)

The Exemplar(s):

- Mohammed
- *Imams*
- Many People of "Ordinary" Faith

7. Sikhism (ca. 25 million adherents worldwide, but concentrated in northwest India, especially the Punjab)

- Founder: Guru Nanak (1469–1539) and nine successive Gurus ("teachers")
- A monotheistic, universalizing religion
- A Sikh is a "disciple" of God who follows the teachings of the ten Gurus
- The last Guru died in 1708
- Beliefs:
  - There is only one God who is creator, sustainer, and destroyer of all life.
  - The same God for all people of all religions.
  - God cannot take human form.
  - The goal of one's life is to break the cycle of death and rebirth and to "merge" with God ("salvation").
  - Salvation may be achieved by overcoming the five cardinal vices (lust, anger, greed, worldly attachment; pride).
  - An emphasis is placed on daily devotion to God.
  - Rejection of all forms of "blind" worship (fasting, yoga, pilgrimage, religious vegetarianism, etc.).
  - Devotees must live "in the world" but keep a "pure" mind.
  - A rejection of all distinctions based around caste, creed, race, gender; equality of women is especially stressed.
  - A stress is placed on charity and community service.
- The Gurus preached a message of love, compassion, and understanding for all human beings and criticized the "blind rituals" of Hindus and Muslims.
- Practices:
  - The very devout decry all forms of violence and undergo a baptism ceremony, after which they become members of a special order; membership in this order requires strict adherence to a Code of Conduct and a prescribed physical appearance (hair is not cut; the wearing of turbans by men; the carrying of a ceremonial sword).
  - Sikhism does not have priests; all Sikhs are considered to be "custodians" of the faith's sacred text.
- Sacred Text:
  - The *Guru Granth Sahib*; this text itself is considered to be the leader of the faith; it was written by the ten Gurus and numerous other authors.
- Cultural Landscapes:
  - The "Gurdwara" (temple); a "Gurdwara" is a structure in which the *Guru Granth Sahib* has been installed; each temple has a community kitchen which serves free meals to people of all faiths;
  - The most significant historical religious center is the Golden Temple at Amritsar, in the state of Punjab in northwest India; it is an inspirational and historic center but not, necessarily, a place of religious pilgrimage

© ppart/Shutterstock.com

The Sikh World-View

The Problem(s):

- Separation from the Divine ("God")
  - This separation comes from giving in to our vices (lust, anger, greed, worldly attachment; pride).
  - This separation results in an endless cycle of death and rebirth in this world.

The Solution(s):

- Salvation
- Becoming One with the Divine ("God")

  - "Merging" with God by breaking the cycle of death and rebirth in this world

The Technique(s):

- Following the Teachings of the Gurus
  - Overcoming the five cardinal vices
  - Love and compassion for all people
  - A rejection of forms of "blind ritual"
  - A rejection of distinctions based on nationality, creed, race, sex, gender, and class

The Exemplar(s):

- Guru Nanak and the Nine Successive Gurus

Feeding the World
By Elizabeth Jordan

Taiga/Shutterstock.com

Figure 114.1 Food is a renewable resource.

We obtain our food from fishing, raising livestock, and growing crops. Due to its ample supply, food is considered a renewable resource—but the means in which we produce it in some parts of the world is unsustainable (**Figure 114.1**).

There are over 7 billion people on the planet, and we produce more calories per day today than we produced in the mid-1900s. And the population is expected to grow to 9 billion by the year 2050. But despite this, people are still not getting enough to eat due to poor distribution. Not getting enough to eat and therefore lacking nutrients is called **food insecurity**.

According to FAO (Food and Agriculture Organization of the United Nations), 821 million people on earth (that is roughly 10%) were undernourished in 2017. All regions in Africa, parts of Latin America and parts of Asia are hit the hardest are the most impacted by hunger and fared worst in this statistic. But most developing countries—especially in Southeast Asia and Latin America—have been successful in actualizing Millennium Development Goal 1 (MDG1) of eradicating worldwide hunger.

Macronutrients are substances that the body needs a large supply of, such as essential fats, proteins, and carbohydrates. **Micronutrients** are substances that the body needs in small amounts, such as iodine, iron, vitamins, and other minerals. A lack of both macro- and micro-nutrients leads to malnutrition among other health problems (an iodine deficiency leads to goiter, a lack of vitamin D and dairy leads to rickets, and a lack of vitamin C leads to scurvy, for example). It is estimated that 2 billion people globally are deficient of important vitamins and minerals in their diet; particularly Vitamin A, iron, zinc, and iodine. This is attributed to a lack of micronutrient-rich foods such as fresh fruits and vegetables.

The more developed countries (MDCs) are big meat-eaters that consume plenty of protein for consumption with the vast number of animals we raise; but raising livestock creates pollution and uses a lot of water. (The least-developed countries (LDCs) typically rely more on fish rather than on meat-based diet.) The consumption of a meat-based diet uses twice as much water as a plant-based diet according to UNESCO Institute of Water Education. And according to International Water Management Institute, meat-eaters use roughly 5000 liters of water a day while people consuming a vegetarian diet use 1–2000 liters. And consider that according to the Food and Agriculture Organization of the United Nations (FAO), in 2014, it was estimated that we produced 312 million tons of meat, with 68 million tons of that being bovine meat![viii] And it does not end there. Much of our fossil fuel is used to raise livestock, and a major negative impact of raising cattle is increased concentrations of methane, which is a greenhouse gas. Livestock releases 80 million metric tons of methane per year, 28% of global methane emissions.[ix] And keep in mind that methane is 20 times more potent as a greenhouse gas than carbon dioxide in a 100 year period and 84 times more potent than carbon dioxide in a 20-year period. Some sources, such as the think tank 'World Watch' rank animal agriculture much higher in terms of greenhouse gas emissions.[x] Their comprehensive research included examining the release of methane, nitrous oxide, and water vapor. Their research concludes that animal agriculture is responsible for nearly half of all global greenhouse gases! The methane released by cows is not only a greenhouse gas; it contaminates, and is a major polluter of soil and groundwater. Animal agriculture is also the major driver for deforestation in the Amazon Rainforest, which acts as an invaluable carbon sink and also provides a major source of Oxygen. Furthermore, up to 56% of the world's freshwater supply goes to feed livestock in the United States. Is this an efficient system, given that 1 billion people are without access to clean drinking water, and over 800 million suffer from food insecurity?

Freer/Shutterstock.com

Figure 114.2 Single crops are known as monoculture.

Most crops in the United States are cultivated through **monoculture**, the harvesting of a uniform type of crop (**Figure 114.2**). This requires lots of irrigation, and our freshwater supply is limited. Consider that it takes 1,800 gallons of water to produce 6.5 pounds of grain. Industrial farming increases the competition for land and water use, which is unsustainable given the growing population.

Source: http://www.worldwatch.org/files/pdf/Livestock%20and%20Climate%20Change.pdf

Monocultures also contribute to large numbers of a specific pest. This is usually counteracted with pesticides. Fertilizers, pesticides, and herbicides all use phosphorus and nitrogen. These chemicals leach into the soil and eventually harm aquatic life and water quality. Pesticides and herbicides have chemicals that also harm beneficial organisms on land and limit terrestrial biodiversity.

Organic farming methods use natural herbicides and pesticides and include **crop rotation** to conserve soil fertility. Organic farming methods also exclude genetic engineering, or the production of **transgenic crops**—crops with DNA from two sources. Crops are usually engineered to be resistant to pests and herbicides, tolerant to drought and salt, as well as the addition of vitamins.

Case Study

To cork or not to cork? That is the question.

A common hot-button debate among many wine connoisseurs is whether to utilize cork *or* modern screw caps—with wine and champagne bottles. The cork controversy has many tiers and environmentalists support both sides of the argument. Cork is technically considered a renewable (and biodegradable) resource if, and only if, it is harvested sustainably. Cork, which matures every 10–20 years, is taken from the outer bark layer of the Mediterranean cork oak (*Quercus suber*). It does not harm or kill the tree if done properly. And in fact these evergreen angiosperms can live up to 300 years, even when harvested.

Although traditionally associated with wine, cork has many other uses such as providing material for furniture and flooring. Widespread cork use, many argue, should be encouraged as it facilitates the conservation of cork oak forests—rich with bio-diversity. These oak forests are home to many endangered, sensitive and endemic species including birds such as eagles and mammals such as the lynx.

LianeM/Shutterstock.com

Box Figure 114.1.1 **Cork Oak**

In addition to its intrinsic and ecological value, such biodiversity also provides economic value. These forest resources are a major component for stabilizing the economy of countries like Portugal and Spain where cork oak groves are found. Using cork oak groves provides jobs to thousands in the region. Those employed for forest-related jobs naturally become stewards of the forest and have incentive to preserve them and to prevent erosion and desertification of the soil.

Cork harvesting gets support from the WWF (World Wildlife Fund established in 1961), a conservation and international fundraising group whose mission is to reduce biodiversity loss worldwide. According to the WWF, "Cork oak landscapes are one of the best examples of balanced conservation and development anywhere in the world. They also play a key role in ecological processes such as water retention, soil conservation, and carbon storage."

Unfortunately, many harvesting practices result in carelessly cut into the vascular layer of the tree— the inner layers that transport nutrients. Ultimately it ends up compromising and killing the oak. Dead cork oaks are of no value to anybody.

Wine production companies often prefer screw caps and to avoid cork taint—parasite that lives in cork oak bark and ruins wine—but is of no risk to human health.

What are your opinions regarding the ongoing cork debate? Are you in favor of using cork or not using cork tops in wine bottles?

inacio pires/Shutterstock.com

Box Figure 114.1.2 **Harvested Cork Oaks in Portugal**

Foods
By Cynthia McKenney, Ursula Schuch, and Amanda Chau

History of Plant Domestication for Food and Other Uses

Plants and other organisms have supported humans to develop and thrive for thousands of years. Our primitive ancestors lived on plant parts such as roots, leaves, seeds, berries, and fruits and supplemented their vegetarian diet with game, fish, and eggs. Humans relied on foraging and hunting until about 10,500 years ago when they began to cultivate plants as crops. This allowed them to store food and have resources during times of crop failure. They could also feed more people that were living in larger settlements or towns. Today there is only a very small population of nomadic people in some deserts and rainforests who forage and hunt for food.

Crop cultivation started with the collection and preservation of desirable seeds, seedbed preparation and planting, plant protection from herbivores and competing vegetation, and gathering, processing, and storing crop products. Rice and soybean remnants in Thailand from the Neolithic period about 10,000 years ago are interpreted as evidence for the first domestication of plants. The grains emmer and barley were found at a site in Iran dating to 6750 B.C. Early Chinese agriculture ventures started in the valleys of the Yellow River and Yangtze River with millet production around 7500 B.C., soon followed by the cultivation of rice. In the New World crops such as corn, squash, chili peppers, and potatoes were found among the earliest crops in the highlands of Central and South America.

Egyptian agriculture is documented in harvest scenes of hieroglyphs, and specimens of seeds and plants in ancient tombs from 5000 to 3400 B.C. The Fertile Crescent, the area from the Mediterranean Sea to the Persian Gulf, is the origin of agriculture. Wheat, barley, and legumes were the first foods cultivated upon which the Egyptian and Sumerian civilizations were built. With some people free to pursue activities other than hunting and gathering food, cultures started to develop. The arid climate, fertile soils, and access to river water brought many advances in agriculture. Written documentation of plants and their properties on papyrus were found dating to 1550 B.C. Greek and Roman scholars wrote about plants and agricultural production.

Important plants during this early agricultural period were wheat, corn, rice, and barley. These grains formed the basis of staple foods and were supplemented with legumes and meat. Different geographical areas specialized in crops appropriate to their environment. In the first century A.D., Romans became the first commercial bakers. Yams were planted as a staple food starting 10,000 years ago in Africa. Lentils and fava beans were cultivated about 6,500 B.C. in the Mediterranean and dates were cultivated in Pakistan. Linen, a cloth produced from flax, was made as early as 5000 B.C. while cotton started to be cultivated in Mexico around

1300 A.D. More than 5,000 years ago silkworms were domesticated in China. They were selected to consume mulberry (*Morus alba*) leaves and to produce precious silk (Lev-etin & McMahon, 2012).

From 1450 to 1650 trade and discovery of plants and plant products played an important role in nutrition, medicine, and the spice trade (How-ell, 2009). During this period gardens became established in different parts of the world. Sugar was introduced to different countries. Many spices were brought from Asia to Europe with different countries vying for dominance of the trade. Spices such as black pepper, nutmeg, and cinnamon became important goods. Vanilla and chocolate came from the Americas and were originally only available to the wealthy. Although paper was invented in 105 A.D. in China, it was not widely available in Europe for another 1,500 years until the printing press was invented. Soon after, more books on plants, including botanical descriptions, medicinal and herbal properties, and agricultural methods were published. From the middle of the 17th century until the middle of the 18th century botanical explorers roamed the different continents and published natural histories including descriptions of the flora about different geographical areas. Plants of economic value such as spices and sugar and plants having medicinal and ornamental interest were moved from their native countries to different countries. Coffee and tea became popular beverages outside their native countries. Subsequent periods are marked by more trade of plants between countries and continents. The rise of significant commercial products such as sugar, pineapple, coffee, and plant nurseries raising plants for sale supported the practice of growing more plants outside their indigenous locations. Advances in science in the last 150 years have led to greater understanding of different plant properties and ways to cultivate and propagate plants. New uses for plants are continuously discovered.

The edible plants our primitive ancestors consumed are different from today's modern plants, which are produced in small- or large-scale agricultural operations. Domesticated plants have been selected over time for specific traits such as larger cob size and larger kernel size of corn, nonshattering heads of grains, and larger tubers, fruits, and seeds. Corncobs from 8,000 years ago were very short and had just a fraction of the 600–1,200 kernels found in a contemporary ear of corn. In their wild form, heads of wheat and rice shattered easily when ripe and scattered their seeds widely to reproduce. Early gatherers selected the seeds of those plants that remained on the head and were nonshattering.

Food security is the basis of a healthy population and today still varies greatly in different areas of the world. In their quest to provide enough food, people have changed natural ecosystems and have altered natural biogeological cycles to where some ecosystems cannot fulfill their natural functions anymore. Hunters and gatherers and early agricultural societies learned about the perils of overharvesting food in one area and degrading the land. Today, slash and burn, clearing vegetation and using the fertile soil for a short time before it is eroded or degraded, is still practiced in some parts of the world.

Food Plants Essential to Humans

Of the more than 300,000 plant species known, scarcely over 100 cultivated species supply more than 90 percent of the energy eaten as food by people (Howell, 2009). Maize, rice, and wheat are the most important food staples, providing more than 60 percent of the calories for humans.

Grasses: maize, rice, Wheat, and other Grasses

Grains or cereal crops that were the basis of ancient agriculture are still the most important food staples for the world today. Maize, rice, and wheat account for more than half of the food calories consumed by the world population (Food and Agriculture Organization, 2010). These grains provide carbohydrates, protein, and essential minerals and vitamins. Other crops from the grass family ranking among the top 35 food crops are barley, sorghum, and millet. Oats and rye are also of importance regionally. Cereals are cultivated worldwide in cool, temperate, subtropical, and tropical climates. Some cereals are warm-season C4 plants and include maize, millet, and sorghum. Cool-season C3 cereals are rice, wheat, barley, rye, oat, and wild rice.

Cereal crops are in the Poacea or grass family and produce a seed packed with energy. The center of the seed, or botanically the dry fruit, consists of the embryo or germ, which is a good source of oils, enzymes, and vitamins, and the starchy endosperm. This is surrounded by the protein-, vitamin-, and oil-rich aleurone layer. The surrounding seed coat is fused to the outer fruit wall and is known as the bran, providing fiber from the grain. The refining process produces foods such as white flour or white rice and removes the chaff or bracts around the grain and the bran and aleurone layer. What is remaining is the endosperm, high in starch and with fewer nutrients compared to whole-grain foods where only the chaff is removed.

Table 115.1	Production of Major Food Crops and Areas of Harvested Land Worldwide, 2010	
Crop	**Production (1,000 tons)**	**Area Harvested (1,000 ha)**
Cereals	2,476,416	693,701
Vegetables	1,044,380	55,598
Starchy roots and Tubers	747,740	53,578
Oil Crops	170,274	269,680
Pulses	68,829	78,311

Source: FAO Production Yearbook 2013. January 2, 2018, from http://www.fao.org/docrep/018/i3107e/i3107e.PDF

The dry grains of cereal crops can be easily transported, stored, and milled. This is an advantage over tubers and starchy fruit such as sweet potato, cassava, yam, and plantain, which have shorter shelf-lives. Cereal grains provide carbohydrates, oils, and many essential vitamins for human and animal nutrition.

The seeds of some broadleaf plants are used like cereal grains, although they are not in the *Poaceae* family. Those plants are called pseudocereals and include buckwheat (*Fagopyrum esculentum*), amaranth (*Amaranthus caudatus*), and quinoa (*Chenepodium quinoa*).

Maize

Maize (*Zea mays*) is native to America and was domesticated from the grass teosinte (*Zea diploperennis*) around 5500 B.C. in Mesoamerica- (Levetin & McMahon, 2012). According to archaeological evidence, cultivation of the grain spread throughout the American continent from Mexico starting about 5,000 years ago. Maize is known as corn in North America and Australia. It is the most widely grown cereal in the world. Corn is consumed as whole grain, coarsely ground or as ground cornmeal in many cooked or baked foods. Corn is converted to high-fructose corn syrup, Bourbon whiskey, cooking oil, breakfast cereals, and cornstarch (**Figure 115.1d**). Corn not only serves as food for humans, it is a major source of animal feed for livestock, the biofuel ethanol, and other industrial products such as plastics, fabrics, and pharmaceutical products. Sweet corn cultivars are grown for human food; field corn is cultivated for animal feed. Overall, about 40 percent of the United States' corn crop is grown for animal feed and 33 percent for biofuel. In contrast, corn for human consumption accounts for only 1.5 percent and for seed corn only 0.2 percent. Corn is also used for industrial, pharmaceutical, export, and other purposes. Ancient people prepared maize with a lime treatment to make niacin or Vitamin B3 available. Without this process a diet with corn as the main staple can cause pellagra, a disease due to vitamin deficiency.

Teosinte is believed to be the ancestor of the modern corn plant, but evidence is not definite. Teosinte is a short grass plant with multiple terminal male inflorescences and several small ears of corn carrying about 6–10 kernels each. Kernels on the ears were surrounded by a hard fruit case and arranged on a spike that shattered upon maturity. Modern corn plants consist of a single stem 2–3 m tall with one broad, long leaf emerging from several nodes along the stem. Corn plants are monoecious with the tassel, the staminate flowers, located at the terminal end of the stem (**Figure 115.1a**). The lateral female inflorescence is a spike and entirely covered by leaves called *husks*. Silks look like soft hair and are elongated styles, with the stigma emerging from the leaves that cover the flowers. Each silk needs to be pollinated to develop into one grain on the ear of corn. The center of the ear is the stem of the female flower and unique to corn among the grasses. Pollen released from the tassel will spread by wind and pollinate nearby silks. Ripened kernels remain attached to the center of the spike, the corncob, and will not shatter (**Figure 115.1b–c**). When the ears are immature they are usually consumed fresh. As they mature they dry out and are unsuitable as fresh food, but are then used for milling and other purposes.

Corn has been used worldwide as a model plant for genetic studies for many decades. Each kernel is pollinated separately and is therefore a different genotype. A whole population is present on one ear of corn. Great variability is found in corn and the crop is easily grown in about 5 months. The large corn chromosomes are easily visible even under a light microscope. One of the major genetic discoveries using multicolored Indian corn (**Figure 115.1c**) were transposable elements or jumping genes for which Barbara McClintock was awarded the Nobel Prize in 1983 for her work in the 1950s. McClintock, an American scientist, studied genes on chromosomes of corn and found the genes responsible for the different colors could move from one chromosome to another. The genome of corn has more than 32,000 genes and has been completely mapped in 2009.

Extensive breeding experimentation has been conducted on maize to produce thousands of modern cultivars with desirable traits to enhance the genetic base of corn. Traits of interest are overall biomass production; kernel size; grain composition such as protein, oil, starch, and water content; sugar content of stems; tolerance to drought, high temperature, insects, and disease; and **aflatoxin** resistance, to name a

© Wattichok Painchiwarapun/Shutterstock.com
(a)

© Tish 1/Shutterstock.com
(b)

© Candace Hartley/Shutterstock.com
(c)

© Elena Schweitzer/Shutterstock.com
(d)

Figure 115.1a–d   (a) Corn plants with terminal staminate inflorescence and female spike developing into an ear of corn. (b) Dried corn ready for harvest. (c) Multicolored corn gets its color from the pigments in the aleurone layer and the pericarp. Yellow corn has the pigment in the endosperm. (d) Sampling of products from corn.

few. Adaptations to different climatic conditions and length of growing seasons are also important to accommodate the wide geographic and climatic range where corn is produced. Genetically modified (GM) corn is widely grown in most countries other than in Europe. GM corn has been engineered for herbicide resistance, insect resistance, and increased content of the naturally low amino acids tryptophan and lysine.

> **aflatoxins:** Compounds in foods such as nuts, corn, or cotton seed caused by the infection with *Aspergillus flavus* or related fungi that produce toxic metabolites.

Rice

Rice (*Oryza sativa*) is the main staple for people in Asia and feeds more than half the earth's population (**Figure 115.2a**). Rice has been cultivated for more than 10,000 years, starting in China. Upland rice is grown in regular fields but makes up only a small percentage of cultivated rice. About 90 percent of rice is grown in flooded fields or paddies where the water is 5–10 cm deep (**Figure 115.2b**). The rice plant is an annual monocot that grows with several stalks to a height of 1.0–1.5 m. Grains are surrounded by bracts and are arranged in a terminal panicle. Once the rice is harvested it has to be dried before being milled where the husk is removed. Today China and India are the top rice-producing countries, although rice is cultivated worldwide in areas where temperatures are favorable.

There are thousands of rice cultivars differing in grain size, shape, color, flavor, and consistency of the cooked product. The two most important subspecies are *indica* and *japonica*. Long-grain rice (*O. sativa* subsp. *indica*) is high in amylose starch and separates into individual grains after cooking. Long-grain rice is grown in tropical and subtropical lowlands, primarily in flooded fields. Short-grain rice (*O. sativa* subsp. *sativa*) is also known as japonica or sticky rice because the grains will adhere to each other when cooked. This rice as well as medium-grain rice is high in amylopectin but low in amylose starch. It is cultivated in upland tropical and temperate areas. Wild rice (*Zizania sp.*), with long grains dark brown to black in color, is a different genus from *Oryza sativa* rice. It is native in the upper Midwest and Eastern Canada and China.

Conventional breeding continues to develop new cultivars with desirable traits related to environmental adaptation, yield, and nutritional improvements. The International Rice Research Institute, devoted to rice research worldwide, has a collection of more than 100,000 rice accessions. The rice genome has been sequenced in 2002, the first of the cereals. Rice has the smallest genome of the cereals, with 430 Mb (millions of base pairs). This makes rice a good model system for genetic research and genetic modification. GM rice has been developed for enhanced nutrition, flood tolerance, insect resistance, and herbicide tolerance. Golden rice synthesizes beta carotene, a precursor to Vitamin A, which is lacking in regular rice. People who rely on rice as their main staple suffer from Vitamin A deficiency, which leads to blindness. Although rice is grown in shallow flooded paddies, the plant is intolerant of being covered entirely by water for more than a few days. GM rice with flood tolerance is now available and will reduce losses to flooding worldwide. Resistance to insects and herbicides in GM crops boosts yields and reduces the input for the rice crop.

Rice is often grown with azolla (*Azolla spp.*) (**Figure 115.2c**), which are several species of aquatic ferns fixing nitrogen. The blue-green algae *Anabaena azollae* has a symbiotic relationship with the weed azolla. The small ferns float on the surface of rice paddies and other aquatic crops and add nitrogen, shade out other weeds, and add organic matter to the production areas.

Rice is often milled, which removes the bran and germ, resulting in white, polished grains. Brown rice, only hulled and not milled, contains more minerals and vitamins, similar to unmilled wheat. The lack of thiamine, Vitamin B1, in white rice causes a nutrient deficiency known as *beriberi*, when people rely on white

© szefei/Shutterstock.com © javarman/Shutterstock.com © Nikita Tiunov/Shutterstock.com

(a) (b) (c)

Figure 115.2a–c (a) Rice plant. (b) The majority of rice is cultivated in flooded fields, often terraced on mountain slopes. (c) The nitrogen-fixing water fern azolla is often cultivated with rice.

rice as their main nutrition source. Rice is **gluten** free, an important attribute for food sought by people with gluten allergies. Although rice is prepared in many ways, it is also made into gluten-free noodles.

gluten: Protein found in the endosperm of wheat and other cereals, making dough elastic and helping it to rise.

Wheat

Bread wheat (*Triticum aestivum*) (**Figure 115.3**) is also known as the staff of life because when ground, mixed with water, and baked, bread has served as a staple for thousands of years. Leavened bread was discovered by Egyptians. They added yeast to the dough, fermenting it and trapping carbon dioxide, which resulted in a light bread. Leavening requires a certain amount of gluten, proteins that make dough elastic. Only wheat and rye flour contain enough gluten for leavening; barley and oat grains lack sufficient gluten to allow the dough to rise.

Grain was originally milled by stones to break down the grain and the whole grain was used for preparing bread. Later on, the milling process improved and steel rollers refined the grain, removing all of the bran and germ and leaving mostly the starchy endosperm. Longer shelf-life of bread resulted from this change because the oils from the germ were taken out. Along with it essential minerals and vitamins were removed, leaving refined wheat flour a nutritionally inferior product compared to the wholegrain type. Today, many grain products are enriched or fortified with the minerals and vitamins lost in the refining process.

There are different types and species of wheat. The first cultivated species of wheat was einkorn (*T. monococcum*), a diploid species cultivated first in Turkey (Bacon, 2008). Emmer wheat (*T. dicoccon*) and durum wheat (*T. durum*), both tetraploid species from a cross between wild wheat and wild goatgrass (*Aegilops sp.*), were the important grains for Mediterranean civilizations 3,000 years ago. Today's bread wheat (*T. aestivum*) and spelt (*T. spelta*) are hexaploids and originated about 6,000 B.C. as a cross between goatgrass (*Aegilops tauschii*) and tetraploid wheat species. Early wheat species are hulled with tough glumes surrounding the grain. After threshing, the spikelets require further processing to remove the hulls. Bread and durum wheat are free threshing; they bear naked fruit yielding the grains after threshing without further need for processing. Other important food grains with hulled fruit are oats, barley, and rye.

Bread wheat accounts for the majority of wheat grown today, which is used for bread, cereals, and pastries. Durum wheat is the second most important wheat species after bread wheat. It is the basis for Italian pasta because of the high protein content. Semolina is the endosperm of the ground durum wheat or other grains

© rodho/Shutterstock.com

(a)

© Orientaly/Shutterstock.com

(b)

Figure 115.3   Wheat is the second most important cereal grain and is grown from extreme northern locations to tropical climates due to many cultivars developed by breeders. Modern cultivars are short to prevent lodging.

after being crushed and sieved. Bulgur, a parboiled whole-wheat product, and couscous, made of wheat semolina, are important staples from durum wheat in North Africa. Breeders have developed thousands of modern wheat cultivars to maximize yields under different environmental conditions and to provide wheat suitable for different purposes. These cultivars vary in protein content, gluten content, adaptability to various environmental conditions, and disease resistance. In the United States red and white wheat are distinguished based on their grain color. Soft and hard wheat differ in protein content. Hard wheat has high protein content and is commonly used for bread whereas soft wheat with lower protein content is favored for cakes and pastries.

Other Grains important to Humans

Sorghum (*Sorghum bicolor*; **Figure 115.4a**) is a C4 grass with grains in a terminal inflorescence. Sorghum is important as food in Africa, India, and Asia. It is used primarily for forage, biofuel, and syrup production in North America. *Millet* is a term for a number of species of small-grain cereals. Pearl millet (*Pennisetum glaucum*; **Figure 115.4b**) is the most important millet grown and has been an important staple in India and some areas of Africa for thousands of years. Among the cereals, millet is most tolerant to arid conditions, poor soil fertility, and harsh environmental conditions.

Barley (*Hordeum vulgare*), similar in appearance to wheat (**Figure 115.4c**), is among the earliest cereals domesticated in the Fertile Crescent. It is used for animal feed and is a main ingredient in beer and other fermented beverages. A very small amount of barley is milled and used for human consumption. Barley production has decreased in the last decades. Leading production areas are Eastern Europe and Canada.

Rye (*Secale cereal*) has been domesticated more recently, only for a few thousand years. Rye is similar in appearance to wheat and thrives in colder, drier regions than wheat, making it popular in Northern Europe. The grain has a lower gluten content, rendering heavier bread such as pumpernickel. Rye is also used as animal feed and for alcoholic beverages. Breeders have crossed rye with wheat, which yielded triticale (× *Triticosecale*) about 100 years ago. Triticale combines the best features of both grains with high protein and lysine content, higher yields, disease resistance, and cold hardiness.

Oat (*Avena sativa*) grains grow in an open panicle with several florets (**Figure 115.4d**). Oats were cultivated more recently and thrive in temperate regions such as northwest Europe and Canada. They are used for food, in beverages, and as horse fodder. Oatmeal is considered a health food due to its cholesterol-lowering properties.

© fotohunter/Shutterstock.com

(a)

© Frontpage/Shutterstock.com

(b)

© jokerpro/Shutterstock.com   © Andrew Koturanov/Shutterstock.com

(c)   (d)

Figure 115.4a–d Other important grains for food, animal feed, and energy are (a) sorghum, (b) pearl millet (Purple Majestic variety), (c), barley, and (d) oats.

Legumes

Legumes are plants in the Fabaceae family used for food, forage, and cover crops. Legume refers to the fruit of a plant in this family, which is commonly called a *pod* and botanically a simple dry fruit that usually dehisces. Peanuts (*Arachis hypogaea*) are an exception in that they are an indehiscent legume fruit. **Pulse** is another term used for legumes and refers to plants harvested for their dry seed. Legumes produce their own nitrogen through *Rhizobium* bacteria in their roots and are rich in protein and oils. These plants are well adapted to arid climates and thrive in hot, dry conditions.

pulse: Dried seeds of legumes.

Beans, peas, lentils, and peanuts are legumes and are an important source of protein, especially for people who choose not to eat or cannot afford animal protein. Beans, peas, and lentils have been cultivated for thousands of years. Several species of beans are grown with the common bean (*Phaseolus vulgaris*), which include haricot, black, red, white, brown or mottled seeds, and are the most cultivated species in the world (Bacon, 2008). Other bean species are the large-seeded broad or fava bean (*Vicia faba*) from the Mediterranean or southwestern Asia; the tepary bean (*Phaseolus acutifolius*) with rich, nutty flavor originating from Central America; and the adzuki bean (*Vigna angularis*) from tropical Asian origins and popular in Japan and China. Chickpeas or garbanzo beans (*Cicer arietinum*) were cultivated in the Fertile Crescent more than 6,000 years ago and are an important staple in India, North Africa, and the Middle East. Chickpeas have the third highest protein content after soybeans and peanuts and are prepared into well-known Middle Eastern dishes such as hummus and falafel. Common beans are cooked and eaten in their pods when young, mature beans are larger and hard and require soaking before cooking. Mung beans (*Vigna radiata*), originating in India, are very small and do not require soaking before cooking. They are used as sprouts; their seeds are cooked or ground for starch to produce vermicelli noodles. Beans are an ingredient in many signature dishes of different cultures and include refried beans, chile, or sweet bean paste in desserts.

Soybeans (*Glycine max*; **Figure 115.5a**) are native to East Asia and have been grown there for 3,000 years. They were introduced to other continents in the 18th century (Bacon, 2008). The United States adopted soybeans as a major crop in 1915 and is now leading world production followed by Brazil, Argentina, and China. Soybeans mature in two to four months under optimum growing temperatures of 68–86°F (20–30°C). Genetically modified soybeans were introduced in 1995 to tolerate the herbicide Roundup. Today, Roundup Ready® soybeans are almost exclusively cultivated in North and South America, but not in European countries. Mapping of the soybean genome was completed in 2010.

Soybeans contain 40 percent protein, the highest protein content among plants, and 20 percent oil, thus making it a very valuable commodity for animal feed, industrial use, and food. Soybeans are traditional staples in the human diet in Asian cultures. They require cooking or fermenting because they are indigestible when consumed raw. Typical food products include tofu, soy sauce, soy milk, and fermented soybean products such as tempeh and miso (Levetin & McMahon, 2012). Soybeans are manufactured into many dairy-type products and are important to people who cannot tolerate lactose in cow's milk. Soybeans are also processed into textured vegetable protein, spun soy fibers flavored and shaped to resemble meat products. Soy products are also valued for the **isoflavones** they contain, phytoestrogens attributed to have several health benefits such as lowering cholesterol and preventing or reducing recurrence of certain cancers. Soy oil is used in many food and industrial applications, including cooking oil, ink, and biodiesel.

isoflavones: Phytoestrogens that have possible health benefits for lowering cholesterol or that are used in cancer treatment.

Peanuts originated in South America and likely were already cultivated in pre-Inca times. Peanuts develop in the ground on an annual herbaceous plant (**Figure 115.5b**) and contrary to their name are not nuts but seeds of a legume in a pod. Flowers grow aboveground and die after pollination. The flower stalk or pedicel turns toward the ground, pushing the developing fruit into the soil. Fruits develop one to four ovules, mostly two per pod within four to five months (Bacon, 2008). Peanuts contain about 25 percent protein, the second highest level after soybeans and more than any nut, and have an oil content of about 45 percent. The nutritional value of peanuts is high, especially in essential nutrients such as several vitamins and minerals. Peanuts are used worldwide for their oil, in sweet and savory dishes, and the majority of peanuts in the United States are made into the popular peanut butter. In the United States, peanuts were grown more widely, especially in the South, based on the work of George Washington Carver (1864–1943), who devised many industrial and culinary uses for peanuts and promoted their cultivation. The leading

© FLarivier/Shutterstock.com   © Sunsetman/Shutterstock.com

(a)   (b)

Figure 115.5a–b   (a) Soy beans and (b) peanuts are important legumes for food because of their high protein content and as oil crops.

peanut-producing country is China. Allergies to peanuts and contamination with aflatoxins can pose health risks through food and animal feed. Peanuts can cause allergies with mild to severe symptoms, affecting 1–2 percent of the population in the United States. Aflatoxins develop when peanuts, but also corn or cotton-seed, are infected by *Aspergillus flavus* or related species, fungi that produce toxic metabolites.

Potatoes, cassava, sweet Potatoes, and other starches

Underground storage organs rich in carbohydrates are an important food source for humans and animals. Potatoes (*Solanum tuberosum*; **Figure 115.6a**) are stem tubers high in starch, originating in the Andean highlands of southern Peru where people domesticated wild plants about 10,000 years ago. This plant became the staple of the Inca civilization in South America. Potatoes were introduced to Spain in the 16th century and became a major food staple throughout Europe about 200 years later. Potatoes are the third largest staple food in the world and about one-third of world potato production is in China and India. There are thousands of potato cultivars known and tubers come in different sizes, shapes, and colors of white, yellow, or purple. Potatoes are a cool-season crop that can be cultivated from sea level to elevations above 4,500 meters. Potatoes are very efficient in using water and produce higher yields per area than cereals.

All aboveground parts of the potato plant are poisonous due to the **glycoalkaloid** solanine. Solanine is found in plants of the nightshade or Solanaceae family, such as tomato and eggplant. Tubers of potatoes are not poisonous, except when they are exposed to excessive light after harvest and develop large green patches, which contain solanine. The nutritional value of potatoes lies in its carbohydrates, fiber, protein, vitamins, and minerals. From its introduction to Europe until the middle of the 19th century, potatoes became the main food for peasants in Ireland. The pathogen *Phytophthora infestans* destroyed stems and leaves of potatoes, killing plants in a short time and causing widespread famine and death among the Irish population from 1845 to 1849. Potato production and use have surged in developing countries in the last 50 years. Potatoes are used fresh and are fried, boiled, or baked. They are also used for their starch as animal feed and seed potatoes and are fermented into alcoholic beverages such as vodka.

glycoalkaloid: Alkaloids in plants from the nightshade (Solan-aceae) family.

Cassava (*Manihot esculenta*) is a woody shrub producing edible tuberous roots (**Figure 115.6b**). It is a crop of great importance to people in tropical and subtropical areas, especially Africa, Asia, and South America. Cassava originated in South America and is widely grown in tropical lowlands. The plants can tolerate a wide range of soil and moisture conditions, making them a popular crop in drier regions. Cassava takes 8–16 months until harvest after stem cuttings are planted. Cassava belongs to the Euphorbiaceae or spurge family, the only food crop in this family of plants containing a white milky sap that can be an irritant. The roots and leaves of cassava are eaten, but proper cooking is required to remove the toxic cyanides. Cassava is classified into sweet or bitter depending on the level of cyanides. The roots contain high levels of carbohydrates and minerals but are low in protein. Cassava is often ground into flour and then cooked with other ingredients or used as tapioca. It is also fried or fermented into alcoholic beverages.

The sweet potato (*Ipomoea batatas*) is a tuberous root vegetable (**Figure 115.6c**). This herbaceous perennial vine in the Convulvulacea or morning glory family is native to Peru. The plants grow well in poor soil with little water and have become an important food staple in Africa and Asia (Bacon, 2008). This crop needs a warm growing season and cannot tolerate frost. The tubers vary in color from beige, yellow, orange, red, and purple. They are prized for high carbohydrate levels, sugar, Vitamin A and beta-carotene, and minerals. Sweet potatoes are used boiled, fried, baked, or ground into flour. George Washington Carver developed more than 100 products from sweet potato, including glue, starch, and dehydrated sweet potato.

True yams (*Dioscorea sp.*; **Figure 115.6d**) are sometimes confused with sweet potato. The plant, an herbaceous, perennial vine with underground tubers, is an important food source in many tropical countries. The white-to orange-colored tubers are rich in starch. Tubers can vary from the size of a potato to over 2 m long and weighing over 40 kg.

Bananas (*Musa acuminata, M. balbisiana*; **Figure 115.6e**) grow from a corm and are herbaceous perennial plants in spite of their shape and size resembling a tree. What appears to be the trunk of the plant are the tightly packed petioles broadening into a sheath below the leaf blade. This **pseudostem** produces one inflorescence with up to several hundred bananas developing from one flower. The stem then dies and new pseudostems emerge at the plant base. Bananas contain high amounts of starch and sugar and are the fourth most important fruit worldwide. Native to Southeast Asia, many banana cultivars are grown in tropical and subtropical regions. Some are consumed raw such as the Cavendish banana popular in North America and Europe; many other cultivars are cooked, fried, baked, or dried. Bananas are a good source of some vitamins and minerals. Fibers from the plant have been used to make textiles and paper.

pseudostem: Structure appearing like a stem, but composed of folded or rolled petioles and leaf blades; found on the herbaceous banana plant.

Fruits, nuts, and Vegetables

Humans require macronutrients such as carbohydrates, proteins, and fats, which are partially supplied by many of the grains and starchy tubers discussed earlier. Vitamins and minerals are essential micronutrients in the human diet, which are also partially provided by these staples. Fruits, nuts, and vegetables are a major source of energy and micronutrients (**Figure 115.7**).

However, they also contribute **phytochemicals,** a group of compounds produced by plants that have biological activity (Higdon & Drake, 2013). Phytochemicals are considered beneficial for human health, but not essential. These compounds include carotenoids from yellow, orange, and red fruits and vegetables; curcumin from the spice turmeric; indole-3-carbinol and isothiocyanates from cruciferous vegetables such as broccoli, kale, and cabbage; resveratrol from grapes, red wine, peanuts, and certain berries; and isoflavones from soybeans. Diets rich in plant-based foods have many health benefits. However, the specific effect of individual phytochemicals to prevent a specific disease or condition or to treat one has not been conclusively proven.

© Madlen/Shutterstock.com
(a)

© KROMKRATHOG/Shutterstock.com
(b)

© mtsyri/Shutterstock.com
(c)

© JIANG HONGYAN/Shutterstock.com
(d)

© T.W. Van Urk/Shutterstock.com
(e)

Figure 115.6a–e (a) Potato plants in the Solanaceae family are toxic except for the tubers. (b) Cassava tubers contain toxic cyanides that are removed by special preparation. (c) Sweet potato and (d) yams are important starch crops. (e) Banana bunches carry many fruits from one flower.

> **phytochemicals:** Compounds produced from plants that have biological activity and are thought to be beneficial but are not essential to human health.

Antioxidants are chemical compounds preventing oxidation of molecules and the production of free radicals. These free radicals can cause reactions damaging cellular DNA, lipids, and proteins or killing cells and initiating disease. The consumption of fresh fruits and vegetables that are rich in antioxidants prevents cardiovascular disease and is thought to prevent several other diseases. Antioxidants found in many fruits, vegetables, and nuts are Vitamins E and C and carotenoids. The antioxidants resveratrol and **flavonoids** are contained in coffee, tea, chocolate, and some spices and foods discussed earlier.

© Adisa/Shutterstock.com

(a)

© szefei, 2014. Used under license from Shutterstock, Inc.

(b)

Figure 115.7 Fruits and vegetables are important for a healthy diet because they provide vitamins, minerals, and phytochemicals, in addition to calories for energy.

antioxidant: Chemical compound preventing oxidation of molecules and production of free radicals, which can damage cells.

flavonoids: Secondary plant metabolites with potential protective properties for human health.

Members of the Rosaceae or rose family are important sources of stone fruit and pome fruit. (Stone fruits are plums, peaches, apricots, and cherries; apples and pears are pome fruit. Aggregate fruit such as strawberries, raspberries, and blackberries also belong to this family). Apples (*Malus × domestica*) are native in Europe, Asia, and North America and are the most consumed fruit worldwide (Bacon, 2008). Thousands of cultivars have been bred and grow well in temperate climates. Apples are eaten fresh, cooked, or used in pastries and for alcoholic beverages such as apple cider, cider vinegar, and apple juice. The European pear (*Pyrus communis*) and the Asian pear (*Pyrus pyrifolia*) also grow well in temperate climates. They are mostly eaten as fresh fruit. Pears can be picked before fully ripening to increase longevity in cold storage. Apples and pears are good sources of fiber and carbohydrates.

Of the stone fruits, apricots (*Prunus armeniaca*) were cultivated 4,000 years ago in their native China. Apricots grow on small deciduous trees in climates with temperate, cool winters. The perishable fruit is eaten fresh but is often dried, poached, or baked in desserts. Peaches are also native to China and were cultivated there for at least 3,000 years. Nectarines have been around for over 2,000 years. Both fruits are excellent for eating fresh when they are ripe from the tree, but also popular baked, poached, or grilled. Cherries originated near the Caspian Sea and sweet cherries (*Prunus avium*) and sour cherries (*P. cerasus*) were cultivated by 300 B.C. Sweet cherries are mostly eaten fresh; sour cherries are cooked and used in many baked pies or other desserts and jams.

Avocados (*Persea americana*) originated in Central America and are cultivated in Mediterranean climates and tropical areas. They are nutritious with high levels of monounsaturated fats, antioxidants, and minerals. This tree fruit is sometimes considered a fruit and sometimes a vegetable. It is eaten fresh and prepared as guacamole, and is added in salads and desserts such as avocado ice cream. Another highly nutritious fruit is the date (*Phoenix dactilyfera*). Dates have been an important food staple for Middle Eastern cultures since prehistoric times (Bacon, 2008). Dates grow on the date palm, which is also widely grown as an ornamental tree. Dates are used in sweet and salty dishes; when dried they contain more than 70 percent sugar. Dried dates are the food of ancient travelers as the fruit contains many essential macro-and micronutrients. Other parts of the plant are used for fiber and wood.

Pineapple (*Ananas comosus*) belongs to the bromeliad family and is valued for their fruit high in carbohydrates, manganese, and Vitamin C. Each fruit is composed of many fruitlets and can weigh from 1–3 kg. Origins of this plant are the South American tropics. Pineapples are used as fresh fruit, juice, and in savory Thai and other Asian dishes. The pineapple represents hospitality and welcome in some cultures.

Citrus fruits are in the rue family or Rutaceae and originated in Southeast Asia. The many citrus species—including oranges, lemons, limes, grape- fruits, kumquats, mandarins, and tangerines—grow on broadleaf evergreen trees or shrubs, some of them with spiny branches. They need a warm subtropical or tropical climate and most cannot tolerate frost. Citrus fruits are treasured for their high juice content and for the essential oils in the exo-carp, which is used as zest to flavor baked goods or other dishes. The sour flavor is typical for citrus fruits, which are often eaten fresh or the pressed juice is used for many culinary purposes. Citrus is high in Vitamin C and flavonoids. Citrus trees are also grown as ornamental plants. In the 17th and 18th century citrus trees were favored by royalty in northern Europe who grew the trees in large containers, moving them into glass houses called *orangeries* to protect them from cold weather.

Important fruit growing on vines are watermelons, kiwis, melons, and grapes. Watermelons (*Citrullus lanatus*) are an annual trailing or climbing plant. Their fruit can weigh up to 20 kg, are composed of 92 percent water and 6 percent sugar, and contain Vitamin C, beta-carotene, and lycopene. Watermelons are native to Africa and were cultivated in ancient Egypt about 1,000 B.C. Watermelons are one of the most popular fruits, botanically they are a pepo, a berry with a mostly red fleshy mesocarp and endocarp and a thick exocarp (rind). Watermelons are the quintessential summer fruit and are consumed fresh or used in salads and chilled desserts. The rind is cooked or pickled in some areas. Different cultivars of *Cucumis melo* are the cantaloupe, honeydew, and muskmelon, belonging to the pumpkin or Cucurbitaceae family like the watermelon. Their wild ancestors are from subtropical and tropical areas in Africa and Asia and they grow well in arid climates. Melons are an important commercial crop worldwide.

Wild ancestors of grapes (*Vitis vinifera*) were growing in the Northern Hemisphere 23 million years ago as evidenced by fossil leaves (Howell, 2009). In ancient Egypt wine was produced in 2400 B.C. and later was brought to Greece where wine production and consumption flourished. Grapes were brought to China by 100 B.C. and to Northern Europe by the Romans. Grapes in North America were cultivated by European settlers who used the native Fox or Concord grape to cross with European varieties. Phyllox-era (*Dactylosphaera vitifoliae*), a tiny, sap-sucking insect feeding on roots and leaves of grapevines, was introduced from America and destroyed major parts of the European wine-growing area, especially France, in the 1860s. The partially resistant Fox grape and other grapes native to North America were used as root stock for grafting desirable cultivars onto or as a parent in hybridization with the European grapes. Grapes are important for today's worldwide wine industry, the fresh market, juice, or dried as raisins (**Figure 115.8a**–b). Grape leaves are used in Middle Eastern cuisine and grapevines are also cultivated as ornamentals for their colorful fall foliage.

Nuts are an important addition to the human food palette, providing protein, unsaturated fats, fiber, vitamins, essential minerals, and antioxidants (Table 115.2). What is called an edible nut is botanically a nut, drupe, or seed (**Figure 115.9a**–c). Coconuts account for more than half of the world nut production, peanuts for one-third, and all other nuts for the remainder (Food and Agriculture Organization, 2010). Coconut flesh developed from the endosperm is eaten fresh or is used dried in many sweet and savory dishes (**Figure 115.9a**). Liquid coconut water is used as a drink, often sold in the coconut with a straw inserted. Coconut milk is the liquid pressed or extracted from the grated coconut meat and is high in saturated fat. Coconut oil is extracted after processing the meat and used in cooking and cosmetics. The remainder of the coconut husk, fronds, and stems are used for growing media, construction materials, fuel, brooms, and dye.

© Nainong/Shutterstock.com

(a)

© Svetlana Lukienko/Shutterstock.com

(b)

Figure 115.8a–b (a) Grapes are used for eating fresh, (b) dried as raisins, pressed as juice, and fermented as wine.

Table 115.2 ## Nuts Used for Culinary Purposes and Their Major Nutritional Properties

Latin Name	Common Name	Fruit Type	Nutritional Properties
Corylus avellana	Hazelnut	Nut	Protein, fiber, unsaturated fat, Vitamin B
Juglans regia, J. nigra	Walnut	Drupe	Protein, unsaturated fat, antioxidants, manganese
Prunus amygdalus	Almond	Drupe	Protein, fiber, unsaturated fat, Vitamins E and B, minerals
Anacardium occidentale	Cashew	Drupe	Protein, carbohydrates, unsaturated fat, essential minerals
Cocos nucifera	Coconut	Drupe	Saturated fat, fiber, minerals
Macadamia integrifolia	Macadamia nut	Seed	Unsaturated fat, vitamins
Pinus spp. (P. edulis, P. pinea, P. koraiensis)	Pine nut	Seed	Protein, unsaturated fat, Vitamins E and B, manganese
Carya illinoinensis	Pecan	Drupe	Protein, unsaturated fat, Vitamin B

Vegetable is a term used to describe plants that are entirely or in part edible for human consumption. Vegetables are usually differentiated from fruit as having a savory flavor, whereas fruit has a sweet flavor. These descriptions are arbitrary and not related to botanical definitions. Vegetables are categorized based on whether they are taxed by law as a vegetable. According to this definition the United States Supreme Court ruled in 1893 that the tomato is considered a vegetable and is taxed as one according to the 1883 U.S. Tariff

Act, although botanically it is a fruit. Mushrooms, for example, are considered a vegetable, although they are not even classified in the plant kingdom. Leafy vegetables are easily recognized and include lettuce, kale, cabbage, spinach, beets, and mustard greens. Common root vegetables are beets, onions, carrots, radishes, potatoes, and sweet potatoes, the last two of which require cooking. Leafy stems of celery and rhubarb are eaten; however, rhubarb stems are the only edible part of the plant, the leaves are poisonous. Tomatoes, eggplant, squash, and pumpkins are fruits. Corn is a grain commonly served as a vegetable. Legumes such as French beans or sugar peas are cooked as a vegetable in the whole while black beans, lentils, and split peas are prepared without the pods.

© mrfiza/Shutterstock.com   © Jakkrit Orrasri/Shutterstock.com   © goghy73/Shutterstock.com

(a)   **(b)**   **(c)**

Figure 115.9a–c   (a) Coconuts, (b) cashews, and (c) almonds are a rich source of fat, fiber, and minerals.

Vegetables are parts of herbaceous plants consumed raw or cooked, with some requiring cooking to make them edible. The different parts of vegetables eaten include leaves, stems, roots, bulbs, fruits, flowers, and flower buds. Vegetables are important for the human diet because they contain fiber, vitamins, essential minerals, antioxidants, and phytochemicals. Diets rich in vegetables and fruits are recommended to prevent or treat some common diseases related to a diet lacking a wide variety of plants. **Vegan** diets rely exclusively on vegetables and other plant-based foods.

vegan: Diets using plant-based food only, excluding also egg and dairy products.

BOX 115.1	Moringa—Another Superfood

Moringa oleifera Lam. is a fast growing tree native to northern India along the Himalayas. Common names for *M. oleifera* include moringa, horseradish tree due to the roots tasting like horseradish, drumstick tree due to its long, narrow shaped seed pods, and benoil tree due to the oil that is extracted from the seeds. Moringa is classified in the family Moringaceae that contains only one genus, Moringa. It is widely cultivated in tropical and subtropical areas of southeast Asia, Central America, and Africa. The plant requires mean annual temperatures of 59°F to 86°F (15°C to 30°C) and annual rainfall between 76 to 225 cm. Long dry periods or cold temperatures stunt leaf growth, lead to leaf abscission, and prevent flowering and fruit set. The plants are well adapted to a wide range of soil conditions and thrive with sufficient irrigation and fertilization.

Different parts of the moringa plant make it versatile as food, medicine, and for horticultural and industrial uses. The leaves contain between 20% and 35% protein on a dry matter basis, and are high in essential amino acids, vitamins A and C, calcium, and potassium. This makes them an ideal source of food to alleviate malnutrition in tropical countries. Moringa powder produced from the leaves is often used as a nutritional supplement in Africa.

Moringa's reputation as a superfood is not only based on the valuable nutritious characteristics but also on the many traditional medicinal uses. Bark and roots of the tree contain isothiocyanate compounds that account for the horseradish scent lending one of the common names of the plant. These compounds have been shown to have antimicrobial and antifungal properties and extracts of the plant have been used to treat tumors and cancer.

The large leaves are tripinnate and are 30–60 cm long. Flowers are creamy white and have yellow stamens. They develop into triangular pods, tapered at both ends, and a length of 30–120 cm. Immature, green pods are consumed as vegetables; once pods mature and turn brown, dry seeds can be processed for oil extraction. Moringa oil is prized for its stability and can be stored for a long time without losing quality. The oil is used for industrial, culinary, and cosmetic purposes. In addition to serving as human food, leaves are also used as supplemental animal fodder.

Moringa trees can be propagated from seed or cuttings. Under optimum conditions, some plants selected for early flowering can produce marketable pods within 6 months, while others may take one to two years to yield a large number of pods. As trees increase in size, they are often cut at a height of 50 to 100 cm and new branches with harvestable pods can develop within 6 months.

Moringa has great potential to expand into a more prominent specialty tree crop in climates where it grows well. The undemanding cultivation requirements and the many benefits from this plant offer great promise as an industrial crop, cultivating on small acreage, or as backyard tree. The evaluation of different Moringa species and continuous selection and breeding of *M. olifeira* for desirable traits further improves opportunities to develop greater production capacity.

Source: Ursula Schuch

(a)

Source: Ursula Schuch

(b)

Figure Box 115.1   Moringa leaves, flowers (left), and immature, green pods (right) are rich in nutrients and popular in vegetable dishes.

Plant oils and sugar

Plants are the source of oils used for many different purposes and have been used by humans for thousands of years. Vegetable oils are used for food preparation, as preservatives, as fuel, in cosmetics, and as lubricants. Vegetable oils have **triglycerides** as their main component. Triglycerides from plants are primarily unsaturated, are liquid at room temperature, and are considered healthy in the human diet (see Appendix). Fat from animals is primarily composed of saturated triglycerides, which are solid at room temperature and potentially less healthy for humans. Oils liquid at room temperature spoil and turn rancid sooner than those solid at room temperature. The process of hydrogenating vegetable oils renders them solid at room temperature, but has negative health effects on human blood chemistry.

> **triglyceride:** Lipid composed of three fatty acids connected to a glycerol molecule; main component of vegetable oils.

Oil from palms and soybeans are extensively used worldwide. Important oils for use in cooking are extracted from rapeseed (canola oil), corn, sunflower seeds, peanuts, olives, coconuts, and sesame seeds. Biodiesel is produced from soybeans, rapeseed, castor, other oil crop plants, algae, and animal fats. Biodiesel is an alternative to petroleum-based fuel and is considered environmentally cleaner. Increased demand for these oil crops cause concern they may compete for land use and food production.

Essential oils are extracted from plants for their volatile aromatic compounds and they contain a specific scent related to the plant. The oils are extracted by distillation or pressing. Mint, eucalyptus, citrus, roses, lavender, almond, and cloves are used for harvesting essential oils. Some plants such as mint or lavender are used entirely for distillation. Peels of citrus fruit are pressed for their oils while rose petals and orange blossoms are distilled. Essential oils are used in perfumes, cosmetics, as flavoring in food or beverages, as scent in many different products, and in alternative medicine such as aromatherapy.

essential oils: Oils harvested from plants for their volatile aromatic compounds.

Sugar is a luxury in the human diet because it is not essential for survival (Bacon, 2008). Major sugar-producing plants are sugar cane (*Saccharum officinarum*) and sugar beets (*Beta vulgaris*) (**Figure 115.10a–b**). The cultivation of sugar cane in the 18th century brought with it slavery, war over the sugar trade, domination of

© Hywit Dimyad/Shutterstock.com

(a)

© butterflydream/Shutterstock.com

(b)

© Simone Voigt/Shutterstock.com

(c)

Figure 115.10a–c (a) Sugar cane, (b) sugar beets, and (c) sweetleaf or stevia satisfy the world's sugar demand.

colonial powers, and forced migration of people. The increase in refined sugar consumption has been linked to increased obesity and several other diseases. Sugar cane cultivation requires a tropical climate, ample water, and labor. Sugar cane provides most of the world with raw, brown, or white sugar, molasses, or rum. Sugar beets are a root crop cultivated in cooler climates. They provide the majority of European table sugar.

Sweetleaf or stevia (*Stevia rebaudiana*; **Figure 115.10c**) is a shrub grown for the high sugar concentration in its leaves. The plant is native to SouthAmerica and the leaves have 30–45 times the sweetness of sucrose, regular table sugar. Stevia contains no carbohydrates and no calories, making it popular for people on diets and those avoiding sugar. Stevia is used widely in many parts of the world but to a lesser extent in Europe and North America. Other sources of sugar are maple syrup, which is the boiled xylem sap from *Acer saccharum* harvested in late winter primarily in southeast Canada and the northeastern United States. The carob tree (*Ceratonia siliqua*) in the legume family produces seeds high in sugar and a chocolate-like flavor. Sugar is also extracted from agave, sorghum, and many other plants used regionally.

Commercial Products and Medicinal Plants

By Cynthia McKenney, Ursula Schuch, and Amanda Chau

Topic 116

Commercial Products

Flavored and Fermented Beverages

Humans have long discovered improvements to beverages to satisfy the need for water. Many flavored beverages such as tea, coffee, and cocoa are based on water flavored with the extract of leaves or beans. Fermented drinks such as beer and wine rely on the fermentation of grains and grapes. These beverages have been used by humans for thousands of years and have become important in some cultural and religious ceremonies. Sharing a cup of tea or coffee is a social ritual many people participate in. Caffeine contained in these beverages is a stimulant, making them a daily habit for many people. For ancient Greeks and Romans wine was an important part of daily life; in colder regions beer was a common beverage in the Middle Ages. Juice extracted from fruit or used for flavoring beverages has been popular for thousands of years. Fruit-flavored waters have come on the market recently.

Tea from the perennial shrub *Camellia sinensis* (**Figure 116.1a**) is the most consumed beverage after water worldwide. Tea originated in East Asia where new growth of leaves and buds are harvested up to 30 times a year. The different types of tea—white, green, oolong, and black—derive their distinct color and flavor from processing, the amount of oxidation the leaves are exposed to before drying stops oxidation. Many different cultivars are grown today in tropical, subtropical, and even temperate climates. Herbal teas do not contain tea, but fruit or leaves from other plants; they are also free of caffeine. Tea has beneficial health effects based on phytochemicals and vitamins that protect from certain diseases.

Coffee is brewed from roasted seeds of the evergreen shrub *Coffea arabica* (**Figure 116.1b**). This species dominates today's coffee trade. *Coffea canephora*, producing the more bitter "Robusta" coffee but being more resistant to disease, plays a secondary role. *Coffea arabica* grows best at middle and higher altitudes in Africa, Southeast Asia, and Central and South America. Robusta coffee thrives at lower elevations where coffee leaf rust (*Hemileia vastatrix*) decimates *C. arabica*. This plant, native to Ethiopia, was first commercially cultivated in the 15th century in Arabia (Bacon, 2008). The Arabian monopoly of the increasing coffee production and trade was broken when viable seeds were smuggled out of the area and the Dutch established coffee plantations in Java. With increasing demand for coffee worldwide, plantations were established in many subtropical and tropical countries and today Brazil and Vietnam are the major coffee producers.

© Iakov Kalinin/Shutterstock.com

(a)

© SOMMAI/Shutterstock.com

(b)

© Sursad/Shutterstock.com

(c)

© mosista/Shutterstock.com

(d)

Figure 116.1a–d (a) Tea, (b) coffee, and (c–d) cocoa are used worldwide for beverages and cocoa for chocolate. The plants contain alkaloids, which are stimulating.

Coffee berries are red, yellow, or purple when they are ripe about 8 months after flowering. Hand labor is necessary to pick only the ripe berries as they mature for best flavor. The flesh of berries is removed and they need to be dried before roasting. For decaffeination, the oils containing the caffeine are extracted from the green seeds with hot water and solvents. The final flavor, body, and color of coffee are affected by the intensity of roasting. The roasted beans are ground and brewed by many different methods into various coffee beverages or are used as flavoring for other drinks or food.

Cocoa is a product of the seeds of the cocoa tree (*Theobroma cacao*; **Figure 116.1c–d**) and is used to make chocolate. Cocoa trees originate in tropical America and were cultivated first around 1000 B.C. The small evergreen trees grow flowers in clusters on their trunks, which are pollinated by flies. Cocoa pods develop to contain about 40 seeds. Seeds and surrounding pulp are taken out of the pod, fermented, dried, and roasted. About half of the roasted seeds are composed of cocoa butter, which is used with milk and sugar to prepare chocolate. Theobromine in cocoa is an alkaloid similar to but milder than caffeine.

Herbs and spices

Humans have used herbs and spices for thousands of years in food preparation to change, add to, or mask the flavor and aroma of dishes. Herbs and spices are generally used in dried form and in very small quantities because their flavor and aroma are so strong (Figure. 116.1a–b); they often are expensive and considered luxury items because they are not essential to human nutrition. Herbs are most often the leafy part of a plant, used fresh or dried for flavoring and garnish of food, as medicine, and for fragrance. Spices are the dried fruit, seed, root, or bark of plants. Spices often have antimicrobial properties and are used for flavoring, in religious ceremonies, for medicinal purposes, and in cosmetics (Table 116.1).

International trade of spices, parts of plants native primarily in tropical and subtropical Asia, became widespread and profitable in the 15th century when maritime European powers brought the desired goods from the Asian continent to Europe. Prominent trading routes brought wealth to the cities lying on the crossroads such as Alexandria in Egypt. During wars over the dominance of the spice trade many indigenous people were killed in Southeast Asia and colonial powers were established over spice monopolies.

Important spices include black pepper, cinnamon, clove, nutmeg, allspice, hot or chile peppers, ginger, vanilla, and saffron (Table 116.1). Many herbs belong to the mint (*Lamiacae*) and parsley (*Apiaceae*) family. They are prized for their **secondary metabolites**, which give the plants their specific flavor. Essential oils are extracted to capture the fragrance, flavor, or aroma of the plants. Herbs also have been used traditionally for medicinal purposes in addition to culinary use.

secondary metabolites: Chemical compounds produced by plants that are not essential for plant survival but assist in primary metabolism and defense mechanisms against pests, pathogens, and herbivores. See also alkaloids.

Paper, Cloth, and Wood

Plants classified as fiber crops provide people with material to make paper, cloth, baskets, and ropes. Fibers are the elongated sclerenchyma cells providing support in plants. They have thick lignified walls and are dead once they mature. Flexible fibers have thinner cell walls with less lignin such as yucca leaves compared to inflexible fibers, which are found in hardwood such as oak. One type of fiber cells, the **bast fibers** or soft fibers, occur as bundles or sheaths around the vascular system in stems, bark, or cortex. Bast fibers are in stems of jute (*Corchorus* spp.), which is used to produce coarse cloth such as burlap, jute rope, mats, carpets, and woven chair coverings. Jute is the least expensive fiber and used second in quantities produced after cotton. It is also useful for its biodegradable properties as a ground cloth to temporarily stabilize soil. Other bast fibers are found in flax (*Linum usistatissimum*), used to produce precious linen fabric, banknotes, and rope; hemp (*Cannabis sativa*) to make rope, cloth, and pulp; and ramie (*Boehmeria nivea*) for fishing nets and fire hoses because of its high strength when the fiber is wet.

bast fibers: Sclerenchyma cells surrounding the vascular tissue or contained in the bark or cortex of fiber plants.

© Elena Schweitzer/Shutterstock.com

(a)

© Sandra Caldwell/Shutterstock.com

(b)

Figure 116.2a–b (a) Spices are almost always used dried in small quantities, many of which originated in tropical Southeast Asia. (b) Herbs are generally the leaves or shoots of plants, often treasured for their aromatic compounds, and are used fresh and dried.

Table 116.1 **Properties of Common Spices and Herbs**			
Latin Name	**Common Name**	**Plant Part**	**Properties and Uses**
Capsicum annuum	Hot pepper	Fruit	Hot- and mild-flavored peppers (bell pepper) are the same species. Capsaicin gives hot peppers the heat, used fresh or dried in many savory dishes; potential medicinal use.
Cinnamomum verum	Cinnamon	Bark	Medicinal and ceremonial use in ancient times, contemporary use for flavoring for sweet and savory dishes and beverages.
Crocus sativa	Saffron	Stigma of flowers	Medicinal use, yellow dye for fibers, culinary use for flavoring and color.
Laurus nobilis	Bay laurel	Leaves	Bay wreaths were symbols of victory in Roman times, for scholars and poets in Middle Ages and later, culinary use in many cooked dishes.
Mentha spp.	Mint	Leaves	Medicinal use, essential oils for fragrance, fresh or dried in beverages, Asian dishes, desserts. Many different flavors such as peppermint, spear-mint, apple mint, and pineapple mint.

(continued)

Table 116.1	Properties of Common Spices and Herbs (*continued*)		
Latin Name	**Common Name**	**Plant Part**	**Properties and Uses**
Myristica fragrans	Nutmeg and mace	Nutmeg seed for nutmeg, pericarp for mace	Used since ancient times for the nutty flavor and warm, sweet aroma in Middle Eastern and Indian dishes, both sweet and savory.
Piper nigrum	Black pepper	Fruit	Produces black, pink, green, and white pepper. Used before refrigeration to improve flavor of salted meat. Universal spice for savory dishes and food flavoring.
Rosmarinus officinalis	Rosemary	Leaves	Medicinal and ceremonial use, signifies remembrance; preservation of meat, condiment, used fresh and dried in savory dishes.
Vanilla planifolia	Vanilla	Pods	Most labor-intensive crop due to hand pollination. Whole processed pods or "beans" used as flavoring in baked goods and desserts.
Zingiber officinale	Ginger	Rhizome	Consumed in ancient China until now for medicinal purposes, used fresh in many Asian cooked dishes and beverages, as pickles, and in desserts, used dried in baked goods.

Fibers from cotton (*Gossypium hirsutum*) are trichomes growing from the seed coat and covering the cotton seeds in the cotton boll or seed capsule (**Figure 116.3a**). They are harvested to produce many different textiles and cloth. Cotton is the most important fiber crop in the world and has been used for 7,000 years. Cotton fibers are well suited for textiles because they are soft and smooth and can be woven into thin or thick fabric, they are naturally white and readily accept dye, and they hold up to repeated washing without breaking.

Leaf fibers are harvested from several plants in the agave (*Agavaceae*) family. Sisal fibers are from *Agave sisalana* and are used for crafts, twine, and rough carpeting. Abaca is a leaf fiber from *Musa textilis* in the banana (Musaceae) family and is manufactured into specialty paper for tea bags, bank notes, and specialty textiles.

Wood fibers are important for pulp production, which is the basis of paper (**Figure 116.3b**). Wood fibers consist of cellulose, hemicellulose, and lignin, with cellulose being the desired component for paper. Fibers or tracheids from conifers are longer and are preferred for paper production over the fibers from hardwood. Wood is used for many other purposes including fuel, timber for construction, furniture, posts, railroad ties, veneers, and plywood, to name a few.

© Steven Frame/Shutterstock.com

(a)

© V.J. Matthew/Shutterstock.com

(b)

Figure 116.3a–b Cotton field ready to harvest. (a) The white fibers in the boll are almost pure cellulose. (b) The spruce logs will be made into pulp, which separates the cellulose fibers from the lignin. Cellulose fibers are used to manufacture paper.

BOX 116.1	Bamboo–Versatile for over 3,000 Years

Bamboo is a fast-growing, woody evergreen grass in the Poaceae family and Bambusoidae subfamily (Farrelly, 1984) (**Figure Box 116.1a**). The more than 1,500 species of bamboo are adapted to diverse environmental conditions from low-lands to high elevation, and humid, hot-to-cold climates. Bamboo is native to Asia, Australia, and North and South America. Bamboo is a flowering monocot and grows either **sympodial** or **monopodial** based on their rhizome structure. Sympodial bamboo species are native to the tropics, sensitive to frost, and grow in defined clumps. Monopodial bamboo, also known as free-standing or running bamboo, grows in colder environments and some species tolerate temperatures as low as −20.2°F (−29°C). Underground rhizomes of monopodial bamboo grow significant distances and form extensive root systems whereas rhizomes from sympodial species are confined and will not spread widely. The root system of bamboo can potentially become invasive, which in some areas is desirable to prevent soil erosion. Buds on the rhizomes emerge aboveground (**Figure Box 116.1b**), giving rise to a culm or stem that reaches its final height within one to four months. The nodes on culms give bamboo its characteristic appearance. Adventitious roots can grow on nodes at lower culm height and thin branches can emerge on nodes further up. The culm diameter remains the same throughout its lifespan. The internodes of culms are hollow and give great strength when culms are used as building materials (**Figure Box 116.1h**). They also allow this shallow rooted plant great resilience in high winds, making the plant a desirable windbreak. Individual culms live from 5 to 20 years. Plant propagation occurs primarily vegetative as flowering of most bamboo species is infrequent, several decades to over 100 years apart. Many bamboo species flower **gregariously**, meaning their vegetative growth stops and all plants start synchronized flowering, which spreads over wide areas and continents, taking one to several years. Many species die after flowering, requiring years to regenerate new culms and mature groves.

sympodial: Growth resulting in lateral branching from secondary axis.

monopodial: Growth of a plant from a single growing point in one direction.

Bamboo has one of the fastest growth rates in the world, making it an ideal plant for sustainable production of many materials. In optimum growing environments bamboo groves can increase biomass by 10–30 percent annually while biomass in a forest increases only 2–5 percent during that time (Farrelly, 1984). Daily height growth of larger species can be 0.1–0.4 m, with the fastest growth recorded on *Phyllostachys edulis* in Japan with 1.2 m in 24 hours. Bamboo varies in height from 0.1 m up to 40 m, with some notable exceptions reaching twice this height. Culm diameter varies as well and larger species can grow to 0.2 m in diameter. Taller, thicker culms grow in warmer, tropical regions.

For more than 2,000 years bamboo played an important role in China and Southeast Asia for its many uses. Bamboo is likely the most versatile plant utilized for commercial, practical, medicinal, and edible purposes, with more than 1,000 uses listed (Farrelly, 1984) **(Figure Box 116.1c–g)**. Bamboo is used to construct buildings, scaffolding, furniture, skin on airplanes, ancient water systems, many different musical instruments, bicycle frames, sandals, matting, and chopsticks. Other uses include needles, arrows, baskets, boats, shelters, laquerware, waterwheels, and as medicine against stomach ailments and respiratory problems. Around the 10th century a Buddhist monk authored a cookbook devoted to recipes with bamboo shoots (Laws, 2010). Contemporary cooks in Asia prepare young bamboo shoots in a variety of different dishes. Strips of bamboo have been used in ancient China to maintain records before paper was invented. The short fiber of bamboo has been used to manufacture paper and rayon fabric. Musical instruments from bamboo include flutes produced since ancient times and more recently the ukulele. Implements used in the Japanese tea ceremony like the whisk and the tea ladle are traditionally made from bamboo. Calligraphy and painting have relied on bamboo brushes. Bamboo is also used as livestock forage for cattle, sheep, and horses, is the primary food source for the giant panda in China, and is consumed by other wildlife. Bamboo plants grow in commercial bamboo forests and as ornamental plants in many areas outside their native range. Bamboo has great significance in Asian cultures, symbolizing longevity, resilience, friendship, strength, and honorability.

© Shi Yali/Shutterstock.com

(a)

© cozyta/Shutterstock.com

(b)

© Toa55/Shutterstock.com

(c)

© paul prescott/Shutterstock.com

(d)

(e)

(f)

(g)

(h)

Figure Box 116.4a–g (a) Bamboo forest in Taiwan; (b) bamboo shoot among larger culms; (c) bamboo scaffolding; (d) bamboo bridge; (e) dish with bamboo shoots; (f) bamboo steamer; (g) weaving of a bamboo basket; and (h) cut bamboo culms showing hollow interior.

Medicinal Plants

Plants have been used for medicinal purposes since ancient times. Ethnobotany studies the relationship people of different cultures have with plants related to many areas of life, one of them the medicinal use of plants. Many different plants are used for healing a great number of diseases and symptoms by using plant parts as a whole, as extracts, as poultice, brewed, or as mixtures of different plant parts. Studies of four groups of Indians indigenous to southern Mexico showed their use of more than 200 species of plants for medicinal purposes (Heinrich, 2000). People from geographically different areas use some of the same species to treat common ailments such as gastrointestinal problems. However, not all species were effective to treat a problem when tested. The compounds that are effective as medicine are often produced by the plant for defensive purposes. They include secondary metabolites such as alkaloids and other compounds that are the basis for modern pharmaceuticals, now often synthesized in the lab instead of being extracted from the plant. Researchers analyze plants from all over the world to discover new medicinal efficacy.

Aloe vera heals burns when the succulent leaves are cut and applied to an injury. Quinine, isolated from the bark of the cinchona tree (*Cinchona officinalis*), is a potent remedy to treat malaria. The bark of willow trees (*Salix* spp.) used to be chewed for pain relief and contains salicin, the ingredient later used to develop the analgesic aspirin. The common ornamental purple foxglove (*Digitalis purpurea*) contains compounds effective in treating irregular heartbeat. Taxol is found in the bark of yew trees (*Taxus brevifolia*) and in the needles of yews and is a powerful drug against breast cancer. Other drugs against cancer are based on compounds isolated from vinca (*Catharanthus roseus*) and star anise (*Illicium verum*). Plants such as opium poppies (*Papaver somniferum*), containing the active alkaloid morphine, and marijuana (*Cannabis sativa*), containing tetrahydrocannabinol (THC), have medicinal value to treat pain and some other ailments. They also affect the central nervous system and can cause hallucinations. Many other plants are used for healing purposes; however, the effectiveness of most has not been validated.

Fisheries and Ghost Fishing

By Elizabeth Jordan

Fifteen percent of the world, mostly comprised of least-developed countries (LDCs), relies on fish as a major dietary protein source. With heart disease, obesity, Type 2 diabetes, and other health-related ailments on the rise, many health professionals and government agencies are recommending a shift from terrestrial meat toward a dietary increase in seafood consumption. This huge demand for fish has escalated to the point at which 70% of the world's fish used as a food source is facing imminent extinction.

There is no international agency governing the oceans, so, unfortunately the most opportunistic fishing methods often win out, preventing sustainable fishing practices. Currently, trawlers are commonly used to catch fish. Trawlers are large nets that drag across the ocean floor and displace many of the species found there—such as mussels and crabs—and damage sea beds. Inadvertently, sharks, rays, and juveniles of many fish species get caught in these nets as **bycatch** and perish. This not only causes a loss in biodiversity but also disrupts the marine food web, thereby threatening dolphins, sea turtles, and sea birds, among many more. Furthermore, derelict fishing gear can get lost in the oceans and continue to "fish," even though there is nobody present to retrieve nets (**Figure 117.1**). Many marine animals accidently become entangled and never make it out of the nets. (Please the case study on Ghost Fishing below.)

Andreas Altenburger/Shutterstock.com

Figure 117.1 Bycatch

The Marine Stewardship Council (**MSC**) is a nonprofit, international agency that works with scientists and fisheries to regulate and set standards to ensure that fishing practices are carried out sustainably. They also ensure that biodiversity and a healthy number of fish are maintained in our oceans, making sure these aquatic environments are ecologically healthy. They reward and promote fisheries that are certified. To be considered certified, fisheries must reduce the number of trawlers, reduce the bycatch of birds and fish, and educate and raise awareness of the impacts overfishing has on the ecosystem.

The problem with sustainable fishing is that it mostly focuses on and target—or the "catch." It also does not take into account the amount of bycatch—the countless animals caught in nets and left to die from lost nets (ghost nets). Since marine animals typically decompose after 3–4 weeks, it is almost impossible to predict the number of animals that are continually caught in nets lost at sea.

An alternative to overfishing is fish farming, or **aquaculture**, which operates more like agriculture than fishing practice. Developing countries rely more on aquaculture than developed nations. According to a report by the International Food Policy Research Institute and World Bank, aquaculture is predicted to provide 75% of the world's fish supply (as of 2009, they provided less than half).[iii] Aquaculture reduces the need to import fish and, if done properly, can decelerate the worldwide overfishing problem (**Figure 117.2**). It provides jobs and improves the economies in coastal areas. In addition, health professionals believe that there could be an overall improvement in health if more people shifted toward a seafood-based diet rather than land animal meat (like beef).

A notable success of aquaculture is the improvement of the Potomac River through oyster aquaculture. The oyster reefs were restored to make room for the oyster farms. The oysters restored overall health by cleaning out the algae from nitrogen pollution.[iv]

Unfortunately, the way that fish farms are currently run causes a lot of ecological dam- age. There is a much higher concentration of fish population inhabiting these farms than one would normally find in the wild. Farmed fish contain the same contaminants and heavy metals found in wild-caught fish. This artificial over-population not only leads to disease and pollution that spread through the aquaculture but also may spread to other natural marine habitats. Furthermore, aquaculture can be ecologically problematic if the farmed fish get out and hybridize with fish in the wild, contaminating their genetic pool. Many farmed fish are not endemic to the area in which they are farmed. The farmed fish are given antibiotics that can not only lead to antibiotic resistance but also spread to the water supply. This could potentially leak out and cause wild fish populations to be exposed, compromising wild populations and contaminating their gene pool.

Vladislav Gajic/Shutterstock.com

Figure 117.2 Fish Farms

Perhaps we need more natural and integrative measures to help solve the problem of over- all health in fish farms. (Wrasse fish swim in and out of the mouths of moray eels in a mutually symbiotic cleaning ritual). It has been recently noted that wrasse fishes are also effective at cleaning out ectoparasites from the mouths of farmed salmon, mutually benefitting both species. Perhaps this mutualistic symbiosis will be a more effective at maintaining health levels and salmon, and keep harmful chemicals out of fish farms. (This is yet another example of why it is important to understand and grasp the interconnectedness and complexity of the natural world, so we can mimic its natural structure to the best of our abilities and thereby reduce our use of synthetic and harmful chemicals.)

Shrimp farms tend to be particularly destructive in other ecological ways to sensitive areas with high biodiversity. They are farmed in mangroves, which are wetlands that harbor much biodiversity and are among the most productive ecosystems. The shrimp farms damage mangroves by polluting and decreasing their natural biodiversity.

Another obstacle to aquaculture is that some fish will not breed in captivity. There are too many variables that go into fish spawning to determine a single variable that is *the* single obstacle to successful breeding, so it is a difficult obstacle to remedy. Also, fish imported to the United States from other countries may be raised under different standards and may use regulatory drugs or chemicals not allowed by the USDA (United States Department of Agriculture).

Furthermore, the productivity of fish farms that cultivate animals with short life spans (such as sardines and shrimp) may be negatively impacted by climate change.[v] Shrimp and sardines eat phytoplankton and algae, small photosynthetic organisms. Anything photosynthetic contains chloroplasts or plastids, subcellular structures that produce chlorophyll (the pigment involved in photosynthesis). Chlorophyll production and photosynthetic rates are greatly disturbed and often erratic and reduced by change in temperatures. This implies a drop in food source for the farmed fish. It also demonstrates the interconnectedness of marine ecosystems and food chain, and how climate change profoundly affects them.

Case Study

GHOST FISHING

Are Ghosts Haunting Our Oceans?

What happens when fishing gear gets lost at sea due to the unexpected wreck of fishing boats or the accidental abandonment of fishing gear? Do nets keep fishing although there is nobody present to man them? It would appear that the sad answer to that is "yes."

Ghost Fishing is a nonprofit organization comprised of volunteer divers and other committed topside support staff. Together, they retrieve lost and abandoned commercial fishing gear at sea. Divers often retrieve thousands of pounds of nets, requiring a good deal of man power to haul up on to the deck. The nets are recycled into various products such as carpet flooring, socks, and bathing suits.[vi]

This agency is aptly called *Ghost Fishing* because these large nets continue to "fish," entangling aquatic life that get caught in them—even though there is nobody there to man or retrieve them. Sometimes this continues over the span of hundreds of years. Sea lions, lobsters, whales, sea cucumbers, octopi, eel, crab, and sea bass are examples of species that have been found caught in nets over the years.[vii]

Source: Tom Boyd

Box Figure 117.1.a Fishing Indiscriminately; Ghost Nets Kill Innocent Marine Life such as this Sea Lion

Some of the aquatic life are found alive and set free. In fact, volunteers save thousands of lives by physically untangling the aquatic life from nets. Sadly, though, other animals come to an untimely death due to their inability to escape. Sea lion bones and remnants of other sea creatures are often found caught in the derelict net.

Saving wildlife can be a dangerous endeavor since the nets are not only a danger to the sea life, but to divers as well. Divers have to be extra careful while maneuvering these thousand-pound nets, lest they become the next casualty of ghost fishing.

In addition to direct-action conservation, *Ghost Fishing* has an outreach program that involves speaking at schools and getting involved in communities to spread awareness. It has published articles in various journals and has been featured in various media outlets. Recent examples include MSNBC's "The Cycle" and features on the Discovery Channel!

To help make more change, *Ghost Fishing* is currently working with lawmakers to help write legislature and change some of the existing laws to enforce that lost commercial fishing gear is reported sooner than later. This could save many more marine lives in the future by ensuring swift and immediate action. For more information, check out www.ghostfishing.org.

Box Figure 117.1.b The incessant demand for seafood is what fuels the problem. If people want to make a difference, they don't need to volunteer—they just have to stop eating sea creatures."—Heather Hamza of Ghost Fishing

Water Crisis
By Elizabeth Jordan

Topic
118

Seventy-one percent of the earth's surface is covered by water of which 97% is found in the oceans, 2% are in glaciers or ice caps, and roughly 1% is fresh water, some of which is available to us as drinking water. The majority of our drinking water supply comes from melting mountain glaciers (that turn into mountain springs), reservoirs, and aquifers. Because such a low percentage of the earth's water supply is freshwater, and also because a high percentage of water from the hydrologic cycle ends up in the oceans (where it is not available to us as freshwater), freshwater is considered a nonrenewable resource. The growing human population (expected to grow almost 50% in the next 50 years!) exhausts the already limited supply. Already, almost 1 billion people on this planet do not have access to clean drinking water, and over 3 million die each year from a water-related disease (**Figure 118.1**).

To solve the problem of water shortage, we will explore several solutions that would provide the growing human population with clean water.

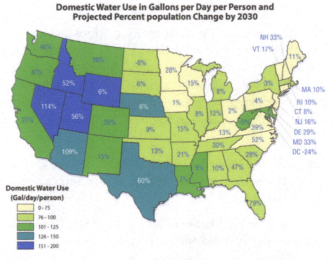

Figure 118.1   Water data from USGS, Estimated Use of Water in the United States in 2005. Table 6, Page 20; population data from U.S. Census Bureau, State Interim Population Projections by Age and Sex: 2004–2030.

Dams

To create freshwater supply, dams, such as the Hoover Dam in Nevada, are built to stop the flow of rivers (or in this case the Colorado River) and create a body of freshwater called a reservoir (**Figure 118.2**). Reservoirs not only provide drinking water, but they also have multipurpose for functions such as flood control, generation of electricity, irrigation, and recreation.

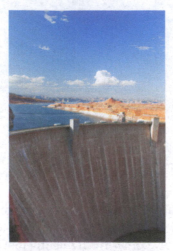

Oleksandr Koretskyi/Shutterstock.com

Figure 118.2   The Glen Canyon Dam, Lake Powell, and the Colorado River

Although multiuse and therefore convenient, dams create a very large land disturbance. As water is dammed up, reservoirs are created on what was often once dry land, forcing people to relocate (if they inhabited the area). This can be a massive under- taking for cities and towns and the residents themselves. Dams also have many other negative ecological impacts such as inhibiting fish spawning and the acceleration of decomposition rates in shallow water, thus increasing the amount of $CO2$ released into the atmosphere.

Water Purification

Municipal water must go through a treatment process which purifies water to ensure it can be reintroduced to the environment. **Potable** water is purified water that is safe to drink. One mechanism used to create potable water is **reverse osmosis**, the process in which water is pushed through a synthetic membrane whose pores are so miniscule that all impurities and contaminants are filtered out. Then it is exposed to UV light which is lethal to most fungal spores, bacteria, viruses, protozoa, algae, and nematode eggs. UV light causes mutations and changes in DNA of microorganisms, making them unable to reproduce. Finally, it is treated with hydrogen peroxide.

Another mechanism involves purifying sewage waste and returning it to a state that is safe for the environment. This process involves primary, secondary, and tertiary stages and can be done locally (septic tanks) or trans- ported to municipal plants (**Figure 118.3**).

Primary sewage treatment uses a screen to filter out and extract water from solid waste. The solid waste that remains is called **sludge** and is eventually dried and used as fertilizer.

Secondary treatment involves adding bacteria to the filtered water in order to break down any remaining organic matter.

irabel8/Shutterstock.com

Figure 118.3 A Sewage Treatment Plant

Tertiary treatment involves using UV light to kill off any remaining bacteria, viruses, and other microorganisms. Finally, is treated with chlorine at which point it becomes potable.

Another purification process involves water being treated by a **desalinization plant** whose purpose is to remove salt from ocean water (**Figure 118.4**). This can be done by either of two mechanisms. The first is to distill the saltwater, and since salt has a different boiling point than water, salt will separate during distillation. Water vapor is collected and cooled to a liquid state and is then used as drinking water. The other desalinization alternative is **microfiltration**, a process in which water is pushed through a membrane-like screen which mechanically separates water from salt.

Kekyalyaynen/Shutterstock.com

Figure 118.4 A Desalinization Plant

Environmental Impacts of Conventional Agriculture
By Terrence Bensel and Jon Turk

Environmental Impacts of Conventional Agriculture

Our modern or conventional agricultural system produces a staggering amount of food at relatively low costs to consumers. Americans spend as little as 7 percent of their income on food, over half of what we spent a generation ago and far less than what people in other countries spend to feed themselves. Despite this success, there are concerns that conventional approaches to agriculture—which emphasize heavy inputs of energy, water, and synthetic fertilizers and pesticides—could impose environmental and health costs on society that are not reflected in the prices we pay for food. This briefing by staff of the U.S. Department of Agriculture Economic Research Service reviews some of the major environmental issues associated with conventional agricultural production.

The following report highlights some of the key areas of concern when assessing the environmental impacts of agriculture, which include soil erosion, water pollution, air pollution, and habitat destruction. Modern agricultural techniques involve extensive plowing and manipulation of soils, and this can result in soil erosion by wind and rain. Eroded soils can reduce farmland fertility, pollute local waterways, and carry fertilizers and pesticides into surface waters. When nitrogen and phosphorous from agricultural fertilizers or animal manure wash into rivers and other bodies of water, they can promote the growth of algae, a process known as eutrophication. This can lead to decreased oxygen levels in the water and the death of many fish and other aquatic organisms. Pesticide runoff, pesticides leaching into groundwater, and pesticide residues on food crops and fruit can also pose health concerns. Agriculture also results in air pollution from a number of sources, including particulate matter from wind erosion and smog formation from agricultural chemicals and emissions of pollutants from farm equipment. Lastly, agriculture often involves the conversion of natural habitats to human uses, and this can have negative impacts on wildlife and biodiversity. Given the seriousness of these impacts it's clear why the debate over biotechnology and genetic engineering takes on such importance. Advocates of this approach argue that it will help address many of agriculture's environmental impacts, but critics argue that it might, in fact, make things worse.

By Staff of the U.S. Department of Agriculture Economic Research Service

Over 440 million acres (19.5 percent of land) is dedicated to growing crops in the U.S., and another 587 million acres (26 percent) is in pasture and range, largely used for domestic live- stock production. Agricultural activities on these lands produce a plentiful, diverse, and relatively inexpensive supply of food and fiber for people here at home and abroad. However, agricultural production practices can degrade the environment.

Transformation of undisturbed land to crop production can diminish habitat for wildlife. Soil erosion, nutrient and pesticide runoff, and irrigation can pollute the air and water, degrade soil quality, and diminish water supplies. the extent and degree of the environmental problems associated with agriculture vary widely across the country. concern over these problems has given rise to local, State, and Federal conservation and environmental policies and programs to address them.

Soil Quality

Soil, as a plant-growing medium, is the key resource in crop production. Soil supports the fundamental physical, chemical, and biological processes that must take place in order for plants to grow. Soil can also function as a "degrader" or "immobilizer" [the ability to break down or hold in place pollutants so that they do not enter groundwater supplies] of agricultural chemicals, wastes, or other potential pollutants, and can mitigate climate change by sequestering [absorbing] carbon from the atmosphere. How well soil performs these functions depends on soil quality. How soil is managed has a major impact on soil quality, and on the potential for various pollutants to leave the field and affect other resources.

Soil quality can be defined as the capacity of a specific kind of soil to function, within natural or managed ecosystem boundaries, to sustain plant and animal productivity, maintain or enhance water and air quality, and support human health and habitation. Soil quality depends on attributes such as the soil's texture, depth, permeability, biological activity, capacity to store water and nutrients, and organic matter content. Soil quality can be maintained or enhanced through the use of appropriate crop production technologies and related resource management systems. Poorly managed fields can lead to soil degradation through three processes: physical degradation, such as via wind and water erosion and soil compaction; chemical degradation, such as toxification [conversion of chemicals into toxic forms], acidification, and salinization; and biological degradation, such as loss of organic matter and decline in the activity of soil fauna. Poor management can also increase runoff of nutrients and pesticides to surface and groundwater systems. thus, soil degradation can have both direct and indirect negative effects on agricultural productivity and the environment. Even on high-quality soils, overuse of chemical inputs can result in soil toxicity and water pollution.

Water Quality

Agriculture is widely believed to have significant impacts on water quality. While no comprehensive national study of agriculture and water quality has been conducted, the magnitude of the impacts can be inferred from several water quality assessments. A general assessment of water quality is provided by EPA's 2002 Water Quality Inventory. Based on State assessments of 19 percent of river and stream miles, 37 percent of lake acres, and 35 percent of estuarine square miles, EPA concluded that agriculture is the leading source of pollution in 37 percent of river miles, 30 percent of lake acres (excluding the Great Lakes), and 8 percent of estuarine waters found to be water-quality impaired, in that they do not support designated uses. this makes agriculture the leading source of impairment in the nation's rivers and lakes, and a minor source of impairment in estuaries. Agriculture's contribution has remained relatively unchanged over the past decade.

Major Agricultural Pollutants

Sediment [naturally occurring material that can wash off of fields] is the largest contaminant of surface water by weight and volume, and is identified by States as the leading pollution problem in rivers and streams and the fourth leading problem in lakes. Sediment in surface water is largely a result of soil erosion, which is influenced by soil properties and the production practices farmers choose. Sediment buildup reduces the useful life of reservoirs. Sediment can clog roadside ditches and irrigation canals, block navigation channels, and increase dredging costs. By raising streambeds and burying streamside wetlands, sediment increases the probability and severity of floods. Suspended sediment can increase the cost of water treatment for municipal and industrial water uses. Sediment can also destroy or degrade aquatic wildlife habitat, reducing diversity and damaging commercial and recreational fisheries.

Nitrogen and phosphorus [two critical plant nutrients] are important crop nutrients, and farmers apply large amounts to cropland each year. They can enter water resources through runoff and leaching [percolate through the ground]. the major concern for surface-water quality is the promotion of algae growth (known as eutrophication), which can result in decreased oxygen levels, fish kills, clogged pipelines, and reduced recreational opportunities. The U.S. Geological Survey (USGS) has found that high concentrations of nitrogen in agricultural streams are correlated with nitrogen inputs from fertilizers and manure used on crops and from livestock waste. EPA reported in its Water Quality Inventory that nutrient pollution is the leading cause of water quality impairment in lakes, and a major cause of oxygen depletion in estuaries.

Consider This

Nitrogen and phosphorous are fertilizers, and when they enter water bodies like lakes and streams, they promote the growth of aquatic plants and algae. Why is this a problem? Design an experiment, using the basic principles of the scientific method, to test how additions of different amounts of nitrogen and phosphorous might affect water quality and wildlife in a body of water.

Eutrophication and hypoxia (low oxygen levels) in the northern Gulf of Mexico have been linked to nitrogen loadings from the Mississippi river. Agricultural sources (fertilizer, soil inorganic nitrogen, and manure) are estimated to contribute about 71 percent of the nitrogen loads entering the Gulf from the Mississippi Basin, and 80 percent of phosphorus loads. The Gulf of Mexico is not the only coastal area affected by nutrients. Recent research by the national oceanographic and Atmospheric Administrations has found that 65 percent of assessed estuaries had moderate to high overall eutrophic conditions, caused primarily by nitrogen enrichment.

Farmers apply a wide variety of pesticides to control insects (insecticides), weeds (herbicides), fungus (fungicides), and other problems (**Figure 119.1**). Well over 500 million pounds (active ingredient) of pesticides have been applied annually on farmland since the 1980s, and certain chemicals can travel far from where they are applied. Pesticide residues reaching surface-water systems may harm freshwater and marine organisms, damaging recreational and commercial fisheries. Pesticides in drinking water supplies may also pose risks to human health. Pesticide concentrations exceeded one or more human-health benchmarks in about 10 percent of agricultural streams examined by USGS as part of the national Water Quality Assessment Program, and in about 1 percent of sampled wells used for drinking water in agricultural areas.

Some irrigation water applied to cropland may run off the field into ditches and receiving waters. these irrigation return flows often carry dissolved salts as well as nutrients and pesticides into surface or ground wat Increased salinity levels in irrigation water can reduce crop yields or d age soils such that some crops can longer be grown. Increased concentrations of naturally occurring to minerals—such as selenium, molybdenum, and boron—can harm aquatic wildlife and impair water-based recreation. Increased levels of dissolved solids in public drinking water supplies can increase water treatment costs, force the development of alternative water supplies, and reduce the lifespans of water-using household appliances. The possibility of pathogens contaminating water supplies and recreation waters is a continuing concern. Bacteria are the largest source of impairment in rivers and streams, according to EPA's water quality inventory. Potential sources include inadequately treated human waste, wildlife, unconfined livestock, and animal operations. Diseases from micro-organisms in livestock waste can be contracted through direct contact with contaminated water, consumption of contaminated drinking water, consumption of crops irrigated with contaminated water, or consumption of contaminated shellfish.

Dan Su Sa/Shutterstock.com

Figure 119.1 A farmer wearing a full-body protective suit sprays crops with pesticides. The level of protection is warranted; many pesticides are known carcinogens, teratogens, and endocrine disruptors for animals and humans.

Air Quality

Ever since farmers began raising animals and cultivating crops, agricultural production practices have generated a variety of substances that enter the atmosphere with the potential of creating health and environmental problems. The relationship between agriculture and air quality became a national issue in the 1930s with the severe dust storms of the **Dust Bowl**. Although dust storms of this magnitude no longer occur in the United States, soil particulates, farm chemicals, and odor from livestock are still carried in the air we breathe. These emissions can harm human health and pollute the environment. Air quality in most rural areas is not a cause for concern, but there are some farming communities where ozone and particulates have impaired air quality to the same extent as in urban areas.

Ammonia is a gas and one of the most abundant nitrogen-containing compounds emitted to the atmosphere. Animal farming systems contribute about 50 percent of the total anthropogenic [man-made] emissions of ammonia into the atmosphere in the U.S. Ammonia is a health hazard to humans and animals in high concentrations. Once in the atmosphere, ammonia is rapidly converted to ammonium particles by reactions with acidic compounds such as nitric acid and sulfuric acid found in ambient aerosols [small, airborne particles]. These ammonium particles can be carried long distances in the atmosphere and contribute to fine particulate pollution and haze. Ammonium is redeposited to the earth's surface by both wet and dry deposition [the process by which particles deposit to the ground; wet refers to rain and dry usually to gravity] contributing to eutrophication of water resources.

Nitrous oxide is another nitrogen compound of concern. It is a greenhouse gas and contributes to ozone depletion. Nitrous oxide forms primarily in the soil during the microbial processes of nitrification [conversion of ammonia to nitrite] and denitrification [conversion of nitrate to nitrogen]. Agricultural sources include manure from livestock farming and commercial fertilizer. Agriculture contributes about 72 percent of total anthropogenic emissions of nitrous oxide in the U.S., mostly from the fertilization of cropland.

Methane is an important greenhouse gas. It is produced by microbial degradation of organic matter under anaerobic conditions. The agricultural sector is the largest anthropogenic source, with livestock production being the major component. Enteric fermentation (within the stomachs of cattle, sheep, goats and other ruminants) and manure management contribute 27 percent of methane emissions in the United States.

Carbon dioxide is the primary greenhouse gas emitted in the U.S., mostly from the combustion of fossil fuels. Carbon dioxide is also a primary input in plant growth. Agriculture can sequester [store] carbon in soils and biomass, thus offsetting greenhouse gas emissions. Carbon entering the soil is stored primarily as soil organic matter. Agricultural soils sequestered an estimated 12.4 million metric tons carbon equivalent in

2004, less than 1 percent of U.S. emissions. Studies indicate that it may be *technically* possible to sequester an additional 89–318 million metric tons of carbon annually on U.S. croplands and grazing lands through various management practices, such as conservation tillage, crop rotations, and fertilizer management. Shifting cropland to grasslands or forest could increase sequestration even more.

Particulates from agriculture result from a variety of activities. Wind erosion can carry soil particles directly into the atmosphere. Many areas west of the Mississippi river experience low average rainfall, frequent drought, and relatively high wind velocities. These conditions, when combined with fine soils, sparse vegetative cover, and agricultural activity, make some western regions susceptible to wind erosion.

Wind erosion can produce short-term levels of particulate pollution in rural areas that exceed urban levels. Particulates from wind erosion can impose costs on those living in affected areas, including cleaning and maintenance of businesses and households, damage to nonfarm machinery, and adverse effects on health. Another source of particulates is open-field burning. Open-field burning is used as a means of removing crop residue after harvest and controlling disease, weeds, and pests. Diesel engines from farm equipment and irrigation pumps are also a source of particulates.

A source of fine particulates (particles smaller than 2.5 microns, also known as PM2.5) is gaseous emissions of ammonia and nitrogen oxides (Nox, or nitric oxide, and nitrogen dioxide). Ammonia and Nox in the atmosphere react with other compounds to form fine particulates, such as ammonium. Fine particulates pose a health risk because they can be inhaled deep into lungs. Fine particulates are also a source of haze, which detracts from views in many popular national parks.

The atmosphere is now recognized as a major pathway by which pesticides can be transported and deposited far from their point of use. Pesticides can enter the atmosphere directly from the spray cloud during application, from evaporation after application, and attached to windborne soil particles. As much as 80 percent of some pesticide applications evaporate. And many of these pesticides across different chemical groups have been detected in the atmosphere. The U.S. Geological Survey found that the most frequently detected pesticides in the atmosphere are DDT, methidathion, diazinon, heptachlor, malathion, and dieldrin. Even though some of these have been banned for years, they continue to be detected.

Wildlife Habitat

Habitat is a combination of environmental factors that provides the food, water, cover, and space that a living organism needs to survive and reproduce. Agricultural land use can benefit some species, harm others, and sometimes do both. Potentially harmful effects of farming include plowing up habitat, farming riparian [along the banks of a river or stream] buffers, fragmenting habitat, diverting water for irrigation, and diffusing agricultural chemicals into the environment. In addition, specialization in agriculture reduces landscape diversity by creating more of a monoculture [growing only one crop]. This reduces the presence of ecological niches, which can limit wildlife populations and biodiversity on farms. Historically, the conversion of native forests, prairies, and wetlands to cropland has diminished wildlife. Habitat loss associated with agricultural practices on over 400 million acres of cropland has been identified as a primary factor depressing wildlife populations in north America. Agriculture is thought to affect the survival of 380 of the 663 species listed by the Federal Government as threatened or endangered in the conterminous 48 States.

Agriculture's negative effects on wildlife need not be permanent. U.S. agriculture is in a unique position with respect to the nation's wildlife resources. The management of land now controlled by U.S. farms and ranches can play a major role in protecting and enhancing the nation's wildlife. In 2002, private farms accounted for 41 percent of all U.S. land, including 434 million acres of cropland and 395 million acres of pasture and range. Farms also account for 76 million acres of forest and woodland, and 17 million acres of nonfederal wetlands. Different types of habitat can be restored or improved through conservation on agricultural lands.

Grassland Habitat

Grasslands constitute the largest land cover on America's private lands. Privately owned grasslands and shrub lands (including tribal) cover more than 395 million acres in the United States. These lands contribute significantly to the economies of many regions, provide biodiversity of plant and animal populations, and play a key role in environmental quality. Grasslands directly support the livestock industry (**Figure 119.2**). They also provide habitat for many wildlife species, reduce the potential for flooding, control sediment loadings in streams and other water bodies, and provide ecological benefits such as nutrient cycling, storage of atmospheric carbon, and water conservation. Grasslands also improve the aesthetic character of the landscape, provide scenic vistas and open space, provide recreational opportunities, and protect the soil from water and wind erosion.

Leonidas Santana/Shutterstock.com

Figure 119.2 **Grasslands support the livestock industries of the surrounding communities and contribute to the overall environmental quality of a region.**

Large expanses of grassland acreage are annually threatened by conversion to other land uses such as cropland and urban development. About half of all grasslands in the U.S. have been lost since settlement, much due to conversion to agricultural uses.

Wetland Habitat

Wetlands are complex ecosystems that provide many ecological functions that are valued by society. They take many forms, including prairie potholes, bottomland hardwood swamps, coastal salt marshes, and playa wetlands. Wetlands are known to be the most biologically productive landscapes in temperate regions. More than one-third of the United States' threatened and endangered species live only in wetlands, and nearly half use wetlands at some point in their lives. Most freshwater fish depend on wetlands at some stage of their lives. Many bird species are dependent on wetlands for either resting places during migration, nesting or feeding grounds, or cover from predators. Wetlands are also critical habitat for many amphibians and fur bearing mammals. Besides supporting wildlife, wetlands also control water pollution and flooding, protect the water supply, and provide recreation.

When the country was first settled there were 221–224 million acres of wetlands in the continental U.S. Since then, about half have been drained and converted to other uses, nearly 85 percent for agricultural uses. Currently, there are about 111 million acres of wetlands on nonfederal lands. About 15 percent are on agricultural lands (cropland, pastureland, and rangeland).

Riparian Habitat

Riparian areas are the zones along water bodies that serve as interfaces between terrestrial and aquatic ecosystems. Riparian ecosystems generally compose a minor proportion of the landscape, but they are typically more structurally diverse and more productive from a wildlife perspective than adjacent upland areas. This is especially true in the arid West. Studies in the Southwest show that riparian areas support a higher breeding diversity of birds than all other western habitats combined. In Arizona and New Mexico, at least 80 percent of all animals use riparian areas at some stage of their lives. Western riparian habitats contain the highest non-colonial avian breeding densities in North America.

Riparian zones also support productive aquatic habitat. They stabilize streambanks, thus reducing streambank erosion and sedimentation. Detritus [non-living organic material, such as fallen leaves] from streamside vegetation provides energy to the stream ecosystem. Vegetation also provides shade, preventing extreme temperature swings that are detrimental to healthy stream ecosystems. Riparian areas also filter out sediment, nutrients, and pesticides in runoff, thereby protecting water quality.

No comprehensive national inventory has been completed on the status and trends of riparian areas. However, NRCS estimates that the conterminous U.S. originally contained 75–100 million acres of riparian habitats and that between 25 and 35 million acres remain.

Implications for Policy

Agriculture has wide ranging impacts on environmental resources. Because of this, it also has the capacity to provide a wide range of environmental services. Understanding the links between agriculture and environmental quality enhances our ability to design programs that best meet the needs of producers and those who value the services the environment can provide.

Adapted from USDA Economic Research Service. 2009 (updated). Environmental Interactions with Agricultural Production: Background. Retrieved from http://www.ers.usda.gov/Briefing/AgAndEnvironment/background.htm

Genetic Engineering
By Elizabeth Jordan

Did You Know?

Monsanto—a Missouri-based bioengineering company specializing in GMO (genetically modified organism) agriculture—developed a genetically modified corn that is resistant to pesticides and herbicides. Their GMO pollen dispersed and hybridized with local family-owned farms (unbeknownst to the family farmers) whose corn was grown organically. These family farmers had saved and replanted seeds the following year, a common agricultural practice.

Asserting that they were protecting their work and innovation, Monsanto sued the local farmers claiming patent infringement. As a blow to small-scale family farmers, Monsanto won the lawsuit. Monsanto's GMOs have been identified as an accidental introduction to many small farm operational seed supplies, which can sometimes result in inviable hybrid seeds.

According to the Public Patent Foundation, family farmers and organizations with organic interests have preemptively sued Monsanto to protect themselves and their interests from Monsanto lawsuits in the future. Their pending lawsuit has set precedence to protect farmers and their organic crops from obsolescence through hybridization with GMOs.

Manipulating Nature

Biotechnology is the science of manipulating living organisms for human needs and benefits. Wine-making and cheese-making are common examples of biotechnology. **Genetic engineering** is the process of *altering the DNA* of a living organism to create products of human interest. **GMOs** are the product of genetic engineering. Reasons for altering DNA include introducing a beneficial trait into an organism or deleting an undesirable trait from an organism.

Some desirable results of genetic engineering include the slow ripening of fruit, higher yield of crops, the inclusion of nutrients not otherwise found in regional crops and staple foods, and the creation of crops resistant to drought, salinity, pesticides, and other conditions that may otherwise decrease agricultural production.

One example of nutritional fortification of foods can be seen with beta-carotene. Deficiency in dietary beta-carotene can lead to blindness. In many developing nations, the rate of blindness is particularly high and has been increasing due to malnourishment. The use of biotechnology has allowed the introduction of beta-carotene.

It is interesting to note that almost 75% of the food in U.S. supermarkets is genetically modified. According to the International Crops Research Institute for the Semi-Arid Tropics (ICRISAT), over 70% of the U.S. crops are transgenic (formed from two or more DNA sources), a much higher percentage than any other country in the world. Some of the most common U.S. GMOs are cotton, soy, and corn, apples, alfalfa, potatoes, and squash (**Figure 120.1**).

cholder/Shutterstock.com

Figure 120.1 **Rows of Cotton**

GMOs are also used to create new drugs and to harvest proteins and hormones such as insulin and growth hormone (GH). Bioengineering also performs bioremediation through microbes (bacterial DNA makes proteins that degrade waste from pesticides and pollution).

How Does Genetic Engineering Work?

Bacteria cells have circular noncoding DNA called **plasmids** that act as storehouses of DNA. Plasmids are able to take DNA from the environment and incorporate it into the genome of its bacterial cell. This is a normal practice of bacteria as it confers beneficial traits such as resistance to antibiotics that may exist in their environment. The contents of the DNA plasmid can be easily manipulated in the lab for our benefit (**Figure 120.2**).

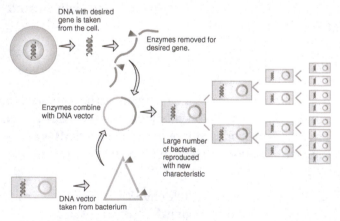

Figure 120.2 **DNA plasmid can be manipulated in the lab.**

An example of what bacterial use of plasmids can do for us is exhibited in how scientists use genetic engineering to create and harvest bovine growth hormone (BGH). BGH increases the yield of milk that a cow can produce. The desired DNA sequence (that confers this increased milk production ability) is isolated from cow cells by using a **restriction enzyme**—a type of DNA scissor that can find a specific sequence of DNA and snip it from the irrelevant DNA. This targets our BGH trait and creates a DNA fragment that can be collected. The bacterial plasmid is cut with the same restriction enzyme so that the BGH DNA fragment is now complementary to the plasmid DNA (meaning the two types of DNA can now be put together). This results in what are called **sticky ends**. The DNA from the two sources is spliced together, and the result is a **transgenic** bacteria. Transgenic means that its DNA comes from two (or more) separate sources.

It is important to note that a plasmid is known as a **vector**, as it essentially carries the trait of interest (BGH) into the bacteria which, in turn, acts as a factory for the protein that will be produced from the trait of interest. The bacteria are cultivated on a growth medium and allowed to produce the protein that our DNA encodes (BGH), and the hormone is extracted and stored for later use.

Pros and Cons of Genetic Engineering

Pros

Genetic engineering can enable crops to grow larger and more colorful, become seedless, ripen slowly, increase yield, become drought, pesticide, and herbicide resistant, and can enable crops to grow on marginal to low-quality soil. In fact, bioengineering performs many of the same agricultural processes that farmers have done for thousands of years (artificial selection for desired traits), but in a much shorter period of time. This is accomplished by inserting the desired gene (that produces a desired trait) instead of taking generations to crossbreed. Traditionally, crossbreeding has been performed with closely related crops species, but bioengineering allows genes to be introduced from many different species of different organisms. In fact, most of the genes bioengineered into crops come from bacteria. Also, genetic engineering helped save Hawaii's papaya industry by helping make them resistant to the papaya ringspot virus, which threatened to wipe them out.

Cons

Criticisms or concerns of GMOs are numerous. One concern is that crops have coevolved over thousands of years with their pollinators (bees, hummingbirds, etc.) and have adapted their own ways to ensure pollination and evade pests. (In fact, spiders have killed more crop pests than any synthetic pesticide.) This can lead to an imbalance or collapse of many populations of organisms in a particular environment by disrupting the food chain. Consider that herbicide-resistant crops often grow more aggressively, requiring more pesticide and herbicide. In fact, the crop ultimately becomes a weed during crop rotation. Plants and crops are difficult, if not impossible, to contain, and GMOs may hybridize with natural, nongenetically modified organisms and disrupt their natural ecological role. Therefore, bioengineering crops may disrupt thousands of years of coevolution and have vast ecological consequences. Furthermore, engineering crops can cause a reduction of genetic diversity of crops. This can make crops more vulnerable to parasites—and situations similar to the Irish Potato Famine for another con: it may make food crops susceptible to antibiotic resistance.

Still another criticism is foul play. A terminator gene has been introduced into some GMOs, making these crops sterile by keeping them from proliferating for a second growing season. An expressed concern is that this feature is fueled by monetary incentive so farmers have to keep buying new seeds each year. This is contrary to the normal practice of families of farmers who may have developed their own family-owned seeds for use. Cross-pollination from a GMO field to a non-GMO is virtually impossible to contain, but can be detected by proprietary seed-producing companies. This situation has in the past caused small business farmers to be sued and ultimately lose property for having their family-based seeds contaminated with a seed of company's genetics.

To address the "pro" that GMOs would be ideal for drought-prone areas: Plants respond in complex and variable ways to drought. This is reflected in the fact that there are multiple genes that code for drought resistance and are expressed at different times and during various stages of the life cycle. This is an adaptive advantage, since droughts vary in how long they last and how severe they are–and at what stages of growth they affect plants. But genetic engineering can only manipulate a few genes at a time, so any one engineered gene may be more successful than others at any given time. But it is difficult to anticipate or predict.

Furthermore, genes that code for drought resistance may have other (sometimes undesirable) effects on crop growth. This is called **pleiotropy**. This interconnectedness and complexity suggests that drought resistance correlates with other aspects of plant growth. Manipulating these genes may carry out the undesirable genetic expression.

Yet another concern is the fact that GMO use in the developing nations has surpassed their use in the developed world. This is harmful to farmers in developing nations who have become increasingly dependent on the United States for their food and seed supply. Farmers in developing nations have lost the knowledge to subsistence farm (farm to survive) and so have become more dependent in growing their food supply.

The final concern of GMOs is the unknown. We still do not know the full extent of side effects from bio-engineering crops, or if the benefits outweigh the costs. Consider that a gene from the bacteria *Bacillus thuringiensis* serves as a natural pesticide for corn and potatoes. It produces a toxin BT that controls caterpillar populations (a common agricultural pest). It has been discovered that the toxin was found in high traces in the crops that people were consuming. Nobody knows the full health side effects of such a toxin in our bodies.

More potential risks include rearranging the plant genome, inadvertently introducing mutations, and development of cancers and viruses—all of which present unknown risks and obvious threats to humans.

Agriculture
By Elizabeth Jordan

Agriculture

Industrial Agriculture

Industrial farming typically involves the use of a **monoculture**, a practice in which one crop is grown at a time in a given area. A uniform chemical herbicide and pesticide (typically synthetic) is used in one fell swoop for monocultures. This yields a high number of crops and costs the least amount of time and money. Industrial agriculture has saved many lives by providing large quantities and varieties of crops at affordable prices. By using synthetic fertilizers, farmers are also able to grow crops in marginal soils; by using pesticides and herbicides, farmers have been able to avoid possible pest outbreaks and famine.

Although efficient, industrial agriculture can be unsustainable and environmentally detrimental, exhausting land resources. It is estimated that we have lost one-third of our topsoil in the last 100 years as a result of our farming methods—particularly monoculture. Because of the resulting poor soil quality, industrial farming relies more and more heavily on nitrogen-rich fertilizers. The nitrogen ultimately leaches into our lakes and streams, resulting in fish kills and **dead zones** (places of low biodiversity). The nitrogen also reacts to form the secondary pollutant nitrous oxide, a greenhouse gas. In addition, many of the pesticides, herbicides, and insecticides used are controversial. Many forests with rich soil and lots of biodiversity are cut down to make room and to be converted to agricultural land (see the case study on palm oil below). Furthermore, agriculture accounts for 14% of global greenhouse gas emissions.[ii]

Case Study

WHY ARE HONEYBEE POPULATIONS DECLINING?

This is a complicated question with many layers, and there may not be a single explanation. But according to Dr. Marla Spivak, an entomologist from the University of Minnesota, monoculture and industrialized farming may be partly to blame. Monoculture, along with a combination of other factors, affects bees by limiting their access to food.

Take into consideration that many honeybees are infected by a widespread parasite that circulates a virus throughout their body, weakening them. This may inhibit their mobility and their ability to effectively visit a food source. To compound the issue, their food source may be too distant for them to access. Why would this be? Before industrialized farming in the United States, farmers grew vegetation such as alfalfa and clover as a cover crop—a good food source for bees. They also enriched the soil with nitrogen. With the onset of industrialized farming, farmers stopped growing these cover crops (moving toward synthetic fertilizers) therefore taking away a bee food source. To add yet another variable to the issue, chemicals found in many synthetic pesticides may disorient honeybees, even further inhibiting them from reaching their food source. Therefore, a combination of parasites, pesticides, and industrialized farming may be responsible for the decline in honeybees.[iii],[iv]

Daniel Prudek/Shutterstock.com

Box Figure 121.1.1

Pest Management

Since the agriculture revolution 12,000 years ago, humans have been using and experimenting with different pesticides to control pests—from sulfur to smoke, from natural oils to ash, from lime to arsenic—and different civilizations have experimented with and employed many varied treatments for pest control.

So what exactly is a pest? A pest is any organism, such as weed, insect, fungus, bird, animal, etc., that does damage to desirable plants, trees, crops, or overall ecosystem health. It may compromise animal or human health, or just be a nuisance.

In the following paragraph, we will examine common (and controversial) types of pesticides used in industrial agriculture today.

Inorganic: Inorganic pesticides are highly toxic substances that typically have high persistence in the environment. They are controversial because inorganic pesticides are often neurotoxins. Examples include arsenic, mercury, and copper.

Organic: Organic pesticides are natural compounds extracted from plants that happen to be toxic to insects. These include phenols, oils from conifers, and nicotine from the tobacco plant.

Fumigants: Fumigants are small molecules that are small enough to penetrate and sterilize the soil and typically produce a vapor or gas. They are also used to control the decay and prevent insect infestations of stored grain, and are also used to control fungal growth in the soil.

Microbial/Biological: These are living cells or organisms, or parts derived from them that naturally deter pests. Examples include fungus, virus, protozoa, ladybugs, wasps, and the BT toxin derived from bacteria (discussed above). They are designated to target certain pests. Some fungi, for example, target and kill specific insects.

Chlorinated hydrocarbons: The chemicals have high persistence and very toxic to organisms, they can bioaccumulate in adipose tissue. Examples include DDT and atrazine.

Organophosphates: These are the most commonly used synthetic pesticide today. They attack the nervous system of some animals by disrupting acetylcholine, a neurotransmitter. Organophosphates are highly toxic and short lived (they do not have high persistence). One example is roundup.

Unfortunately through natural selection, many pests have become genetically resistant to traditional pesticides that once effectively controlled their population. In addition, although effective at saving crops and therefore lives, some pesticides can become controversial due to their toxicity. They also compromise overall ecosystem health (DDT for example).

One way to control pest populations while addressing these above concerns is a strategy called **Integrative Pest Management (IPM)**. The goal of IPM is to maximize pest control by employing many various integrative methods, which effectively minimize the use of pesticides, and adjusting the use of each method accordingly. The decisions regarding which method to use is based on assessing the environment and using common sense and sound judgment to determine the best solution. For maximum effectiveness and to determine the best methodology, it is necessary to know the ecosystem and the pests involved. Monitoring the site is also imperative for long-term effectiveness. Let us examine some of the integrative methods below.

Biological or natural controls for pest populations is the strategy that mimics nature by introducing the pests' natural population controls, such as parasites, competitors, disease, inclement weather patterns, and **parasitoids** (insects such as wasps and mites that use other insects to complete their life cycle, ultimately killing them). It employs the use of pheromones to disrupt their life cycle, growing healthy crops that can resist attacks. Another method is *pesticide or chemical control*. Using alternative methods, such as fungicides, herbicides, and insecticides, pest population can be reduced. *Cultural control* employs crop rotation, agroforestry, strip farming, and other agricultural practices that are used to confuse pests on where its food is found. (See below paragraph on sustainable agriculture for more detail.) And finally, *mechanical* control involves dealing with the pest directly. This involves screens for keeping pests out, or trapping rodents.

Organic Farming
By Elizabeth Jordan

When you are at the supermarket, you may see stickers that say "certified organic." The USDA (United States Department of Agriculture) implanted the standards for "certified organic" as recently as 2002. Since that time, the organic market and consumer demand for it has increased steadily and profoundly. Testament to this is the amount of organic food available at supermarkets, co-ops, and the increased popularity of farmer's markets.

Organic farming typically involves farming that is least invasive to agricultural land, relies on renewable resources, and is least invasive to soil (**Figure 122.1**). These standards are met at every stage of production, from the fields to the market place. For instance, **polyculture** (growing more than one crop at a time in a given area) is practiced instead of monoculture. It also involves planting cover crops, such as alfalfa, clover, pea plants, and other legumes that put nitrogen into the soil and help maintain soil nutrients. Alfalfa may also reintroduce bee populations, as it is a food source. Organic standards apply to livestock as well. Organic meat has no hormones or antibiotics added to it.

There are 15 soil nutrients required from crop growth and shortage of any one of these can compromise crop yield. It can take 1000 years to create one centimeter of soil, and by the year 2050, agriculture yield must increase by 60% globally. If soil was managed more sustainably, it could yield 58% more food.[v] So let us examine more ways in which soil management and agriculture can be more sustainable:

Crop-rotation and **no-till cultivation** are also implemented in organic farming. Crop rotation involves growing different crops each year, which maintains soil fertility, and also enables crops to evade pests naturally. This prevents pests from lingering after finding their food source. No-till cultivation eliminates the use of tractors and is less invasive to the soil than till cultivation. The result is reduced water evaporation and soil erosion. Another common application is using green manure as a fertilizer.

SOIL LAYERS

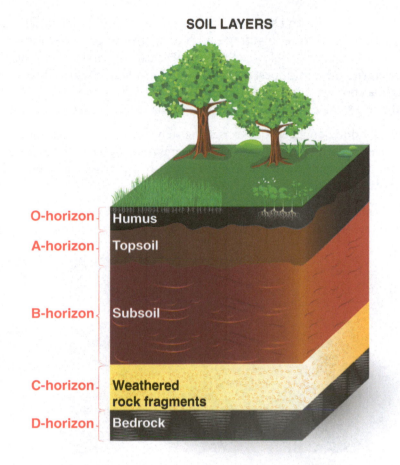

O-horizon — Humus

A-horizon — Topsoil

B-horizon — Subsoil

C-horizon — Weathered rock fragments

D-horizon — Bedrock

Designua/Shutterstock.com

Figure 122.1 Soil: A Mixture of Living and Nonliving Material

Natural pesticides and herbicides are used in organic farming, and the practice does not include GMOs. U.S. crops have to be only 95% organic to be considered "organic" by the FDA, and 70% organic to say it has been made with "organic ingredients." According to the *World of Organic Agriculture*, as of 2010, 35 million hectares of farmland are organic worldwide, with strongest growth being reported in Latin America and Europe.[vi]

Tran Van Thai/Shutterstock.com

Figure 122.2 Terraced Fields in Vietnam

Alternatives to conventional pesticides and herbicides include planting marigolds (a natural herbicide), scalding hot water to kill off pests, polyculture, planting mulch to control weeds, crop rotation to confuse pests and disrupt their cycles, mechanical control (weeding and physical barriers), pheromone and hormone disruptors, introducing insectaries or natural pest predators, and introducing naturally occurring microorganisms. Trap crops such as bok choy are often grown to steer pests away from food crops. The following farming techniques are designed to maximize agriculture use on land that would otherwise not be functional.

Terrace farming is typically done on steep slopes, cutting steps into the slopes in order to maintain water and to keep the ground covered with vegetation (in order to prevent runoff). Some of the ancient rice terraces that were carved into hills and mountains in countries, such as China, the Philippines, and Vietnam, are still in existence today (**Figure 122.2**).

Contour farming involves planting crops to go along with the gentle rolling slopes of hills, instead of orienting them perpendicular to inclines. This reduces soil erosion and maximizes surface area for use (**Figure 122.3**).

Rosliak Oleksandr/Shutterstock.com

Figure 122.3 Contour Farming

Strip farming (or alley cropping) involves growing crops in rows to keep pests confused and to maintain soil fertility. Rows of crops are also typically flanked by tall vegetation in order to block excess wind, thereby preventing erosion. It also provides shade which reduces evaporation, maintaining soil moisture (**Figure 122.4**).

David Aleksandrowicz/Shutterstock.com

Figure 122.4 Strip Farming

Agroforesty is a diverse, highly functional and integrative approach to agriculture that combines crops with shrubs, trees, and livestock. The benefits include a very biodiverse eco-system that may elude pests and curb climate change by absorbing more carbon than traditional agricultural practices. It also avoids the practice of converting forests into monoculture agriculture, thus maintaining soil and biodiversity.

Subsistence Farming. The majority of the world's agricultural land is neither managed nor produced by industrial agriculture, but by 500 million families worldwide and distributed all over the globe! In fact, these 500 million families produce most of the world's food.[vii] These 500 million people engage in **subsistence farming**—farming to eat, or growing one's own food. It is important that subsistence farming maintains its status as the majority of food production. For one, it is important for food security, so individuals, especially in developing countries, do not rely on exterior sources for food. By growing their own crops, they hold on to valuable knowledge and maintain independence. Second, subsistence farming helps protect the natural environment by avoiding the adverse environmental effects that come with industrial faming (**Figure 122.5**).

EcoPrint/Shutterstock.com

Figure 122.5 Subsistence Farming in South Africa

Case Study

THE REAL PRICE OF PALM OIL: SPOTLIGHT ON ENDANGERED ORANGUTANS

Kjersti Joergensen/Shutterstock.com

Figure 122.6

Existing for roughly 20 million years since the Miocene era, Tropical Rain-forests are among the oldest and bio-diverse ecosystems on the planet. They also happen to be home to our arboreal primate relatives, the orang-utans. We split with this great ape from our evolutionary tree 10–12 million years ago and share roughly 97% of our DNA with them. The whites surrounding their irises give them a familiar, almost human gaze, reminding us of our recently shared evolutionary past.

Their home range includes the Tropical Rainforests of Borneo, Sumatra, and Southeast China. They are the only great ape to spend most of their time—about 95% of it—in the canopies of trees. Their birth intervals are roughly 7–9.3 years, which are the longest of any mammal. This long gap between births means that maternal care is substantial and mothers place a lot of time and energy into child rearing. In fact, orangutan babies nurse for up to 7 years and are carried by their mothers until they are 5 years old. The bond between mother and baby is strong and impenetrable (**Figure 122.6**).

Rich Carey/Shutterstock.com

Figure 122.7 **Deforestation in Borneo**

Sadly though, our red ape relatives are in big trouble. 90% of their rainforest habitat range— from Sumatra to Borneo to lower China—has been disturbed or degraded, most of it in the last 20 years. Because of mass destruction to their native habitats, orangutans are now in the IUCN-listed Endangered Species category and in dangerous risk of going extinct in the near future.

Human activity is responsible for most of this disturbance and destruction. Legal and illegal logging, mining, and unsustainable palm oil plantations all cut down and destroy the rainforest and therefore the home ranges of orangutans. Our unsustainable use of palm oil, however, remains the biggest contributor to rainforest destruction in this region (**Figure 122.7**).

Palm oil is found in many products that we use every day: cookies, chips, laundry detergent, ice cream, candles, cosmetics, toothpaste, etc. The demand for palm oil has caused plantations to increase in size—sometimes illegally. Palm oil needs specific soil that is found in rainforests, so monoculture plantations are created in areas that were once rainforest. Making room for palm oil means slash and burn agriculture—this involves cutting down the forest and burning the debris and litter that remains. A 2018 Purdue University study concluded that palm oil labeled as 'sustainable' is also harmful to the environment and responsible for deforestation.

This has driven many orangutans out from their canopy home and toward great danger. Often times they end up burned, shot, starving, motherless or dehydrated and too often near death.

Currently, there is moratorium on rainforest clearing for palm oil plantations in Indonesia and strict laws in place regulating their size. But such laws are difficult to enforce and corporations and plantation owners often take the risk since great wealth comes from palm oil. But clearly, something must be done.

In response to the dire situation of orangutans, Dr. Birute Mary Galdikas founded the Orangutan Foundation International (OFI). Dr. Birute Mary Galdikas works in the Rainforests of Borneo. She was chosen by Dr. Leakey to study orangutans and has been doing so since 1971. Along with Jane Goodall and Dianne Fosse, she is known as a "trimate," one of the three notable women doing research on great apes.

OFI works with Indonesian Government to rescue, rehabilitate and reintroduce orangutans into the wild. They are often quarantined, given medical care, and eventually returned to the forests where they are reintroduced once they become healthy and strong enough to live independently. OFI also helps patrol the intact rainforest to prevent further expansion of rainforests or mining or illegal logging.

But they still need support and action from the rest of the world to save our beloved orangutans. What we can all do to help: Assist with tree planting programs in southeast Asia (and everywhere for that matter); Come see orangutans at Camp Leakey in Borneo; Promote ecotourism in Borneo and stay in touch with OFI (www.orangutan.org); and most importantly, significantly reduce palm oil use or cut it out altogether.

Source: Sustainable palm oil may not be so sustainable.

Saving the Sustaining Earth

By Elizabeth A. Jordan

"Green" is all about providing conditions that keep our environment and our quality of life healthy; therefore, any green discussion would be incomplete without addressing the **social**, **working**, and **political** conditions that affect our environment and our lives.

Did You Know?

Until the 1960s, there was not considerable momentum for wildlife and biodiversity preservation, at least it was not the nexus of global concern that it is today. Dialogue and environmental concern slowly gained traction during the 1960s as human population growth reached its biotic potential and inevitably placed strain on natural resources. With enough people understanding that the state of the natural environment is closely entwined with the quality of human life, preservation efforts expanded on both local and global scales.

In 1964, the *National Wildlife Preservation System* in the U.S. was implemented, protecting Federally managed land and working with the Federal Agencies (such as the National Parks and U.S. Wildlife Fish and Game) that manage them. In 1969, *NEPA*—an act by Congress to improve environmental quality—was passed. In 1970, the *EPA* (Environmental Protection Agency)—the Federal Organization through which all Environmental Legislature moves—was created. In 1971, *Greenpeace* was established. Starting as an anti-war protest, Greenpeace is comprised of a group of volunteers who consider themselves peaceful warriors fighting to protect the environment. Nonviolently, they also expose those who commit offenses against the natural world. In 1972, *UNEP* (United Nations Environment Programme) was created. UNEP is the UN's advocate for stewardship of the environment, promoting education and sustainability. In 1972, the *Noise Control Act* was passed. Enforced on a national level, it serves to protect the health and well-being of people, wildlife and the environment by promoting a noise-free environment. In 1973, *The Endangered Species Act* was passed, protecting species that are about to go extinct and their habitats. 1973 *CITES* (The Convention on the International trade in Endangered Species of Fauna and Flora) was created. It is an international treaty protecting wildlife from poaching and ensuring their well-being during international sales and trade. *The Fishery Conservation and Management Act* of 1975 protects the U.S. coast from foreign fishing—up to 200 miles out. In 1980, *The Alaska National Interest Lands Conservation Act* was passed—designating large parts of Alaska as National Forests, Parks, and Wildlife Refuges. 1980s and beyond brought us of an ivory ban, a whaling ban, the *Montreal Protocol* (to protect the ozone layer), and *Kyoto Protocol* (to reduce greenhouse gas emissions). The collective determination and gestalt effort of these organizations has brought environmental and wildlife concerns to the forefront of global awareness.

Many of our major environmental problems can be traced back to overpopulation—and with good reason! As of 2019, there are over 7.5 billion people (according to census.gov). This number is staggering considering that there were only just more than 1.5 billion people on the planet in the year 1900. This considerable growth in such a short time exhausts natural resources that do not have time to regenerate before they are repeatedly harvested. It also generates waste and pollution that causes many environmental, ecological, and health problems.

Urban Landscapes of the World

By Tim Anderson

The basic structure of the world- economy, especially with respect to relative levels of development and under-development are the core, semi-periphery, and periphery. Global patterns of urbanization, especially during the post- industrial era of the late twentieth and early twenty- first centuries, within this world-systems context will be discussed. Although urbanization has been a phenomenon associated with many cultures for thousands of years (since the Neolithic Revolution), the highest rates of urban growth have material-ized during the most recent industrial and post-industrial eras. As a result, more people are now living in cities than ever before in human history. Where are the most urbanized places in the world? Where are urban populations growing the fastest? What are the potential consequences of high rates of urban growth in different regions of the world-economy? What are some of the distinguishing characteristics of cities and urban landscapes in the post- industrial era? How do urban landscapes differ in the various regions of the world-economy?

Global Patterns Of Urbanization

A logical place to begin this discussion is with an overview of some general global patterns of urbaniza-tion. The most urbanized populations today are in core and semi-peripheral regions. Indeed, of the regions of the world in which 70 percent or more of the population lives in urban areas, all are either in the core or semi-periphery of the world-economy:

- North America (The United States and Canada)
- Mexico
- Northern and Western Europe
- South America (with the exception of the Andean highlands)
- Australia and New Zealand
- Japan and South Korea
- Parts of Southwest Asia and North Africa (Libya, Saudi Arabia, Israel, Jordan, United Arab Emirates)

At the same time, however, data compiled over the last 10 years indicate that urban populations are growing the fastest in the periphery and parts of the semi-periphery. The overall global trend in the post-industrial era, then, has been stagnant or negative rates of urban growth in the core but very high rates of urban growth in the semi-periphery and periphery. The growth of urban populations in the semi-periphery and periphery is cause for concern because these economies can ill afford the social and economic pressures resulting from ever-increasing urban populations. Such pressures might include:

- Housing—to shelter newly arrived migrants (most such migrants in the semi-periphery and periphery have moved from rural areas to urban areas in search of jobs)
- Food—urban dwellers do not produce food, but rather consume food produced in rural areas by fewer and fewer farmers
- Natural resources—clean air and water
- Public services—water and sewage services, trash collection, communication and transportation infra-structures, security (police), social services
- Social problems—crime, ethnic conflict, economic inequalities, unemployment
- Jobs—to provide a living for newly arrived migrants from the rural countryside

For an urban region to function smoothly, each of these factors must be addressed or provided. While similar problems plague all cities worldwide, including those in core regions, core economies are much better equipped to handle large urban populations. In many parts of the periphery and in parts of the semi-periphery, such services are woefully inadequate at best and nonexistent at worst. Large and growing urban populations, then, present such areas with myriad issues and problems, and only add to the many economic and social problems that afflict these areas of the world-economy.

The growth of large cities in the semi-periphery and periphery of the world- economy during the late industrial and early post-industrial eras is illustrated in the following list of the world's largest metropolitan areas in 2018, according to the United Nations:

World's Fifteen Largest Cities (metropolitan area) 2018		
City	**Population**	**Location**
Tokyo, Japan	37.4 million	Core
Delhi, India	28.5 million	Periphery
Shanghai, China	25.6 million	Semi-Periphery
São Paulo, Brazil	21.7 million	Semi-Periphery
Mexico City, Mexico	21.6 million	Semi-Periphery
Cairo, Egypt	20.1 million	Semi-Periphery
Mumbai, India	20.0 million	Periphery
Beijing, China	19.6 million	Semi-Periphery
Dhaka, Bangladesh	19.6 million	Periphery
Osaka, Japan	19.3 million	Core
New York City, USA	18.8 million	Core
Karachi, Pakistan	15.4 million	Periphery
Buenos Aires, Argentina	15.0 million	Semi-Periphery
Chongqing, China	14.8 million	Semi-Periphery
Istanbul, Turkey	14.8 million	Semi-Periphery

As late as the mid-twentieth century, nearly all of the most populous cities in the world were located in Europe or North America. Today, however, of the 20 largest cities in the world, only six are located in core regions of the world- economy. Eight are in semi-peripheral locations, and six are in the periphery. Thus, while large cities are associated with the wealthiest countries in most people's minds, it is clear that large urban populations are increasingly phenomena of the semi-periphery and periphery as well. Indeed, experts speculate that Mexico City will overtake Tokyo as the world's largest metropolitan area by 2025 and that only two or three cities in core regions (Tokyo-Yokohama, New York, and perhaps Osaka) will remain on the list of the 20 largest cities in the world.

The Nature Of Cities

Why do cities exist in the first place? What advantages are offered by the agglomeration of large populations in a certain place? What functions do cities perform? The answers to these questions are extremely complex and may differ from place to place, not only within the same country, but within the different regions of the world-economy as well. But, we can begin to understand the nature of cities by pointing out a few basic caveats concerning urban regions about which most experts agree:

1. Cities perform certain basic economic functions.
   This is the reason that cities exist at all: for the efficient performance of functions that a population could not perform adequately or efficiently if it were randomly dispersed through space. For example, producers are nearer to the consumers of their products, and workers are nearer to their places of employment. Time, money, and efficiency are saved by the agglomeration of people in space.

2. Cities function as markets.
   This has been the case since the earliest Neolithic Revolutions. As such, they have close reciprocal relationships with their rural hinterlands. For example, cities consume food that is produced primarily in rural areas. Cities are also the places where raw materials from rural areas (such as agricultural products or natural resources) are processed into consumable goods. Cities dispense goods and services not only for their own urban populations, but for rural populations as well.

3. Cities tend to be located strategically.
   Cities are most commonly located at certain advantageous sites:
   - Sites that offer security and/or defense, such as a hilltop or island (many Neolithic and Medieval cities occupy such sites)
   - Sites that are economically advantageous, such as a **head of navigation site** on a river, a river fording or portage site, or a railhead site

4. Cities function as central places.
   The idea of cities as "central places" stems from the work of the German geographer Walter Christaller, who in 1933 published a theoretical study concerning the distribution of service centers. Christaller wanted to understand the theoretical spatial patterns that would result when rural residents traded with a central market town providing goods and services. Human geographers call the results of this study **Central Place Theory**. The theory has been applied in many different places around the world to more fully understand urban patterns and appears to have stood the test of time in terms of its explanatory value. Christaller was concerned with why cities are located where they are, why some cities grow while others do not, and why there is an apparently non-random pattern concerning the location of cities relative to other cities. Central Place Theory holds that the importance of a market city is directly related to its centrality—the relative importance of a place with respect to its surrounding region. The central concept of Central Place Theory concerns the *range* of a good or service—the distance that people are willing to travel to obtain a certain good or service. Some goods and services, such as bread or food in general, are **low-range goods**, those for which people are not willing to travel far to obtain. Others, such as cars or furniture, are **high-range goods**, for which people are willing to travel long distances to buy. Christaller argued that the *centrality*, or relative importance, of a market town or city is directly proportional to the types of goods and services offered there, and that a natural hierarchy of size will arise with respect to market locations based upon what types of goods or services are offered there. This central place hierarchy ranges from **low-order places** that offer only low-range goods, to **high-order places** that offer both low- and high-range goods and services. Central Place Theory thus gives us insight into the functions that cities perform, why some cities grow and others do not, and why there are many small central places but only a few very large central places.

Urban Landscapes of the Core

By Tim Anderson

The following sections list the distinguishing characteristics of urban regions in selected locations in the core, semi-periphery, and periphery of the world- economy:

The core regions of the world-economy contain some of the largest cities in the world. Several of these cities developed into the nodes or "command and control centers" of the world-economy and remain so today. These so-called **world cities** are distinguished by the following characteristics:

- Financial centers—head offices, stock market locations
- Command and control centers of world capitalist economy—corporate headquarters of multi-national firms
- Political centers
- Cultural centers
- Nodes of international linkages

Western European Cities

Medieval Origins

One of the most striking features of most western European cities today is the juxtaposition of pre-industrial and post-industrial features in the region's urban landscapes. Many western European urban centers can trace their origins to the medieval era (from the tenth to the fifteenth century), when they emerged as high-order places in the central place hierarchy due to economic and/or political importance. The siting and the pre-industrial landscapes of many western European cities reflect these medieval origins, as well as their early roles as transportation centers and economic and political centers. Many of the largest western European cities, for example, are located either on the coast or on a large, navigable river. Some occupy early head-of-navigation sites (London, for example) or river crossing sites (Frankfurt, Germany). Others occupy defensive sites, such as hill-top or island locations, and are often associated with the sites of medieval-era castles and fortifications (Edinburgh, Scotland and Heidelberg, Germany, for example). Another medieval feature of many western European urban landscapes is the seemingly haphazard arrangement of streets, reflective of a lack of centralized city planning and growth by accretion over long periods. Only since the nineteenth and twentieth centuries have many western European cities begun to enact city-planning schemes to alter ancient medieval urban city plans. The presence of straight, ceremonial boulevards, such as the Champs Elysées in Paris and Unter den Linden in Berlin, reflect this kind of centralized city planning.

Abundant "Green" Spaces

Also reflective of the more modern trend toward centralized urban planning is the presence of relatively large areas set aside for public use. Such include pedestrian zones, public markets, and parks. Many western European cities are surrounded by large tracts of forests and parks located at the urban-rural fringe on the outskirts of urban areas; these tracts are collectively known as **greenbelts**.

Dense Public Transportation Networks

Modern western European cities are characterized by well-developed, efficient urban transportation networks. These networks usually include bus, tram (streetcar), subway, and light rail transportation.

"Low" Profiles

Compared to many North American cities, most large western European cities have relatively few tall skyscrapers. With the exception of London, Paris, and Frankfurt, all of which have downtown central business districts with prominent skyscrapers, most European cities have a comparatively low landscape profile comprising many square miles of multi-unit housing structures and retail services.

Emerging Post-Industrial Multiethnic Cities

Although most western European states emerged from the colonial era as very strong nation-states with ethnically homogeneous societies, the late twentieth and early twenty- first centuries have been characterized by significant influxes of ethnic minorities, primarily from southern and eastern Europe, Asia, and Africa. From Moroccans and Algerians in Spain and France, to Indians and Pakistanis in England, to Serbs and Africans in Germany and the Scandinavian countries, the larger urban centers of western Europe have received the largest number of these immigrants; their presence has fundamentally altered the ethnic makeup of the region such that most western European societies can now be characterized as fundamentally multiethnic. Increased immigration has altered European societies in many positive ways, but it has also not occurred without some significant social issues. For example, many of these new immigrants are refugees escaping political and economic turmoil in their home countries, and they arrive in European cities without jobs or housing. Given the immigrants' need for jobs and housing, some cities have experienced significant strains in dealing with large influxes of unemployed immigrants. In response, many cities have constructed several immigrant housing developments in specific areas set aside for such housing, often on the outskirts of urban regions. This practice has created **ethnic enclaves**, neighborhoods that are numerically dominated by specific ethnic groups, with businesses, such as restaurants and retain shops, catering to those groups.

North American Cities

Changing Forms of Transportation

Historically, the most significant factor that has influenced the structure and size of North American cities has been changing dominant forms of transportation over time. In the pre-industrial era—from the colonial period until the late nineteenth century—most cities were oriented toward pedestrian and/or animal transportation. These pedestrian cities tended to be rather compact, with zones of varying land use forming concentric circles around a **central business district** nucleus dominated by retail services and high-rent residential housing. During the early industrial period, between roughly 1880 and 1940, urban transportation came to be dominated by streetcars, subways, and railways. As a result, cities expanded dramatically in size as commuters could now live further from the central business district due to faster and more efficient forms of public transportation. Zones of varying urban land use expanded outward from the central business district along these public transportation arteries. These "streetcar cities" came to resemble a wheel, with the central business district as the hub and railway and streetcar lines as the spokes radiating outward from the center.

After World War II, the widespread use of the automobile as the dominant mode of transportation engendered even more radical changes in the size and shape of American cities. With the construction of interstate highway systems, commuters began to live further and further away from the central city, leading to the development of intense **suburbanization** at the urban-rural fringe dominated by upper-income residential housing and services catering to that income group. Cities expanded dramatically in size such that most American urban regions can now be characterized as comprising multiple "cities within cities," covering hundreds of square miles and connected via a dense and efficient network of large highways.

Spatial Differentiation Based on Ethnicity, Race and Income

One of the most distinctive characteristics of North American cities today is the development of conspicuous sectors of varying residential land use that reflect societal differences in ethnicity, race, class, and income. This has led to the development of distinctive ethnic and class-based neighborhoods within American urban regions. This trend began in the nineteenth and early twentieth centuries with the immigration of millions of Europeans, especially those coming from southern and eastern Europe who settled in the large industrial cities of the Northeast, and the migration of hundreds of thousands of African-Americans from rural areas of the South to industrial cities of the North. Many of these immigrants and migrants settled among one another in distinctive ethnic neighborhoods near central business districts, downtown areas that were abandoned by upper-income whites in favor of suburban locales at the urban-rural fringe.

Zoning

Zoning refers to the detailed urban land-use planning that city governments in the United States undertake; it is yet another distinguishing characteristic of American cities. Through the use and enforcement of zoning laws, city governments have the power to authorize and enforce what types of economic activities can take place in certain areas and what kind of structures can and cannot be built in certain areas. Such areas are said to be "zoned" for certain activities or kinds of structures. Such zoning laws have had a significant impact on the spatial differentiation of American cities with respect to both residential and business land use.

Gentrification

Gentrification refers to the revitalization of formerly abandoned properties in the central business district of American cities, a trend that is increasingly characteristic of American cities in the post-industrial era of the late twentieth and early twenty-first centuries. As upper-income whites moved to suburban areas from the central city during the 1950s, 1960s, and 1970s, and as warehousing and light industry activities also moved to urban peripheral regions during the same period, downtown areas in many American cities fell into disrepair. To revitalize downtown areas and to entice suburbanites back to the central business district, city governments began to support the e orts of wealthy investors in purchasing and revitalizing formerly abandoned downtown properties. These gentrification schemes often involve the construction of pedestrian malls dominated by expensive restaurants and specialty shops that cater to upper-income customers. While these activities have given many downtown areas a second life and contributed to economic revitalization, gentrification does not occur without some social costs. For example, as downtown areas were abandoned during the era of rapid suburban growth, they were often repopulated by lower- income residents and recent immigrants. Because such residents lack the political and economic power of wealthy investors and developers, gentrification schemes often result in such residents being forced to move.

Urban Landscapes of the Semi-Periphery and Periphery

Topic 126

By Tim Anderson

Latin American Cities

An Iberian Colonial Imprint

The most conspicuous urban landscape features of Latin American cities reflect the Iberian (Spanish and Portuguese) colonial imprint that is common throughout the region, from the southwestern United States in the north to the southern tip of South America. Spanish colonial goals were focused on the expansion of empire, the expansion of Christendom, and the extraction of valuable natural resources such as gold and silver, and distinctive urban landscape features reflect these colonial goals. For example, to accomplish these goals, the Spanish instituted a centrally planned and ordered network of urban centers that was built upon pre-existing networks of Native American towns. By law and in practice, all Spanish colonial towns were constructed on a rectilinear grid of streets oriented to the cardinal directions surrounding an open, public *plaza*. Almost all colonial towns were associated with presidios, forts of garrisoned military troops that exerted political and military control, and cathedrals staffed by Jesuit priests who were charged with converting Native Americans to Christianity. Other Iberian landscape features common in Latin American cities include Spanish architectural features, such as adobe construction and red tile roofs.

Spatial Differentiation Reflecting Strong Class Differences

Like other semiperipheral areas of the world-economy, Latin America is a region characterized by relatively intense social stratication based upon class, race, ethnicity, and income. The urban landscapes of the region reflect this stratication. In contrast to American cities, the wealthiest members of Latin American societies often live very near city centers, in elite sectors or neighborhoods. These elite sectors are surrounded by distinctive neighborhood sectors according to race, ethnicity, and income. The poorest members of Latin American societies, especially those that are homeless, live in so-called **squatter belts** in urban-rural fringe areas on the outskirts of cities in very poor conditions devoid of urban services such as running water, electricity, and sewage and trash disposal.

Southeast and South Asian Cities

A Western European Colonial Imprint

In contrast to Iberian colonial goals, the goals of western European colonial powers (such as Holland, England, and France) in South Asia (e.g., Pakistan, India, and Sri Lanka) and Southeast Asia (e.g., Vietnam, Myanmar, and Malaysia) were decidedly merchant capitalist in nature. That is, profit based on the establishment of privately financed plantations specializing in the production of tropical and subtropical agricultural products (such as tea, coffee, sugar, rubber, and spices) was more important than the expansion of the empire. After politically securing a colonial area, private companies typically established plantations in interior areas and warehousing and port facilities on the coast, often at the mouth of a major river. These port cities, which existed prior to European colonialism, came to be remade into European colonial outposts with distinctive European urban landscape features. Such features included European architectural styles employed in the construction of public buildings and the dwellings of European plantation managers, retail services catering to a European clientele, and European schools and churches.

Residential Segregation Based on Income and Class

As is the case in most urban centers around the world, South Asian and Southeast Asian urban residential sectors reflect differences in income and class. Colonial port cities (such as Mumbai and Calcutta in India and Hanoi in Vietnam) are usually characterized by three distinct types of residential zones: 1) an elite, European sector surrounding old warehousing facilities near port zones where European colonial managers and civil servants lived, worked, shopped and sent their children to school; 2) a sector of low-income housing also near historical port and warehousing zones numerically dominated by lower and middle-class workers; and 3) a sector on the outskirts of cities dominated by low-income landless families. These sectors, called **shantytowns**, resemble the squatter belts that can be found on the outskirts of many Latin American cities, and like squatter belts, they lack basic services such as running water, public sewage systems, and electricity.

Urbanization
By Elizabeth Jordan

In the year 2007, for the first time ever, there were a greater percentage of people living in urban areas than in rural areas. Consider that in the year 1950, roughly 30% of the world's population lived in urban areas. As of 2014, 54% of the world's population lived in urban areas. By 2050, it is estimated that 66% of the world' population will be living in urban areas, cities and metropolises.

The world's most urbanized countries are North America, South America, Europe and the Caribbean. Roughly 90% of the world's most rural areas are in Asia and Africa, but these areas are also urbanizing quite rapidly.

Tokyo is the world's most populated city with 38 million people (as of 2014). The second most populated city is Delhi with 25 million people, followed by Mumbai, Mexico City, and Sao Paolo, each with roughly 25 million people.

From 2010 to 2015, MDCs had an urbanization rate of .3%, while LDCs had an urbanization rate of 2.9%. From this data, we see that less developed nations are urbanizing at a much greater rate than more developed countries. (As of 2015, MDCs contain 1.3 billion people, projected to be 1.3 billion in the year 2030. LDCs contain 6.9 billion people as of 2015, but are projected to have 8.3 billion people as of 2030).

With urbanization increasing, and many people moving to urban areas, there are questions and concerns regarding such growth. In particular, there are three factors that place high stress on sustainable urban development, including *environmental protection, social development*, and economic development. These factors are all considered with consistent and predictable growth. Unfortunately, rapid and unplanned urbanization makes sustainable development in all of these areas a big challenge.

Consider that urbanization is often linked to higher literacy rates; longer life expectancies; lower fertility; and more mobility throughout various geographic areas. It also typically means better health services, better public transportation, and more participation in political and cultural endeavors. However, unplanned urban expansion often leads to pollution, environ- mental degradation, as well as unsustainable production and consumption of goods.[ii]

Can Microcredit Help Curb Population Growth and Eradicate Poverty?

Let us recall that education and employment of women are both factors that decrease birth rate. Women who are employed and financially independent tend to give birth to fewer children and therefore have a greater chance of breaking the cycle of poverty. **Microcredit** is a system that involves lending money to the poor, and, in many cases, mostly women, who have no collateral in many areas of the world. This loan is usually a couple of hundred dollars, suitable or commensurate with the need of recipients which is usually modest relative to U.S. standards. The loan is enough to help get a business started and generate income, but not an overwhelming amount so loan recipients cannot get out of debt. The goal is to help recipients become financially secure, breaking the cycle of poverty, and potentially strengthen the economy and stabilize population growth.

Microlenders also offer financial assistance and knowledge on how to get a small business started and generate income, ensuring that payback rates are high. There are several micro-lending banks including BancoSol in Bolivia and Women's World Banking (WWB) which only offers financial assistance to women.

Dr. Muhammed Yunus, an economist from Bangladesh who spent several years teaching and studying in the United States, returned home to his native land only to find the people impoverished. Wanting to help his people get out of debt, his efforts included microlending and soon materialized into the Grameen Bank in 1983. Its goal is both social and economic, to help people living in extreme poverty become financially independent. This practice of microlending set a precedent for the international community and became a more global endeavor. (Recognized for his selfless actions, Dr. Muhammed Yunus (**Figure 127.1**) and the Grameen Bank won a Nobel Peace Prize in 2006.[iii]) The Grameen Bank aspires to help achieve the first Millenium Development Goal, to eradicate extreme poverty (MDG 1).[iv]

lev radin/Shutterstock.com

Figure 127.1 Nobel Laureate Muhammed Yunus

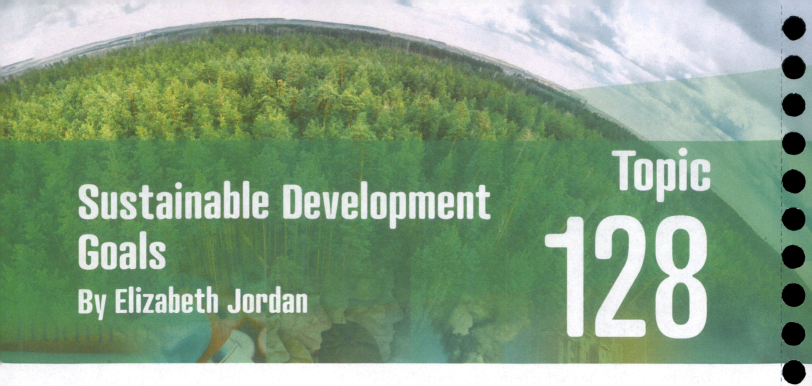

Sustainable Development Goals

By Elizabeth Jordan

Getting Involved

Although the government has passed laws to protect the environment, private citizens have assumed a sense of personal responsibility and taken measures to live sustainably and reduce waste on their own.

A noteworthy trend is to use the natural landscape and native vegetation, as opposed to the water-thirsty English gardens that homeowners once preferred. Given water demands and arid climates, homeowners and cities have started growing native gardens with only vegetation native to that given area, minimizing the use of water that would otherwise get to lawns and water-thirsty plants.

Many homes and office buildings are now being built to maximize use of natural energy and sun-light (**Figure 128.2**). This is known as **green architecture**. Norman Foster designed the Gherkin building in London—a prime example for green architecture. This structural masterpiece was specifically designed with a curved exterior to maximize its own ventilation and to become less reliant on air conditioning. Its glass exterior and open floor plans maximize ambient sun- light, so it is less reliant on electrical lighting.

Katherine Welles/Shutterstock.com

Figure 128.1 **A LEED Certified Building**

A popular trend of the past 25 years has been to build homes as large as possible. But a new trend is taking hold and that is to build smaller homes for quality instead of quantity. Interior designers and architects seek to capitalize on this trend, and books are published on how to maximize space and use renewable energy to meet the home's electrical needs. Reclaimed lumber and recycled copper are often used.

LEED (Leadership in Energy and Environmental Design) **certification** is a voluntary program whose participants (buildings and home owners) get certified by an objective third party as being considered green. This program has now spread to 135 countries.

LEED homes and buildings have become a desired commodity because they are designed to reduce waste, use local resources, conserve water and energy, lower greenhouse gas emission, and provide an overall healthy environment to its occupants (**Figure 128.1**). Highly sustainable buildings earn a platinum status and now there are even 'living buildings' that are carbon neutral!

Figure 128.2 **Green Architecture**

Chevron Corporation's building in Louisiana is LEED certified. The College of William and Mary has constructed LEED certified buildings as well as many new buildings in the University of California (UC) system. The UC system has estimated that it has already saved millions of dollars with its sustainable practice.

The World Takes Action

In 2002, an UN-sponsored coalition of businesses, foundations, heads of state, and dignitaries decided to start a campaign and take action to achieve eight Millennium Development Goals (MDGs) originally by 2015, and now beyond. Various fundraisers, committees, and summits have been held to spearhead the cause and reach its set goals. The eight MDGs are listed below:

1. Eradicate extreme hunger and poverty
2. Achieve universal primary education
3. Promote gender equity

4. Reduce infant mortality
5. Improve maternal health
6. Combat HIV/AIDS and other diseases
7. Ensure environmental sustainability
8. Establish global partnership

Every year, an MDG report is generated to provide yearly assessments of how much progress the globe has achieved and what strides it is taking to reach its MDGs. Although there is still a lot of focus, determination, and hard work ahead, the following are encouraging statistics surrounding the progress of each of the Millennium Development Goals:[iii]

Because of the great success of the UN's Millennium Development Goals for 2000, the United Nations spearheaded a new initiative that built on the success of the MDG's; a global rally cry to protect the environment, end poverty, and ensure a quality life for all of the planet's inhabitants and adding the planet as a priority. From this initiative came the 'Sustainable Development Goals' or SDGs.

The SDGs are as Follows:

SDG1. End poverty

SDG2. Zero Hunger; fight food insecurity, promote sustainable agriculture and healthy nutrition

SDG3. Health and Well-being for everybody

SDG4. Inclusive and equitable education

SDG5. Gender equality and the empowerment of women

SDG6. Clean water for all, and sustainable water management

SDG7. Affordable, renewable, clean energy is available to everybody

SDG8. Sustainable work growth and employment

SDG9. Build resilient infrastructure and sustainable and inclusive industrialization

SDG10. Reduce inequality within and among countries

SDG11. Sustainable cities and communities

SDG12. Promote responsible consumption and production

SDG13. Combat climate change and take climate action

SDG14. The conservation and sustainable use of the oceans and all marine ecosystems

SDG15. Promote sustainable terrestrial ecosystems by halting land degradation, biodiversity loss, sustainable use of terrestrial ecosystems and sustainable forest management

SDG16. Promote peaceful societies with sustainable development

SDG17. Strengthen global partnership for sustainable growth (source www.undp.org)

Historical Outbreaks

By Ingrid Ukstins and David Best

Epidemics are the rapid spread of infectious disease to a large number of people in a given population within a short period of time. In the past, these diseases were usually contained within a region, but with today's mobile society, they can spread rapidly across countries and continents to produce a **pandemic**, which occurs over a wide geographic area such as multiple continents or worldwide, and affects an exceptionally high proportion of the population. An example of an epidemic would be the spread of measles within a local community or a university student body. A pandemic would be much more widespread and create havoc for millions of people across continents. Examples of pandemics include the bubonic plague that overtook Europe in the mid-1300s or the Spanish influenza outbreak of 1918. In the case of the Spanish flu, which started in 1918 and lasted until 1920, up to 40% of the world's population at that time was infected, or 500 million people. The mortality rate was 10% to 20%, with 25 million people dying in the first 25 weeks, and up to 50 million people dying in all. This flu pandemic was particularly bad because it killed healthy young adults instead of those with already weakened immune systems. Diseases such as cancer or congestive heart failure, although killers of many people, are not contagious and hence do not produce epidemics or pandemics.

> **epidemic:** The sudden occurrence of a highly-contagious disease or other event that is clearly in excess of normally expected numbers.
>
> **pandemic:** A term applied to a highly-contagious disease that spreads throughout the world.

Historical Outbreaks

Several notable pandemics have affected large civilizations or the entire world. Among the earliest recorded was the Plague of Athens in 430 BC through 426 BC (**Figure 129.1**):

> In the winter following the first year of the war, morale had fallen considerably in Athens. It was at the year's public funeral (held annually for men who had fallen in battle in the course of the year) that Pericles pronounced the famous funeral oration that is so often quoted as summing up the greatness of Periclean Athens (Thuc.2.34–46). Pericles' speech was an encomium on Athenian democracy and it provided the high point of Thucydides' account of the war. It is immediately and dramatically followed in his account by the description of the plague which struck the city in the following summer, as the Spartans again invaded Attica. Crowded together in the city as the result of Pericles' strategy, the Athenians fell victim to the virulent sickness that was spreading throughout

the eastern Mediterranean. People died in large numbers, and no preventive measures or remedies were of any avail. It has been estimated that a quarter, and perhaps even a third, of the population was lost. The plague returned twice more, in 429 and 427/6, and Pericles himself died during this time, probably as a result of the disease.

Source: From http://www.indiana.edu/~ancmed/plague.htm. Used with permission.

In 2006, Manolis Papagrigorakis and others reported in the *International Journal of Infectious Diseases* that analysis of dental pulp extracted from the teeth in bodies in a mass grave in Athens, Greece, contained evidence that the people mostly likely died from typhoid fever.

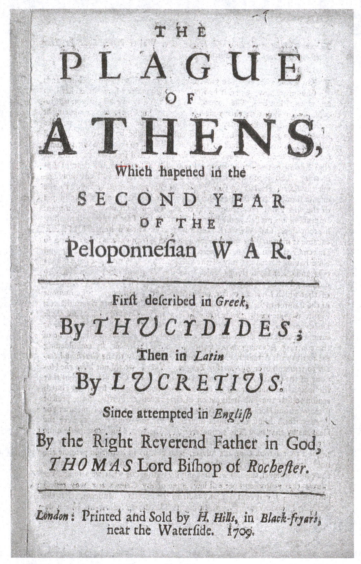

Source: Wellcome Collection.

Figure 129.1 The Plague of Athens was one of the first epidemics to be recognized in ancient history. It impacted much of the Eastern Mediterranean.

The Antonine Plague, which lasted from AD 165 to AD 180, was most likely smallpox that had been brought back from the eastern Mediterranean. It is estimated that about 25 percent of those infected died—almost 5 million people. In a second outbreak (AD 251 to AD 260), as many as 5,000 people died each day in Rome and the surrounding area.

The bubonic plague first made its appearance when a pandemic outbreak occurred in AD 541. The Plague of Justinian began in Egypt and surfaced in Turkey the following year. At its worst in Constantinople, it killed

10,000 people each day and wiped out close to 40 percent of the population of the city. At the end of the outbreak, 25 percent of the entire population of the eastern Mediterranean region had died.

Other well-known pandemics include the Black Death outbreak of the mid-1300s, a series of influenza pandemics that extended from the late 1800s until the most recent worldwide outbreak in 1968–1969, and typhus, which had its first outbreaks in the late 1400s and recurred intermittently until the Second World War. More details on all these disease pandemics will be provided in the following sections that address past and present biological hazards.

Natural Disasters and Disease

By Ingrid Ukstins and David Best

Dangerously Unprepared for Future Deadly Pandemics

Researchers estimate that birds and mammals harbor anywhere from 631,000 to 827,000 unknown viruses that could potentially leap into humans—these are the sources for future epidemics and pandemics. Ebola jumped from bats to humans in Guinea, West Africa, when a boy was infected in December of 2013. Genetic studies found that the virus was spreading quickly and mutated to form a strain that replicated faster, spread faster, and was significantly better at killing its human host. Treating and containing fast-spreading, deadly diseases is challenging even with the best medical facilities. In response to just 10 cases of Ebola in 2014, the U.S. spent $1.1 billion on domestic preparations, including $119 million on screening and quarantine. A severe 1918-style flu pandemic would cost the United States an estimated $683 billion dollars, according to the nonprofit Trust for America's Health. The World Bank estimates that global output would fall by almost 5 percent—totaling some $4 trillion—from another global influenza pandemic, and warns that we as a planet are 'dangerously unprepared' for future deadly pandemics. According to Jim Yong Kim, president of the World Bank speaking in 2015, the international response to the Ebola crisis was belated and disorganized. We need countries, corporations and donors working together to prepare for rapid response to outbreaks and emergencies, with stronger health systems, improved monitoring and supply transportation, and fast-acting medical teams.

Natural disasters are catastrophic events that have serious health, social and economic impacts. Deaths related to disasters, especially rapid-onset disasters like earthquakes, tsunami or floods are usually due to blunt force trauma, crush-related injuries, or drowning. The number of dead bodies in disaster-affected areas increases concerns about disease outbreaks, although human remains do not pose a significant risk for epidemics unless the deaths are from cholera or **hemorrhagic fevers** (such as the Ebola virus, **Figure 130.1**). The main risks for outbreaks after disasters are due to population displacement and overcrowding, and the availability of safe drinking water, sanitation facilities, and healthcare services. Disasters can greatly increase the risk of epidemics, especially in crowded situations such as refugee camps or temporary living areas for people displaced by natural disasters (**Figure 130.2a**). Often poor sanitary conditions and the loss of a reliable water supply lead to explosive outbreaks of diarrheal disease, such as a large cholera epidemic that sickened 16,000 people and killed 276 in West Bengal in 1998 after flooding contaminated the water supply. Hepatitis A and E are also associated with lack of access to safe water and sanitation. Overcrowding facilitates the transmission of communicable diseases, and is common in displaced populations. After the eruption of Mt. Pinatubo in the Philippines in 1991 more than 18,000 people came down with measles. In addition, standing water from heavy rainfall or river flooding, cyclones or hurricanes can increase breeding sites for mosquitoes that transmit disease. Malaria outbreaks after flooding are common. Biological hazards can quickly develop and have far-reaching, devastating effects at all levels of our population.

> **hemorrhagic fever:** A viral infection that causes victims to develop high fevers and bleeding throughout the body that can lead to shock, organ failure and death.

A **biological hazard** (also called a **biohazard**) is any organism or substance that has the potential to threaten the health of animals or the environment (**Figure 130.2b**). These can be in the form of a living and/or replicating pathogen such as a bacterium, virus, **protozoan**, **parasite** or **fungus**. A different type of hazard is material produced by chemical waste or nuclear accidents; this can also become a primary cause of biological disasters.

> **biological hazard (biohazard):** A biological substance or condition that poses a threat to the health of living organisms, primarily that of humans.
>
> **protozoa:** Microscopic one-celled free-living or parasitic organisms that are able to multiply in humans.
>
> **parasite:** An organism that lives on or in an organism of another species, called the host, and benefits by deriving nutrients at the host's expense.
>
> **fungus:** A single-celled or multicellular organism that gets their food from decaying matter, such as yeast, mold, and mushrooms.

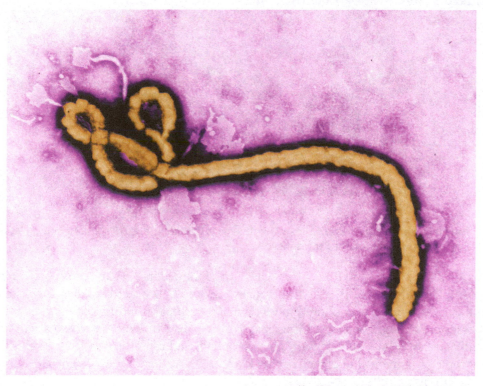

Image created by CDC microbiologist Frederick A. Murphy.

Figure 130.1   **This colorized transmission electron micrograph shows the shape of an Ebola virus virion.**

© Sk Hasan Ali/Shutterstock.com.

Figure 130.2a Overcrowding, poor sanitation, unsafe drinking water and lack of medical care all contribute to the spread of disease in displaced populations fleeing natural disasters.

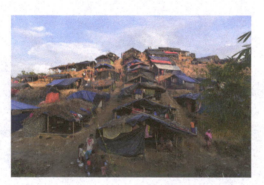

© Kaspi/Shutterstock.com.

Figure 130.2b The biohazard symbol, developed by Dow Chemical Company in 1966, contains four circles representing the chain of infection: agent, host, source and transmission.

The Centers for Disease Control and Prevention (CDC) was founded in 1992 and is one of the major operating components of the Department of Health and Human Services in the United States. It tracks and responds to global health threats, and has monitored more than 300 outbreaks of different diseases in 160 countries. Infectious illnesses are thought to cause 15 percent of all deaths worldwide. **Bacteria** are microscopic creatures that usually consist of one cell, contain no chlorophyll, and reproduce by simple cell division. **Viruses** are ultramicroscopic organisms that contain either RNA or DNA, surrounded by a protein case. Viruses can only multiply within living cells.

bacteria: A tiny, single-celled organism that reproduces by cell division, having a shape similar to a rod, spiral, or sphere, and has no chlorophyll.

virus: Acellular, non-living infectious particles of either DNA or RNA associated with various protein coats; these only multiply in living cells.

Nature of Diseases

A **pathogen** is any agent that is capable of causing a disease. Most pathogens are microorganisms. **Microbes** infest a host, which serves as the source of energy and nutrition necessary for them to survive. Microbes have the ability to sustain their existence by not killing the host immediately, which would terminate the particular disease—and the microbes. Many of Earth's 6.7 billion people live in areas that are constantly threatened by outbreaks of highly communicable diseases, such as malaria and tuberculosis (**Figure 130.3a**).

pathogen: An agent, such as a bacterium or virus, that causes a disease.

microbe: A microscopic living organism, including bacteria, protozoa, fungi, algae, amoebas and slime molds. Viruses are also called microbes. Some microbes are pathogens, capable of causing disease.

The malaria parasite that infects humans has existed for up to 100,000 years, but increased with the development of settlement and agriculture 10,000 years ago. Both ancient Greece and the Roman Empire suffered from malaria, which means bad air in ancient Italian, and was associated with swamps and standing water (**Figure 130.3b**). Malaria parasite species infect birds, reptiles and other mammals, including other primates such as chimpanzees. Bubonic plague has been documented since before the first millennium and has persisted for almost 1,500 years. Several diseases, including TB, have become resistant to early treatment techniques and are now more difficult to combat.

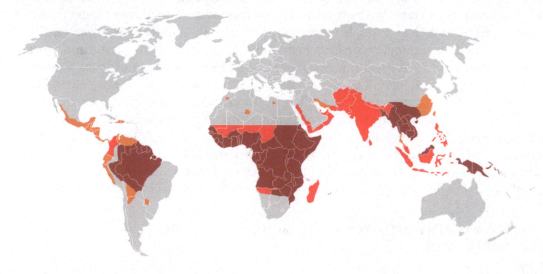

Percherie CHU de Roun.

Figure 130.3a   In 2015 malaria infected between 350 and 550 million people and killed 438,000 according to the World Health Organization. Death tolls have dropped by 60% since 2000, when almost a million people died, due to the use of insecticide-treated mosquito nets. Malaria is still endemic around the equator in tropical and subtropical regions that have stagnant water that mosquito larvae need to mature. Areas in dark red have malaria resistant to multiple treatments, red areas host chloroquine-resistant malaria, orange areas do not have resistance to treatments, and gray areas do not yet host malaria.

CDC; Global Health, Division of Parasitic Diseases and Malaria.

Figure 130.3b The malaria parasite is transmitted to humans by the host female *Anopheles* mosquitoes which infect people while they are feeding on them for blood.

Epidemics and Pandemics

Epidemics are the rapid spread of infectious disease to a large number of people in a given population within a short period of time. In the past, these diseases were usually contained within a region, but with today's mobile society, they can spread rapidly across countries and continents to produce a **pandemic**, which occurs over a wide geographic area such as multiple continents or worldwide, and affects an exceptionally high proportion of the population. An example of an epidemic would be the spread of measles within a local community or a university student body. A pandemic would be much more widespread and create havoc for millions of people across continents. Examples of pandemics include the bubonic plague that overtook Europe in the mid-1300s or the Spanish influenza outbreak of 1918. In the case of the Spanish flu, which started in 1918 and lasted until 1920, up to 40% of the world's population at that time was infected, or 500 million people. The mortality rate was 10% to 20%, with 25 million people dying in the first 25 weeks, and up to 50 million people dying in all. This flu pandemic was particularly bad because it killed healthy young adults instead of those with already weakened immune systems. Diseases such as cancer or congestive heart failure, although killers of many people, are not contagious and hence do not produce epidemics or pandemics.

epidemic: The sudden occurrence of a highly-contagious disease or other event that is clearly in excess of normally expected numbers.

pandemic: A term applied to a highly-contagious disease that spreads throughout the world

Types of Diseases
By Ingrid Ukstins and David Best

Bubonic Plague

Bubonic plague primarily affects rodents, but it is transmitted to humans by flea piercings (we incorrectly refer to these as bites, but fleas and mosquitoes do not have teeth) (**Figure 131.1**). The disease is caused by a bacterium called *Yersinia pestis*, is highly communicable, and spreads very rapidly. Symptoms include a fever, chills, and painful swelling of the lymph glands called buboes—hence its name. Red spots develop on the skin and these eventually turn black from subdermal dried blood caused by internal bleeding, giving rise to the name **Black Plague**, or the **Black Death**. The Black Death swept through Asia, Europe and Africa in the 14th century, killing an estimated 50 million people and up to 60 percent of the European population. Plague doctors were hired by towns to treat the victims; they often lacked medical training and rarely cured their patients. Starting in 1619 plague doctors began to wear beak-like masks filled with aromatic herbs, straw and spices in the hopes of protecting themselves from the contaminated air which they believed caused the plague (**Figure 131.2**).

> **bubonic plague:** A rare, but fast-spreading, bacterial infection caused by *Yersinia pestis;* transmitted in humans and rodents by infected flea piercings; symptoms include headaches and painful swelling of lymph nodes.
>
> **Black Plague (Black Death):** An epidemic of bubonic plague that killed more than 20 million people in Europe during the mid-fourteenth century.

© Cosmin Manci/Shutterstock.com

Figure 131.1 The Oriental rat flea is the host for bubonic plague, the plague forms a biofilm in the flea's gut and when the parasitic flea in turn feeds on a host the plague bacteria infects the wound.

Paul Fürst, Der Doctor Schnabel von Rom, 1658.

Figure 131.2 The plague doctor mask had a beak-like front that was filled with herbs and spices because people believed at that time that breathing in contaminated air was the way to transmit disease.

Following the pandemics that were recorded before AD 550, including the first recorded cases of bubonic plague, there was a lull in worldwide diseases until 1347, at which time the Black Plague resurfaced in China and found its way to Italy. Sailors on several Italian trade ships were dying from plague as the ships pulled into Sicily. Within a year, the plague had made its way through western Europe and into England (**Figure 131.3**). Here, it received the name "Black Death" (Box 131.1). Outbreaks slowed in the winter because fleas were dormant, but they had a resurgence each spring, when deaths increased dramatically. Although Ziegler (1991) chronicles an ongoing debate about the population of Europe and the number of people who died, it is fairly well established that by 1353, more than 20 million Europeans—almost one-third of Europe's population—had perished. This loss of life represented the estimated population growth that Europe had experienced in the previous 250 years.

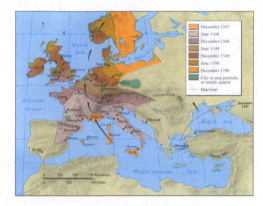

Source: http://www.odu.edu/~mcarhart/hist102/images.htm

Figure 131.3   Spread of the Black Plague across Europe.

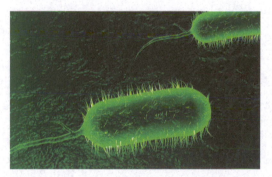

© Creations/Shutterstock.com

Figure 131.4   Cholera is a bacteria found in salt water or brackish water that, when ingested, can cause diarrhea and vomiting in the host within several hours to two to three days of ingestion.

Cholera

Cholera results from an intestinal infection by the bacterium *Vibrio cholerae* that could be present in drinking water or contaminated food (**Figure 131.4**). It is spread when people come in contact with feces from others who are infected, a problem in areas without proper sanitation facilities or safe water supplies. Extensive breakdown of infrastructure when cholera is present can produce epidemic conditions. Floods and earthquakes easily destroy a community's ability to handle its sanitation needs. There are documented cases of people contracting cholera from ingesting raw shellfish from coastal regions and brackish rivers, areas that are often tainted by raw sewage, especially when cyclonic storms hit populated coastal areas.

cholera: An acute intestinal disease caused by the bacterium *Vibrio cholera;* often found in contaminated drinking water and food.

Symptoms of cholera range from almost nonexistent to severe, the latter affecting approximately 5 percent of those who become infected. Vomiting and extremely water-laden diarrhea create a rapid decrease in body fluids that brings on dehydration and shock. Death is immediate without treatment.

Table 131.1 shows pandemics and concentrated outbreaks of cholera that have occurred since the first outbreak in India in 1816–1826. By the late 1880s, most developed countries understood the need for a clean reliable water supply, which helped reduce the pervasive nature of the disease in most Western nations.

After the 2010 earthquake in Haiti, which displaced approximately 1.5 million people, a cholera epidemic began to ravage the internally displaced persons camps. Studies have traced the outbreak to faulty sanitation in a United Nations peacekeeping force camp. Estimates suggest that more than 9,000 people have died of cholera, but a study by Doctors Without Borders suggests that this is severely underreported and may be up to a factor of three greater than that reported during the most intense early stages of the epidemic. As of 2018, the cholera epidemic is still ongoing in Haiti but the number of cases has dropped dramatically. About 40 percent of the population in Haiti does not have daily access to clean water and less than 25 percent has regular use of a toilet, according to Pan-American Health Organization and World Bank, making Haiti vulnerable to recurring strikes of cholera during future disasters such as hurricanes.

Table 131.1	Major Outbreaks of Cholera Throughout the World Since 1816	
Dates	**Localities**	**Comments**
1816–1826	First in Bengal, to India in 1820, then to China and eastern Europe	
1829–1851	Began in Europe, then London, Canada, and New York in 1832; west coast of North America in 1834	
1852–1860	Mainly Russia	More than 1 million deaths, including composer Peter Tchaikovsky
1863–1875	Prevalent in Africa and Europe	
1866	North America	
1892	Hamburg, Germany	City water supply contaminated, killing more than 8,500
1899–1923	Russia	
1961–1966	First in Indonesia, then to Bangladesh, India, and Russia	
1994	Rwanda refugee camps	48,000 cases resulted in almost 24,000 deaths
2010	Haiti earthquake displaced persons camps	>800,000 cases and more than 9,000 deaths

Source: https://www.asm.org/index.php/microbelibrary/367-news-room/iceid-releases/93626-influenza-vaccine-while-not-100-effective-may-reduce-the-severity-of-flu-symptoms

Influenza

Influenza viruses produce the flu, a disease that affects millions of people every year (Figure 131.5). Symptoms include a fever, headache, nasal congestion, muscle aches, a sore throat, and a general loss of energy and appetite. A case of the flu varies from being a mild illness to one that can be quite severe and lead to death in those with low immunity, especially the young and the aged. An annual vaccination is moderately effective against the disease, reducing your risk of developing an influenza-related illness by about 60 percent, and may also lessen your symptoms if you do get the flu, according to the American Society for Microbiology.

> **influenza:** An acute viral infection that affects the respiratory system; this can occur as isolated, epidemic, or pandemic in scale; symptoms include fever, headache, nasal congestion, and general lethargy.

The Centers for Disease Control and Prevention (CDC) report that every year in the United States, on average,

- Five to 20 percent of the population get the flu.
- More than 200,000 people are hospitalized from flu complications.
- About 36,000 people die from flu infections.

Pandemics of influenza have struck approximately three times every century since the 1500s. Recurrence intervals range between 10 and 50 years for these major outbreaks, beginning with the one that infected people in Africa and Europe in 1510. A more widespread episode occurred in 1889 and 1890, when influenza was reported in Russia in May 1889 and spread to North America by December of that year. In only five months, this pandemic had overtaken India and South America, and it finally reached Australia in the early spring of 1890.

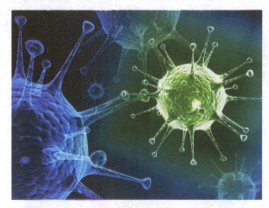

© mathagraphics/Shutterstock.com

Figure 131.5   Three-dimensional structure of seasonal influenza virus from electron tomography.

The 1918 influenza pandemic killed more people than World War One, which was just winding down. It is cited as the most devastating epidemic in recorded world history, and more people died of influenza in a single year than in the entire Black Death from 1347 to 1351. Although the strain began in the central United States, this episode became known as the "Spanish flu" or "La Grippe" because the spread of the disease received more press coverage in Spain, which was not involved in World War I and hence news in that country was not censored.

The supposed cause of this episode was from a mutated swine virus that affected soldiers at Camp Funston, Kansas, on March 11, 1918. Within two days, more than 500 people were sick, many of whom contracted pneumonia. Within a week, the flu had spread to every state in the United States and in the next two months the flu outbreak was worldwide. Interestingly, the disease usually peaked within a few weeks of its initial onset in a given area.

In the end, there were nearly 20 million cases in the United States, from which almost 1 million people died. It infected 28 percent of all Americans, and a fifth of the world population. The worldwide death toll has been estimated to range between 25 and 40 million people. Overall, most deaths occurred in people between 20 and 45 years old, generally considered the healthiest portion of a population. Besides a shortage of medical workers and medical supplies, there were also shortages in coffins, morticians and gravediggers. The Red Cross created a special committee to mobilize all resources to fight the flu, and even President

Woodrow Wilson suffered from the flu in early 1919 while negotiating the Treaty of Versailles that ended the World War.

Smallpox

Smallpox is an acute, contagious, and sometimes fatal disease caused by the variola virus (an orthopoxvirus), and marked by fever and a distinct, progressive skin rash. The only defense is prevention by vaccination, as no adequate treatment exists. Infected people develop raised bumps filled with fluid that appear on the skin covering the body. Historically, there were major episodes of outbreaks in the New World that killed several million people. In the twenty-first century, the United Nations reported that more than 300 million people worldwide have died from smallpox. The last documented case in the United States was in 1949 and the last case worldwide was in Africa in 1977. In 1980, the disease was declared eradicated following worldwide vaccination programs, and no cases of naturally occurring smallpox have happened since. However, in the aftermath of September 11, 2001, the CDC reports that the US government is taking precautions to be ready to deal with a possible but unlikely bioterrorist attack using smallpox as a weapon. Adequate vaccine supplies exist to inoculate everyone in the United States against the disease.

smallpox: An acute viral disease that was once a major killer but has been eradicated; symptoms include headaches, vomiting, and fever, followed by a widespread skin rash that eventually permanently scars the skin.

Centers for Disease Control and Prevention. Photo by James Gathany.

Figure 131.6 Body lice are parasitic insects that live on the body, and in the clothing or bedding of infested humans. They spread infections rapidly under crowded conditions where hygiene is poor and there is frequent contact among people. The dark mass inside the abdomen is a previously ingested blood meal.

Typhus

The CDC reports that **typhus** generally occurs only in communities and populations with poor sanitary conditions and crowding, in which body louse infestations are frequent (typically seen in refugee and prisoner populations, particularly during times of wars or famine; **Figure 131.6**). Typhus also occurs sporadically in cooler mountainous regions of Africa, South America, Asia, and Mexico, especially during the colder months when louse-infested clothing is not laundered and person-to-person spread of lice is more frequent.

> **typhus:** An acute infectious disease with symptoms of high fever, severe headaches, and a skin eruption; common in wartime, famines, or catastrophes, it is spread by lice, ticks, or fleas.

The four main types of typhus are (1) *epidemic typhus*, (2) *Brill-Zinsser disease*, (3) *endemic* or *murine typhus*, and (4) *scrub typhus*. Scrub typhus was a major problem in World War I as it was associated with trench warfare that was prevalent in Europe. Typhus was a major cause of death during World War II, killing those in POW camps, ghettos, and Nazi concentration camps, including Anne Frank, at the age of 15. The newly discovered insecticide DDT was used as a lice killer and helped avert major post-war epidemics.

Symptoms of epidemic typhus include chills and fever, headache, vomiting, and a rash that generally begins in the trunk area. If these conditions remain untreated, older adults who are infected can experience a death rate as high as 60 percent. Children usually recover well from epidemic typhus. Death rates for Brill-Zinsser disease, endemic or murine typhus, and scrub typhus are generally less than 1 percent. The best prevention for any form of typhus is to avoid the carrier insects and to use insect repellents and good hygiene.

Yellow Fever

Another viral disease that is transmitted between humans by mosquitoes is **yellow fever**. Symptoms are similar to other tropical diseases in that infected people experience fever, headache, muscle pain, jaundice, and nausea. One's pulse can slow down also, but all these symptoms generally vanish in a few days. There can be a sudden return of more serious conditions that include bleeding from the nose, eyes, mouth, and in the stomach. Kidney failure follows and usually death occurs within 10 to 14 days.

> **yellow fever:** Disease caused by a vector-transmitted virus; symptoms include high fever, headaches, jaundice, and often gastrointestinal hemorrhaging.

An epidemic of yellow fever struck the young United States in the late 1700s. From August to November 1793, between 4,000 and 5,000 people perished from the disease that at the time was unexplained. Philadelphia, then the new nation's capital, was the hardest hit in terms of deaths. The government came to a halt because so many workers were either afflicted with the disease or left the city to seek a safer area. A recurrence in 1798 killed many who were involved with forming the new government of the United States.

Historically, yellow fever was a major disease in equatorial regions and was one of the key impedances to the construction of the Panama Canal in the early 1900s. The research of William C. Gorgas (**Figure 131.7**), who also played a key role in the quest to eliminate malaria in the region, was paramount in controlling outbreaks of yellow fever. Gorgas contracted yellow fever as a young soldier and became immune to it in his later life. He also contracted typhoid fever in 1898.

U.S. Army.

Figure 131.7 Major General William C. Gorgas, Surgeon General of the US Army, eradicated yellow fever and reduced malaria in Panama.

Agricultural and forestry workers in South America and Africa are most affected by yellow fever as they are exposed to mosquitoes in their workplace, fields, and forests. Moist savanna regions in Central and West Africa are primary areas when the rainy season produces conditions that foster mosquito larva growth.

Yellow fever is now very rarely a concern to travelers but they must have a vaccination to prevent the disease. A single vaccination provides immunity for 10 years or more, with inoculation of a booster dose required every additional 10 years. If people are able to avoid mosquito piercings through the use of protective clothing, mosquito nets, and insect repellent, they can minimize the chance of contracting the disease. As the world's climate becomes warmer, more localities will experience increased infestation of mosquitos. This could result in an increase in yellow fever and other mosquito-borne diseases, especially in developing nations.

HIV infection and AIDS

Human immunodeficiency virus (HIV) is a retrovirus, or a type of of virus that use RNA as its genetic material, which can lead to **acquired immunodeficiency syndrome (AIDS)** in humans. This disease is considered a pandemic and was first diagnosed in 1981 in the United States although cases dating back to the late 1950s and 1960s are believed to be the earliest known AIDS infections. There is no cure or vaccine. The disease is thought to originate in non-human primates from Africa and to have been transferred to humans in the early 20th century.

> **human immunodeficiency virus (HIV):** An RNA (ribonucleic acid) virus that causes an immune system failure and can lead to AIDS.
>
> **acquired immunodeficiency syndrome (AIDS):** A serious disease that results from an infection with HIV; it is spread through direct contact with contaminated bodily fluids.

The virus attacks the body's immune system (specifically the CD4 or T cells) and, over time, so many of these cells are destroyed that the body can't fight off infection or disease. People diagnosed with HIV may become infected with life-threatening diseases called opportunistic infections, which are caused by microbes such as viruses, bacteria, or fungi that usually do not make healthy people sick. As the HIV infection progresses it increases the risk of common infections like tuberculosis as well as tumors. These late-stage symptoms, often accompanied by weight loss, are referred to as AIDS.

More than 78 million people have been infected with HIV and 35 million people have died from AIDS-related illnesses in the last 35 years (**Figure 131.8**). As of 2016 about 36.7 million people are living with HIV infections. There were 1.8 million new infections in 2016, down from 3.1 million in 2001, according to UNAIDS, a joint United Nations program on HIV and AIDS that involves 11 UN organizations and has a mission to end the AIDS epidemic by 2030. AIDS related deaths have fallen by 48 percent since their maximum in 2005, and tuberculosis remains the leading cause of death among people living with HIV. The greatest occurrence of HIV and AIDS is in eastern and southern Africa, 68 percent of all HIV cases occur there and it is believed to infect about 5 percent of the adult population. This is a blood-borne disease that is transmitted through unprotected sexual contact with infected partners. It is also possible to become infected through the use of nonsterile needles or, on extremely rare occasions, by tainted blood received through transfusions. HIV can be transmitted by women to their babies during pregnancy or birth. Approximately one-quarter to one-third of all untreated pregnant women infected with HIV will pass the infection to their infants. HIV can also be spread to babies through the breast milk of mothers infected with the virus.

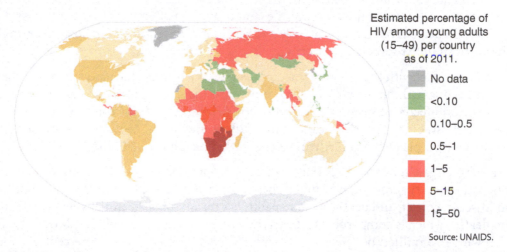

Estimated percentage of
HIV among young adults
(15–49) per country
as of 2011.

- No data
- <0.10
- 0.10–0.5
- 0.5–1
- 1–5
- 5–15
- 15–50

Source: UNAIDS.

Figure 131.8 **The estimated amount of the population of young adults ages 15 to 49 who were infected with HIV as of 2011, shown per country.**

The incidence rate of AIDS is highest among homosexual men, but education programs in high-incidence countries have helped stem the spreading rate. Recent data, however, show an increase in the Americas and in Asian countries.

As reported by the CDC, almost 700,000 cases of AIDS have been reported in the United States since 1981. As many as 1.2 million Americans may be infected with HIV, one-fifth of whom are unaware of their infection.

Symptoms of an HIV infection are not evident when someone is first infected. Within one or two months, flu-like symptoms appear and may include a fever, fatigue, headaches, and swollen lymph nodes. Because these generally disappear within a few weeks, they are usually mistaken for some other viral infection. It is during this period that people are highly infectious, especially through unprotected sexual contact.

Medical workers have noticed that the more severe or persistent symptoms may take 10 years or more to be evident, at which time the disease is well-established in the individual. The time it takes for the symptoms to be clearly defined varies with each person.

Much progress has been made toward the treatment of HIV and AIDS during the past decade. Advances and new drug treatments have improved the survival rate. The most current information concerning HIV and AIDS will be found on appropriate Internet sites sponsored by major medical institutions such as the CDC and National Institutes of Health (NIH).

Hepatitis

Hepatitis is a liver disease and the most common cause worldwide is viruses. Several forms of hepatitis (designated A, B, C, D, and E) exist. Hepatitis A and E are both passed from person to person through contact with fecal material of infected people. These are common in developing countries with poor sanitation, and do not lead to chronic hepatitis.

> **hepatitis:** A liver disease caused by a virus; four different forms exist; symptoms include fever, jaundice, fatigue, liver enlargement, and abdominal pain

The most widespread of the five types of hepatitis is Hepatitis B (HBV), which affected about 343 million people worldwide in 2015, compared to 114 million with hepatitis A and 142 million with hepatitis C. In comparison, alcoholic hepatitis affects about 5 million people. Its symptoms include jaundice, fatigue, abdominal pain, nausea, and pain in the joints. One-third of the world's population is infected with HBV, making it the number one infectious disease at the present time (and of all time). Death from chronic liver disease occurs in approximately 25 percent of all cases, usually as a result of liver cancer or cirrhosis (scarring) of the liver. Most deaths occur in underdeveloped countries.

HBV is spread by contact with the blood of infected people, which enters the bloodstream of the noninfected person. Unprotected sexual contact with infected people and the sharing of needles and other drug paraphernalia will put the uninfected at severe risk of contracting Hepatitis B. HBV can produce a lifelong infection and can cause cirrhosis of the liver, liver cancer, liver failure, and death. A vaccine has existed since 1982 and is recommended for people between birth and age 18 if they have any likelihood of contracting the disease.

The World Health Organization reports that as of 2017, 325 million people are living with chronic hepatitis B virus or hepatitis C virus infection. Many of these people lack access to testing and treatment, and are at risk for liver disease, cancer, and death. Viral hepatitis is a major public health challenge, and in 2015 caused 1.34 million deaths, which is comparable to deaths caused by tuberculosis and HIV. Mortality from hepatitis is increasing, with 1.75 million people infected with HCV in 2015. Globally, 84 percent of children born in 2015 got the three hepatitis B vaccine doses, according to the WHO, and the proportion of children under 5 years of age with the infection has fallen to 1.3 percent.

Hepatitis C is contracted through contact with blood of an infected person. Hepatitis B and C are the two main types out of the five different infections, and are responsible for 96% of all hepatitis deaths. A primary means is by shared use of needles and other sharp objects related to drug use. Hepatitis D needs HBV present to exist. Hepatitis E, which is transmitted in similar ways to HAV, rarely occurs in the United States. Hepatitis A, B and D are preventable with immunizations, and chronic viral hepatitis can be treated medically, although it results in more than a million deaths each year.

Leprosy

Leprosy (Hansen's disease) is an infection caused by slow-growing bacteria called *Mycobacterium leprae* that attacks the skin and peripheral nerves and can have many other manifestations. Two forms of the disease exist: paucibacillary Hansen's disease is the milder form that causes one or more pigmented lesions under the skin. The multibacillary form is recognized by symmetric skin lesions, nodules, thickened dermis, and nasal mucosa which produce nasal congestion and nose bleeds. *Mycobacterium leprae* is a rod-shaped bacterium that spreads very slowly and mainly infects the skin, nerves, and mucous membranes.

> **leprosy (Hansen's disease):** A chronic, mildly infectious disease which affects the nervous system, skin, and nasal regions; it is characterized by skin ulcerations and nodules.

The first description of symptoms similar to leprosy comes from writings in India from 600 BC, and it is mentioned in both the Old and New Testaments in the Bible. It is uncertain how the disease is spread but researchers believe that humans contract the bacillus through respiratory droplets by coughing or other contact with fluid from the nose of an infected person. Contrary to common belief, it is not highly contagious and most of the human population, about 95 percent, is not susceptible to infection, according to the Health Resources and Services Administration, US Department of Health and Human Services. It is curable with multi-drug therapy.

The number of leprosy cases was in the tens of millions in the 1960's, this has dropped dramatically to about 175,000, but there are still about 210,000 new cases diagnosed worldwide every year (**Figures 131.9**).

© fivepointsix/Shutterstock.com

Figure 131.9   **Leprosy patient at Munger Leper Colony, Munger, India.**

Malaria

Since the beginning of history, malaria has killed half of the men, women, and children who have lived on the planet. It has outperformed all wars, all famines and all other epidemics. Until World War II it still accounted for 50 percent of the business at most cemeteries.

(Nikiforuk, 1991)

Malaria (derived from Italian: *mala aira*, meaning bad air) was so named because it was first thought to have been caused by bad air. It is a mosquito-borne infectious disease caused by a parasitic protozoan single-celled micro-organism that has affected humans for more than 50,000 years (**Figure 131.10**). Malaria has the cyclic symptoms of body chills, muscular pain, flu-like illness, and a high fever, which first appear 10 to 15 days after infection and then may recur months later if not properly treated. Different strains exist, and one type can cause more serious problems, damaging vital organs including the heart, kidneys, lungs, or brain, thereby producing death. The female *Anopheles* mosquito is the reservoir for malaria and is the **vector** that infects people by means of puncturing the skin. Male mosquitoes feed on nectar from flowers and pose no threat of transmitting the disease.

malaria: An infectious disease caused by the bacterium *Plasmodium falciparum*, which is transmitted by the female *Anopheles* mosquito; symptoms include high fever, chills, and sweating.

vector: An agent (for example, a flea or mosquito) that can spread parasites, a virus or bacteria.

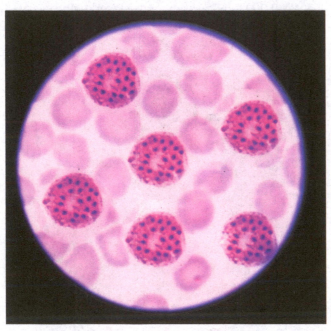

© toeytoey/Shutterstock.com

Figure 131.10 A Plasmodium parasite from an infected mosquito. The mosquito host transmits the parasite into a vertebrate host—the secondary host—and is the transmission vector for malaria.

Mosquitoes infected with malaria are less fertile and less long-lived than uninfected mosquitoes, but vector biologist Janneth Rodrigues and colleagues at the National Institute of Allergy and Infectious Diseases have found that many mosquitoes are malaria-resistant and are able to fight off the infection upon exposure. This may be useful in helping fight malaria, which can kill more than a million people every year, most of them children and pregnant women, according to the World Health Organization.

Most often found in Central and South America, Africa, and South Asia, malaria is a serious disease in those regions. This was especially true in the 1800s, when the British Army had large numbers of soldiers stationed throughout the tropics. The British government recognized that if the disease, and more importantly, its cause and source, could be identified, Great Britain would continue to dominate the regions it controlled in South Asia.

Sir Ronald Ross, a former major and surgeon in the British Army (**Figure 131.11a**), studied the malaria parasite in birds and recognized its connection with mosquitoes, although he did not recognize that malaria was transmitted by the female *Anopheles* mosquito. In 1880, Charles L. A. Laveran, a French army surgeon stationed in Algeria (**Figure 131.11b**), was the first to notice parasites in the blood of a patient suffering from malaria. For his discovery and continuing work, Laveran was awarded the Nobel Prize in Physiology or Medicine in 1907.

Source: CDC.

Figure 131.11a   **Sir Ronald Ross.**

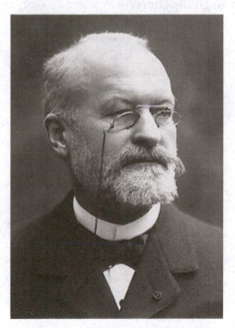

Source: Wikimedia Commons.

Figure 131.11b   **Charles Louis Alphonse Laveran.**

In rare instances, the disease can be transmitted if people come in contact with infected blood. It is also possible for an infected pregnant woman to pass malaria to her fetus. However, it is a myth that malaria can simply be passed to anyone who comes in close contact with a carrier.

Today, malaria infects more than 200 million people each year, mostly young children in sub-Saharan Africa. Every two minutes, malaria kills a child under the age of five. Over three billion people live in areas at risk of malaria transmission, in 106 countries and territories. This depends on climatic factors such as temperature, humidity, and rainfall. Malaria requires tropical and subtropical areas where mosquito hosts can survive and breed, and where the malaria parasite can complete its growth cycle in the mosquito. This can't happen at temperatures less than 20ºC (68ºF). In warm countries, close to the equator, malaria transmission may be year-round and more intense, compared to cooler regions or those with seasonal temperature or moisture fluctuations. No vaccine is currently available for malaria, so preventive drugs must be taken continuously to reduce the risk of infection. Unfortunately, these prophylactic drug treatments are too expensive for most people living in the endemic areas of the tropical and subtropical regions of the world.

The most cost-effective way of preventing malaria is to sleep under an insecticide treated bed net to protect against mosquito bites, which typically happen between 10 pm and 2 am. Each net costs about $2, lasts for three to four years, and protects, on average, two people (**Figure 131.12**). Since the year 2000, more than one billion nets have been given out in Africa alone. However, a WHO study from 2016 suggests that mosquitoes may be developing an immunity to the insecticide used in these nets, and malaria increased by five million cases—an estimated 216 million people infected—from the year 2015 to 2016.

Malaria was prevalent in Central and equatorial South America and created a major challenge during the early stages of construction of the Panama Canal. In 1906, more than 26,000 workers were involved with the project. Malaria and yellow fever hospitalized more than 80 percent of them during some stage of their employment. By the time the workforce had swelled to more than 50,000 people in 1912, fewer than 12 percent of them were hospitalized. Through the efforts of William C. Gorgas, malaria was significantly reduced as a biological threat in the region.

© Tuttoo/Shutterstock.com

Figure 131.12   Inexpensive, insecicide-treated bed nets help prevent malaria.

Malaria is less likely to occur in developed countries where public health measures are better established. Although the United States and western Europe have eradicated malaria, these areas still harbor *Anopheles* mosquitoes that can transmit malaria, and reintroduction of the disease is a constant risk. The high levels of deaths from malaria has put the strongest known rates of selective pressure on the human genome since the beginning of agriculture, and humans may be developing a genetic resistance to it. Miguel Soarez and Ana Ferreira of the Gulbenkian Institute of Sciences in Portugal showed, in 2011, that people carrying the gene for sickle-cell anemia disease are protected from malaria, not by protecting from infection, but by preventing the disease from taking hold because of changes to hemoglobin that sickle-cell gene carriers have in their blood.

Tuberculosis

"My sister Emily first declined. The details of her illness are deep branded in my memory, but to dwell on them, either in thought or narrative, is not in my power. Never in all her life had she lingered over any task that lay before her, and she did not linger now. She sank rapidly. She made haste to leave us. Yet, while physically she perished, mentally she grew stronger than we had yet known her. Day by day, when I saw with what a front she met suffering, I looked on her with an anguish of wonder and love. I have seen nothing like it; but, indeed, I have never seen her parallel in anything. Stronger than a man, simpler than a child, her nature stood alone: The awful point was, that while full of ruth for others, on herself she had no pity; the spirit was inexorable to the flesh; from the trembling hand, the unnerved limbs, the faded eyes, the same service was exacted as they had rendered in health. To stand by and witness this, and not dare to remonstrate, was a pain no words can render.

Two cruel months of hope and fear passed painfully by, and the day came at last when the terrors and pains of death were to be undergone by this treasure, which had grown dearer and dearer to our hearts as it wasted before our eyes. Towards the decline of that day, we had nothing of Emily but her mortal remains as consumption left them…

We thought this enough: but we were utterly and presumptuously wrong. She was not buried ere Anne fell ill. She had not been committed to the grave a fortnight, before we received distinct intimation that it was necessary to prepare our minds to see the younger sister go after the elder."

Charlotte Bronte, in a preface for the new addition of *Wuthering Heights* and *Agnes Grey* in 1850, a year after the authors, her sisters Emily and Anne, died, on December 19th, 1848 and May 28th, 1849, respectively.

Called 'consumption' in Victorian times, tuberculosis is one of human history's greatest killers, responsible for a billion deaths over two centuries. It has been found in human remains from 5000 BC, Egyptian mummies from 2400 BC (**Figure 131.13**), ancient Greece—Hippocrates identified it as the most widespread disease of his age in 460 BC—and medieval Europe. In 1882 Robert Koch discovered that bacterial infection causes tuberculosis (TB), and that air and lung fluid secretions, such as coughing, from victims contains live bacteria and can infect those around them. TB can affect the skin, the bones, or—the most common—lungs, which was called consumption for the wasting away late-stage victims experienced, along with night sweats, fevers, chills and coughing. TB is estimated to be responsible for 20 percent of deaths in 17th century London and 30 percent in 19th century Paris. Consumption has been romanticized in music and film, from Mimi dying in Puccini's *La Boheme* to Satine in *Moulin Rouge*. Emily Bronte died at the age of 30 from TB, as did her brother and sister. *Wuthering Heights* is filled with characters who develop TB and transmission from one character to another is used as a plot device to show the nature of their relationships.

© Andrea Izzotti/Shutterstock.com

Figure 131.13   An Egyptian mummy from the British Museum that shows spinal decay from tuberculosis.

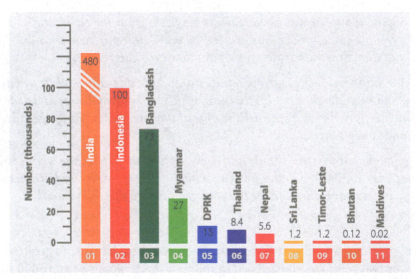

Source: World Health Organization.

Figure 131.14   Ranking of South-East Asian countries by TB incidence, according to the World Health Organization Regional Office for South-East Asia 2017 annual report.

While fresh air was considered a treatment for TB, and in-patient hospitals called sanatoriums were set up to treat patients, 80 percent of those who contracted TB died because there was no cure. This changed in 1944, when the first critically ill TB patient was cured with the antibiotic streptomycin. Rates of the disease dropped throughout the world, and in the mid-1980's the American Medical Association predicted that TB would be eradicated by 2010.

Yet by 1985 the decline in TB cases stopped and then began to rise. The drug regime developed to treat TB needs to be taken continuously, in regular doses, for six to eight months. Most patients do not follow through, because of the cost the inconvenience, or because they feel better and stop taking medication early. This leaves antibiotic-resistant microbes in the host that have turned into treatment-resistant strains of TB—it takes three years to develop a new antibiotic, and three months to develop a new drug-resistant strain of microbe. Tuberculosis and HIV have formed a new combination that is deadly—immunocompromised HIV victims represent an additional 1.4 million cases of TB each year. The WHO now calls TB a 'fire raging out of control' (**Figure 131.14**).

Two forms of TB exist: latent and active. One-third of the world's population is thought to be infected with TB, but if a person has latent TB, they are asymptomatic and the bacteria are being carried around but are not transmittable (**Figure 131.15**). Approximately 10 percent of latent TB cases transform into active TB. People with latent infections should seek treatment to prevent them from developing active infections later in life. Active TB spreads through the body of the carrier and can be transmitted to others if the carrier's lungs are infected. TB is the number one cause of death from an infectious disease and people with HIV/AIDS are 16 to 27 times more likely to develop TB than those without.

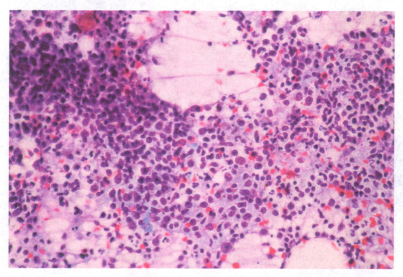

© Komsan Loonprom/Shutterstock.com

Figure 131.15   **The tuberculosis bacteria in mucus.**

In 2015, the World Health Organization reported more than 4.7 million new TB cases, resulting in 1,945 deaths per day—equivalent to nine passenger planes crashing each day. Of those new cases, only about 2.5 million were treated. TB death rates are highest in Africa, India, Indonesia and Bangladesh (**Figure 131.16**). Between 2000 and 2014, improvements in diagnosis and treatment saved 43 million lives worldwide. The current rate of decline in TB is 1.5 to 2 percent a year, which is not enough to meet the WHO's 'End TB' target of the year 2030. TB is not just a biomedical and public health problem, it's association with poverty complicates efforts. To meet the WHO target, TB decline has to reach 10 percent a year to 2020, and 17 percent a year to 2025.

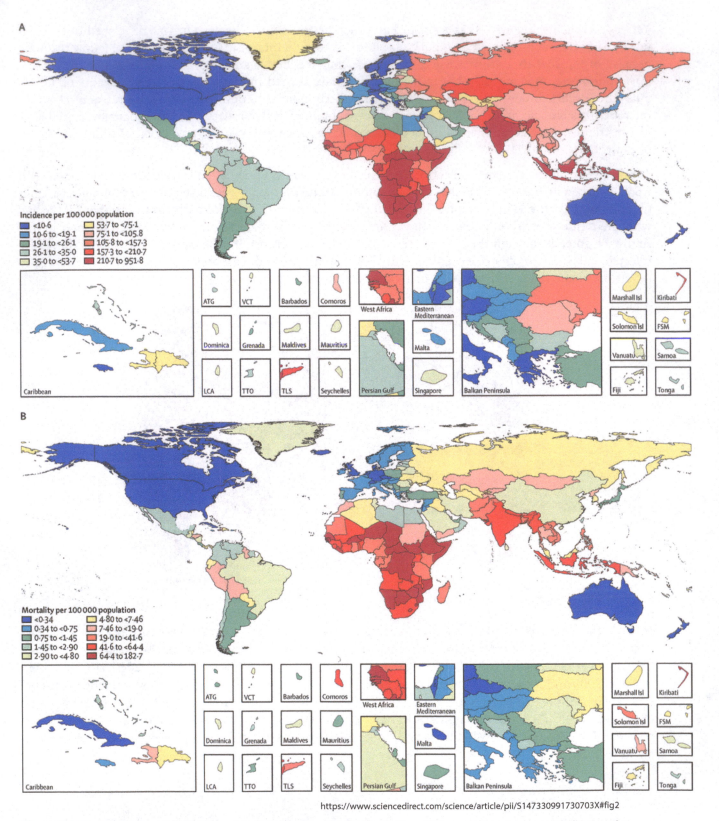

Figure 131.16   For the year 2015, these maps show the age-standardized rates of tuberculosis occurrence in HIV-negative individuals (top map) and the mortality rates (bottom map) per every 100,000 people. This research, published in The Lancet by the Global Burden of Disease Tuberculosis Collaborators, was funded by the Bill and Melinda Gates Foundation.

Typhoid Fever

The bacterium *Salmonella typhi* is the cause of **typhoid fever**, which is is common worldwide, affected 12.5 million people and resulted in 149,000 deaths in 2015. Approximately 400 cases are diagnosed each year in the United States, three-quarters of which are attributable to international travelers. The bacteria grow in the intestines and blood, and the disease is spread by eating or drinking food or water contaminated with the feces of an infected person, which can occur through poor sanitation or poor hygiene. Symptoms, which vary from mild to severe, start six to 30 days after exposure and include a high fever, weakness, abdominal pain, constipation and headaches. Some people develop a skin rash with rose colored spots. Some people are carriers without having any symptoms—they can still transmit the disease.

> **typhoid fever:** An illness spread by contamination of water, food, or milk supplies with *Salmonella typhi*; symptoms include fever, diarrhea, stomach aches, and rash.

Although typhoid fever is rare in developed countries, it is still very prevalent throughout Latin America, Asia, and Africa. Typhoid occurs most often in children and young adults between the ages of five and 19 years old. The rates of typhoid fever in developed countries declined during the first half of the 20th century because of the implementation of vaccines and improvements in public sanitation and hygiene, including the chlorination of public drinking water in 1908. In developed countries, typhoid fever rates are low at about 5 cases per million people per year.

William Wallace Lincoln, son of former US president Abraham Lincoln, died of typhoid in 1862, and the English novelist Arnold Bennett drank tap water in a Paris restaurant to prove it was safe, despite being warned by the waiter, and died in 1931 from the typhoid he contracted. Mary Mallon, a cook who worked in New York City between 1900 and 1907, was the most notorious carrier identified for typhoid, even though she was not the most destructive—three deaths have been directly linked to her but some estimates suggest she caused up to 50 fatalities.. She became the first person to be recognized as a healthy individual who was clearly documented to be a carrier of typhoid fever. As a cook, she ended up infecting 53 people with typhoid fever through her dessert dish—peaches and ice cream. "Typhoid Mary," as she was nicknamed, continually denied her involvement in transmitting the disease and also refused to stop work as a cook, sneaking back under a false name to continue working. She was arrested and permanently quarantined at Riverside Hospital for 26 years total, before dying of pneumonia at age 69.

A vaccine can prevent up to 70 percent of cases over two years, and is recommended for people travelling to areas where the disease is common. Other preventative measures include drinking only clean drinking water, better sanitation, and rigorous handwashing. Food preparers should be checked for infection as they can be asymptomatic but still transmit the disease. The risk of death is as high as 20 percent if not treated, and antibiotics can treat the infection once it is acquired, but strains of typhoid have been developing resistance which makes treatment and complete recovery more difficult.

Five medical researchers who examined their causes and effects on humans won Nobel Prizes for physiology or medicine (Table 131.2). Their work has wide-ranging and beneficial effects for humanity, as do the contributions of countless other scientists and medical personnel who, with less recognition, have greatly expanded our current knowledge of these and other diseases.

Table 131.2	**Nobel Prizes for Physiology or Medicine Awarded for Research into Various Major Diseases**		
Year	**Name**	**Country**	**Achievement**
1902	Sir Ronald Ross	Great Britain	Recognizing the role of mosquitoes in spreading malaria
1907	Charles L. A. Laveran	France	Discovering the role of parasites in blood as related to the transmission of malaria
1928	Charles Nicolle	France	Researching the causes of typhus
1948	Paul Mueller	Switzerland	Recognizing that DDT killed mosquitoes
1951	Max Theiler	South Africa	Developing a vaccine for yellow fever

Emerging Infectious Diseases

By Ingrid Ukstins and David Best

Emerging infectious diseases are infections that have recently appeared in a population or ones that have had rapid increase in their geographic range or number of cases, or diseases that threaten to increase in the near future. These can be caused by four different factors:

1. New detection of an unknown or previously undetected infectious agent
2. Known infection agents that have spread to new locations or new populations
3. Previously known disease agents whose role in specific diseases has just been recognized
4. Re-emergence of disease agents whose incidence of disease declined in the past, but has re-appeared. These are classified as re-emergent infectious diseases.

The WHO warns that infectious diseases are emerging at an escalating rate—since the 1970s we have discovered about 40 new infectious diseases. Increased travel, increased population density, and contact with wild animals all increase the potential for emerging infectious diseases to become global pandemics. Many of the following emergent diseases are considered by the WHO to be both a dire epidemic threat and in need of urgent funding in research and development of clinical solutions.

Viral Hemorrhagic Fevers

Viral hemorrhagic fevers (VHF) are a group of illnesses produced by five families of RNA virus and have the common features of causing a severe, multi-system syndrome—by affecting many organs, damaging blood vessels, and impacting the body's ability to self-regulate. They are often accompanied by hemorrhage, or bleeding. Some can cause mild symptoms but many, like Ebola or Marburg, result in severe, life-threatening illness and death. Specific VHF diseases are usually limited to the geographic region of the animal that hosts the virus. Viral hemorrhagic fevers are spread in a variety of ways, some may be transmitted through a respiratory route, and the ability to disseminate by aerosol, along with their severity, means they have the potential to be weaponized for biowarfare.

Crimean-Congo Hemorrhagic Fever

This virus is typically spread by tick bites or contact with livestock carrying the disease, as well as contact with infected people. Infection occurs most frequently in agricultural workers, slaughterhouse workers, and medical personnel. Death rates are up to 40 percent, and outbreaks occur in Africa, the Balkans, the Middle East and Asia.

Ebola Hemorrhagic Fever

Fruit bats are believed to be the carrier of the Ebola virus, and can spread the virus without being affected by it (**Figure 132.1**). It spreads by direct contact with body fluids like blood, or items contaminated by body fluids, from an infected person or animal. Eating infected bushmeat (wild animals) may expose people to the virus; animals can become infected when they eat infected animals or eat fruit partially eaten by infected bats. People who have recovered from Ebola continue to carry the virus in their semen or breast milk for several weeks and up to nine months afterwards, and can infect others. Death rates from Ebola are extremely high—from 25 percent to 90 percent of those who are infected—with an average of about 50 percent mortality. As soon as two days after exposure flu-like symptoms start, followed by vomiting, diarrhea, rash, decreased kidney and liver function, and then by internal and external bleeding—vomiting or coughing up blood, internal bleeding and blood in the stool, and bleeding into the eyes.

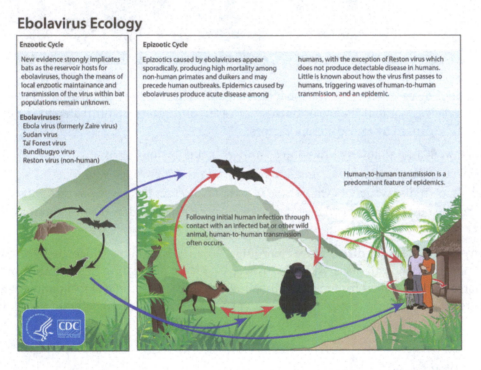

Source: CDC.

Figure 132.1   **The Ebola virus host and transmission cycle from wild animals to humans.**

Two simultaneous outbreaks in 1976 were the first time Ebola was identified, one in South Sudan and one near the Ebola River in the Democratic Republic of the Congo (then known as Zaire), which is where the name comes from. A total of 602 people were infected and 430 died. The largest outbreak currently documented was in West Africa from 2013 to 2016—Guinea, Liberia and Sierra Leone had 28,661 suspected cases and 11,310 reported deaths, although the WHO believes these numbers were underestimated. Ten percent of the dead were healthcare workers.

Lassa Hemorrhagic Fever

Lassa fever is relatively common in West Africa, with about 300,000 to 500,000 cases each year. Rats are the disease vector and you can contract the virus through exposure to their urine or feces; it can then be spread from person to person. Most of those who contract the virus do not develop symptoms, which are similar to Ebola, although a complication from the disease is that about a quarter of those who survive develop deafness, which can improve over time. Death rates are about one percent, with about 5,000 deaths each year, although in epidemics death rates can reach 50 percent.

Marburg Hemorrhagic Fever

Marburg was first described in 1967 from an outbreak in Marburg and Frankfurt, Germany, and Belgrade, Yugoslavia, where 31 cases were identified and seven people died when workers were exposed to infected tissue from grivet monkeys at a German biopharmaceutical company. The virus is carried, and can be transmitted, by fruit bats and also between infected people or other animals via body fluids, broken skin, unprotected sex, or exposure to infected tissue. Prolonged exposure to mines or caves inhabited by bats carrying the disease may also result in infection. Disease progression and symptoms are similar to Ebola. The largest outbreak so far was in Angola in 2004 to 2005, with 252 cases and 227 deaths (**Figure 132.2**). The most recent outbreak, in Uganda in 2017, was successfully controlled within weeks of detection, and resulted in only three deaths, according to the WHO. It was Uganda's fifth Marburg outbreak in 10 years.

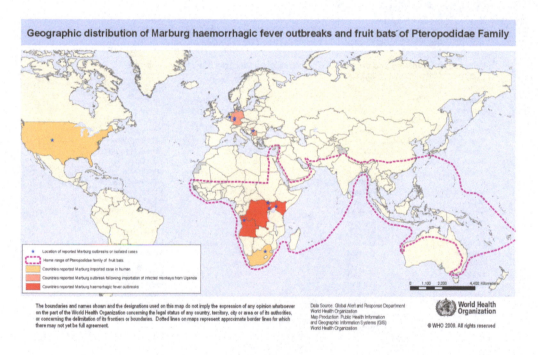

Source: WHO.

Figure 132.2   Geographic distribution of Marburg hemorrhagic fever outbreaks and fruit bats of Pteropodidae Family.

Chikungunya Disease and Dengue Fever

Chikungunya disease is caused by a virus that is related to Dengue fever; both are tropical diseases that are transmitted by mosquitoes. In the past, both were restricted to tropical areas around the Indian Ocean, but with modern climate change, tropical diseases such as these are spreading to new areas and can have significantly greater impact (**Figure 132.3**). In 2007, there was a mysterious outbreak of illness that caused fever, exhaustion and severe bone pain in Italy, which was later identified as Chikungunya. By 2014 outbreaks were reported from Europe, Asia, Africa, the Caribbean, Central and South America, and North America—initially only in Florida, but as of April 2016 cases have occurred as far north as Alaska. Chikungunya has now been identified in 45 countries and infects three million people each year (**Figures 132.4a and b**).

Countries and territories where chikungunya cases have been reported*
(as of April 22, 2016)

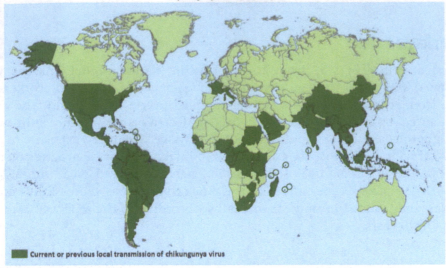

■ Current or previous local transmission of chikungunya virus

*Does not include countries or territories where only imported cases have been documented. This map is updated weekly if there are new countries or territories that report local chikungunya virus transmission.

Data table: Countries and territories where chikungunya cases have been reported

AFRICA	ASIA	AMERICAS	
Benin	Bangladesh	Anguilla	Nicaragua
Burundi	Bhutan	Antigua and Barbuda	Panama
Cameroon	Cambodia	Argentina	Paraguay
Central African Republic	China	Aruba	Peru
Comoros	India	Bahamas	Puerto Rico
Dem. Republic of the Congo	Indonesia	Barbados	Saint Barthelemy
Equatorial Guinea	Laos	Belize	Saint Kitts and Nevis
Gabon	Malaysia	Bolivia	Saint Lucia
Guinea	Maldives	Brazil	Saint Martin
Kenya	Myanmar (Burma)	British Virgin Islands	Saint Vincent & the Grenadines
Madagascar	Pakistan	Cayman Islands	Sint Maarten
Malawi	Philippines	Colombia	Suriname
Mauritius	Saudi Arabia	Costa Rica	Trinidad and Tobago
Mayotte	Singapore	Curacao	Turks and Caicos Islands
Nigeria	Sri Lanka	Dominica	United States
Republic of Congo	Taiwan	Dominican Republic	US Virgin Islands
Reunion	Thailand	Ecuador	Venezuela
Senegal	Timor	El Salvador	
Seychelles	Vietnam	French Guiana	**OCEANIA/PACIFIC ISLANDS**
Sierra Leone	Yemen	Grenada	American Samoa
South Africa		Guadeloupe	Cook Islands
Sudan	**EUROPE**	Guatemala	Federal States of Micronesia
Tanzania	France	Guyana	French Polynesia
Uganda	Italy	Haiti	Kiribati
Zimbabwe		Honduras	New Caledonia
		Jamaica	Papua New Guinea
		Martinique	Samoa
		Mexico	Tokelau
		Montserrat	Tonga

Source: CDC.

Figure 132.3 Countries and territories where chikungunya cases have been reported* (*as of April 22, 2016*).

Dengue Fever is also a tropical, mosquito-transmitted viral infection that has epidemic potential, but is not classified as an emergent disease because there are active major disease control and research networks, and mechanisms for intervention, already in place. About half of the world's population is at risk for Dengue Fever—3.9 billion people in 128 countries—and severe Dengue is a leading cause of serious illness and death among children in some Asian and Latin American countries. Incidences of Dengue have risen dramatically around the world in the last several decades, and according to the WHO up to 390 million people are infected each year, of which 96 million develop mild to severe symptoms of the disease. Not only are the number of cases increasing, but severe outbreaks are are also occurring, and the threat of Dengue in places like Europe is real—outbreaks have been reported in France, Portugal, and Croatia. In the United States cases have been

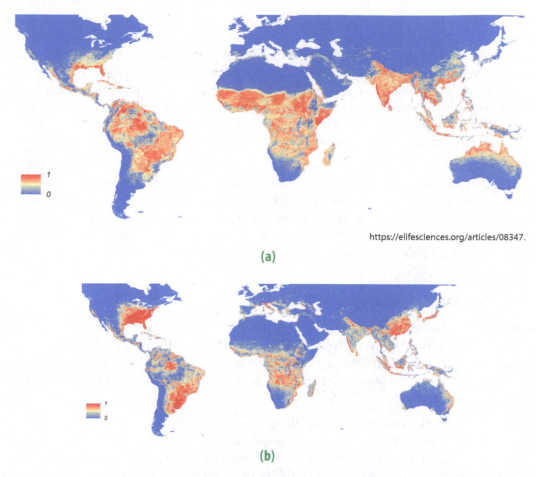

https://elifesciences.org/articles/08347.

(a)

(b)

Figure 132.4   Global maps of the distribution of *Aedes aegypti* and *Aedes albopictus* (top and bottom maps, respectively), mosquito species that host and spread Dengue and Chikungunya Fever. This study by Moritz Kraemer and colleagues, published in Ecology, Epidemiology and Global Health in 2015, shows the distribution of these mosquito species to be the widest ever recorded—they are now extensive in all continents, including North America and Europe.

reported from Florida and Hawaii. In 2016 there were large Dengue outbreaks worldwide, with more than 2.38 million in the Americas.

Zika Fever

In early 2015, a Zika Fever epidemic which originated in Brazil spread through South and North America. The 2015–2016 outbreak was picked up by social media, and an analysis of tweets on February 2, 2016 showed that there were 50 tweets per minute posted about Zika. The World Health Organization declared this outbreak a public health emergency of international concern.

Zika is a virus hosted by monkeys, and transmission to humans was rare before the current pandemic began in 2007. Zika is spread among monkeys, and to humans, by a mosquito species most active during daytime. Related virus that infect the same mosquito species, such as Dengue, are known to be intensified by urbanization and globalization, and this is thought to be a factor in the current spread. The potential risk of Zika is dependent on the distribution of the mosquito vector that transmits it, and the global distribution of this species is the most extensive ever recorded—across all continents including North and South America and Europe—and expanding due to global travel and trade. This mosquito species may be adapting to colder climates and has been found in Washington D.C., where they can now survive through the winter, according to Dr. David Severson, expanding their range from the tropical and subtropical areas they are commonly found. Zika can also be transmitted from men and women to their sexual partners, most cases are from infected men to women.

Most people infected with Zika have few or no symptoms, and those that develop symptoms have fever, joint pain, headache, and rash. Symptoms last about a week and no deaths have been reported due to the initial infection. A much greater problem with Zika is the risk of transmission from a pregnant mother who becomes infected to her unborn child. The disease can then cause birth defects such as underdevelopment in the fetal brain leading to microcephaly. Adults who have been infected may have increased risk of developing Guillain-Barre Syndrome, a rapid-onset muscle weakness caused by the immune system damaging the central nervous system. This can be life-threatening in about 8 percent of victims, especially those who develop weakness of the breathing muscles or abnormalities in heart rate or blood pressure.

Because of the evidence for a link between Zika infection and birth defects and neurological problems, the Centers for Disease Control issued a travel alert in early 2016 advising pregnant women to consider postponing travel to areas with ongoing Zika Virus cases. The advice was updated to caution pregnant women to avoid these places entirely, and several countries in Latin America and the Caribbean—Colombia, Ecuador, El Salvador and Jamaica—issued warnings for women to avoid pregnancy until more is known about the virus and its impact on fetal development.

Animal Influenza Viruses (Zoonotic Influenza)

Humans can be infected by avian, swine and other animal-based influenza viruses such as bird flu and swine flu. These kinds of infections are mainly from direct contact with infected live or dead animals or contaminated environments, and thus far the viruses have not mutated to acquire the ability to effectively transmit themselves among humans. Symptoms of animal-based flu virus infections—called zoonotic influenza— range from mild flu-like symptoms to death.

Flu viruses infect many different species of animals, such as birds, seals, pigs, cattle, horses, and dogs, in addition to humans. While flu viruses are adapted to the specific species that hosts them, they can jump to other species, such as the Avian flu outbreak of H5N1 in 1997 which has been linked back to a goose (Figure 132.5). So far, no zoonotic influenza has acquired the ability of sustained transmission among humans. Species such as pigs, which can be infected with swine, avian and human viruses, allow the virus genes to mix and can facilitate the creation of a new virus which people may have little to no immunity from. The diversity of

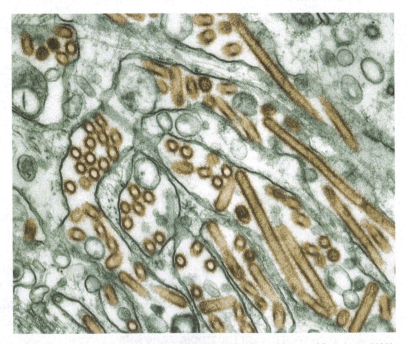

Source: Centers for Disease Control and Prevention's Public Health Image Library image #1841.

Figure 132.5 **Transmission electron micrograph of the avian flu virus, shown in yellow.**

zoonotic influenza viruses is 'alarming' according to the WHO, and necessitates strengthening surveillance and pandemic preparedness planning.

avian flu (H5N1): An acute viral disease of chickens and other birds (except pigeons) capable of being transmitted to humans; first noticed in the Far East, its symptoms include fever and lack of energy.

Prion Diseases

Prion diseases are part of a class of transmissible spongiform encephalopathies (TSE) that are characterized by a malformed protein molecule that produces clumps in the brain. Bovine spongiform encephalopathy (BSE or mad-cow disease) is a TSE, as is scrapie, a nervous system disease in goats and sheep, and chronic wasting disease in deer. Human consumption of a diseased animal can lead to the spread of TSE among humans. Once contracted, prion diseases are always fatal.

prion disease: An infliction that attacks the brain and nervous system, and disrupts normal protein activity within the neural cells.

TSEs are spread by prions, which are only protein material, unlike other diseases which are spread by agents with DNA or RNA, such as virus or bacteria. Transmission occurs when a healthy animal consumes tainted tissue from one with the disease. BSE spread as an epidemic in cattle in the 1980's and 1990's because cattle were fed processed remains of other cattle, a practice which is now banned in many countries. People who ate the infected cattle also contracted prion disease, called variant Creutzfeldt-Jacob disease (vCJD), which is a fatal brain disease with an extremely long incubation period of many years (**Figure 132.6**). Prions can't be transmitted through the air, by touch, or most other forms of casual contact. They can be transmitted through contact with infected tissue, body fluids, or contaminated medical instruments. Normal sterilization procedures such as alcohol, boiling, acid, or irradiation do not destroy the prions. Infected brains that have been sitting in formaldehyde for decades can still transmit spongiform disease.

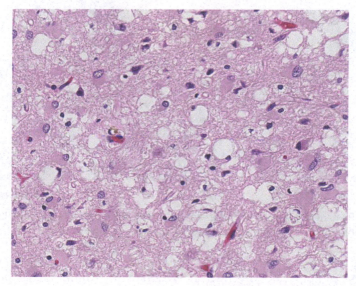

Source: CDC Public Health Image Library #10131.

Figure 132.6 This brain tissue has been magnified 100 times and shows prominent spongy texture in the cortex and loss of neurons—the patient was diagnosed with the prion disease Creutzfeldt-Jakob (vCJD). This was first diagnosed in the UK in 1996 and results from eating cows infected with bovine spongiform encephalopathy (BSE or 'mad cow' disease), which is the same agent responsible for the outbreak of vCJD in humans.

Included in the human category of TSEs is Kuru, a rare brain disorder that reached epidemic proportions in the mountains of New Guinea in the 1960s. Natives called Kuru the 'laughing disease' and thought it was caused by sorcery. Symptoms include shaking, dementia and uncontrolled crying and laughter. It incubates for about 10 years and victims die within 12 months of developing symptoms. People suspected of being sorcerers and inflicting kuru on others were murdered brutally, by biting their trachea and using clubs or stones to crush their genitals. As a result of rituals involving mortuary cannibalism among some of the indigenous tribes, kuru spread to a large percentage of the population, but it has now essentially disappeared due to government intervention.

SARS and MERS

SARS (severe acute respiratory syndrome) and MERS (Middle East respiratory syndrome) are infectious respiratory diseases caused by coronaviruses, which have a crown-like appearance in microscopic view, and can both be deadly to humans (**Figure 132.7**). SARS is characterized by severe pneumonia-like symptoms, and is transmitted from person to person through respiratory droplets produced by sneezing or coughing and through direct contact with a surface contaminated with respiratory droplets. It was first identified in Guangdong Province, China, in November of 2002, and by July 2003 there were 8,098 cases in 37 countries resulting in 774 deaths. No cases of SARS have been reported since 2004.

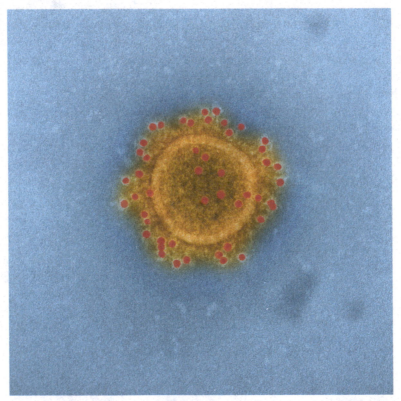

Source: NIH NIAID.

Figure 132.7   Middle East Respiratory Syndrome Coronavirus (MERS-CoV) is a viral respiratory illness first reported in Saudi Arabia in 2012. Both MERS and SARS (Severe acute respiratory syndrome) belong to a family of viruses called coronaviruses.

> **severe acute respiratory syndrome (SARS):** A respiratory disease of unknown source that first occurred in mainland China in 2003; symptoms include fever and coughing or difficulty breathing; it is sometimes fatal.

The SARS epidemic was brought to the public spotlight in February of 2003, when an American businessman flying back from China got extremely sick on the airplane to Singapore. The flight made an emergency landing in Vietnam and the victim died. Several of the medical professionals who treated him in the hospital also died, and Dr. Carlo Urbani identified the risk and alerted the World Health Organization; he later died of SARS also.

The source of the virus was traced to cave-dwelling horseshoe bats in Yunnan Province, China, in late 2017 (**Figure 132.8**). Antibiotics are ineffective since SARS is a viral disease, and there is no vaccine—isolation and quarantine are the most effective means of preventing the spread of SARS. Because SARS is most infectious in extremely sick patients, which usually occurs during the second week of illness, quarantine can be highly effective if infected people are isolated in the first few days of their sickness.

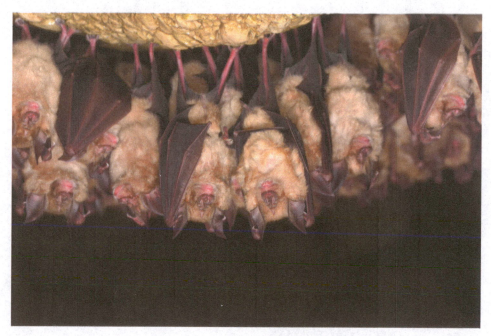

© All-stock-photos/Shutterstock.com

Figure 132.8   Cave-dwelling horseshoe bats like these, living in Yunnan Province, China, are thought to be the source of the SARS epidemic that killed 774 people.

MERS was first identified in the Arabian peninsula in 2012, and MERS is similar to SARS because it is a severe respiratory illness that can be fatal to humans, with a death rate of from 30 to 40 percent. MERS is also thought to originate in bats, but was passed to camels which then infected humans (**Figure 132.9**). Spread between humans requires close contact with an infected person, and because of that, so far the risk to the global population is considered to be relatively low. As of February of 2018, the WHO has identified 2,143 confirmed cases, with 750 MERS-related deaths in 27 countries. The two largest localized outbreaks so far have been in Saudi Arabia in April of 2014, with 688 people infected and 282 deaths, and in South Korea in May of 2015. The first case there was a man who had visited Saudi Arabia, and as of June 2015—when it was finally contained—a further 186 cases of infection and 36 deaths occurred there.

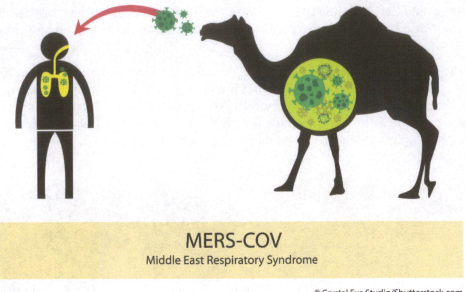

MERS-COV
Middle East Respiratory Syndrome

Figure 132.9   MERS is transferred from infected camels to humans through close contact.

Minimizing Biological Hazards

By Ingrid Ukstins and David Best

One means of reducing the spread of insect-borne diseases is the use of insecticides. Perhaps the most widely used has been DDT (dichloro-diphenyl-trichloroethane), first synthesized in 1874 by a German chemistry student as a thesis project. In 1939, its ability to kill insects was discovered by the Swiss scientist Paul Müller (Nobel Prize in physiology or medicine in 1948). Several countries used it during World War II to rid areas of lice that carried typhus. Following the war, it was used in numerous countries to kill mosquitoes that carried malaria. However, the use of DDT had profound effects on many forms of wildlife, including the embryos of bald eagles and peregrine falcons. Its use in the United States was banned in 1972. However, it is still used in many tropical regions to curtail the spread of mosquitos (**Figure 133.1**).

Figure 133.1   Thai man fogging DDT to kill mosquito, and control malaria, as well as other mosquito borne diseases.

An example of the eradication of a disease in a developed country occurred in 1951 when malaria was officially declared no longer a threat in the United States. This happened after a four-year plan had been carried out to spray millions of homes and their surroundings to kill the carriers. As late as 1947, 15,000 cases of malaria were reported in the United States; by 1951 there were none.

Safe water sources and clean living and sanitary environments will help minimize many diseases that are associated with poverty. Diseases can move with people who are sick whenever they are trying to avoid a pending natural disaster, such as an imminent volcanic eruption. We must all be vigilant to prevent the spread of diseases that can become a challenge to our very existence. Global warming will enhance the spread of diseases that thrive in warmer temperatures. Malaria and yellow fever will spread when mosquitoes are able to expand their habitats across the globe.

Zombie Apocalypse Preparedness

By Ingrid Ukstins and David Best

Zombie Apocalypse Preparedness

Zombies are reanimated corpses with a taste for brains. Their cause is often attributed to a virus passed through an infected human host by bites, or wounds that come in contact with infected bodily fluids. The Zombie Survival Guide identifies the source as the virus solanum, Harvard Medical School psychiatrist Dr. Steven C. Schlozman diagnosed Night of the Living Dead zombies as suffering from Ataxic Neurodegenerative Satiety Deficiency Syndrome, attributed to an infectious agent released when a radioactive space probe exploded in Earth's atmosphere. Other (fictional) zombie plagues have been blamed on mutations in prions, mad-cow disease, measles and rabies. No matter the cause, the preparation is the same—have an emergency kit in your house, ready for any disaster, zombie-based or otherwise (**Figure 134.1**). Emergency kits, based on CDC guidelines, should include:

- Water (1 gallon per person per day)
- Food (non-perishable items)
- Medications (7 day supply of prescription and non-prescription)
- Tools and Supplies (utility knife, duct tape, battery powered radio and spare batteries)
- Sanitation and Hygiene (household bleach, soap, towels)
- Clothing and Bedding (a change of clothes for each person in the household, blankets)
- Important documents (copies of driver's license, passport, and birth certificate, plus other important paperwork)
- First Aid supplies (for basic cuts and lacerations, won't help with zombie bites)

You should also have a household emergency plan, listing who you would call and where you would go in case of emergency. The first step is to identify the types of emergency that might occur in your area, pick a meeting place in case you need to flee your home—one right outside your house for sudden emergencies and another outside your neighborhood in case you are not able to get to, or return to, your home. Make a list of all emergency contacts—police, fire departments, out-of-state relatives. Have this printed out, in case your phone does not work. Finally, plan an evacuation route in case you need to leave home. Have multiple routes in case one road or highway is blocked.

Source: CDC.

Figure 134.1 CDC guidelines for Zombie Apocalypse preparedness are also useful for other disasters.

Race, Ethnicity, Nationality

Topic 135

K.L. Chandler

Race, Ethnicity, Nationality

Geographers classify groups of people into different categories, not to be demeaning nor socially divisive but with the objective of studying them. Historically, geographers described **race** as distinctive, inherited physical traits shared by members of a people group who were connected regionally. However, scientific research over the last two decades has shown only minute differences between races, and today it is accepted that only based on self-identification can race be determined. The nine racial groups that were historically defined by region were African, Asian, Australian (Aboriginal people), Caucasian, Indian (South Asia), Indigenous American, Melanesian, Micronesian, and Polynesian.

Unlike race which historically looked at inherited physical characteristics, **ethnicity** is identified by shared cultural universals like language, history, societal norms, traditions, and religion. Many times, ethnicity stems from genealogy and ancestral identification, and should not be confused with race. Topics of ethnicity have become popular with the many DNA testing kits that are available to the general public. These tests supposedly help determine ancestry however they are based on only single genotyping instead of the entire genome, and generally only gross generalities can be determined. Differing from ethnicity, **nationality** is the legally recognized citizenship of and declared allegiance to a particular political state or nation, whether by birth or through **naturalization** (the legal act to become the citizen of a country other than where one was born) (**Figure 135.1**).

© Evgenia Parajanian/Shutterstock.com

Figure 135.1 North Korea has been reported as the political state that is the most difficult for foreigners to gain citizenship, on the other hand, Canada has a reputation for easier admission.

Development and Underdevelopment in the World Economy

By Tim Anderson

One of the most useful aspects of world-systems analysis with respect to understanding the human geography of the world and its division into developed and underdeveloped realms is its delimitation of the spatial, political, and economic structure of the current capitalist world-economy. According to Wallerstein, the capitalist world-economy that has developed since the mid-fifteenth century has three primary structural features:

1. A single world market that operates within the context of a capitalist ideology and logic; this logic and ideology affects economic decisions throughout the entire world-system.

2. A multiple-state system in which no single state (country) dominates totally and within which political and economic competition among various states is structured and defined.

3. A three-tiered economic and spatial structure of stratification in terms of the degree of economic development and underdevelopment of the world's various states. This textbook employs this basic three-tiered structure as the primary tool, as a model and a theory, for understanding basic differences in the human geography of the world today and in the past. In place of terms like "developed world" and "developing world," or "first world" and "third world," we will use the terms *core, semi-periphery, and periphery*, which are employed in world-systems analysis to differentiate between regions of varying economic development in the world. This text argues that an understanding of the differences between these regions is fundamental to a meaningful appreciation of the differences in such things as the nature of cultures, the structure of economies, political organization and population issues, and problems around the world.

Table 136.1 employs widely available socio-economic statistics for selected countries to illustrate general socio-economic differences between core, semi-peripheral, and peripheral regions. While the terms *core and periphery* are employed to indicate geographical location in common usage, in world-systems analysis, they do not *necessarily* refer to a state's geographical location on the globe. Rather, they are used to refer to a state or region's "location" in the world-economy, that is, on the "inside" (core), on the "outside" (periphery), or somewhere in the middle (semi-periphery) of the world-economy in terms of relative control. Those states in the core control and direct the world-economy, while those in the periphery are most often controlled and directed by the core states. Further, history shows us that it is clear that core and periphery are not geographically or temporally static— states move up and down within this hierarchy over time at a non-constant rate. A recent excellent example of this rule is the former Soviet Union. Before its breakup in late 1991, one could argue (using socio-economic data) that the Soviet Union was a core state. Since the breakup, however, economic conditions have deteriorated to the point that Russia is now clearly a semi-peripheral state. Further examples would include several countries in Southeast Asia, such as Indonesia, Thailand, Malaysia, and Singapore. Peripheral states until the 1980s, they all are now part of the semi-periphery due to surging manufacturing economies and greater political and economic stability.

Core States

The **core states** of the world-economy today include most states in northern, western, and southern Europe; the United States and Canada; Japan; and Australia and New Zealand. It is from here that the world-economy is directed and where its primary command and control centers, such as New York, London, Paris, Frankfurt, and Tokyo, are located. Western Europe emerged as the first core region of the capitalist world-economy in the mid-fifteenth century and remained the primary core until the twentieth century, when the United States, Canada, Japan, and Australia and New Zealand joined its ranks. The wealth of the core stemmed at first from highly efficient agricultural production and control of merchant capitalism through long-distance sea trade in valuable tropical agricultural products (such as tea, coffee, sugar, tobacco, cotton, and spices) during the era of colonialism. Portugal, Spain, the Netherlands, and the United Kingdom were the successive **hegemonic powers** (dominance in economic, political, military, and cultural world affairs) of this early world-economy from the mid-fifteenth century until the early twentieth century (Figure 1.3).

Table 136.1 World Demographic/Socio-Economic Indicator Data, 2020

COUNTRY	POP	BR	DR	NI	IM	TRF	<15	>65	GNI/PPP
World	7,723	19	7	1.1	31	2.3	26	9	$16,885
More Dev.	1,272	10	10	0.0	4	1.6	16	19	$46,188
Less Dev.	6,501	20	7	1.4	34	2.5	28	7	$10,814
Least Dev.	1,062	33	7	2.6	49	4.1	40	4	$2,923
Selected Core States									
Germany	83.3	10	12	-0.2	3.2	1.6	14	22	$55,980
Japan	126.0	7	11	-0.4	1.9	1.3	12	29	$43,010
USA	329.9	12	9	0.3	5.7	1.7	18	16	$63,780
Norway	5.4	10	8	0.3	2.1	1.5	17	18	$70,530
Spain	47.6	8	9	-0.1	2.6	1.3	15	19	$40,570
Selected Semiperipheral States									
Russia	146.7	11	13	-0.2	5.1	1.6	18	15	$27,840
South Africa	59.6	20	9	1.1	22.0	2.3	29	6	$12,530
Costa Rica	5.1	13	5	0.8	8.3	1.6	22	9	$18,670
Argentina	45.4	17	8	0.9	8.8	2.3	24	12	$22,470
Mexico	127.8	17	6	1.2	11.0	2.1	27	7	$19,870
China	1,402	10	7	0.3	9.0	1.5	17	13	$15,320
Malaysia	32.8	16	5	1.0	7.0	1.8	23	7	$27,200

Table 136.1 **World Demographic/Socio-Economic Indicator Data, 2020** (*continued*)

COUNTRY	POP	BR	DR	NI	IM	TRF	<15	>65	GNI/PPP
Saudi Arabia	35.0	15	3	1.2	12.0	2.0	25	3	$49,200
Selected Peripheral States									
Pakistan	220.9	28	6	2.2	62.0	3.6	36	4	$5,110
Laos	7.2	23	7	1.6	52.0	2.8	32	4	$7,410
Bolivia	11.6	22	6	1.6	29.0	2.8	31	7	$8,640
Rwanda	13.0	32	5	2.7	28.0	4.0	40	3	$2,070
Nigeria	206.1	37	12	2.5	67.0	5.3	44	3	$5,040
Malawi	19.1	34	7	2.7	40.0	4.2	44	3	$1,070

POP = Population (millions)
BR = Births/1000 persons/year
DR = Deaths/1000 persons/year
NI = Natural Increase/annum (%)
IM = Infant Mortality Rate (deaths of infants under 1 per 1,000 live births)
TFR = Total Fertility Rate (average number of children born to a woman during her lifetime)
<15 = Percentage of population less than 15 years of age
>65 = Percentage of population greater than 65 years of age
GNI/PPP = Gross national income in purchasing power parity/population (US $) (2018)
Population Reference Bureau, Population Reference Bureau, 2020 World Population Data Sheet

The United States emerged as the primary hegemonic power after World War II. From this period until the early 1970s, the wealth of the core was based primarily on dominance in industrial capitalism focused on heavy industry (cars, ships, chemicals, manufacturing, and the like). The United States, Canada, the Soviet Union, Germany, and Japan all emerged as the major players in world industrial production during this era. With the exception of the Soviet Union, all of these states remain in the core today. Since the early 1970s, post-industrial restructuring in the core economies has taken place, with a switch away from a reliance on heavy industry to a focus on service industries and information technologies. This era is generally called the era of *globalization*, which is characterized by a pronounced international division of labor.

In comparison with semi-peripheral and peripheral locations, the core states all have highly developed economies that are oriented toward service industries and post-industrial technologies, such as information technology and computer software production (Figure 1.4). These advanced economies enjoy a very high standard of living, high gross national products, and per capita incomes, and are generally the "richest" countries in the world. Other characteristics of the core states include stable, democratic governments with large militaries; low infant mortality rates and high life expectancies (measures of relative health and availability of adequate health care); low birth rates, fertility rates, and rates of natural increase (all signs of very low population growth rates); and large urban (as opposed to rural) populations.

Peripheral and Semi-Peripheral States
By Tim Anderson

The **peripheral states** of the world-economy today are located primarily in Sub-Saharan Africa, South Asia (Pakistan, India, and Bangladesh), and parts of Southeast Asia (e.g., Papua New Guinea, Cambodia, Laos) and some parts of Middle and South America (e.g., Guatemala, Bolivia, Haiti). The peripheral states all have several things in common. Historically, they are all former colonies of core states and, accordingly, are generally located in either subtropical or tropical areas. During the era of merchant capitalism and colonialism, these areas were assigned a specific role in the world-economy by the dominant core powers: to be producers of valuable tropical agricultural goods such as sugar, tea, coffee, cotton, and spices. As colonies, these areas lost political sovereignty to colonial governance by the core powers, and traditional economies and societies were replaced by a colonial economy focusing on plantation agriculture linked heavily with the core. The core powers employed slavery and peonage as the primary form of labor control in the periphery. Most of these former colonies regained political sovereignty after the end of the colonial era, beginning in the late nineteenth century, but have struggled since then to rebuild economies and societies that were heavily disrupted by colonialism.

Today, the economies of the periphery remain largely agricultural. The majority of the populations in these countries rely on subsistence farming, low-wage labor on agricultural plantations, or low-skilled, low-wage service positions in urban areas. Compared with the core and the semi-periphery, living standards are relatively low, as are per-capita incomes—these areas are among the "poorest" countries in the world. Peripheral regions are plagued by myriad problems today: weak, inefficient, and often corrupt governments; political instability; ethnic conflict; weakly developed economies and a lack of public services (such as regular electric service, availability of clean water, and public sewage services) taken for granted in core regions. High infant mortality rates and relatively low life expectancies belie weakly developed health-care systems and a susceptibility to epidemics and chronic problems with diseases such as malaria and AIDS. While core states have very low rates of natural increase (or even negative population growth in some countries), the peripheral populations are the world's fastest growing populations. Compared with the core and semi-periphery, the peripheral states have very high birth and fertility rates and "young" populations—a significant proportion of these populations is younger than fifteen years of age. If core populations can be described as "old" with stable growth rates, then peripheral populations can be characterized as "young" and growing rapidly.

Semi-Peripheral States

The **semi-peripheral states** today include many of the countries of Middle and South America (e.g., Mexico, Costa Rica, Chile, Brazil, and Argentina), Eastern and Southeastern Europe (e.g., Russia, Poland, Bulgaria, and Hungary), Southwest Asia (e.g., Turkey, Saudi Arabia, and Iran), and Southeast and East Asia (e.g., Indonesia, Malaysia, Thailand, China, and South Korea). Other semi-peripheral outliers include such countries as South Africa and much of Saharan Africa (e.g., Tunisia, Algeria, and Libya). As the term implies, the semi-periphery occupies a middle place in the hierarchy of development in the world-economy. If one considers socio-economic indicators and demographic data, it is clear that the data for semi-peripheral countries are midway between the extremes of the core and periphery—per capita incomes, birth and fertility rates, and rates of natural increase in the semi-periphery are neither the highest nor the lowest in the world, but rather fall somewhere in between. Accordingly, many world-systems analysts characterize semi-peripheral countries as not the richest, but certainly not the poorest countries in the world.

Many, but not all, of these countries are also former colonies. But, since the 1960s, most of them have managed to achieve some amount of economic and political stability, largely with loans from the World Bank and foreign aid from the core countries. As a result, however, many are saddled with large debts to the World Bank and banks in the core. Economic stability in these countries was largely achieved through a focus on heavy industry and manufacturing as central features of the economy under the authority of rather strong (and sometimes corrupt and heavy-handed) central governments. At the same time, most semi-peripheral economies still depend heavily on the agricultural sector. Indeed, one of the most characteristic aspects of the semi-periphery is a mixed economy dependent upon agriculture (largely for export to the core), heavy industry and manufacturing, and a small but growing service sector. Social stratication in semi-peripheral societies reflects this mixed economy: a large rural agricultural lower class; a relatively large, urban blue-collar manufacturing class; and a small, wealthy urban professional class.

This intense social stratication—a vast difference between the richest and poorest members of society—is one of the hallmarks of the semi-periphery. According to Wallerstein, the most pronounced and acute class struggle occurs in the semi-periphery. This class struggle is often accompanied by chronic political and economic instability. The semi-periphery is also the focus of periodic restructurings of the world-economy during times of economic stagnation, which provide the necessary conditions for this restructuring. For example, the semi-periphery usually is most adversely affected by crises in the world-economy. During the Industrial Revolution in the late eighteenth and early nineteenth centuries, for example, traditional agricultural and artisan economies in places such as Germany and Ireland (part of the semi-periphery at that time) were upset by the changes wrought by industrialism in Great Britain. Many farmers and artisans who could no longer make a living at home moved to core regions like Great Britain and the United States to take jobs in urban factories, thus supplying the core with a needed industrial workforce. In today's world-economy, a similar situation is occurring in the semi-periphery. This time around, traditional economies are being reordered by the current restructuring usually called globalization. This restructuring involves the outsourcing of manufacturing jobs by multinational firms from the core to the semi-periphery, especially in the textile industry (the manufacturing of clothes, shoes, and the like). At the same time, this economic reordering has again resulted in traditional economies being upset and phased out. One of the consequences of this has been a renewed large-scale migration of low-skilled farmers and laborers from the semi-periphery (Latin America, East and Southeast Asia, Southwest Asia) to the core (Western Europe, North America, Australia) (Table 137.1).

This restructuring has resulted in a pronounced **international division of labor** characterized by economic specialization in each of the three regions of the world-economy. Peripheral economies are dominated by subsistence agriculture, plantation agriculture, and natural resource extraction, all mainly for export to the semi-periphery and core. While local, low-wage labor is employed in the production of these resources and products, the capital and management is often controlled from or by the core in the form of multi-national corporations. Semi-peripheral economies specialize in small-scale commercial agriculture, heavy industry and manufacturing (steel, chemicals, etc.), and textile production, the latter for export primarily to the core.

Core economies are highly diversified but are primarily service-based. That is, most workers are employed in the service sector of the economy, which includes everything from retain sales to real estate, banking, health care, education, government, and high-tech industries such as computer software production. Commercial agriculture is an important part of the economy in all of the core countries, but relatively few people make a living wholly as farmers (typically less than 10 percent of the population). While heavy industry and manufacturing was the mainstay of the industrial economies of the core from World War II until the 1970s, employment in this sector of the economy and its overall importance to the economies of the core have declined dramatically over the past 20 years.

Instances of Global Hegemony in the Modern World-System (ca. 1450-Present)

Hegemonic Power	Periods of Hegemony	Basis of Dominance	Major Conflicts	Major Competitors
Portugal	1494-1580	Navigation/Sailing Technology	The Italian Wars	Spain, England; Holland; France; Ottoman Empire
Dutch Republic "Holland"	1580-1688	Navigation; Banking/Credit Colonialism	Anglo-Dutch Wars The Glorious Revolution Thirty Years' War	Spain, England; France; Holland Habsburg Empire
England	1688-1792	Naval Superiority; Colonialism Agricultural Productivity Early Industrial Revolutions	Napoleonic Wars French Revolution American Revolution	France; England; United States
Great Britain	1815-1914	Industrial Productivity/ Innovation Railroads; Iron; Coal; Textiles	World War I Franco-Prussian War	Great Britain; France; United States, Russia; Germany
United States	1945-???	Industrial/Commercial Capitalism Banking; Petrochemicals; Military;	World War II, Cold War "War on Global Terrorism"	United States Germany; USSR; Japan; China European Union

Source: George Modelski, *Long Cycles in World Politics* (University of Washington Press, 1987)

Table 137.1 Instances of Global Hegemony in the Modern World-System

World Political Geography
By Tim Anderson

For most people, the term political brings to mind things like elections, political parties, and vigorous partisan debates about various issues that dominate the airwaves and that are so much a part of our current popular culture. While these concepts are each certainly political in nature, in an academic sense, political refers largely to the structure and function of governments, issues related to territoriality, and power structures in various types of societies. Political geography, then, is the sub-field of human geography that involves the analysis of the spatial expression of these issues. How do people govern themselves, and how has this changed over time? How are governments structured in different parts of the world-economy? What are the spatial characteristics of different types of political organization around the world?

Types Of Historical Political Organization

The world-systems model of social and political change over time. World-systems analysts argue that, historically, there have been only three types of political, social, and economic organizations: mini-systems, world-empires, and the capitalist world-economy. We can more fully understand such historical change by coupling the world-systems model with descriptions of societal structure borrowed from the fields of anthropology and political geography. Anthropologists and political geographers generally identify five different types of political organization that have occurred throughout human history, four of which can still be observed in societies today:

Band Societies

A band society is one in which the there are no formal positions of power and in which members of the society are united by ethnicity, cultural traditions, and kinship. Bands are usually quite small, perhaps only a few dozen extended families, and order is based in and around these extended nuclear families. Such societies exhibit no formal political claim to territory, with the exception, perhaps, of a claimed hunting territory. Strong territorial identity, however, is often associated with cultural identity. Until the Neolithic Revolution, most human societies were organized in such a manner, but few examples survive today. Band societies are limited to a few populations in southwest Africa, parts of tropical Southeast Asia, and some tropical rainforest regions such as the Amazon River basin of South America.

Tribal Societies

A tribal society is comprised of a few, perhaps many, bands of people united by common descent, linguistic similarity, and cultural values and traditions. Political leadership in these societies usually is transitory and determined by virtue of perceived courage, bravery, or wisdom. Tribes are largely egalitarian in nature with respect to the communal use of resources and formalized class structures. Tribes usually claim a home territory, but the defense of this territory is rarely undertaken with organized military power. Nevertheless, as is the case with band societies, tribal societies exhibit a very strong identity with a specific territory that is often conceived of as a people's homeland. Many societies around the world today can be described as tribal in nature. The vast majority of these societies are located in the periphery and in some parts of the semi-periphery of the world-economy: much of sub-Saharan Africa, parts of tropical Southeast Asia, parts of Southwest Asia, and parts of Middle and South America. In world-systems analysis, both band and tribal societies are considered as mini-systems in which production and exchange (mode of production) is largely egalitarian and reciprocal in nature. Tribal societies today, however, are also part of the capitalist world-economy, whose economies and societies are increasingly influenced by it.

Chiefdoms

A chiefdom describes a feudal social and economic order in which a powerful royal and aristocratic elite in a centralized control center controls the production and redistribution of agricultural products from different parts of a claimed political territory. World-systems analysis refers to this organization as the redistributive-tributary mode of production. Leadership in such societies is hereditary (by blood birth), and leaders often claim special divine authority to rule. Chiefdom societies are highly stratified by royalty and occupation, and social rank is largely determined by birth. Agricultural surpluses are generated by coercion of a large peasant class through peonage, serfdom, or slavery. These surpluses are then collected, controlled, stored, and redistributed to the rest of the society by the royalty and aristocracy living in a central urban control center. Chiefdoms usually claim large territories, from which natural resources and agricultural products are extracted, and raise large, organized militaries to defend these territories by force. The central geographical feature of the societies is the city-state. World-systems analysis refers to this type of political and social organization as a world-empire. Today, there are no surviving examples of chiefdoms. They were at one time, however, found all over the world, emerging after the Neolithic Revolution in the various so-called culture hearths: Mesopotamia, the Nile Valley, lowland Middle America, West and Southeast Africa, northern China, and the Indus Valley. This political and social organization also describes the situation in feudal Europe and Japan.

States

A state is an independent political unit occupying a defined, well-populated territory, the borders of which are recognized by surrounding states and militarily defended. All of the countries of the world today are in this sense states. This type of political organization represents a significant departure from that of bands, tribes, and chiefdoms for territory; cultural or ethnic affiliation is the basis of organization. Most of the world's states are multi- ethnic and multi-cultural in nature, and thus, they are not defined by a certain culture, language, or religion, but rather by place, by a territory. This basis of political organization developed in several areas around the world as feudal orders ended, especially in Europe and East Asia.

All states have a government, within which political institutions of the state function to exert control over the state's population and territory. Through such political power, governments are empowered to impose laws, exact taxes, and wage wars. The structure of this empowerment is basically found in two different forms of internal state political structure today. In unitary states, governmental power and authority is centralized in a very strong central government operating from the state's capital city. The vast majority of the nearly 200 states in the world today are unitary states. A handful of states today, however, are federal states, in which governmental power and authority is vested in several different levels. There is thus a hierarchy of power from the national level (federal governments) to the regional level (state or provincial governments) to the local level (city governments). Authority and control in states are vested in governments, but for those governments to have political legitimacy, they must have some sort of ideology behind it that unites disparate groups within the society. Brute force, through the use of the military for example, may work in the short run, but without some central integrating philosophy behind it (such as freedom or democracy in the United States), a state's government can lose legitimacy and find its right to govern questioned by those it governs.

Nation-States

The idea of the nation-state emerged in Europe in the eighteenth and nineteenth centuries. By definition, a nation-state is a state (a territory) that is inhabited by a group of people (a nation) bound together by a general sense of cohesion resulting from a common history, ancestry, language, religion, and political philosophy. A nation in this sense refers not to a country, but to a group of people, and a state refers to a political territory. The ideal of the nation-state, then, combines these two concepts. As such, this type of political organization involves very strong allegiance to nationality and to territory on the part of the nation-state's citizenry. This political ideal probably first emerged after the Industrial Revolution in Europe as improved communication and transportation technologies enabled more efficient control of large territories. All of the major European colonial powers developed a strong nation-state ideal of political organization in the eighteenth and nineteenth centuries and later exported this ideal all around the world as European power and influence grew very strong during the colonial era.

Core-Periphery Patterns Of Political Cohesiveness

Today, all of the countries of the world-economy are politically organized as a state, but most aspire to the ideal of the nation-state. In practice, however, there are very few true nation-states as the term is defined above: one nation of people living in and claiming a defined political territory as its homeland. Why is this? To begin with, even in the strongest nation-states there are always threats to national cohesion. Economic inequality, racial and/or ethnic hostilities and injustices, or perceived disenfranchisement on the part of certain groups in a society may threaten national ideals. But, the biggest obstacle to the ideal of the nation-state is the increasingly globalized world that has emerged since the Merchant Capitalist Revolution. Large-scale migrations (some voluntary, some involuntary) during this era, which continue today, have resulted in the creation of states that are fundamentally multi-national in nature, plural societies in which a variety of ethnic and national groups count themselves as citizens.

Some of the strongest (politically speaking) of such multi-national states have developed a strong sense of nationality in spite of the plural nature of the society. In some instances, this occurred by happenstance as a result of a group or groups of people occupying a large territory over a long period (e.g., the European nation-states). In other instances, strong central governments sought to foster nationality overtly through public education systems and the development of a strong sense of patriotism (e.g., the United States, the Soviet Union, and China).

At the other end of the spectrum, multi-national states without a central organizing principal can sometimes degenerate into civil war or ethnic conflict in which various nations of a state struggle for political power (the former Yugoslavia and present-day conflicts in Africa, for example). At the same time, nations of people without their own state often engage in violence to achieve their own state and thus their own political power. Ongoing examples of such conflicts include the struggle of Palestinians, Kurds, and Basques to forge independent governments and states. Most such struggles, civil war, and political instability in general occur today in the periphery and parts of the semi-periphery of the world-economy. While the core states are not without their conflicts, such states usually have very strong central governments that may seek to quell such conflicts. Ethnic problems and conflicts are also usually worked out in the core through democratic processes or through public debate, both of which are relatively peaceful methods compared to the civil wars and military coups that are common in the periphery.

Political cohesiveness in most states today is in uenced by two main types of forces working in society. Centripetal political forces are those that tend to bring together disparate groups in multi-ethnic, multi-national states. Such forces might include:

- Nationalism—identification with the state and acceptance of its national goals, ideals, and way of life
- Iconography—symbols of unification (flags, national heroes, rituals and holidays, patriotic songs, royalty, etc.)
- Institutions—national education systems, armed forces, state churches, common language
- Effective state organization and administration—public confidence in the organization of the state, security from aggression, fair allocation of resources, equal opportunity to participate, law and order, efficient transportation and communication networks
- On the other hand, centrifugal political forces are those that tend to destabilize a society and pull disparate groups apart in multi-ethnic and multi-national societies. Examples of centrifugal forces might include:
- Internal discord and challenge to the authority of the state, which can lead to political devolution in which national or ethnic groups seek to form separate political authority
- Ethnic separatism and regionalism—this is often seen in states where disparate populations have not been fully integrated (nations without states); this can lead to Balkanization in which multi-national states break apart along ethnic lines
- Trouble integrating peripheral locations—this is especially a problem where disparate rural populations are located far away from the capital
- Social and economic inequality—this is most often seen in multi-nation states where the dominant group is seen to exploit minority groups in terms of control of wealth and social services, etc.

Political Geography

The idea of a **state** is a political concept. It is a recognized boundary under the jurisdiction of a governing body. Unlike other nations around the world that refer to themselves as political states, the people groups of the United States of America use the term *country*, referring to the partitioned subregions as states. Political geography recognizes territoriality as the control of a geographic area and its people, stemming from European colonialism, which impacted most of the known world.

A **nation-state** is a self-governed political entity, occupied by people with a shared language, history, or culture. True nation-states are homogenous, with the majority of the population's sharing the same ethnicity, like Japan or the Koreas. However, it is estimated that only 3% of the world's recognized nation-states today are true nation-states.

Multinational states are sovereign (or self-governed) political entities, composed of multiple nations. Their diversity can serve to be both unifying and contentious at times, in the quest to ensure that the joined nation groups are represented with equality. These political states function most successfully when **centripetal forces**, elements that bind a group together, have greater influence than **centrifugal forces** that divide and cause conflict. In extreme cases, multinational states can become so fractionalized (splintered) that the entity becomes a **failed state**, where a government loses control over part or all of its territory, as befell Syria.

A **stateless nation** is an ethnic group that is not the majority population of any state and is minimalized and marginalized in society (**Figure 139.1**). Civil wars often result when these minoritized people groups rise, and **balkanization**, the breaking or dividing of a political state into one or more new political states, can occur. Some political states unify to create **supernational organizations** that affect or have jurisdiction over more than one nation, like the European Union, which has varied control over twenty-seven European states.

© Peter Hermes Furian/Shutterstock.com

Figure 139.1 The Kurdish people, a stateless nation, reside across borders in southwest Asia.

Continents and Regions
K.L. Chandler

The Four Corners of the Earth

The phrase "the four corners of the Earth" is believed to have originated from a religious prophet in the 8th century BC. The phrase led to the use of an idiom in the English language that came to mean "the far reaches of the Earth." In historical context, the four corners of the Earth came to represent the 15th centuries' known continents – Africa, Asia, Europe, and America. The latter, America, was recognized as one continual landmass in the northwest and the southwest quadrants of the world.

The southern continents of Antarctica and Australia were not yet known to exist, although interestingly there had been some debate centuries earlier that the landmasses in the northern hemisphere needed to be counterbalanced by similar-sized landmasses in the southern hemisphere. This led to many cartographers adding a *Terra Australis Incognita*, or "unknown lands of the south" to the bottom of their maps regardless that no explorer could validate such a land.

Ultimately, the southern landmass of Australia would be found by the Dutch in 1606, while in 1820 Russian explorers would discover an Antarctic ice shelf, and in the same year, a British explorer would spot Antarctica's Trinity Peninsula, which lies south of the Drake Passage. By the 19th century nearly all the world's land had been found, claimed, and charted, with the exception for a few Arctic and Antarctic islands that would be established later.

© Peter Hermes Furian/Shutterstock.com

Figure 140.1 **The traditional seven continents.**

Continents

Today, in the United States the Earth's landmasses are organized into seven continents: Africa, Antarctica, Asia, Australia, Europe, North America, and South America. However, many political states do not share the same categorization of a seven-continent theory. Other theories include:

- Six-continent theory: Africa, America (combining North and South America), Antarctica, Asia, Australia, and Europe
- Five-continent theory: Africa, America, Antarctica, Australia, and Eurasia (combining Europe and Asia)
- Four-continent theory: Afro-Eurasia (combining Africa, Europe, and Asia), America, Antarctica, and Australia

Regions

Given the size of the continental landmasses, geographers split them into regions for ease of studying them. Three facets help determine how a region is defined – formal, functional, and perceptual.

Formal Regions

A formal region is defined by a boundary line, it is generally understood that the areas within the border share certain components, such as cultural or physical traits, laws, and government. (**Figure 140.2**) Sometimes, borders don't change, but the name does. Southern Africa's Swaziland became eSwatini in 2018. The country of Macedonia in Eastern Europe became the Republic of North Macedonia in 2019.

© Chintung Lee/Shutterstock.com

Figure 140.2   In 1414, Henry V of England was the first to create a document to help people prove their identity and nationality when traveling to foreign lands. The term *passport* was not used until 1540.

Functional Regions

A region can also be identified by its function, which forms the area's central portion and radiates outward, affecting the areas that surround it. The functional region or its reputation may not extend to the edges of the region, and it may not have a stronghold throughout its entirety, but it is a way the area can be identified and discussed. Detroit is known as the Motor City, and many communities in its metropolitan area developed as a result of people's relocating to work in the automotive industry.

Perceptual Regions

Sometimes called popular or vernacular regions, perceptual regions are tied to a person's impression of an area. It could be based on an emotional experience, stereotype, prejudice, or reputation that may or may not reflect the truth. Perceptual regions can be broad, like Tornado Alley, or closer to home, as in *the wrong side of the tracks*. Both of these references convey an impression, rather than a formal region, based on what a person knows or thinks she knows.

Economic Geography
K.L. Chandler

Topic 141

Economic Geography

Economic geography is the process of organizing political states into groups to study how they experience economic change over time. Various economic and political organizations have specific demographic thresholds (e.g., infant mortality rates, literacy rates, GDP per capita, etc.) that political states attain to earn a development score. These scores can vary, but typically they are used to divide political states into one of the following three categories:

- **more developed country (MDC)**, having a four-tier income system that includes high income, upper middle income, lower middle income, and low-income brackets
- **less developed country (LDC)**, having two income levels that include the poor masses and the elite rich
- **newly developing country (NDC)**, having the two income levels of a less developed country but including an emerging middle-income tier

© Wara1982/Shuttershock.com

Figure 141.1   The first paper bills were used by the Chinese during the Tang Dynasty (A.D. 618-907).

Economic Indicators

There are several different indicators to determine the economic condition of political states, including **gross domestic product (GDP)**, which is the total value of goods produced and services provided within a political state's borders during one year. This indicator offers valuable insight into that political state's economic health. The abundance of its manufactured products and its excess raw materials, coupled with the importing of goods and materials that cannot be obtained within its borders, directly impact the political state's balance of trade and the value of its currency. Sometimes GDP by **purchasing power parity (PPP)** is considered an economic indicator, as it allows for contrasting political states through a comparison of cost of living by GDP. Another indicator, **gross national income (GNI)** gives the total income earned by a political state's people and businesses, regardless of whether they are located in the political state or abroad. Either number can be considered at the per capita level by dividing the total market value (GDP) or the total income (GNI) by the total population. This per capita view expresses how a political state looks on paper. China's GDP, for example, is high, but, because its population is also large, its income per capita for the year 2020 was just over 8,000 USD; the United States' GDP per capita for 2020 was 53,240 USD. Economists often give a threshold of 12,000 USD as a developing country's benchmark.

Other economic indicators can be natural resources, demographics, and geographic location. LDCs usually have higher birth rates than MDCs, and life expectancy is usually longer in MDCs than LDCs. Literacy rates and levels of income also indicate the economic status of a political state.

The abundance or absence of natural resources within a political state's boundaries by itself does not reliably determine its economic success. Some political states have limited or no natural resources but prosper by combining what they do have with imported raw resources. Others may be rich in gold, diamonds, or oil but languish because another political state or private entity has assumed the management of those resources, and the general population does not reap their benefits.

Economic well-being can also depend on physical location. A **landlocked** state, without a coastline or access to a seaport, may be disadvantaged in conducting trade or procuring supplies but protected from invasion or certain natural disasters that threaten coastal states or islands. At the same time, an island state may have many ports, but common international trade routes could limit its proximity to trading partners. A political state's ability to provide for itself via the land and its capacity for technological infrastructure can propel or cripple its economic potential, too (**Figure 141.1**).

Economic Systems

There are four main economic systems that most political states are categorized; these economic systems are typically organized by society or government for the production and distribution of goods, services, and resources. The first is the **traditional (subsistence) economic system,** which is common in LDCs, with local family or community units' producing enough to meet survival needs on a day-to-day basis, as in the case of nomadic herders. Surplus production rarely occurs.

The second economic system is a **market (commercial) economic system**. In this system, private companies hold great power in a competitive market, which can lead to neo-colonialism and maintain the divide between rich and poor. Many governments attempt to regulate trade by instituting controls to reduce the influence of private companies. The third economic system is where the central government controls the production, distribution, and pricing of goods and services called a **command (planned) economic system**. Commonly known as a "communist economy," this rigid system does not thrive for extended periods because the government concentrates its focus on the wealthiest sectors or contributors, to the detriment of the general population. The last is the **mixed economic system**, which combines the market (commercial) system with the command (planned) model, creating a dual economic system that keeps industries in the private sector and puts public services under the government's purview in an effort to achieve balance. However, the government's influence often permeates the system, assuming more liberties over time.

The Classification Of Economic Activities

Economists identify five different types of economic activities, usually referred to as sectors of an economy. This classification of the various sectors of an economy will be employed to compare and contrast economic activities, modes of production, and social relations of production in the various regions of the capitalist world-economy of today. The primary sector of an economy refers to activities related to the extraction of natural resources. This includes fishing, hunting, lumbering, mining, and agriculture. The secondary sector describes so-called heavy or blue-collar manufacturing industries that involve the processing of raw materials (usually natural resources) into finished products. These industries include such activities as steel production, automobile assembly, the production of chemicals, food-processing industries, and paper production.

The following three sectors are defined as service industries, those businesses that provide services to individuals and the community at large. The tertiary sector refers to financial, business, professional, and clerical services, including retail and wholesale trade. The quaternary sector of an economy describes jobs and industries that involve the processing and dissemination of information, as well as administration and control of various enterprises. These jobs are often described by the term white collar and consist of professionals working in a variety of industries such as education, government, research, health care, and information management. Finally, the quinary sector refers to high-level management and decision making in large organizations and corporations.

Factors in Industrial Location

The location of various types of industrial activities is influenced by a variety of geographical factors. These factors are at work not only at the regional scale, but also at the national and global scale. Among these factors are the following:

- The Costs of Production
  - **Geographically fixed costs**—costs relatively unaffected by location of the enterprise (e.g., capital, interest)
  - **Geographically variable costs**—costs that vary spatially (e.g., labor, land, power, transportation)
- Capitalist Ideology and Logic
  - Since the goal of almost all industries is the minimization of costs and the maximization of profit, the location of an industry is most likely to be where the total costs of production are minimized.

- Complexity of the Manufacturing Process
  - The more interdependent a manufacturing process is, the more its costs of production are affected by location (e.g., steel production).
- Type of Raw Materials Involved in the Manufacturing Process
  - Raw materials that are bulky and heavy, perishable, or undergo great weight loss or gain in processing have the greatest effect on siting
  - Examples: pulp, paper and sawmills; fruit and vegetable canning; meat processing; soft drink canning
- Source of Power
  - Important when a source of power is immovable (e.g., the aluminum industry)
- Costs of Labor (Wages)
- The Market for the Product
  - **Market orientation**—placing the last stage of a manufacturing process as close to the market for that product as possible; products that undergo much weight gain during the manufacturing process (e.g., soft drink canning and bottling)
  - **Raw material orientation**—locating the manufacturing plant as close to the raw material that is used as possible; usually applies in industries that use very heavy or bulky raw materials (e.g.. paper production)
- Transportation Costs

  - Water—the least expensive for of transportation for bulky and heavy goods
  - Rail—also relatively inexpensive for bulky and heavy goods but with less flexible routes
  - Trucking—relatively expensive but carries the advantage of very flexible routes
  - Air—the most expensive form of transportation; employed usually for transporting very valuable goods or those that are time- sensitive (e.g., overnight mail service)

Economies of The Semi-Periphery And Periphery

The economies of the periphery and parts of the semi-periphery are dominated by primary economic activities. The vast majority of people in the periphery, as illustrated in the socio-economic data in Table 1.1, live in rural areas and make their living from subsistence farming. Subsistence farming systems involve agricultural activities that are undertaken not necessarily for money profit, but rather for daily sustenance. Three main types of subsistence agricultural systems are dominant in the peripheral regions of the world-economy today:

Shifting Cultivation

Shifting cultivation, sometimes called slash-and-burn agriculture, involves a complex set of farming practices employed in tropical wet regions where environmental conditions (heavy annual rainfall and very poor soils) are delicately balanced. This ancient practice probably dates back to the earliest stages of the Neolithic Revolution and represents a fairly sound ecological solution to the vagaries of life in such an environment. Shifting cultivation cannot support very large populations because tropical soils do not support the kinds of crops that can feed very large populations (such as grain crops). Rather, it involves the use of low-technology tools such as machetes, hoes, digging sticks, and re to harvest crops such as bananas, taro, cassava, and manioc. Very small groups (bands and tribes) practice a nomadic or semi- nomadic lifestyle and practice shifting cultivation in which a small area is cleared of brush with a machete, covered to allow it to dry, and burned, which fixes nitrogen into the soil. Plants that reproduce vegetatively (like those listed above) are then planted in the ashes. After one or two growing cycles, the soils are exhausted and the group moves on to another site. This lifestyle is practiced mainly in tropical rainforest regions today by a relatively small number of people in locations such as the Amazon and Orinoco basins of South America, parts of Java and other Indonesian islands, parts of tropical central Africa, and parts of Middle America and the Caribbean islands.

Pastoral Nomadism

Pastoral nomadism, sometimes called extensive subsistence agriculture, describes the practice of following or hunting herds of game or herding domesticated animals. It is practiced by large numbers of people in tropical grassland environments (savannahs) in central and east-central Africa and some mid-latitude grasslands regions in central Asia. Pastoral nomadism involves an almost wholly nomadic lifestyle that is extensive in its use of land since livestock like cattle, sheep, and goats require much land per animal to thrive. Contrary to popular belief, most pastoral nomads each meat very infrequently because the animals they herd are the main source of wealth and income. Such societies are usually tribal in nature, and the lifestyle involves seasonal movements to greener grazing lands. As such, there are few permanent settlements in the areas where pastoral nomadism is the dominant economic activity.

Intensive Subsistence Farming

Intensive subsistence farming refers to the subsistence production of a variety of grains and vegetables on small, permanent plots that are farmed intensively the year round. This is possible because such farming activities are undertaken mainly in wet regions of the sub-tropics. The most important of these regions and the most important crops are: China, India, and southeast Asia (Vietnam, Thailand, Cambodia, Laos, Myanmar), where rice is the most important crop; Middle America, where beans and corn are dominant; Southwest Asia from Syria to Pakistan, where wheat and rice are most widely grown; and central Africa, where millet, sorghums, and peanuts are very important. This is not to say that these are the only crops grown in these areas. In fact, intensive subsistence agriculture usually involves the production of a large variety of fruits, grains, and vegetables, but at a relatively small scale. Intensive subsistence agricultural systems employ relatively low-technology tools and innovations such as animal-drawn plows, manure for fertilizer, terracing systems, and not a small amount of human muscle power. These farming systems are undertaken by literally millions of people in much of the periphery and semi-periphery. Indeed, rice, the most important grain crop in the world, sustains at least half of the world's population on a daily basis.

Plantation Agriculture

A final important type of agricultural system undertaken primarily in the periphery and semi-periphery of the world-economy today is plantation agriculture. While all of the other forms of agriculture in these regions of the world-economy are subsistence in nature, plantation agriculture is a form of commercial farming in which the main goal is the maximization of profit per unit area of land under cultivation to sell the products in the international marketplace. Plantation agriculture involves the commercial production of tropical and sub-tropical products, such as tropical fruits, sugar, coffee, tea, and cocoa. A plantation is a large farm on which usually only one crop is grown. Most plantations are owned and managed by large multi-national corporations in the core but employ local low-wage labor. The capital input, as well as the main market for the crops grown on such plantations, is in the core regions of the world-economy. Thus, most plantation products are grown explicitly for export to the core. A British and Dutch innovation, the idea of the plantation dates from the colonial era when British, French, and Dutch companies established sugar plantations using African slave labor in the Caribbean. Plantations were also established in other colonial areas to produce a variety of valuable tropical agricultural crops: tea in India, spices in Indonesia, and coffee in Africa and parts of Middle and South America. Today, most plantations are found in the Caribbean, Central America, and tropical central Africa.

Manufacturing

The globalization of the post-industrial era has created a global economy that is punctuated by an international division of labor in which the periphery and the semi-periphery play an increasingly important role. In the semi-periphery, the post-industrial era ushered in an era in which manufacturing activities have become an important element of most economies. Semi-peripheral economies today are characterized by a mix of subsistence and commercial farming (mostly plantation agriculture) and light and heavy manufacturing activities that employ low-wage, low- skilled urban workers. Most manufacturing enterprises and plantations, however, are owned and managed by multinational corporations in the core —the United States, Japan, and Europe. This pattern has developed over the past 30 years because of evolving service economies in the core and the movement of manufacturing activities out of the core in search of lower costs of production, especially wages. Some of the industries most affected by this transition are textiles (clothing and shoes), inexpensive retail goods (toys, for example), and automobile assembly. Thus, an international division of labor punctuated by multinational companies in the core has evolved into what characterizes it today: high-wage service sector labor in the core; low- wage, low-skilled secondary sector (manufacturing) jobs in the semi- periphery; and low-wage, low-skilled primary sector jobs (subsistence agriculture and plantation enterprises) in the periphery.

Economies of The Core

Primary Activities

Commercial agricultural activities in the core regions of the world-economy are punctuated by several distinguishing characteristics. First, agriculture employs a very small number of people, but it is nevertheless very productive and remains an important part of core economies. But, agriculture has undergone immense changes in the core over the past 50 years: the number of farmers has drastically declined, the average farm size has grown substantially, and many agricultural activities are increasingly being controlled by large corporations (this is known as agribusiness).

Farming in the core today is characterized by high inputs of capital, heavy mechanization, the heavy use of hybrid crop and animal varieties, chemical fertilizers, and relatively low inputs of human labor. While such innovations have put many farmers in the core out of work or made them redundant, agricultural production in the core is the highest in the world in terms of crop and animal yields per unit area.

Five main types of farming systems dominate the agricultural economies of the core. First, commercial dairying involves the production of milk and milk products, primarily for large urban markets. While fresh milk production is located in the urban hinterlands of most large metropolitan areas in the core (because it is perishable), the production of milk products like cheese, yogurt, and butter is concentrated in the northern United States and northern and Alpine Europe. Second, market gardening, sometimes called "truck farming," involves the commercial production of fruits, vegetables, or other specialized crops, again mainly for large urban markets. Market gardening is concentrated along the Atlantic coast of the United States from New Jersey to Florida to Texas, parts of coastal California, northwestern Europe (especially the Netherlands), parts of coastal Japan, and parts of Australia and New Zealand. Third, mixed livestock and grain farming refers to the production of livestock for human consumption alongside the production of grains (mainly corn and wheat) for use as livestock fodder. This describes the agricultural systems dominant in the Great Plains (wheat and cattle) and Midwest (corn, soybeans, hogs, and cattle) of the United States, the Pampas of Argentina (wheat, corn, and cattle), much of central Europe (sorghum, corn, cattle, and hogs), and the interior grasslands of Australia (wheat and sheep). Fourth, livestock ranching involves the commercial production of livestock (mainly cattle) for large urban markets and for international export. Ranching takes place mainly in semi-arid grassland environments of the mid-latitude regions of the core: the interior West of the United States, interior Australia, southern Brazil, parts of Argentina, and parts of Spain and Greece. Finally, Mediterranean agriculture involves the production of highly specialized Mediterranean crops like grapes, figs, dates, olives, and citrus fruits. Such activities are undertaken in regions around the world with the distinctive Mediterranean climate (hot, dry summers and cool, wet winters): southern California, the circum-Mediterranean, southwest Australia, parts of southwestern Africa, and parts of Chile.

Service Activities

The economies of the core regions of the capitalist world-economy were built upon a factory organization of labor and a reliance on heavy manufacturing industrial activity during the nineteenth and early twentieth centuries. But, the last 30 years have witnessed a vast shift in how core economies are structured. In search of lower costs of labor, many large corporations have spearheaded the development of an international division of labor in which manufacturing activities (and manufacturing jobs along with them) have been moved to semi-peripheral locations. Today, the vast majority of the population in core regions is employed in the service sector of the economy, especially the tertiary and quaternary sectors. Core economies are also increasingly information-based: some of the largest companies in the world, like AT&T and Microsoft, are in the business of perfecting and selling the access and dissemination of information.

This business has resulted in revolutionary changes in the industrial landscape of the core countries and the geography of industry and manufacturing in these places. The once-dominant industrial centers such as the Great Lakes region in the United States, the British Midlands, and the German Ruhr district have now become "rust belts." Entirely new manufacturing centers have now developed in places in which the main resource for new information technologies, human brainpower, is nearby. Examples of these new manufacturing regions include Silicon Valley in northern California, the Research Triangle of North Carolina, and the "Golden Triangle" of central Texas. In Europe, such centers have been developed in many areas of eastern Europe, such as eastern Germany, the Czech Republic, Hungary, and Poland.

China's Belt and Road Initiative in Africa
Cathleen Fritz

The continent of Africa is rich in oil, iron ore, timber, gold, diamonds and many other natural resources. Over the last two decades Africans have seen a new "scramble" for economic development opportunities in their continent. The "scramble" is between many African countries and China. This is a complex issue because the continent of Africa finally cut their ties to being colonized. The decolonization of Africa began in the 1950's and ended in the late 1970's. They had been dominated by Europe in the nineteenth and twentieth century. Much of Africa, except Ethiopia and Liberia, experienced European dominance. They fought hard to build democracies and stand on their own again. They intended to end being colonized forever.

Now? China is appealing to many African countries to build…Build…BUILD! The quest that is discussed by many political scholars and economists is whether Africa is being recolonized…Again. China has built many infrastructures like trains, roads, bridges, and seaports for shipping and trade, water and sewer systems to support migration to the cities, energy systems to power the new cities and corporations, and broadband internet to link all people, corporations, and governments to the economic development. China's Belt and Road Initiative in Africa is meant to serve the people of Africa and China. Who is funding all of these projects? China!

Read a few of the initiatives and then decide whether the African countries are benefitting from this growth as much as China.

1. China has funded two railways projects; Addis Ababa Light Rail Transit and Ethiopia-Djibouti Railway.
2. China built an African Union headquarters in Addis Ababa Ethiopia.
3. China is building a West African Union (ECOWAS) headquarters in Abuja Nigeria.
4. China is paying Guana for bauxite exploration and mining.
5. China is building a hydropower plant in Dondo, Angola and Guinea.
6. China is building Special Economic Zone in the Congo.
7. China is building an oil refinery in Nigeria.
8. China is building a cement factory in Zambia.
9. China is building a residential district, an industrial zone, schools, a university and recreational centers in Egypt.
10. China built a new parliament building Harare Zimbabwe.

All of these investments have cost China billions of dollars but the investment may be worth it if Africa generates financial gains for China. Is all of this building creating financial gains for Africa too? Research. Decide (**Figure 143.1**).

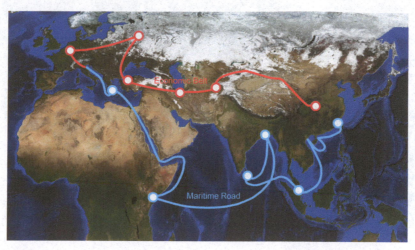

© YIUCHEUNG/Shutterstock.com

Figure 143.1   China's Belt and Road Initiative connects Africa, Europe and Asia.

Four Revolutions
By Tim Anderson

Topic
144

Four significant "revolutions" in human history have occurred that have successively shaped and reordered the world's economies, political geographies, cultures, and landscapes. Each revolution represents a clear break with what had come before concerning dominant modes of production, forms of labor control, social relations of production, dominant technologies in use, and the types of natural resources exploited. Each revolution resulted in a significant increase in the amount of power that could be harnessed per person per year. Further, each revolution resulted in substantial changes in the ways in which human beings conceived of the world around them, how humans perceived of themselves in the world and their place in it, and how certain natural resources could be used and exploited. Over time, the effects of these revolutions resulted in the alteration of the world's cultural landscapes into what they are today. The cultural landscapes of today reveal these past changes. They reveal changing cultural traditions, ideals, and values. They reveal social, political, and economic struggles that have taken place over time and space. The world's cultural landscapes are, in short, a palimpsest of the last 12,000 years of human history.

The Neolithic Revolution

What Was It?

The first revolution to significantly alter the ways in which human beings viewed and used the natural world around them was what anthropologists refer to as the **Neolithic Revolution**. This term refers to the period in human history about 10,000–12,000 B.C.E. when the first large-scale urban settlements began to appear, together with concomitant changes in the structure and nature of societies, modes of production, and social relations of production. The most important feature of the Neolithic Revolution, however, and what engendered most of these changes, was the domestication of plants and animals. **Plant and animal domestication** refers to the gradual genetic change of plants and animals through selective breeding, such that they become dependent upon human intervention for their reproduction (**Figure 144.1**). Certain plants and animals, or certain varieties of plants and animals, were selected for particular qualities, such as taste or caloric value or nutritional value, while other varieties were left behind. These selected varieties of plants and animals were nurtured, protected, and cared for in gardens and fields, and they continuously reproduced. Over many generations, this caused genetic change, resulting in plants and animals with distinctive characteristics found to be helpful to human beings (again, characteristics such as taste or resistance to pests and drought). Today, this process occurs in the form of scientific genetic breeding of plants and animals carried out at laboratories and universities.

Without a doubt, the most important and far-reaching result of plant and animal domestication was the advent of agriculture on a scale that had not been seen previously. Before the advent of large-scale agriculture, the vast majority of societies all over the world could be described, in world-systems analysis language, as mini-systems, with social relations of production and social and political organizations that were tribal in nature. Exchange in these mini-systems was basically reciprocal, social classes were weakly developed, and their geographic extent was very small, amounting to little more than a claimed homeland or hunting territory. Agriculture, however, wrought social and economic changes that changed the world forever. Where it occurred, much more complex societies quickly developed, with highly stratified societies, vast increases in food production, and despotic forms of political organization that claimed and defended large territories by force. That is, where plant and animal domestication first occurred, where agriculture first originated, mini-systems gave way to the development of world-empires.

Why and Where Did It Occur?

Two primary theories exist about when, where, and why agriculture first originated. The first, what we might call the orthodox theory of domestication, is agreed upon by most historians and anthropologists. The second was proposed by the geographer Carl O. Sauer in the mid-twentieth century. The orthodox theory, that which is most widely accepted by social scientists and for which there is the most archaeological evidence, argues that plant and animal domestication occurred independently in several areas of the world at about the same time in history—from about 8,000 to 12,000 B.C.E.—through a process of **independent invention**.

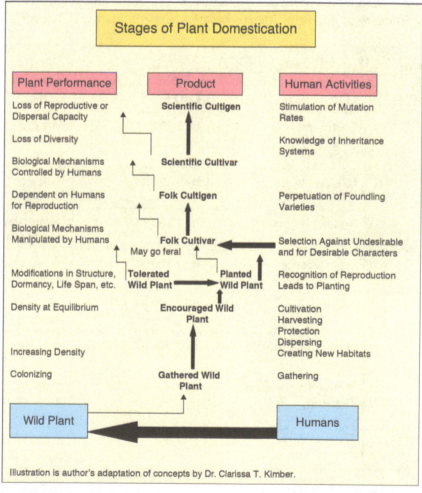

Illustration is author's adaptation of concepts by Dr. Clarissa T. Kimber.

Source: Timothy G. Anderson

Figure 144.1 **Stages of Plant Domestication**

That is, populations in different parts of the world "invented" agriculture on their own without contact from outside populations, without learning it from others. Most experts argue that some sort of ecological stress—overpopulation or climate change, for example—forced these populations to move from a nomadic hunting and gathering type of lifestyle to a more sedentary lifestyle more and more dependent upon agriculture to feed burgeoning populations.

This revolutionary change did not happen overnight, or even in a few generations, but rather most likely occurred over hundreds of years and over many human generations. But, where it did occur, in the following so-called **culture hearths** of innovation and invention, societies began to look very different, the hallmark of which was the first urban civilizations. Table 144.1 lists some of the earliest culture hearths and gives dates by which world-empire types of civilizations had developed. The dates for the domestication of various plants and animals are not given here, but radiocarbon dating of plant and animal remains suggests that plant domestication occurred as early as 12,000 to 15,000 years ago in Southwest Asia and the Nile Valley and 5,000 years ago in Middle America and the Andean Highlands of South America.

Another less widely accepted theory of the development of agriculture was proposed by the American geographer Carl O. Sauer in his book *Agricultural Origins and Dispersals* in 1952. Sauer suggested that agricultural first developed in Southeast Asia as early as 20,000 years ago and from there diffused to the other areas of the world. This theory, therefore, is one based on the idea of **cultural diffusion** from an original hearth area to all other regions of the world. Sauer reasoned that agriculture would have developed first in an area where the richest array of different kinds of useful plants naturally occurred, most likely a tropical region, and in a place where people were not under any kind of ecological stress. Such populations, Sauer argued, would not have been forced to invent agriculture, but rather would have had ample time over many generations to experiment with growing different varieties of useful plants. This theory, however, is at most

Table 144.1 Major Culture Hearths

CULTURE HEARTH	LOCATION	DATE	DOMESTICATES
Mesopotamia	Southwest Asia	by 5,500 B.C.E.	wheat, barley, rye, grapes, oats, cattle, horse, dogs, sheep
East Africa	Nile Valley	by 3,300 B.C.E.	barley, coffee, cotton, millet, wheat
Incan	Andean Highlands	by 2,500 B.C.E.	potato, tomato, llama, guinea pig, alpaca
Mediterranean	Crete, Greece	by 2,500 B.C.E.	barley, grapes, goats olives, dates, garlic
Indus Valley	Pakistan	by 2,300 B.C.E.	wheat, cattle, dog, rye, sheep, horse
North China	Huang He Valley	by 2,200 B.C.E.	soybeans, buckwheat, cabbage, barley, plum
Southeast Asia	Vietnam, Cambodia, Thailand	by 1,500 B.C.E.	bananas, chicken, pig, tea, dog, rice, taro, water buffalo, yams
Meso-America	Mexico, Guatemala	by 1,250 B.C.E.	maize, beans, taro, chili peppers, dog
West Africa	Ghana, Mali	by 400 B.C.E.	arrowroot, millets, pig, rice, oil palm,

a thought experiment, and little archaeological evidence supports it. It also presupposes that populations in certain parts of the world were not "advanced" enough to invent agriculture on their own and that they therefore must have been shown how by invading populations having such knowledge. For these reasons, Sauer's theory is not widely held to be true.

What Changes Did It Engender?

An agricultural lifestyle had the effect of radically altering the nature and structure of societies and economies in each of the various culture hearths. Aside from new innovations and techniques that increased crop yields and food production, such as crop rotation and large-scale irrigation projects, the most revolutionary change was the development of highly stratified, agriculturally based societies with codified class structures, a class-based division of labor, and a redistributive-tributary mode of production. Anthropologists refer to such societies as *chiefdoms*, and in essence, this describes a *feudal* social and economic order. In the world-systems perspective, Wallerstein refers to these societies as *world-empires*.

At one end of the socio-economic spectrum in these societies, an agricultural underclass comprised up to 90 percent of the population. This peasant underclass produced agricultural surpluses, the storage and redistribution of which was controlled and directed by an aristocratic (title by blood birth) royalty at the other extreme of the socio-economic spectrum. These rulers also directed the building of vast agricultural projects, such as irrigation schemes (in North China and the Nile Valley), and the construction of monumental structures (such as the Great Wall and the pyramids). Such rulers (known as kings, emperors, pharaohs, and the like) are often called **god-kings** by anthropologists because they often claimed to possess supernatural powers and to have achieved communion with the gods. These rulers surrounded themselves with other elite classes, such as scribes and priests. These ruling elite lived in a central urban location known as a city-state, the basic political unit of world-empires. The presence of distinctive elite classes points to the development of other revolutionary changes first witnessed in each of these culture hearth that set the stage for the development of vastly more complex societies and economies that forever changed the ways in which humans interacted with the natural world around them:

- Written languages to keep records of agricultural surpluses and the redistribution of surpluses
- The development and use of calendars to track the seasons and to predict planting and harvesting times
- The development of an organized military under the command of the ruling elite to defend a claimed political territory
- Significant population increases as a result of more stable food supplies

The Merchant Capitalist "Revolution"

What Was It?

Although it is not always called a "revolution," the development of merchant capitalism in northwestern Europe in the mid-fifteenth century and its diffusion around the world over the ensuing 400 years engendered changes in societies, economies, and cultures that shaped the modern world-economy more than any other event. If we define **capitalism** as the production and exchange of goods and services for private money profit, then most historians would agree that it has existed for millennia in many places around the world (**Figure 144.2**). **Merchant capitalism**, however, arose in only one region of the world at a specific point in history—Holland and England in the mid-fifteenth century. Merchant capitalism was an altogether new version of capitalist production and exchange because it was based heavily on long-distance sea trade in exotic products from the tropics and subtropics.

Because such products were rare and costly to acquire (it took five years to travel by ship from Holland to the southeast Asian spice islands in the sixteenth century), the high relative value of such trade meant great wealth to the countries and individuals that controlled it. Before the advent of merchant capitalism, most trade was merely regional in nature and used land-based caravans or shipping routes that plied coastal regions. Merchant capitalism involved the use of new sailing and navigation technologies to strike out into the oceans out of sight of land on risky overseas ventures in search of rare tropical goods for which Europeans were willing to pay high prices (such as sugar, cotton, tea, coffee, pepper, and other spices). In short, merchant capitalism ushered in the era of colonialism and global trade that radically altered the world forever. It was the first step toward creating the "globalized" world that we live in today.

Why and Where Did It Occur?

Why did merchant capitalism develop in a relative backwater spot in the world at the particular time that it did? Compared to the great civilizations of China, South Asia, and Southwest Asia, Europe in the early Middle Ages lagged far behind in terms of technology and scientific know-how. Earlier, traditional theories attributed the development of merchant capitalism in northwest Europe to such things as Protestant Christianity (Weber's "Protestant work ethic") or European racial superiority. While some still hold Weber's theory to be of explanatory value, nobody still believes that Europeans were somehow "better" or "smarter" than other world civilizations and "invented" capitalism due to this superiority. Instead, most historians today argue that a series of events happened first in Europe, affecting societies there in more drastic ways than others and resulting in a fundamental reordering of European societies and economies.

The most important of these events was most likely the **Black Death**, the Bubonic plague that ravaged Europe from 1340 to 1440. The Black Death struck elsewhere around the world, but nowhere with such far-reaching consequences. Anywhere from one-third to one-half of the population of Europe died during this period. Most of those that died came from the peasant classes, in a feudal society, that part of the population with the lowest caloric intake and the highest susceptibility to such respiratory diseases. This loss had drastic repercussions for European economies because with so much of the peasantry gone, labor was now in very short supply. As a result, those peasants that remained acquired something that they had never possessed before—bargaining power over their aristocratic landlords. Peasants began to demand something for their labor, either payment in kind or money wages. In short, the Black Death precipitated a **crisis of feudalism** in Europe. The ancient feudal organization of labor and production could no longer keep up with the demand for food, especially as populations began to increase rapidly again after the plagues subsided. World-systems analysts argue that the replacement of a feudal organization to societies and economies with a different form of production and exchange based on individual money profit—capitalism—was the solution to this crisis in the feudal order.

Along with the reordering of economies came a fundamental reordering of societies as well, the most important aspect of which was the emergence of a new class of people—a middle class. Because production was now based on profit and competition, those that could produce the most reaped the most profit. This spurred technological innovations, especially in agricultural production. The use of steel plows, the draining of marshlands, and crop rotation schemes, for example, led to higher agricultural surpluses than ever before. This reordering also meant that more and more of the European population did not have to farm because agriculture had become more efficient and more productive. Many people began to move to towns and cities and became shopkeepers, merchants, and artisans (weavers, blacksmiths, millers, tailors, bakers, etc.). Over ensuing generations, this new middle class rose in social, economic, and political importance, and the middle class began to pass on wealth to successive generations. Such dynastic, wealthy urban merchant families came to dominate the social and economic life of coastal towns in northwestern Europe, especially in Holland and England, by the sixteenth and seventeenth centuries. By the late sixteenth century, these powerful urban merchants began to sponsor overseas trading ventures to newly "discovered" tropical regions to supply the new and growing demand for rare, expensive tropical agricultural products by the expanding middle class. **Colonialism**—the political, economic, and social control of tropical peripheral regions by European core powers using coerced labor—was invented by northwestern European urban merchant capitalists to more efficiently supply this demand. It was this new merchant capitalist colonial order that created the European-centered capitalist world-economy with its characteristic tripartite global economic geography of core, semi-periphery, and periphery.

What Changes Did It Engender?

In summary, the development and expansion of merchant capitalism based on long-distance sea trade resulted in the following revolutionary developments:

- The replacement of feudal, agricultural societies with economies dominated by merchant capital interests involved in long-distance sea trade
- The invention and domination of colonialism
- The birth and dominance of a middle class of urban merchant entrepreneurs
- The creation of a capitalist world-economy dominated by a few nation-states in Europe
- The creation of a "core" of relatively high economic and technological development and a "periphery" of relatively low economic and technological development
- The creation of the use of varying labor control systems in the different part of the world-economy: slavery in the periphery, wage labor in the core, and tenant farming in the semi-periphery
- The increased use of nonhuman and non-animal sources of energy; a drastic increase in the amount of energy harnessed per capita, per year
- The initiation of a capitalistic logic in the world-economy; since capitalism rewards innovations that make production cheaper, it engenders constant technological innovation

The Industrial Revolution

What Was It, and Where Did It Occur?

The **Industrial Revolution** represents a third major break with the past with respect to far-reaching societal changes around the world, especially in the core countries of the world-economy during the nineteenth and twentieth centuries; their wealth was built on the backs of economies focused on heavy industry and manufacturing. Most significantly, the Industrial Revolution ushered in vastly different methods and modes of industrial production, systems of labor control, and social relations of production. These changes occurred first in the Midlands of central England beginning in the early-to mid-eighteenth century but spread very quickly to the rest of Europe (Belgium, Holland, France, Germany, and Russia) and North America (especially the northeast United States) by the mid-to late-nineteenth century. The Industrial Revolution ushered in the Machine Age.

The Industrial Revolution radically altered societies and economies in two important ways. First, modes of production were revolutionized with the introduction and application of new labor-saving technologies and machines. The earliest of these new technologies was the steam engine, first put to use in the British Midlands running pumps to remove excess water from coal mines; but, the technology was soon put to use running all types of machines that heretofore had employed human or animal power, such as mills and looms. The first industry to be completely revolutionized by the machine age was the textile industry. Steam engines that spun and wove yarn into textile goods replaced traditional cottage textile industries in the rural countryside of Europe, where such goods had for millennia been produced by hand. The mode of production that came to dominate regions that became industrialized was the urban factory staffed by low-wage, low-skilled workers running steam-driven machines.

Many of these workers were women and children from the rural hinterlands of emerging industrial cities who came to the cities in search of work because traditional ways of making a living were being radically altered. This migration represents the second major change wrought by the Industrial Revolution—a new form of labor control characterized by the factory organization of low-wage, low-skilled laborers. Such laborers lived in housing, often built by the factory owners, very near to the factories in which they worked. Long working hours (production could take place around the clock with the advent of gas lighting and, later, electricity), poor living conditions, and exposure to hazardous environmental pollutants (coal smoke, chemicals) characterized the lives of the earliest factory workers. Workers often performed the same menial tasks hour after hour in cramped and hot conditions. In the United States, the industrialist Henry Ford mastered the factory mode of production and labor control with the innovation of the assembly line, where workers performed the same task hour after hour as cars traveled down an assembly line (this mode of production is often called **Fordism**).

The factory mode of production built around urban factories staffed by low-skilled wage laborers and powered by petroleum-based fuels (first coal, then oil products like gasoline) revolutionized production. The countries that industrialized first, like Great Britain, Germany, and the United States, soon outpaced and outproduced their competitors. Although industrialization came to different countries at different times, the economies of all of the core powers of today were built upon heavy industrial production between the mid-eighteenth century and 1960. Industrialization brought immense wealth and power to these core countries, especially in the early twentieth century when steel production, shipbuilding, and automobile manufacturing became the hallmark of the core economies. Many semi-peripheral areas did not industrialize until the 1960s and 1970s, while large-scale industrialization has yet to appear in most areas of the periphery. The semi-peripheral economies hope that a strong manufacturing and heavy industry base will also bring the same wealth that it brought to the core economies, but it remains to be seen whether this will happen or not.

The Post-Industrial (Post-Modern) Revolution

The **Post-Industrial Revolution** describes revolutionary changes in societies, economies, and cultures that have taken place since the early 1970s, predominately in the core regions of the world-economy. But, because changes that take place in the core affect all regions of the world-economy, this revolutionary period in which we now live is fundamentally transforming societies, economies, and cultures all over the world. This period has seen the development of new modes of production, the advent of so-called "new information technologies" such as the internet (these have ushered in the "computer age"), and a pronounced global division of labor in the world-economy.

Table 144.2 Historical Stages of World Capitalism Since 1750

Era	Form of Capitalism	Hememonic State/ Challengers	Economic Policy	Capital	Dominant Commodity	S
1750-1810	Mercantilsm	**England** United States France	Imperialism	Merchant Capital	Textiles	Abs Mon
1810-1870	Liberalism	**Great Britain** United States Germany	Liberalism	Industrial Capital	Light Industry	Nat Stat

(continued)

Table 144.2 Historical Stages of World Capitalism Since 1750 (*continued*)

Era	Form of Capitalism	Hememonic State/ Challengers	Economic Policy	Capital	Dominant Commodity	S
1870-1930	Imperialism	**United Kingdom** United States France German Italy	Imperialistic	Finance Capital	Heavy Industry	Imp
1930-1990	Late Capitalism	**United States** United Kingdom France Italy Soviet Union Japan	Liberalism	State-Monopoly Capital	Durable Consumer Goods	Wel Stat
1990-???	Neoliberalism	United States China United Kingdom Germany Russia	Imperialistic	Multinational Capital	Information	Reg

Kojin Karatani, The Structure of World History (Duke University Press, 2014) p. 272

Perhaps the most significant development during this latest revolution is the process of **globalization** in the world-economy, directed from the core by multi-national conglomerates employing rapidly evolving forms of information technology to increasingly expand the scope of *interconnectedness* among the parts of the world-economy. Such *interconnectedness* has led not only to a fundamental restructuring of the world-economy. This post-industrial or "postmodern" age has also ushered in an era in which people and places all over the world are linked and tied together to a degree never before seen in human history. Rapidly evolving forms of mass communication, faster and cheaper forms of travel between continents, and large-scale international migrations are all in part responsible for such international linkages.

In terms of modes of production, a central feature of the post-industrial era has been a restructuring of the world-economy characterized by globalization and a marked international division of labor. In the core regions of the world-economy, this restructuring involved a move from economies based on heavy industry and manufacturing between World War II and the early 1970s, to service-based economies from the 1970s until the present. Manufacturing jobs and manufacturing-based industries (such as steel production and automobile assembly plants), once the cornerstone of core economies, have increasingly moved to the semi-periphery and, in some instances, to peripheral locations. This process may be called **global outsourcing**, and it has occurred primarily due to lower costs of labor (wages) in semi-peripheral and peripheral locations such as Mexico, Indonesia, Malaysia, and much of Latin America. This geographical outsourcing has led to the movement of manufacturing-based industries and jobs (especially those with high costs of labor, such as textile production and automobile assembly) to such semi-peripheral locations. In the core regions themselves, a new mode of production, called "just-in-time" or **post-Fordist production**, has replaced traditional factory organizations of labor. So-called "blue collar" low-wage and low-skilled jobs have rapidly declined in number with the decline in the number of manufacturing industries. In their place, a new kind of worker has emerged: highly skilled, high-wage labor in which brains rather than brawn matter the most. This is especially true in the new information technologies, such as in software production and computer assembly, which are an increasingly important part of the core economies in the post-industrial era.

In the end, the post-industrial era has led to a reinforcement of the core—semi-periphery—periphery structure of the world-economy and made the differences between them greater than ever before. That is, life in the core for the average person is vastly different than it is in the periphery. Today, the core controls the flow of wealth and information in the post-industrial world-economy. The core contains the nodes, or nerve centers, of the world-economy (places such as New York, London, and Tokyo), and it is here that the largest multi-national companies in the world are located. Core economies are primarily service-based and information-based in nature. The semi-peripheral regions today are the places where much of the world's manufacturing now takes place. Agriculture is still an important part of these economies, especially the production of plantation products for export to the core. A small but expanding service sector is also characteristic of many of the semi-peripheral economies. Peripheral economies remain largely dependent upon traditional subsistence agricultural economies and plantation agriculture for export to the core. The service and manufacturing sectors of the economy in the periphery are both still rather weakly developed.

Finally, **post-modernism** has significantly altered traditional ways of life and forms of artistic expression such as art, literature, music, and poetry, as well as philosophy and other scientific endeavors. These alterations have been most conspicuous in the core regions of the world-economy, but their effects have trickled down to the semi-periphery and periphery as well, albeit to a lesser extent. Post-modern expression is characterized by a lack of faith in absolute truths, a mélange of forms and styles, a rejection of order, and a deconstructionist ideology in which traditional ways of articulation are continually questioned. This post-modern "condition" has resulted in a reordering of societies, economies, and modes of production, especially in the core, and is characterized by the following conditions:

Increasing globalization of the world-economy

- The development of a "frenetic" international financial system
- The development of, and reliance upon, new information technologies
- A world-economy more and more reliant upon the ow of information
- A world-economy that is increasingly illegible to the average person; interconnections are so complex that the world is harder to comprehend, global capitalism is harder to "locate"; a world of confused senses and order
- A world that is increasingly hyper-mobile; a world-wide informational economy with telecommunication technology as its foundation; a "space of flows" that dominates sense of place; a perception of the world through the medium of information technologies
- A world increasingly effected by time-space compression; a marked increase in the pace of life; a seeming collapse of time and space that affects our abilities to grapple with and comprehend the world